Biofuel and Biorefinery Technologies

Volume 6

Series editors

Vijai Kumar Gupta, Department of Chemistry and Biotechnology, Tallinn University of Technology, Tallinn, Estonia
Maria G. Tuohy, School of Natural Sciences, National University of Ireland Galway, Galway, Ireland

This book series provides detailed information on recent developments in biofuels & bioenergy and related research. The individual volumes highlight all relevant biofuel production technologies and integrated biorefinery methods, describing the merits and shortcomings of each, including cost-efficiency. All volumes are written and edited by international experts, academics and researchers in the respective research areas.

Biofuel and Biorefinery Technologies will appeal to researchers and post-graduates in the fields of biofuels & bioenergy technology and applications, offering not only an overview of these specific fields of research, but also a wealth of detailed information.

More information about this series at http://www.springer.com/series/11833

Meisam Tabatabaei · Hossein Ghanavati
Editors

Biogas

Fundamentals, Process, and Operation

Springer

Editors
Meisam Tabatabaei
Biofuel Research Team (BRTeam) /
Agricultural Biotechnology Research Institute of Iran (ABRII)
Agricultural Research, Education and Extension Organization (AREEO)
Karaj
Iran

Hossein Ghanavati
Agricultural Biotechnology Research Institute of Iran (ABRII)
Agricultural Research, Education and Extension Organization (AREEO)
Karaj
Iran

ISSN 2363-7609 ISSN 2363-7617 (electronic)
Biofuel and Biorefinery Technologies
ISBN 978-3-030-08431-8 ISBN 978-3-319-77335-3 (eBook)
https://doi.org/10.1007/978-3-319-77335-3

Printed on acid-free paper

This Springer imprint is published by the registered company Springer International Publishing AG part of Springer Nature
The registered company address is: Gewerbestrasse 11, 6330 Cham, Switzerland

Preface

This book is about biogas production using anaerobic digestion technology, with an emphasis on waste utilization/valorization. Biogas is arguably the most commercialized type of biofuel. Nevertheless, there has been a renewed interest in biogas production from waste-oriented feedstock worldwide and in particular in the developing countries. A number of parameters including increasing energy demands and worsening environmental conditions are among the main driving factors of this surge in interest in biogas. The present book, which is the six book in the series on Biofuel and Biorefinery Technologies, offers a comprehensive reference guide to biogas production from different waste streams by internationally recognized experts in the field of biogas production form both academia and industry. The 18 chapters cover various aspects of anaerobic digestion technology from the basics, i.e., microbiological aspects to prominent parameters governing biogas production systems, as well as major principles of their operation, analysis, process control, and troubleshooting. In addition, major issues affecting biogas production, including the type of feedstock, pretreatment techniques, production systems, design and fabrication of biogas plants, as well as biogas purification and upgrading technologies are comprehensively reviewed and discussed. Providing in-depth and cutting-edge information on central developments in the field, 'Biogas: Fundamentals, Process, and Operation' also addresses the application of advanced environmental and energy evaluation tools including life cycle assessment (LCA), exergy, techno-economics, and modeling techniques. The book is intended for all researchers, practitioners and students who are interested in the current trends and future prospects of biogas production technologies.

It is expected that the present volume on biogas would assist both the scientific and industrial communities in further developing this industry worldwide. We are thankful to authors of all the chapters for their efficient cooperation and also for their readiness in revising the manuscripts. We also would like to extend our appreciation to the reviewers who in spite of their busy schedule assisted us by evaluating the manuscripts and provided their critical comments to improve the manuscripts. We sincerely thank Dr. Vijai Kumar Gupta and Dr. Maria G. Tuohy and the team of Springer Nature for their cooperation and efforts in producing this book.

Karaj, Iran

Meisam Tabatabaei
Biofuel Research Team (BRTeam) /
Agricultural Biotechnology Research Institute of Iran (ABRII)

Hossein Ghanavati
Agricultural Biotechnology Research Institute of Iran (ABRII)

Contents

Chapter 1
Waste Management Strategies; the State of the Art

Alireza Ghasemi Ghodrat, Meisam Tabatabaei, Mortaza Aghbashlo and Solange I. Mussatto

List of Abbreviations

MSWM	Municipal Solid Waste Management
MSW	Municipal Solid Waste
SWM	Solid Waste Management
WM	Waste Management
MRF	Materials Recovery Facility
RDF	Refuse-derived Fuel
SRF	Solid Recovery Fuel
AD	Anaerobic Digestion
LCA	Life Cycle Assessment
EU	European Union
e-waste	Electronic Waste
EPA	Environmental Protection Agency
ABS	Acrylonitrile butadiene styrene
HIPS	High impact polystyrene

A. Ghasemi Ghodrat
Biofuels Research Team (BRTeam), Karaj, Iran

A. Ghasemi Ghodrat
Arta Sepehr Kavian Co., Tehran, Iran

M. Tabatabaei (✉)
Biofuel Research Team (BRTeam) / Agricultural
Biotechnology Research Institute of Iran (ABRII), Agricultural Research,
Education and Extension Organization (AREEO), Karaj, Iran
e-mail: meisam_tabatabaei@abrii.ac.ir; meisam_tab@yahoo.com

S. I. Mussatto
Novo Nordisk Foundation Center for Biosustainability, Technical
University of Denmark, Building 220, 2800 Kongens, Lyngby, Denmark

M. Aghbashlo (✉)
Department of Mechanical Engineering of Agricultural Machinery,
Faculty of Agricultural Engineering and Technology, College
of Agriculture and Natural Resources, University of Tehran, Karaj, Iran
e-mail: maghbashlo@ut.ac.ir

M. Tabatabaei and H. Ghanavati (eds.), *Biogas*, Biofuel and Biorefinery
Technologies 6, https://doi.org/10.1007/978-3-319-77335-3_1

PC	Polycarbonates
PVC	Polyvinyl chloride
PET	Polyethylene terephthalate
ASTM	American Society for Testing and Materials
TFS	Transfrontier shipment
VOCs	Volatile Organic Compounds
C/N ratio	Carbon to Nitrogen Ratio
CH_4	Methane
CO_2	Carbon Dioxide
N_2	Nitrogen
H_2	Hydrogen
H_2S	Hydrogen Sulphide
NH_3	Ammonia
VS	Volatile Solids
BOD	Biological (Biochemical) Oxygen Demand
COD	Chemical Oxygen Demand
UNIDO	United Nations Industrial Development Organization
t	Tonne
BMP	Biochemical Methane Potential
OM	Organic Material
LAC	Latin America and Caribbean

1.1 Waste Management; A Conceptual Approach

Maximizing resource (material and energy) recovery and minimizing environmental impacts such as contribution to the global warming are two of the very first and important objectives in the solid waste management (SWM) sector, which is considerably developed over the past century (Habib et al. 2013). In fact, the way solid wastes have been managed over the history of human civilization has faced a tremendous level of change, shifting from the focus on public cleansing of the cities to modern waste management strategies. This happened mainly due to the lifestyle changes and swift industrialization process all around the world, which has led to the introduction of new materials and the consequent changes in the types and composition of the generated wastes (Christensen 2011).

It is crucial to note that the definition of the term "waste" is totally subjective and depends on various factors including time, location, income level, state, and personal preference (Table 1.1). In another words, culture, climate, religious, ethnic background, as well as economic abilities could affect what would be defined as waste. For instance, The European Union (EU) defines waste as "any substance or object which the holder discards or intends or is required to discard." This is also supplemented with various examples of items and materials that can be considered as waste within a long list entitled "European Waste Catalogue" (Christensen 2011). Similarly, the term "management," based on the Basel Convention, means

Table 1.1 Influential factors on the definition of the term "waste" adopted from Christensen (2011)

Time	• During scarcity, e.g., war time and embargo, repairing an item may become economical, since buying a new version may be costly or hard to achieve
Location	• For example, the feasibility of using food wastes for animal feeding in rural *vs.* urban areas
State	• Regarding an item's state (price, age, and type of damage), it may be repairable
Income Level	• The higher the income, more food or other stuffs would be likely to be discarded
Personal Preference	• Waste to an individual may not be regarded as waste to another individual

collection, transport, and disposal of hazardous wastes or other wastes, including after-care of disposal sites (UNEP 2014).

As an example of a type of waste and its subsequent relevant issues, electronic waste (e-waste) would be a good option. E-waste, as one of the rapidly growing waste pollution problems worldwide, can seriously contaminate the environment and threaten human health by a variety of toxic substances. Many protocols for this type of waste have been introduced across countries focusing on management, disposal, as well as reprocessing and reutilization of these wastes as raw materials. Overall, developing eco-design devices, proper collection of e-wastes, safe recovery and recycling of materials, disposal of e-wastes by suitable techniques, raising awareness of the impacts of e-waste, and forbidding the transfer of used electronic devices to developing countries are the dominant factors to be considered to accomplish a successful e-waste management. In spite of that, heavy movements of hazardous wastes, especially e-wastes into Asia, notably India, China, and Pakistan have been observed contrary to the instructions set forth by international protocols, e.g., The Basel Convention (Pariatamby et al. 2015; Kiddee et al. 2013). Almost 5% of the total waste volume generated globally is contributed by e-wastes and, according to the Environmental Protection Agency (EPA), 30–40 million personal computers are estimated to reach their "end-of-life" each year. This means that a huge amount of hazardous materials is ready to be added to the environment, while a variety of valuable materials and minerals can be recovered. There are a number of companies around the world utilizing proper technologies for recovering largely ferrous metals, aluminum, copper, circuit boards, as well as plastics, acrylonitrile butadiene styrene (ABS), high impact polystyrene (HIPS), polycarbonates (PC), and ABS-PC from e-wastes (CP Group 2017).

In the year 2012, approximately 1.3 billion t of MSW were generated globally, and this figure is expected to rise to approximately 2.2 billion t by the year 2025

(Rajaeifar et al. 2017). Regarding the environmental impacts, soil, water, and especially air are prone to be enormously influenced by the unsafe disposal of wastes (Pawłowska 2014). Groundwater pollution at landfills, air quality affected by gaseous emissions through incineration, as well as metals remained in soil and crops after the utilization of MSW-oriented compost are some of the examples of contaminations caused by unsafe SWM. Such consequences have led to the implementation of much more strict regulations and laws in the waste management sector to meet the concepts as sustainability (Christensen 2011).

Sustainability, defined as "development that meets the needs of the present without compromising the ability of future generations to meet their own needs," (Commission 1987) has environmental, social, and economic dimensions with a focus on long-term issues. As a matter of fact, this definition states that each generation has to take the responsibility of their very own-generated problems and try to solve them with the help of local solutions. To do so, there are quite a few protocols covering different aspects of waste management including The Basel Convention, The Montreal Protocol, The Kyoto Protocol, and The Aarhus Convention along with a number of powerful tools such as LCA introduced to perform feasibility studies.

Moreover, considering the above-mentioned issues as well as the complexity and high expenses of waste management in the modern days, new strategies and systems have also been introduced to this sector (Christensen 2011). Among the most important strategies in SWM throughout the world, considering the waste hierarchy (Fig. 1.1), 3R—"reduce, reuse, and recovery"—is one-of-a-kind, frequently used

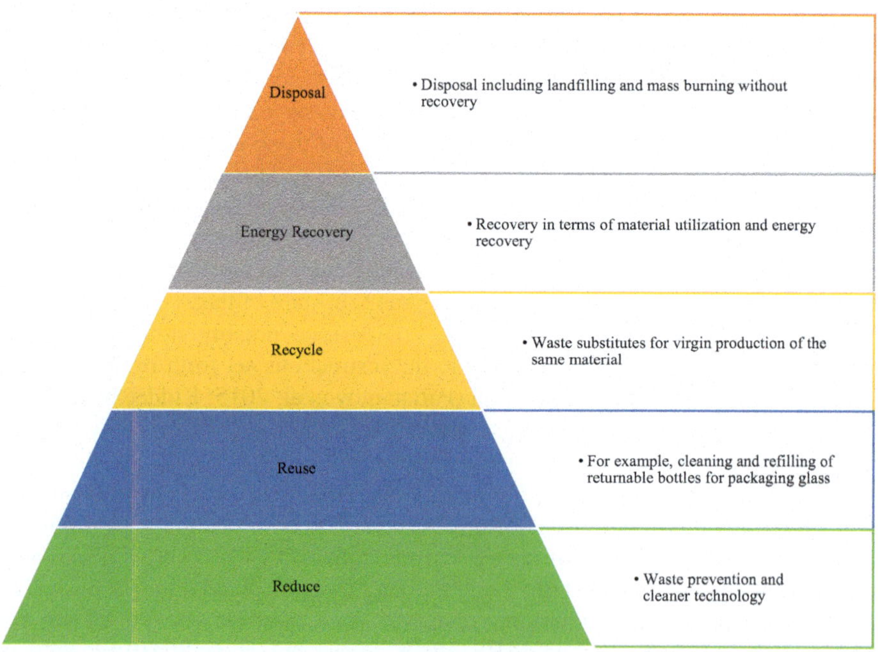

Fig. 1.1 Waste hierarchy adopted from Christensen (2011) and Richards and Taherzadeh (2016)

by the Western world and some parts of the Asia (especially Japan) since early 1980s (Richards and Taherzadeh 2016). MSW through the implementation of a systematic management can serve as a precious resource for different purposes (World Energy Council 2013). It is worth mentioning that resource (material and energy) recovery, as an important step in the waste hierarchy, implies not only the utilization of waste to produce materials and harvest different forms of energy carriers, but also the efforts in the context of avoiding environmental impacts from production of raw materials and simultaneously waste disposal (Christensen 2011). It is also critical to highlight waste collection as well, which contributes a considerable part of WM expenses (usually about half of the costs of a typical waste management system). In better words, within a comprehensive WM system, all factors from the point of waste collection to final disposal have to be considered (Dubanowitz 2000).

Waste management systems can be divided into six different categories namely Landfilling, Composting, MRF, AD, Incineration, and RDF/SRF. Each system has its own characteristics with a wide range of Waste-to-Energy (WtE) technologies offered around the world. In general, WtE technologies can be defined as any waste treatment processes that create energy from a waste source in any forms of energy carrier, i.e., electricity, heat, or transportation fuels (World Energy Council 2013). Based on a report by World Energy Council, increase in the amount of generated waste, high costs of energy, growing concerns of environmental issues, and restricted landfilling capacities are the summarized main drivers for the growth in WtE market in the past decades (World Energy Council 2013). In 2013, the global WtE market faced a growth of 5.5% with respect to its preceding year and reached a value of 25.32 billion USD. Among the various WtE technologies, thermal energy conversion was at the top and accounted for 88.2% of total market revenue in the same year (World Energy Council 2016).

It should be highlighted that while a system with a particular technology is suitable for a region, it may lead to a disaster for another region. Therefore, a comprehensive investigation on different influential factors including demographic, meteorological, and social background, as well as industrial zones, water, and electricity grid availability has to be conducted prior to the decision-making step by well-educated experts.

1.1.1 Global Status

The degree of industrialization, life style, local climate, and economic development are the prominent influential factors on MSW generation rates. As a rule of thumb, the greater the population, the higher the economic development, and the higher the rate of urbanization, all will lead to a higher rate of municipal solid waste production in addition to the change in its composition and treatment technologies (World Energy Council 2013). In this section, population (in million), total MSWs generation (in million tons) and MSWs generation per capita (kg person^{-1} day^{-1})

as well as changes in the contribution of different MSWs treatment options in various parts of the world (i.e., The United States of America (USA), EU-27, Australia, Japan, Iran, Africa, Middle East, East Asia and the pacific region, Eastern Europe and central Asia, China, as well as Latin America and the Caribbean) have been graphically presented (Figs. 1.2, 1.3, 1.4, 1.5, 1.6, 1.7, 1.8, 1.9, 1.10, and 1.11).

As it can be seen from the above figures, various strategies are applied in different regions and countries depending on their distinct local conditions. For example, a 4.7% increase in the amount of incinerated waste (i.e., from 72.02% in 1990 to 76.72% in 2010) was recorded in Japan. On the other hand, the total generated waste in Japan was decreased by 9.8%, while MSW generation per capita was also reduced from 1.11 kg person^{-1} day^{-1} to 0.97 kg person^{-1} day^{-1} over the same time period. These promising improvements could be attributed to technological developments along with implementation of appropriate waste management strategies, laws, and regulations (Rajaeifar et al. 2017).

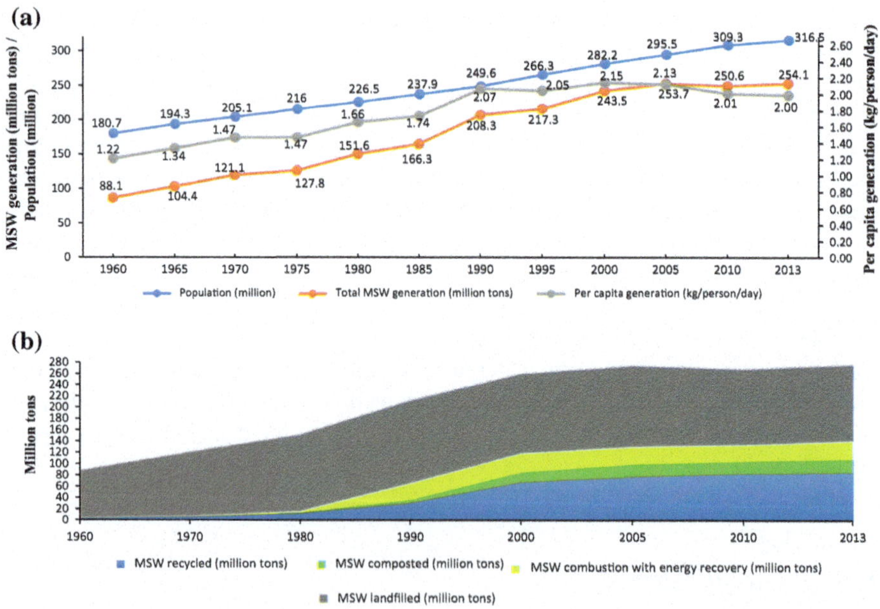

Fig. 1.2 **a** Population (in million), total MSWs generation (in million tons) and MSWs generation per capita (kg person^{-1} day^{-1}) in the USA (1960–2013). **b** Changes in the contribution of different MSWs treatment options in the USA (1960–2013) (Rajaeifar et al. 2017). With permission from Elsevier. Copyright © 2017

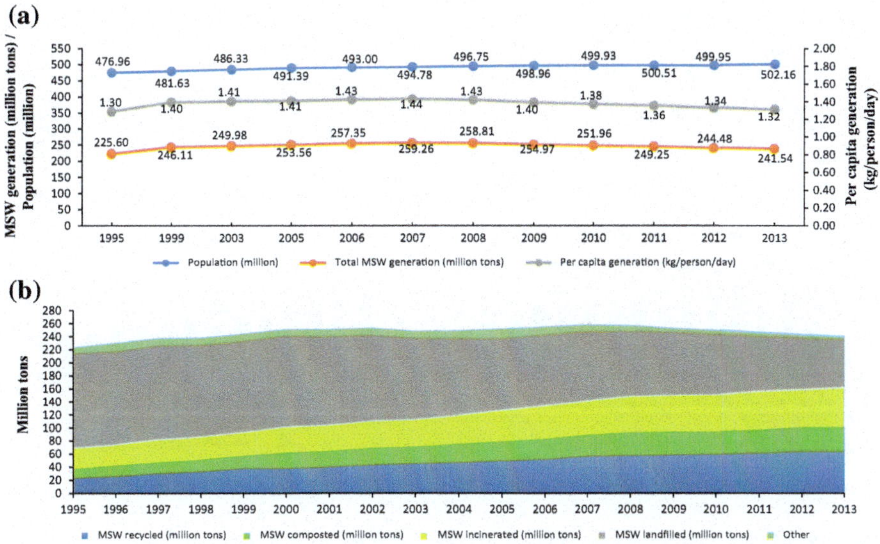

Fig. 1.3 **a** Population (in million), total MSWs generation (in million tons) and MSWs generation per capita (kg $person^{-1}$ day^{-1}) in the European Union (EU)-27 (1995–2013). **b** Changes in the contribution of different MSWs treatment options in the European Union (EU)-27 (1995–2013) (Rajaeifar et al. 2017). With permission from Elsevier. Copyright © 2017

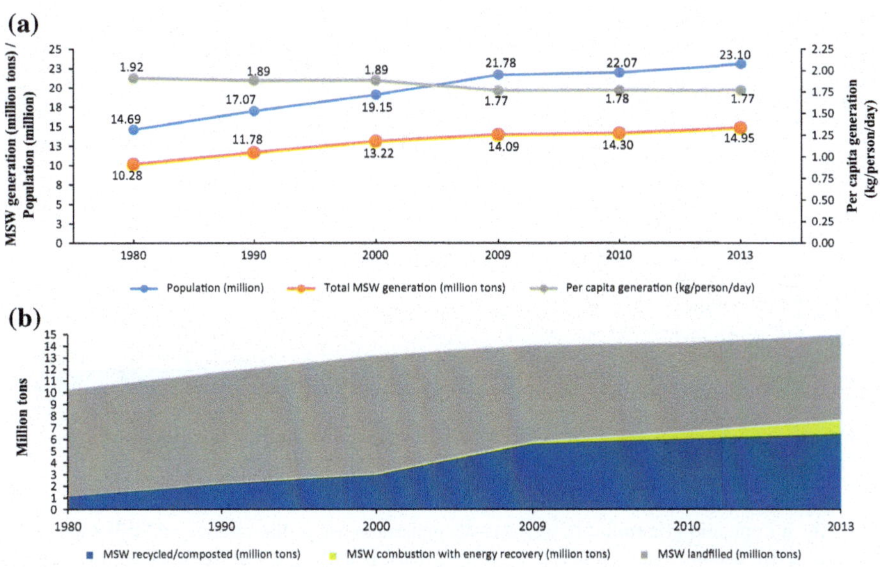

Fig. 1.4 **a** Population (in million), total MSWs generation (in million tons) and MSWs generation per capita (kg $person^{-1}$ day^{-1}) in Australia (1980–2013). **b** Changes in the contribution of different MSWs treatment options in Australia (1980–2013) (Rajaeifar et al. 2017). With permission from Elsevier. Copyright © 2017

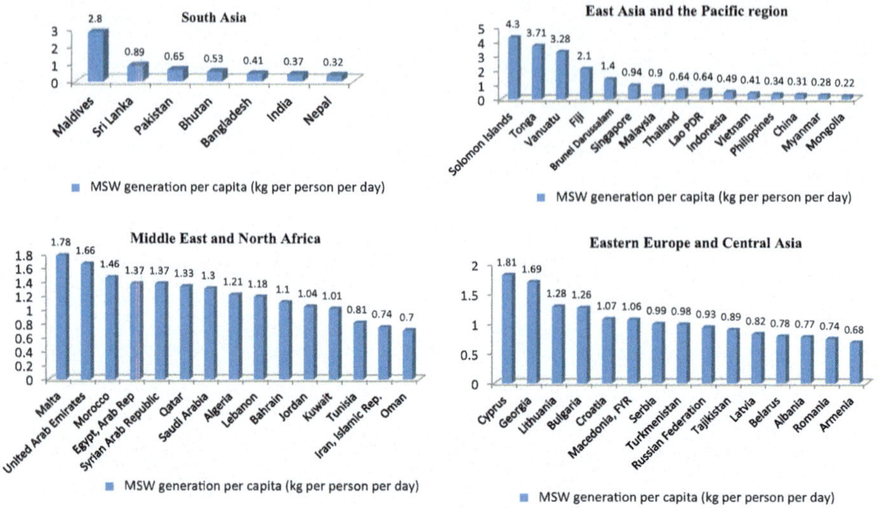

Fig. 1.5 MSWs generation per capita in different regions in Asia (Rajaeifar et al. 2017). With permission from Elsevier. Copyright © 2017

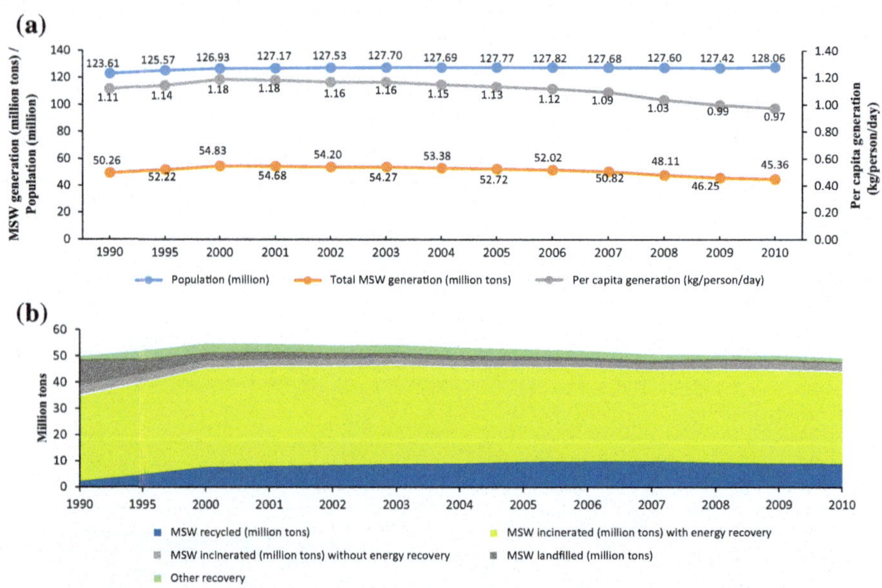

Fig. 1.6 **a** Population (in million), total MSWs generation (in million tons) and MSWs generation per capita (kg person^{-1} day^{-1}) in Japan (1990–2010). **b** Changes in the contribution of different MSWs treatment options in Japan (1990–2010) (Rajaeifar et al. 2017). With permission from Elsevier. Copyright © 2017

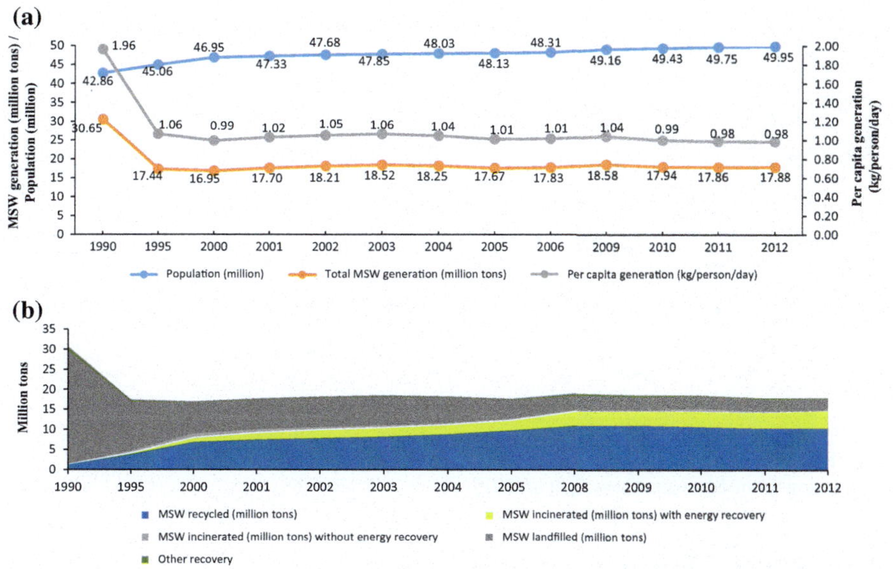

Fig. 1.7 **a** Population (in million), total MSWs generation (in million tons) and MSWs generation per capita (kg person^{-1} day^{-1}) in South Korea (1990–2012). **b** Changes in the contribution of different MSWs treatment options in South Korea (1990–2012) (Rajaeifar et al. 2017). With permission from Elsevier. Copyright © 2017

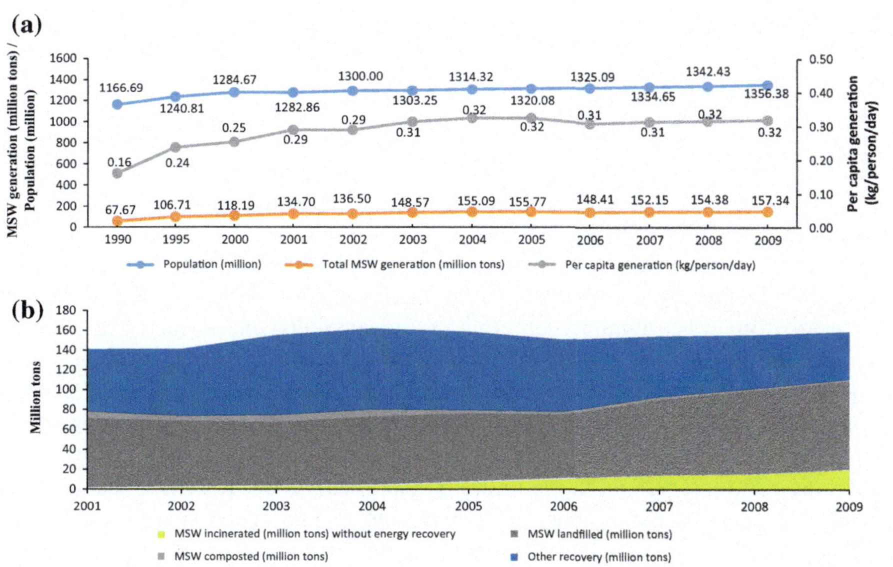

Fig. 1.8 **a** Population (in million), total MSWs generation (in million tons) and MSWs generation per capita (kg person^{-1} day^{-1}) in China (1990–2009). **b** Changes in the contribution of different MSWs treatment options in China (2001–2009) (Rajaeifar et al. 2017). With permission from Elsevier. Copyright © 2017

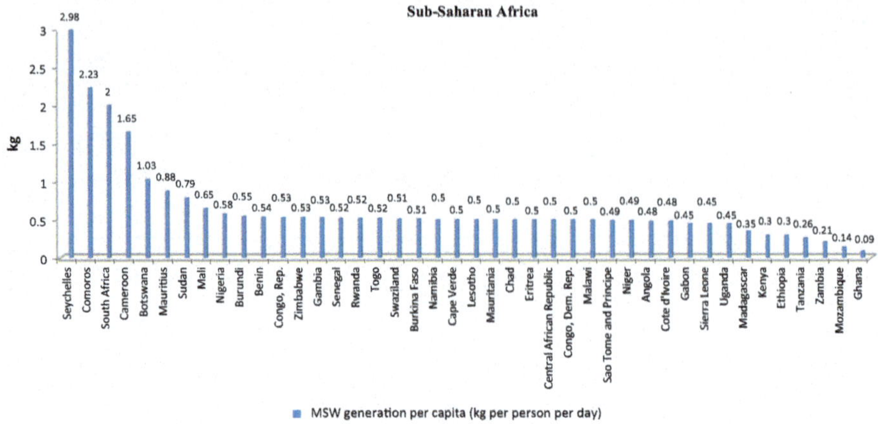

Fig. 1.9 MSWs generation per capita in the Sub-Saharan Africa (Rajaeifar et al. 2017). With permission from Elsevier. Copyright © 2017

1.2 Waste Management Strategies

Generally, MSW technologies are divided into two main categories, namely, mechanical and biological treatment, and thermal treatment. Each one of them is also classified into some subcategories as presented in Fig. 1.12.

1.2.1 Materials Recovery Facility (MRF)

MRF, as a critical and vital step in MSWM strategies, consists of three main stages of separation, processing, and storing, aimed at maximizing the quantity of the processed recyclables. It also targets consistent production of clean products from heterogeneous materials containing some levels of contamination with the highest possible revenue in the market. From environmental point of view, material recovery from waste within such contexts substantially offsets the environmental burdens attributed with resource extraction. Based on a study, it is estimated that every t of MSW is responsible for the extraction of about 71 t of upstream materials (Zaman 2016). MRF separates and processes the accepted materials through different operational units and, at the end, stores them as raw materials for remanufacturing and reprocessing in the future (Dubanowitz 2000; Kessler Consulting Inc 2009). In fact, it is the primary systematic and technological step in a particular MSWM strategy and can be considered as the feed supplier of the other waste management systems, e.g., incinerator. Figure 1.13 illustrates the sequence of developing an MRF facility for separating MSW as feedstock.

The choice between manual and mechanical separation techniques is an important issue in the operation of such facilities. With regard to the high labour

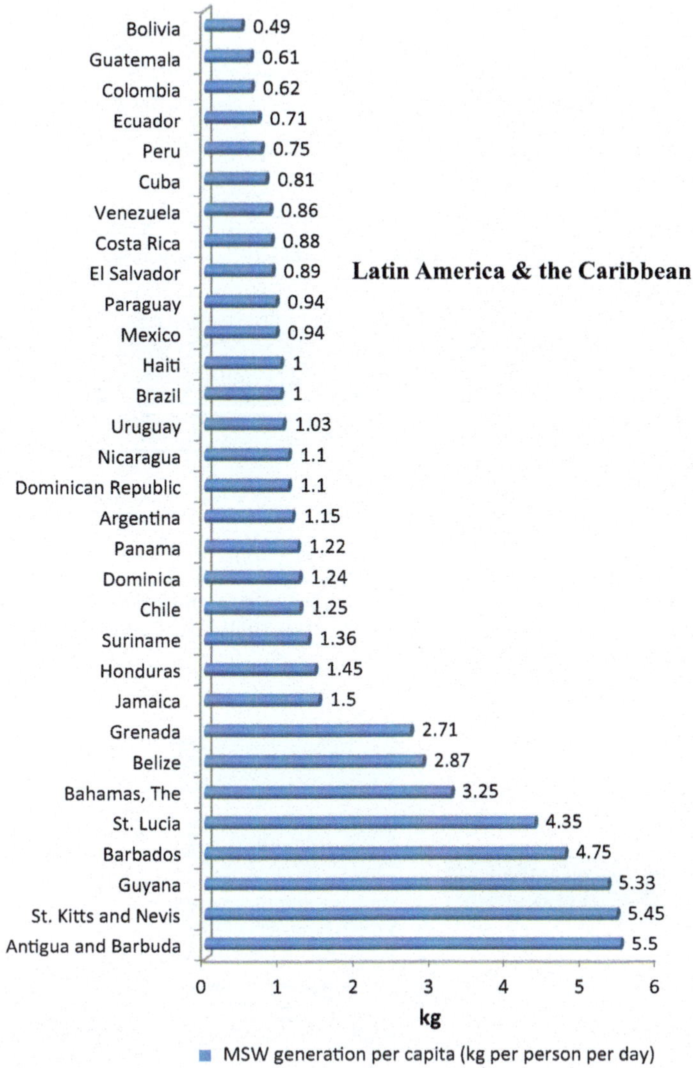

Fig. 1.10 MSWs generation per capita in LAC region (Rajaeifar et al. 2017). With permission from Elsevier. Copyright © 2017

costs, the amount of rejected materials, processing rates, adjustability and flexibility to new waste streams, the level of health and safety risks, and separating difficult-to-detect materials (e.g. PVC and PET), automated processing is a much more cost effective choice. However, given the potential of manual sorting in producing higher quality material recovery, automated sorting is usually accompanied with manual sorting in some units. The types of entering materials, the final quality, the inputs and outputs of each subsystem, and the distinguishing

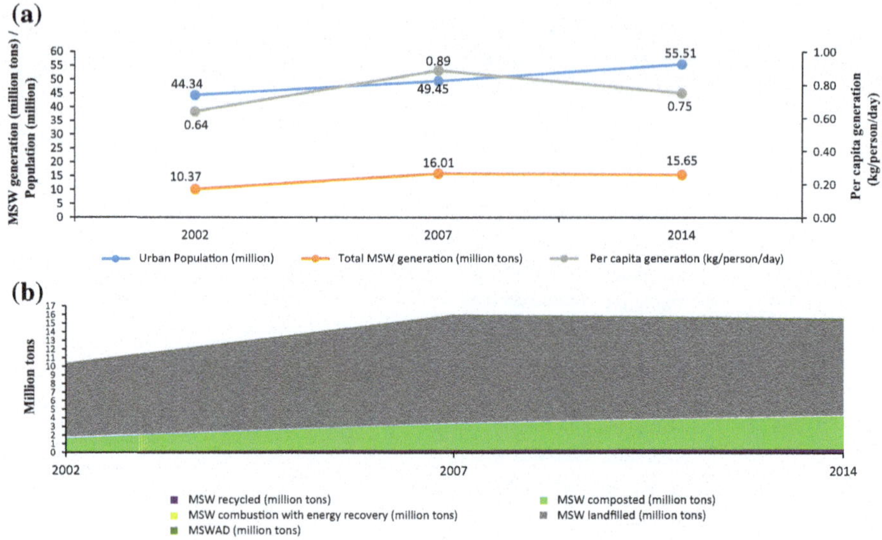

Fig. 1.11 **a** Population (in million), total MSWs generation (in million tons) and MSWs generation per capita (kg person^{-1} day^{-1}) in Iran (2002–2014). **b** Changes in the contribution of different MSWs treatment options in Iran (2002–2014) (Rajaeifar et al. 2017). With permission from Elsevier. Copyright © 2017

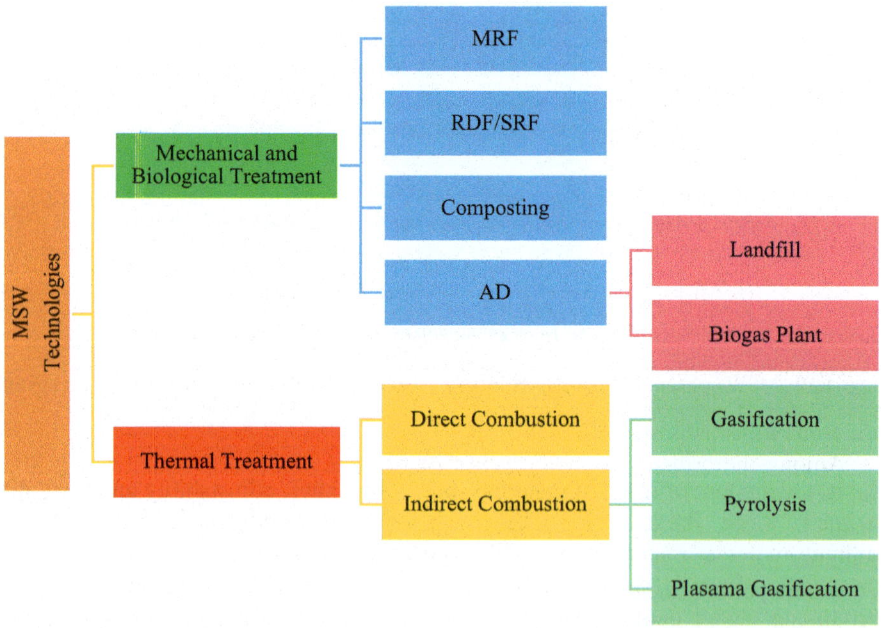

Fig. 1.12 MSW technologies

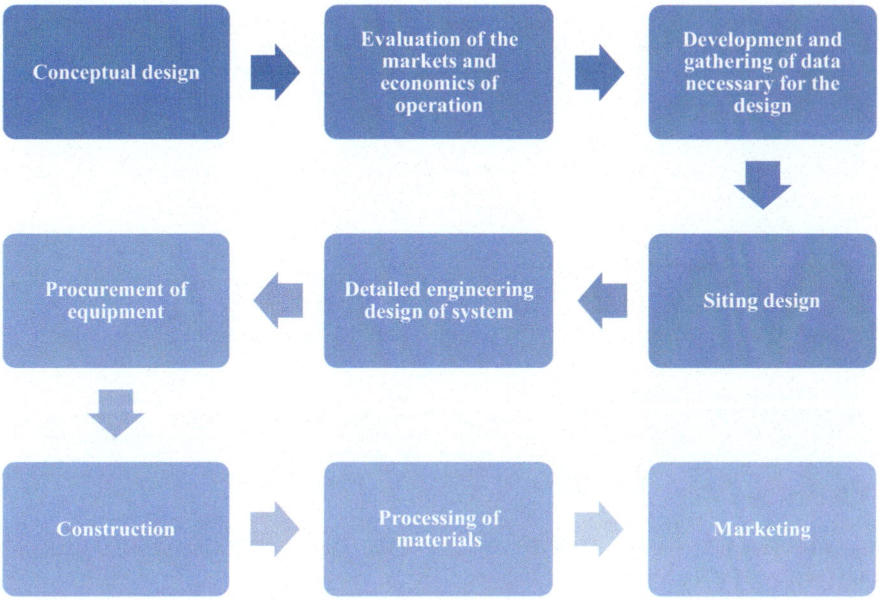

Fig. 1.13 The sequence of developing an MRF facility for MSW as feedstock adopted from Dubanowitz and Themelis (2000)

characteristics of the desired products are the major considerations before designing such unit operations. Overall, an automated MRF may consists of many unit operations with each equipped with various high-tech equipment. These equipment can be any of the followings: (1) Conveyor System; (2) Ferrous Metals Separation; (3) Screening; (4) Air Classification; (5) Non-ferrous Metal Separation; (6) "Detect and Route" Systems, which itself consists of Glass Separation, Plastic Separation, and Paper and Carton Separation; (7) Size Reduction; as well as (8) Compactors and Balers (Dubanowitz 2000).

As an influential factor in designing MRFs, the condition of the input materials will significantly affect the configuration of the processing line. This means that the inflow materials' condition, or the manner by which wastes are collected, will determine the costs and resource utilization of the MRF, as well as its building layout and equipment. In general, MSW can be collected and introduced in four different ways as presented in Fig. 1.14.

Among the critical considerations in developing an MRF unit is to conduct a preliminary investigation on the current recycled materials market and the financial status in a region of interest. This means that a basic requirement for planning new facilities, or for evaluating existing ones, is the simulation of their technical and economic performance (Cimpan et al. 2016). A well-designed MRF unit can cut the municipality's expenses to an acceptable extent by separating the wastes in one stage. Based on a report, the city of Los Angeles faced a 140% increase in the amount of collected waste due to the shift from two-stream to single-stream

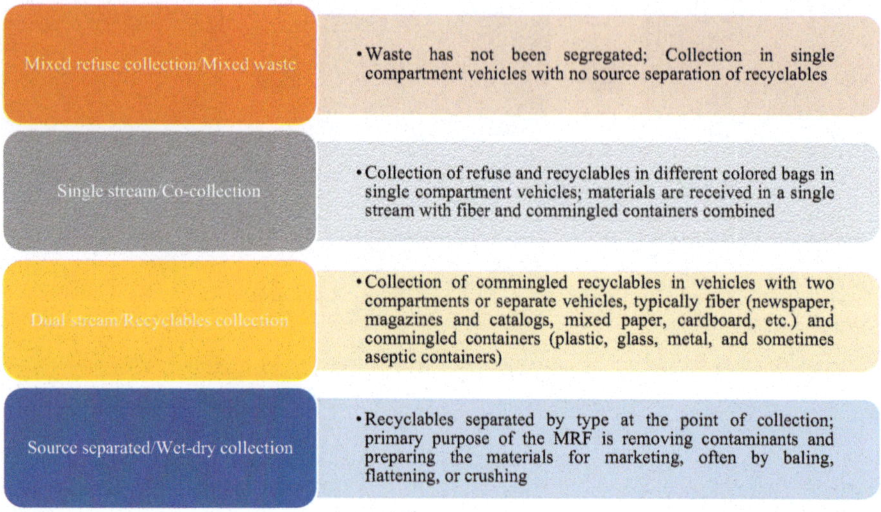

Fig. 1.14 Four different ways of MSW collection before entering MRF adopted from Dubanowitz and Themelis (2000), Kessler Consulting Inc (2009)

collection scheme in which a highly automated MRF was used. This was followed by a 25% reduction in the collection expenses. In general, by increasing the level of automation, higher speed of operation, lower costs, and higher quality of recovery could be achieved (Dubanowitz 2000).

The extent to which each country recycles its generated wastes depends on various factors including legislations, availability of finance, technological availability, cultural habit-building practices, etc. Among the top 10 recycling countries around the world, the highest rate of recycling belongs to Austria, where 63% of all waste is diverted from landfills. The other following 8 countries are Germany with 62%, Taiwan with 60%, Singapore with 59%, Belgium with 58%, South Korea with 49%, United Kingdom with 39%, Italy with 36%, and France with 35%. The last country is the United States of America which, in the year 2014, produced about 25% of the world's generated waste while only recycled 34% of this huge quantity of wastes (World Bank 2010; Aid 2015; General Kinematics 2016). From another point of view, the higher the landfill tipping fees, the higher the chance of recycling becoming economically feasible as a waste management practice. For instance, between 1985 and 1992, the national average landfill tipping fee increased by more than 500% in the northeaster region of the United States. This substantial increase together with an increased reliance on costly and contentious waste exportation made recycling as an economical and proven approach for waste management (Dubanowitz 2000). Nowadays, the use of systems featuring a variety of equipment, from screens, to optical sorters, to cutting-edge electrical solutions are the state-of-the-art technologies to meet the highest quality standards. It should also be noted that an MRF facility can be designed extremely automated, but, as mentioned earlier, the higher the automation, the higher the capital cost as well (Advancedmrf 2017; CPG Group 2017).

1.2.2 Refuse-Derived Fuel (RDF)—Solid Recovery Fuel (SRF)

In order to mitigate the devastating consequences of landfilling along with an efficient utilization of the energy contained in waste, RDF and, its new version, i.e., SRF, have been introduced as strategy in MSWM scenarios to be used in power plants, cement kilns, and other combustion plants. The creation of RDF dates back to the time of the energy crisis in the 1970s. In spite of the different definitions offered for RDF and SRF across different countries, based on Italian decrees (Ragazzi and Rada 2012), the RDF and SRF are defined as follows (Rotter 2011; Ragazzi and Rada 2012):

> RDF is fuel derived from municipal solid waste through treatments aimed to the elimination of substances hazardous for combustion and to guarantee an adequate lower heating value (LHV), and to comply with the technical norms for its characterization.
>
> SRF is the solid fuel prepared (means processed, homogenized and up-graded to a quality that can be traded amongst producers and users) from non- hazardous waste to be utilized for energy recovery in incineration or co- incineration plants and meeting the classification and specification requirements laid down in CEN/TS 15359.

In fact, the two fuels are termed based on their characteristics. Nowadays, the terminology RDF is known as unspecified waste after a basic processing to increase the calorific value and usually refers to the segregated, high calorific fraction of MSW, commercial or industrial wastes (Rotter 2011). SRF, as newer terminology, refers to non-hazardous waste, utilized for energy recovery, and is more homogeneous as well as less contaminated than the generic RDF (Garg et al. 2007). Figure 1.15 shows different unit operations in an RDF production plant.

Based on a classification by the American Society for Testing and Materials (ASTM), RDF is divided into seven categories, depending on the type of processing and not based on chemical or physical parameters (Rotter 2011). As an important advantage, RDF/SRF can be shipped, under transfrontier shipment (TFS) regulations, across countries as an energy carrier (Clarity Environmental 2017). This type of energy carrier can be co-combusted in cement kilns plants, in which up to 40% of their firing thermal capacity can be provided using high calorific waste fuels, co-combusted in coal fired boilers (lignite or hard coal), or mono-combusted in RDF-fired boilers (grate firing or fluidized bed technology) with the aim of district heating or steam and electricity for industries (Rotter 2011).

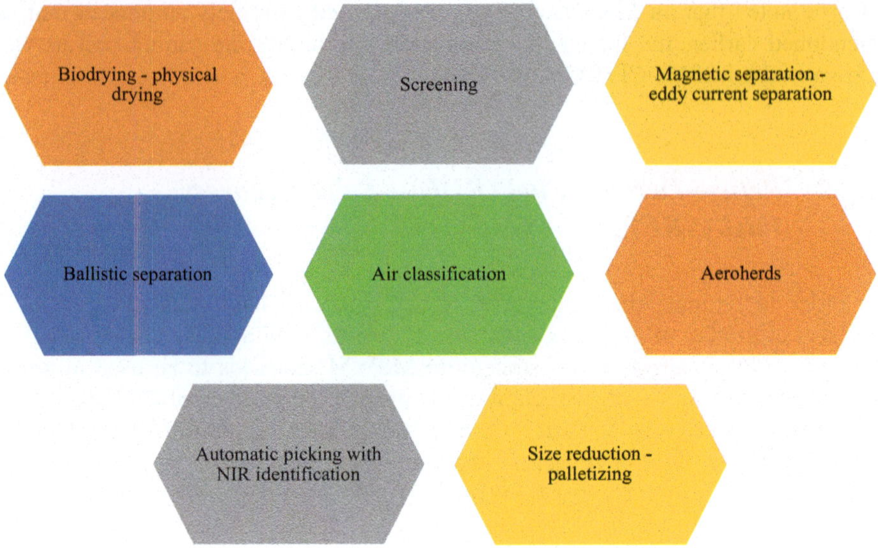

Fig. 1.15 Various unit operations in RDF production. Adopted from Christensen (2011)

1.2.3 Landfill

Landfilling, i.e., dedicated use of land for disposing waste in an engineered facility, is still the predominant and widespread concept for the MSWM of the waste generated by about 7.5 billion of the global population. This prevalence is mainly due to its being the most cost-efficient method of waste disposal, it does not mean that this technology is associated with low environmental risks though. In fact, water, soil, and particularly air are prone to be contaminated by the deposition of wastes in landfills. More to this, a great deal of concern is about its long-term negative impacts on the future generations, since the decomposition of organic materials (OM) under anaerobic conditions takes place at a low rate. Therefore, an appropriate design, considering the type of waste that has to be landfilled together with various standards, conditions, and regulations, should be implemented (Christensen 2011; Pawłowska 2014; Richards and Taherzadeh 2016).

It would be wise to implement a resource recovery facility, moving toward a more sustainable society, even if landfilling is the only option (Richards and Taherzadeh 2016). By constructing and implementing an engineered collection system along with the utilization of complex bio-chemical conversion processes (including different phases like Initial Adjustment, Transition Phase, Acid Phase, Methane Fermentation, and Maturation Phase), biogas, as an energy carrier, can also be harvested from landfills directly (World Energy Council 2013). Typical major biogas composition in a landfill site is: CH_4: 47.7; H_2O: 20; H_2S: 2.4; and CO_2: 29.6 (Fehr 2010).

Technologies related to landfilling can be categorized into five distinct types, namely, aerobic, semi-aerobic, hybrid, anaerobic, and landfill as deposit of inert waste. More in-detail explanations about each type of landfilling technology, as an approach to minimize the impacts of landfills on the atmosphere and the environment, can be found in Pawłowska (2014). The aim of the complex system of interrelated components and sub-systems of a landfill is to break down and stabilize disposed wastes over a long period of time. Each of these types of landfills can to some extent address different concerns including disease vectors such as flies, mosquitoes, cockroaches, rats, and other pests, as well as groundwater contamination because of leachate production (Dhamija 2006).

Each of the above landfill types may be considered for a region based on its local conditions; however, among them, the semi-aerobic bioreactor (known as Fukuoka method due to its first implementation in the city of Fukuoka in Japan in 1975) is one of the best choices in designing a landfill with low capital and operational costs, while meeting the regulations and expectations. Low degree of technical demand, machines, devices, and ease of operation and maintenance, decrease in the load of waste water contamination by quick drainage of waste water, contribution to the prevention of Global Warming by control of the discharge of methane gas, early stabilization of landfill ground by promoting waste bio-degradation, wider alternatives of material for construction, and lower cost of construction are the other advantageous of Fukuoka landfilling method (Global Environment Center Foundation 2006; Fukuoka Municipal Government 2010).

Nowadays, developing sustainable techniques and technologies to enhance the stabilization of landfills and to harvest energy from landfills are the major objectives of modern landfilling. For instance, landfill reclamation, a treatment operation and perhaps the most sustainable manner in operating a landfill, has been utilized in a few regions. Many other practices may also be applied in landfilling to achieved the aforementioned objectives (Krook et al. 2012; Ritzkowski and Stegmann 2012; Pawłowska 2014; Townsend et al. 2015; Wolfsberger et al. 2016).

1.2.4 Compost

Composting, the biological decomposition of organic matter under aerobic conditions, is an excellent and valuable waste management technology. In recent years, composting of MSW has received much attention as a way of ameliorating the soil's physicochemical properties and improving the biological responses of cultivated lands. Eradicating several food-borne diseases (caused by bacteria, viruses, and parasites), providing nutrients for crop production, avoiding methane production and its release to the atmosphere, conserving moisture in soil, improving soil conditions for better crop growth, producing a product that can improve plant growth, reducing runoff and erosion, as well as minimizing landfilling or incineration of waste are the other common benefits of composting practiced by both with the developing and developed countries (Epstein 2011; Srivastava et al. 2016).

There are eight influential factors in composting operation viz. turning frequency, temperature, C/N ratio, moisture content, electrical conductivity, aeration, pH, and particle size. More to this, oxygen and moisture, as the two prominent operational parameters, together with temperature and nutrients, especially carbon and nitrogen, affect the rate of decomposition of the organic matter during composting and are required to be maintained at an optimum level. It has been proved that these operational factors are interconnected. For instance, turning frequency affects total nitrogen, pH, moisture content, carbon to nitrogen (C/N) ratio, dry matter, total carbon, and temperature of composting piles, and, as another example, the higher the O_2 concentration, the lower the concentration of organic acids in the compost leading to a rapid decomposition of the acids. More in-detail information can be found in numerous literatures (Epstein 2011; Onwosi et al. 2017).

Various technologies have been introduced for composting organic materials. There may also be different classifications, among which the concise generic classification tabulated in Table 1.2 is widely approved. In order to choose the most appropriate system, many factors should be considered, that is, economics and cost,

Table 1.2 The generic classification of composting technologies and systems. Adopted from Epstein (2011)

Static Systems	*Passively Aerated Windrows*	Relies on convective air to provide oxygen and to achieve favorable temperatures and stabilization; uses perforated pipes open to the atmosphere; feedstock with a bulking agent is piled over the pipes; not an approved U.S. Environmental Protection Agency (USEPA) method for pathogen reduction for the use of sewage sludge or biosolids; used as a low-cost technology by farmers for composting animal wastes
	Forced Aeration —Static Pile	Originally developed using negative air, i.e., suction, leading to reduction in odors by sucking the air through pipes (negative aeration) and filtering the air into a biofilter; currently utilized using positive air, i.e., forcing the air through the pile, leading to head loss reduction and unrequired external biofilter as the advantages; availability of numerous configuration, e.g., totally open/enclosed
	Bin/Container/ Bag/Tunnel	Principally applied to small facilities; can be very effective in odor control; usually ventilated and are horizontal; different in the way these are loaded, unloaded, and ventilated; mostly used for relatively low volumes of feedstock and where the location is sensitive to odors; require a mixing and final preparation of the product through screening or other techniques
	Silo/Vertical Reactors	Principal problems were excessive compaction, poor aeration, and difficulty in extracting the material; currently are not being built and many have been discontinued

(continued)

Table 1.2 (continued)

Turned/ Agitated Systems	*Windrow*	Essentially operated outdoors; uses turning system where the machine straddles the windrow and agitates the material; attributed with a great deal of emissions; odors can be a significant problem; major advantages are large volume of material it can handle and excellent quality of mixing and pulverizing the material; varies in width and height depending on the equipment used; windrows are 1.5–2.7 m (5–9 feet) high and 2.7–6.1 m (9–20 feet) wide with spaces considered in between for the turning machine; aeration is provided primarily through convective airflow; turning not only provides mixing, but also improves porosity and breaks up the particles
	Drum/Kiln	Have been used in many facilities in the Europe, but to a very limited extent in the United States; uses elongated drums to mix the solid waste and biosolids; mixture is then composted in an agitated bin system; limited temperatures are obtained as well as limited biological degradation of the feedstock; retention time in the drum varies with the technology; stabilization of the compost may be needed; the drum does not lead to complete composting; retention in drums is usually from 24 h to 7 d, depending on manufacturer specifications; additional composting and curing are usually done in aerated bays or windrows
	Agitated Bed	Numerous variations of the agitated bed; horizontal systems using turning machines, paddles, or other turning devices; principally used in the United States for composting biosolids; all are enclosed

location, amount of materials to be handled, type of feedstock, as well as state, country, or local regulations (Epstein 2011).

Sewage sludge, biosolids, septage or night soil, manure, animal mortalities, food waste, yard waste, MSW, industrial wastes, and military wastes are the different types of feedstocks in composting. Numerous factors such as feedstock source and ratio used, toxic compounds, the composting design, maturation length, and procedure adopted during the process of composting are the determinative factors in the quality of the compost obtained from MSW (Epstein 2011; Srivastava et al. 2016).

It has been observed that odours or gas emissions, lack of uniformity of compost maturity index, leachate generation, and subsequent concerns about potential diseases, bioaerosols, or impacts of chemicals, raised by citizens, are the most important operational obstacles facing composting operations. With this in mind, the major focus has been shifted from the utilization of compost and its importance in horticulture, erosion control, plant pathology management, and other uses to emissions and their control over the past decades. In another word, composting has evolved into a more sophisticated technology with environmental and public health aspects as the main focus. More specifically, odour management, volatile organic compounds (VOCs) reduction, and bioaerosols management are the technological points with greater emphasis (Epstein 2011).

Currently, state-of-the-art bioreactor (biological air treatment) design, new indices for determining compost maturity, developing the means to harness heat from composting process as bioenergy, modelling of gas compounds removal and microbial structure analysis, developing technologies related to odour treatment/ control (use of additives), use of inexpensive pre-treatment processes and genetically modified strains as microbial inoculum, as well as moving toward more cost-effective and efficient processes are the cutting-edge research fields and developments (Onwosi et al. 2017).

1.2.5 Biogas

Biogas, the product of the complex biochemical decomposition of organic materials, mainly consisting of 60–70% methane (CH_4), 30–40% carbon dioxide (CO_2), together with the other gases, i.e., nitrogen (N_2), hydrogen (H_2), hydrogen sulphide (H_2S), ammonia (NH_3), as well as water vapour. It is produced through an AD process by consortia of bacteria and archaea. In another word, it is a complex microbial process occurring naturally in oxygen-free environments and is considered as one of the most efficient methods for conversion of biomass to CH_4. The process may be divided into four steps viz. hydrolysis, acidogenesis, acetogenesis, and methanogenesis. A wide range of materials including agricultural wastes, MSW, food waste, industrial waste and wastewater, as well as crops may be considered as feedstock for the AD process (Rapport et al. 2008, 2012; Ullah Khan et al. 2017).

Each of the above materials has their own potentials for biogas, or more specifically biomethane, production. Volatile solids (VS) content, biological (biochemical) oxygen demand (BOD), chemical oxygen demand (COD), C/N ratio, and presence of inhibitory substances are among the most important feedstock parameters to be considered. Not only do the feedstock characteristics affect the performance of AD processes, but also do many factors including reactor design and operational conditions, either by process enhancement or inhibition. Biogas production potential should also be investigated through one or some of the various methods as a crucial step in designing a biogas plant (Jingura and Kamusoko 2017). These methods are broadly divided into two categories namely experimental and theoretical methods. More in-detail information regarding the subcategories of the methods used can be found in Table 1.3.

Many design options have been proposed for AD systems including wet, dry, thermophilic, mesophilic, batch, continuous, single-stage, and multi-stage configurations. However, the process itself is divided into two general categories viz. wet and dry, or, in another word, the AD process is applied to feedstocks ranging from highly liquefied to the ones with high solid contents (e.g., MSW). Likewise, the AD system can also be divided into batch or continuous and single-stage process or

Table 1.3 Different methods for determination of biomethane potential of feedstock*

Method type			Description	Advantages	Disadvantages
Experimental method	*BMP Test (Biochemical Methane Potential)*	Conventional	• Mixing an organic feedstock with an inoculum in distinct operational conditions as the general principle • Physically quantify the gas produced by manometric or volumetric method • The biogas composition is determined by GC • Various technical approaches and experimental sets, e.g., Specific Methanogenic Activities (SMA) test, Anaerobic Biogasification Potential (ABP) test, are used • Influential parameters temperature; pH; inoculum; substrate/inoculum ratio (S/I); particle size; stirring intensity; headspace flushing	• Easy to use • Inexpensive • Repeatable	• Time wasting • Resource consuming
		Automatic	• Automatic Methane Potential Test System (AMPTS) was developed by the Bioprocess Control Sweden Company • Utilizes the basic principle of the conventional BMP test • Methane production is directly measured on-line by means of liquid displacement and buoyancy method • A new version developed is called Biogas Activity Monitoring (BAM) • Influential parameters temperature; pH; inoculum; substrate/inoculum ratio (S/I); particle size; stirring intensity; headspace flushing	• Uses less labour • The equipment is inexpensive • Provides high quality and adequate quantity of data	• Require sound systems

(continued)

Table 1.3 (continued)

	Spectroscopy	The Envital Kit	• Recently introduced as a rapid assay based on fluorescence • The Envital® kit is capable of estimating anaerobic biodegradability of sewage sludge in early stages of development • Results are ready in 48 h • Uses a fluorescent redox indicator	• Rapid • High-through put characterisation of more than 32 samples simultaneously in 48 h • Quickly answers operational requests	• Still in early stages of development • Needs more time for validation
		Near-infrared	• Useful tool for quantitative prediction of compounds in pharmaceutical, food and agricultural industries • Recently emerged as a simple and cheap alternative to several laboratory methods for the quantification of BMP • Is used in conjunction with sophisticated chemometrics	• Rapid • Chemical-free • Easy to use (once calibrations have been developed) • Non-destructive	• Machines are too expensive • Calibration is less accurate than wet chemistry • Small calibration sizes can lead to overconfidence • Measurements outside the range of calibration samples are invalid
		Fourier Transform Mid-infrared	• Suitable for in-line determination of volatile fatty acids (VFA), alkalinity, COD, and TOC • Requires the interpretation of the obtained spectra which is more difficult with NIR spectroscopy due to overlapping overtones and combination bands • FTIR-photoacoustic spectroscopy (FTIR-PAS) is the modified version • Creates a thermal wave from the vibration of molecules as a result of the infrared and the sample interface	• Relatively fast and simple to use • Sensitive and requires small amount of sample • Non-destructive method • Universal method: the instrument and software readily available and can be utilised for routine analysis • Multiple sample analysis: can test samples in the form of liquid, gas, powder, solid or film • Relatively cheap as compared with many other methods • Provides qualitative as well as quantitative data	• A single sample requires background scans and many scans due to variations in the spectra caused by environmental factors surrounding the FT-IR spectrophotometer • May require standardization, extensive data collection and skills in chemometric analysis of spectra

(continued)

Table 1.3 (continued)

Theoretical method	*Chemical composition*	• Applicable in cases where elemental composition of the substrate is unknown • Can be done economically within a short period of time • A more rapid and cheaper method than the BMP test	• Rapid • Cheap • Useful in cases where access to • Laboratory facilities is restricted	• The accuracy of each method presumes complete degradation of OM, yet the actual digestibility is usually 27-76% • The BMP is over-estimated • Several inhibitions may occur during the digestion process, and are not considered in these methods • Requires a lot of measurements which time consuming and costly
	Chemical oxygen demand	• COD indirectly measures the amount of organic matter • Can be applied to estimate the CH_4 yield of biomass • Based on the assumption that 1 mol of methane requires 2 mol of oxygen to oxidize carbon to carbon-dioxide and water		
	Elemental composition	• Applied to calculate theoretical BMP • Consists of different formulas, e.g., the Buswell formula—based on the assumption that OM (e.g., $C_nH_aO_b$) is completely degraded to CH_4 and CO_2, the modified Dulong formula—based on energy value of the feedstock that is estimated from its elemental composition		

**Source* Jingura and Kamusoko (2017)

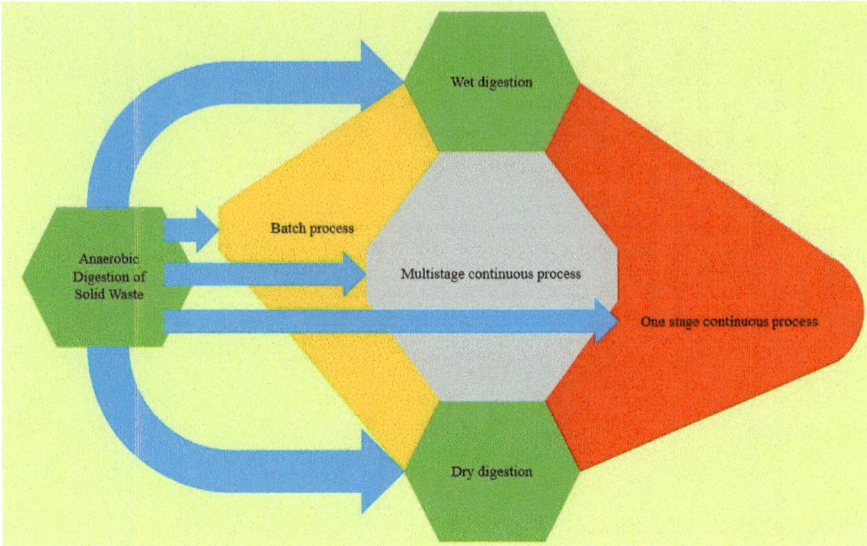

Fig. 1.16 Various AD methods adopted from Richards and Taherzadeh (2016)

two/multistage processes (Richards and Taherzadeh 2016; Rapport et al. 2012). There are numerous companies around the world (European at the top) providing technologies, equipment, and services to the biogas sector. An up-to-date list of these active companies can be found in the report published by Energietechnik et al. (2016). Figure 1.16 presents a holistic overview of the current biogas production methods in the world.

1.2.6 Combustion; Direct and Indirect

Incineration, waste combustion with the goal of disposing waste fractions that cannot be recycled or reused, has been practiced and developed over more than a hundred years. The main objective has evolved from reducing waste volumes and hygienic problems to state-of-the-art waste-to-energy plants accompanied by extensive processes and emission control systems. A key factor in determining the feasibility of generating energy from waste is its heating value, which is expressed as lower and higher heating values (Christensen 2011; Richards and Taherzadeh 2016). A thorough review upon various methods in determining the heating values can be found in Christensen 2011. Table 1.4 shows different routes of waste combustion with their in-detail specifications.

Additionally, an important issue in case of incineration is the public perception about the technologies used which has to be taken into account. This perception is significantly different among various countries around the world, i.e., people in

Table 1.4 A general overview of thermal treatments, i.e., direct and indirect combustion technologies*

Thermal Treatment	Reactor/process type	Description	Advantages	Disadvantages
Direct combustion	Stoker/Grate Furnace	• Conventional mass burn incinerator • Many different designs • Keeps a fuel bed on top of a grate while letting primary air pass through the grate from beneath • Appropriate for waste by use of a sloping reciprocating grate • Typically, a grate can consist of 2–4 modules in a series and 1–2 modules in parallel • 1 MW/m^2 is the usual order of specific heat rate released from the grate; normally, about 60% of the total combustion air is supplied as primary air through the grate • Accompanied by Flue Gas–Cleaning System including particle precipitation, CO control, scrubbers for HCl and SO_2 removal, NO_x removal		–
	Fluidized Bed (FB)	• Consists of a bed of sand (or similar inert material) at the bottom of the combustion chamber • Two main categories: (1) bubbling fluidized bed (or simply fluidized bed)—in the order of 20 MWth is the preferred choice for moderately sized boilers, (2) circulating fluidized bed • No moving parts and, therefore, lower investment cost • Higher running operational cost due to the need for fresh sand and a more homogeneously crushed fuel • Accompanied by Flue Gas–Cleaning System including particle precipitation, CO control, scrubbers for HCl and SO_2 removal, NO_x removal	–	–
	Rotary Kiln	• Consists of a layered burning of the waste in a rotating cylinder • The energy efficiency may not exceed 80% • The possibility of being joined with a moving grate (moving grate as the ignition part and the rotary kiln as the burning out section) • More maintenance is required	–	–

(continued)

Table 1.4 (continued)

			• Rarely used for new MSW incineration plants with high heating value wastes • Common for burning waste with special characteristics, e.g. hazardous/chemical waste where confinement of the waste matters or for low heating value waste such as animal waste/carcasses; may be utilized in gasification • Accompanied by Flue Gas–Cleaning System including particle precipitation, CO control, scrubbers for HCl and SO_2 removal, NO_x removal		
Indirect combustion	*Gasification*	Fixed Beds (Downdraft, Updraft, and Slagging Fixed Beds)	• Generally used for lower throughputs resulting from difficulties in having a large diameter without causing the gases to be channeled • Based on the direction of the gas flow mostly consists of two kinds: updraft (in the opposite direction to the solid material) or downdraft (in the same direction to the solid material) • Downdraft system is more complex, requires more control, leads to high quality produced gas, and is devoid of damaging the gas turbine • Updraft gasifier handles a greater variation in the feedstock quality well (e.g., heat content and moisture content), produces a gas with rather high amounts of tars, as a stand-alone unit has a high energy efficiency • Slagging fixed beds operate at high temperatures for inorganic material—the material is melted and forms a molten solution (is often quenched before further treatment)—as a disadvantage, lime and coke are required to be added to the gasifier not only to control the viscosity of the smelt but also to ensure full reduction conditions in the lower section while keeping the temperature sufficiently high—Nippon Steel and JFE are the two main suppliers	• Recovering chemical energy in the waste as hydrogen and/or other chemical feedstocks rather than converting this energy into hot flue gases • Generally better energy efficiency; lower possibility of corrosion • Less need for flue gas cleaning: better quality and smaller volumes of flue gas • Influential in CO_2 capture • Potentially lower emissions of dioxins • Better quality of solid residues • Utilized in high-temperature processes • Low fuel load in gasification units; potentially lower costs	• Relatively homogeneous fuels are required • Troubles with slagging, tar production, and contaminants in the produced gas are common • Complicated to be controlled (despite the theoretical possibility) • Numerous waste related pyrolysis and gasification technologies only applicable to specific fuel types in small scale • Unable to compete with modern waste regarding overall energy conversion efficiencies
		Fluidized Beds (Bubbling and Circulating)	• Having a rapid and effective heat and mass transfer within them that distributes the fuel and increases mixing • The fuel must be pretreated • A bed material (usually sand) is required		

(continued)

Table 1.4 (continued)

			• Depending on the velocity of the gas divided into circulating and bubbling • Ebrara, Kabelco, and Hitachi Zosen are suppliers of bubbling fluidized bed gasifiers for treating waste		
	Plasma	Plasma Gasification	• It is generated at temperatures exceeding 2000 °C and is generally created by an electric arc • Dissociation of gas molecules starts at about 2000 °C and subsequently at temperatures above 3000 °C, they become ionized by lose of electrons • All tars will be eliminated • Alter NRG, Gasplasma® (Advanced Plasma Power, Swindon, UK), Plasco (Plasco Energy Group, Kanata, ON, Canada), and CHO Power (Europlasma, Oudenaarde, Belgium) are companies currently working with plasma in small-scale waste gasifiers (not yet been used commercially on a large scale)		
	Pyrolysis	Slow Pyrolysis	• Low heating rate of the solid material • The residence time of the solids is in the order of hours • Mild treatment and low entrainment of material into the gas phase are guaranteed • Low temperature (around 500 °C), thereby requiring a longer residence time and giving a solid char with a higher amount of oxygen and hydrogen • Lower energy demand • Less violent reaction during gas devolatilization • More carbon-rich char and less tars in the gas phase result by applying higher temperatures (above 700 °C)		
		Fast (Flash) Pyrolysis	• To produce bio-oil at approximately 510 °C with a proper feeding rate • To explore the secondary cracking of tar at longer residence times • Induces the presence of waxy materials in the liquid products • The higher the temperature, the lower the liquid yield • Currently no large-scale plant is in operation		

*Source Klein (2002), Christensen (2011), Bosmans et al. (2013), Chen et al. (2015), Richards and Taherzadeh (2016) some countries consider incineration plants as a safe and clean waste treatment technology reducing fossil fuel consumptions, while others might think of these plants as major contributors to air pollution, climate change, and public health threats.

Pyrolysis oil and gas, the possibility of recycling the solid materials (i.e., char and metals) after separation are the opportunities offered by pyrolysis. Likewise, production of a clean synthesis gas that can be used in gas turbines or gas engines is the main opportunity offered by gasification. Other advantages include possibly lower emission levels, further reduction in the formation of possible toxic substances (such as dioxins and furans) due the possibility of applying high temperatures and the presence of a high degree of vitrification (slagging), possibility of using the inert produced materials in construction or roads.

On both direct and indirect combustion techniques, research activities aiming at optimizing the processes involved are in progress, especially with a focus on environmental concerns. In case of gasification, it has been used together with ash melting with the goal of achieving very low emissions and increasing the use of solid waste. In the same way, coupling industrial pyrolysis facilities with gasification and combustion stage equipped with gas scrubbing devices are the current state-of-the-art developments (Chen et al. 2015; Panepinto and Zanetti 2017; Richards and Taherzadeh 2016).

1.3 Feasibility Study

Nowadays, MSWM systems consist of various options including materials collection, MRF, composting, combustion, and landfilling, that is, they are highly integrated (Dubanowitz 2000). In order for having an efficient systematic MSWM, a thorough investigation upon various on-going systems, conditions, and policies of the targeted area has to be implemented. This investigation has to cover the collection system (inspection on the overall efficiency of the current system mainly from economical point of view), waste producing sources, demographic and meteorological profiles, social influential parameters, hygienic conditions, water availability (surface and groundwater), electricity distribution and grid accessibility, physical, chemical, and heating value analysis, as well as on-going and future regulations.

More specifically, a given investigation should include an inspection on:

- the collection system to possibly implement new strategies for a more economical system together with lower negative environmental impacts;
- waste producing systems to specify an appropriate fee for every particular producer regarding its type of waste and also targeting illegal producers especially in developing or undeveloped countries;

- demographic and meteorological profiles including immigration rate, precipitation profile, sunny days, wind roses, temperature profile, and the climate for future considerations;
- social influential parameters including acceptability of new technologies among the people or the level of their knowledge about waste management in general for future considerations;
- hygienic conditions including the amount of health-care or hospital waste and the number of centers;
- surface and ground water accessibility for if a particular place is suitable for a particular waste management system;
- electricity distribution and the grid accessibility for selling the possible generated electricity in the future;
- physical, chemical, and heating value analysis of whole generated waste as the most important factor in determining the best scenario;
- on-going and future regulations for the chosen technologies whether or not there is a discrepancy between the regulations and the chosen systems.

In case of MSW standards, there are a few standards, among which ASTMs are more acceptable across countries. Some of these standards are ASTM D4979-12 for physical description screening analysis in waste, D5231-92(2016) for determination of the composition of unprocessed MSW, and D5681-16a for waste and waste management. The complete list of ASTM standards in waste management can be found in ASTM (2017).

In the following subsections, two of the most important must-do investigations, i.e., LCA and financial feasibility, will shortly be discussed.

1.3.1 LCA

Grown to be a major tool to evaluate the environmental performance of products and services, LCA is now utilized for economic analysis of all kinds of activities, from cradle-to-grave, i.e., from resource extraction, manufacturing, transport, wholesale and retail, to use and end-of-life management. Covering approximately all the environmental stressors that contribute to all the problems facing mankind, from resource depletion, climate change, smog formation, acidification, eutrophication, to noise, ecological toxicity, biodiversity loss, and human health (e.g., cancer) as well as non-cancer effects makes this analysis invaluable for decision makers (Hauschild et al. 2018).

Moreover, considering various distinct policies, regulations, and social as well as economic circumstances across countries, LCA is a vital and critical tool to estimate and compare the environmental impacts of waste management strategies (Jeswani and Azapagic 2016). For instance, based on a comparative LCA of five different MSWM scenarios in Iran by Rajaeifar et al (2015), landfilling combined with composting, a conventional but fading MSW management practice in Iran, was the worst scenario; however, the combination of AD with incineration was suggested as the most environmentally-friendly procedure (Rajaeifar et al. 2015).

1.3.2 Financial Feasibility

Financial feasibility, as an important and critical step in accessing the practicality of a proposed project, has to be conducted in order to find an in-detail cash flow in the project. As it is depicted in Fig. 1.17, many factors from two major costs subcategories, that is, investment and operational costs, have to be analysed carefully. In case of conducting the analysis, a few software have been developed, among which COMFAR III EXPERT is among the most promising ones.

In fact, COMFAR III EXPERT (Computer Model for Feasibility Analysis and Reporting) is a tool that has been developed by United Nations Industrial Development Organization (UNIDO), based on the experience, recommendations, comments, and needs of more than 7000 users in 160 countries to solve industrial problems, investment analysis, etc. Since its release, the software has been upgraded yearly to meet the technical developments as well as users' requests (UNIDO 2002).

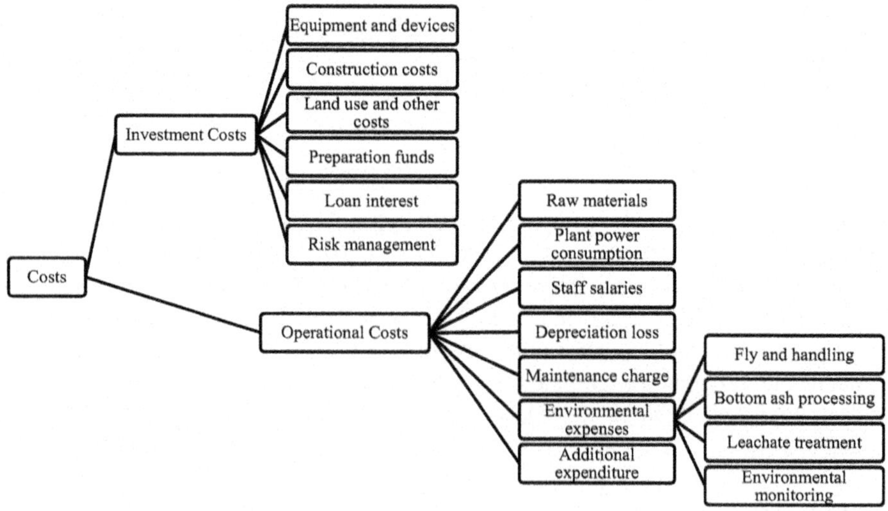

Fig. 1.17 The cost structure of WtE plants (Zhao et al. 2016). With permission from Elsevier. Copyright © 2017

1.4 Conclusions

Over the past century, the term "waste management" has taken a growing level of attention mainly due to the lifestyle changes and swift industrialization process all around the world. From economic and environmental points of view, waste, as a subjective definition, has become a valuable source of various materials, while would be a curse considering especially its negative environmental impacts. In order to have an optimal and efficient management system, building a scenario is a critical step. Within a scenario, various strategies could be applied to the whole system, i.e., a better and optimized collection system along with an efficient WtE system. WtE systems must be chosen by carrying out a thorough investigation of the local conditions of a targeted region. A system with a particular technology may be suitable for a region, while it may lead to a disaster for another region. Ultimately, the scenario can help a wide range of audience, from governments and companies to non-governmental organizations such as environmental protection agencies, to set their long-term objectives logically.

References

Advancedmrf (2017) Available at: http://www.advancedmrf.com/. Accessed 1 Jan 2017

Aid P (2015) Recycling Rates Around the World. Planet Aid. Available at: http://www.planetaid.org/blog/recycling-rates-around-the-world. Accessed 9 Jan 2017

ASTM (2017) Waste Management Standards, ASTM International. Available at: https://www.astm.org/Standards/waste-management-standards.html. Accessed 1 Jan 2017

Bosmans A et al (2013) The crucial role of Waste-to-Energy technologies in enhanced land fi ll mining: a technology review. J Cleaner Prod 55:10–23. https://doi.org/10.1016/j.jclepro.2012.05.032 (Elsevier Ltd)

Chen D et al (2015) Reprint of: Pyrolysis technologies for municipal solid waste: a review q. Waste Manage 37:116–136. https://doi.org/10.1016/j.wasman.2015.01.022 (Elsevier Ltd)

Christensen TH (ed.) (2011) Solid Waste Technology and Management. Wiley

Cimpan C et al (2016) Techno-economic assessment of central sorting at material recovery facilities e the case of lightweight packaging waste. J Cleaner Prod 112:4387–4397. https://doi.org/10.1016/j.jclepro.2015.09.011 (Elsevier Ltd)

Clarity Environmental (2017) RDF Export. Available at: http://www.clarity.eu.com. Accessed 1 Jan 2017

Commission W (1987) Report of the World Commission on Environment and Development : our Common Future Acronyms and Note on Terminology Chairman â€™ s Foreword. New York, USA

CP Group (2017) Available at: http://www.cpmfg.com/material-recovery-facility/e-waste-recycling-equipment/electronic-waste-recycling-equipment/. Accessed 1 Jan 2017

CPG Group (2017) Available at: http://www.cpgrp.com

Dhamija U (2006) Sustainable Solid Waste Management. In: Unnisa SA, Rav SB (ed). CRC Press

Dubanowitz AJ (2000) Design of a Materials Recovery Facility (MRF) For Processing the Recyclable Materials of New York City's Municipal Solid Waste Design of a Materials Recovery Facility (MRF) For Processing the Recyclable Materials of New York City's Municipal Solid Waste

Energietechnik ETW, Blanch A, Evans,T (2016) Biogas Technology Suppliers Directory 2016. (88)

Epstein E (2011) Induatrial composting: environmental engineering and facilities management

Fehr M (2010) The Threshold Target Approach to Waste Management in Emerging Economies : Pragmatic, Realistic, Appropriate

Fukuoka Municipal Government (2010) Fukuoka Method Type Semi-aerobic Landfill. Environment Bureau, Hong Kong

Garg A et al (2007) Wastes as Co-Fuels: the Policy Framework for Solid Recovered Fuel (SRF) in Europe, with UK Implications. Environ Sci Technol 41(14):4868–4874

General Kinematics (2016) Top 10 Recycling Countries From Around the World, General Kinematics. Available at: https://www.generalkinematics.com/blog/top-10-recycling-countries-around-world/. Accessed 9 Jan 2017

Global Environment Center Foundation (2006) Semi-aerobic landfill (Fukuoka method). Available at: http://nett21.gec.jp/waste/data/waste_l-1.html. Accessed 9 Mar 2017

Habib K, Schmidt JH, Christensen P (2013) A historical perspective of Global Warming Potential from Municipal Solid Waste Management. Waste Manage 33(9):1926–1933. https://doi.org/10.1016/j.wasman.2013.04.016 (Elsevier Ltd)

Hauschild M, Rosenbaum RK, Olsen S (eds) (2018) Life Cycle Assessment. Springer International Publishing. https://doi.org/10.1007/978-3-319-56475-3

Jeswani HK, Azapagic A (2016) Assessing the environmental sustainability of energy recovery from municipal solid waste in the UK (in press). Waste Management. Elsevier Ltd, Amsterdam

Jingura RM, Kamusoko R (2017) Methods for determination of biomethane potential of feedstocks: a review. Biofuel Research Journal 4(2):573–586. https://doi.org/10.18331/BRJ2017.4.2.3

Kessler Consulting Inc (2009) Material recovery facility technology review. Available at: http://www.nswai.com/DataBank/Reports_pdf/reports_aug15/MATERIALS%20RECOVERY%20FACILITY%20TECHONOLOGY%20REVIEW.pdf

Kiddee P, Naidu R, Wong MH (2013) Electronic waste management approaches: An overview. Waste Manage 33(5):1237–1250. https://doi.org/10.1016/j.wasman.2013.01.006 (Elsevier Ltd)

Klein A (2002) Gasification: an alternative process for energy recovery and disposal of municipal solid wastes, New York. Available at: http://www.seas.columbia.edu/earth/wtert/sofos/klein_thesis.pdf

Krook J, Svensson N, Eklund M (2012) Landfill mining: a critical review of two decades of research. Waste Manage 32(3):513–520. https://doi.org/10.1016/j.wasman.2011.10.015 (Elsevier Ltd)

Onwosi CO et al (2017) Composting technology in waste stabilization: on the methods, challenges and future prospects. J Environ Manage 190:140–157. https://doi.org/10.1016/j.jenvman.2016.12.051 (Elsevier Ltd)

Panepinto D, Zanetti MC (2017) Municipal solid waste incineration plant: A multi-step approach to the evaluation of an energy-recovery configuration. Waste Manage. https://doi.org/10.1016/j.wasman.2017.07.036.(Elsevier Ltd)

Pariatamby A et al (2015) Municipal Solid Waste Management in Asia and the Pacific Islands. South Asia Economic J. https://doi.org/10.1177/139156140901000108

Pawłowska M (2014) Mitigation of Landfill Gas Emissions. CRC Press

Ragazzi M, Rada EC (2012) RDF/ SRF evolution and MSW bio-drying. Waste Manage Environ. https://doi.org/10.2495/WM120191 (WIT Press)

Rajaeifar MA et al (2015) Comparative life cycle assessment of different municipal solid waste management scenarios in Iran. Renew Sustain Energy Rev 51:886–898. https://doi.org/10.1016/j.rser.2015.06.037

Rajaeifar MA et al (2017) Electricity generation and GHG emission reduction potentials through different municipal solid waste management technologies: A comparative review. Renew Sustain Energy Rev 79(April):414–439. https://doi.org/10.1016/j.rser.2017.04.109 (Elsevier Ltd)

Rapport J et al (2008) Current anaerobic digestion technologies used for treatment of municipal organic solid waste, California Integrated Waste Management Board. Available at: http://scholar.google.com/scholar?hl=en&btnG=Search&q=intitle:Current+Anaerobic+Digestion+Technologies+Used+for+Treatment+of+Municipal+Organic+Solid+Waste#0, www.calrecycle.ca.gov/publications/Documents/1275/2008011.pdf

Rapport JL et al (2012) Anaerobic Digestion technologies for the treatment of Municipal Solid Waste. Int J Environ Waste Manage 9(1/2):100–122. https://doi.org/10.1504/IJEWM.2012.044163

Richards T, Taherzadeh MJ (eds) (2016) Resource Recovery to Approach Zero Municipal Waste Edited by CRC Press

Ritzkowski M, Stegmann R (2012) Landfill aeration worldwide: Concepts, indications and findings. Waste Manage 32(7):1411–1419. https://doi.org/10.1016/j.wasman.2012.02.020 (Elsevier Ltd)

Rotter S (2011) '8.7 Incineration : RDF and SRF—Solid Fuels from Waste. Solid Waste Technol Manage pp 487–501. https://doi.org/10.1002/9780470666883.ch32

Srivastava V et al (2016) Biological response of using municipal solid waste compost in agriculture as fertilizer supplement reviews in environmental science and bio/technology. Springer, Netherlands. https://doi.org/10.1007/s11157-016-9407-9

Townsend TG et al (2015). Sustainable Practices for Landfill Design and Operation. Sustainable Practices for Landfill Design and Operation. https://doi.org/10.1007/978-1-4939-2662-6

Ullah Khan I et al (2017) Biogas as a renewable energy fuel—A review of biogas upgrading, utilisation and storage. Energy Convers Manag 150(August):277–294. https://doi.org/10.1016/j.enconman.2017.08.035

UNEP (2014) 'Basel Convention on the Control of Transboundary Movements of Hazardous Wastes and their Disposal'. UNEP

UNIDO (2002) COMFAR III Expert, October

Wolfsberger T et al (2016) Landfill mining: Development of a cost simulation model. Waste Manage Res. https://doi.org/10.1177/0734242X16628980

World Bank (2010) 'Waste Generation', Urban Development Series-Knowledge Papers, pp. 8–12

World Energy Council (2013) World Energy Resources: Waste to Energy, World Energy Council. Available at: https://www.worldenergy.org/wp-content/uploads/2013/10/WER_2013_7b_Waste_to_Energy.pdf

World Energy Council (2016) World Energy Resources: Waste to Energy. Available at: https://www.worldenergy.org/wp-content/uploads/2013/10/WER_2013_7b_Waste_to_Energy.pdf

Zaman AU (2016) A comprehensive study of the environmental and economic benefits of resource recovery from global waste management systems. J Cleaner Prod. https://doi.org/10.1016/j.jclepro.2016.02.086. (Elsevier Ltd)

Zhao X gang et al (2016) Economic analysis of waste-to-energy industry in China. Waste Manage 48:604–618. https://doi.org/10.1016/j.wasman.2015.10.014. (Elsevier Ltd)

Chapter 2
Feedstocks for Biogas Production: Biogas and Electricity Generation Potentials

Johannes W. A. Langeveld and Eric C. Peterson

2.1 Introduction

The development of bioenergy offers major possibilities for the reduction of Greenhouse Gas (GHG) emissions and fossil fuel dependency. Agricultural and industrial biomass residues have been identified as promising feedstocks that bring fewer risks with respect to competition for food and/or affecting natural resources. The amount of residues available for energy production, the way in which they should be converted, and the organisation of emerging bioenergy chains remain subjects of debate. On the other hand, the land area needed for economic and sustainable production of bioenergy crops has been discussed while no final conclusion as to their potential has been reached. Given the favourable perspective of the conversion of biomass residues and other organic materials into bioenergy, it comes as no surprise that the production of biogas is constantly growing. The number of biogas installations currently exceeds 35 million, mostly comprised of household installations in Asian countries like China and India. Larger farm digesters are mostly found in industrial countries in the Europe and North America (Langeveld et al. 2016).

Anaerobic Digestion (AD) can convert dry as well as wet feedstocks from a range of sources. Feedstocks used in AD range from crops residues including stalks, leaves, roots, seeds and seed shells, to urban and animal wastes. Agro-industrial plants such as sugar refineries and palm oil extraction plants also generate sustainable sources of residues and are regarded as promising for waste valorisation platforms such as AD. The main product of AD, biogas, can be stored before it is

J. W. A. Langeveld (✉)
Wageningen, Biomass Research, Costerweg 1D, 6702 AA Wageningen, The Netherlands
e-mail: hans@biomassresearch.eu

E. C. Peterson
Universidad Icesi, Calle 18#, 122-135 Cali, Colombia
e-mail: ecpeterson@icesi.edu.co

M. Tabatabaei and H. Ghanavati (eds.), *Biogas*, Biofuel and Biorefinery Technologies 6, https://doi.org/10.1007/978-3-319-77335-3_2

used to provide heat, electricity, injected into a natural gas grid, and/or utilized in the chemical industry. This makes it a valuable resource of the bioenergy sector, whether as a stand-alone energy source or part of an integrated food-, animal feed- or biofuel production chain. AD can convert high volume, low value, and low energy-density feedstocks including the organic fraction of municipal solid waste (MSW), crop and field residues, food industry effluents, processing residues, and livestock manure. This is achieved in relatively simple and safe installations, generating a co-product, i.e., digestate, which could be used as biofertilizer to provide plants with nutrients and to increase the organic fraction of the soil.

Production of bioenergy is projected to increase significantly to reach 108 exajoules (EJ) in 2030. This is twice the current level and it would represent 20% of the total primary energy supply and 60% of the final renewable energy use (Nakada et al. 2014). Notwithstanding this potential, development of production capacity remains below expectations. Feedstock availability, conversion techniques, production costs, and financial support are the main subjects of debate (Langeveld et al. 2016).

There is a large variation in composition and quality of the feedstocks offered to AD managers and owners. In particular, residue streams can vary widely in terms of dry matter, proteins, and fats contents while other compositional elements may seriously limit their biogas production potential. This can be a major barrier for managers who need to make considerable efforts to ensure that the installation is fed with feedstock of sufficient quality (Langeveld et al. 2010). While it is not realistic to expect the quality of feedstocks to be always known in advance, it is important to realise that the viability of the operation may depend on the predictability of biogas yield potential. For this reason, major efforts are made to ensure better predictability of biogas yields from feedstocks (Amon et al. 2007a, b). In line with that, an increasing number of companies are offering biogas yield potential assessments.

The information presented in this chapter outlines the current state of biomass utilization for AD and biogas production. More specifically, the aim of the chapter is to encompass biogas activities from differing geographic regions (the Europe, Africa, North and South Americas, and Asia), with a focus on region-specific substrates as examples, describing substrate availability, current AD activities, and future trends. The chapter also summarizes biogas potential of these varied substrates, while considering the emerging trends of using the digested materials as fertilizer.

2.2 Feedstock Types

To understand the correlation between a given feedstock and its potential biogas yield, several factors should be considered. Evaluation of physicochemical properties of feedstocks including moisture content and available organic materials is necessary to characterize them. For instance, total solids (%TS) refers to the overall amount of solids available in a sample, while volatile solids (%VS) refers to the organic fraction of %TS available for biogas production, as characterized by overnight combustion of over dried samples at 550 °C. Finally, fresh matter (FM) is an important quality, as this represents the actual amount of materials fed into digesters by operators. Each value has important ramifications, and offers important considerations for characterizing a given feedstock. In more details, %VS provides an accurate comparison of feedstocks by allowing normalization of biogas per g VS while TS is an additional important consideration, as low TS values, such as in liquid substrates, indicate a highly diluted sample, which results in reduced biogas per unit fresh matter, whether solids or liquids. It should be noted that systems such as methane energy value system (MEVS) predict methane yields based on indicators including crude protein, crude fat, cellulose, hemicellulose, and nitrogen contents (Amon et al. 2007a, b), while a similar, more basic feedstock system is used for the evaluation of liquid feedstocks. Additional indicators may refer to nutrient composition expressed as ratios (C/N, C/P, N/P), or trace elements.

In addition to that, additional compositional considerations need to be addressed as well. For instance, readily hydrolysable feedstocks such as manures and liquid effluents, as well as solid plant biomass, can contain lignocelluloses, which are significantly more resistant to microbial hydrolysis necessary for biogas production. In better words, due to the presence of both cellulose and lignin, digestion of woody residues are significantly restricted, as the crystalline nature of cellulose and the presence of lignin result in slow hydrolysis rates, requiring specialized enzymes (i.e., cellulases and laccases, respectively). Nevertheless, various forms of pre-treatment can increase lignocellulosic biogas yield potentials. The subsequent subsections describe both non-lignocellulosic and lignocellulosic feedstocks, their sources, and how they differ in composition.

2.2.1 Non-lignocellulosic Feedstocks

The global population is projected to increase by one billion people within the next twelve years, reaching 8 billion in 2025, and 9.6 billion in 2050 (United Nations 2013). An increasing number of people are living in cities, which are also growing in size, and in fact, more people are now living in cities than in rural areas. This phenomenon is affecting how food and materials are provided to consumers, and the way waste is generated. Over the last decades, food production systems have developed an increased integration between agricultural production and other economic activities. This has led to an emergence of large scale, complex agro-industrial chains. Expectations are that this trend will continue during the coming years (FAO 2013). Residues of large agro-industrial plants are increasingly targeted by private companies as sources of animal feed, biological compounds (proteins, fats), biobased products, and nutrients, while effluents and by-products can eventually be sourced as biogas feedstocks.

Major sources of biogas feedstocks are found in processing units of main food industries including starch, protein, oil, beverage, meat, and cereal production

Table 2.1 Availability and composition of non-lignocellulosic residues generated by livestock production systems, urban areas, and food processing units

Residue	TS %FM	VS %TS	Availability	References
Liquids				
Potato effluent	4	90	5.4 m^3/tonne potato	FNR (2010) and Hung et al. (2006)
Palm oil mill effluent	3.1	86	0.7–0.9 m^3/tonne FFB	Langeveld and Quist-Wessel (2014) and Sri Rahayu et al. (2015)
Slaughtering waste	15	80	500 kg/tonne meat	Patinvoh et al. (2017) and Verheijen et al. (1996)
Vinasse	1	90	0.85–1.28 m^3/ tonne sugarcane	Janke et al. (2015)
Cattle slurry	11	82	7.3 tonne/head/y	FNR (2010)
Pig slurry	7	86	1.8 tonne/head/y	Matulaitis et al. (2015)
Coffee wastewater	2.3–5.5	52–76	3.67 m^3/tonne raw coffee berries	Syarief et al. (2012), Campos et al. (2014) and Montenegro et al. (2014)
	TS %FM	VS %TS	Availability	References
Solids				
Food residues	20	92	55 kg/person/y	Yong et al. (2015)
Chicken manure	40	75	73 kg/head/y	FNR (2010)
Cattle manure	25	76	7.3 tonne/head/y	FNR (2010)

TS Total Solids, *VS* Volatile solids, *FM* Fresh matter

chains and therefore, these industries could potentially serve as large sources of energy. Typical availability of urban and agro-industrial residues is presented in Table 2.1, which includes liquid effluents from sources such as potato starch, palm oil, or coffee processing, while also including solid wastes such as manures and MSW. It should be noted that it is important to consider TS for effluents which are extremely low in dry matter.

Food processing waste varies between 250 and 800 kg per tonne of raw food but shows large variations in composition. Biodegradable urban waste can be as high as 70 kg per person per year; the organic fraction of MSW is biodegradable, which is sometimes defined as organic waste, or 'bio-waste'. While MSW in many cases is landfilled, increasing efforts are made to valorise this potentially valuable feedstock, which can be used for high quality compost and/or biogas chains. Livestock provides another source of feedstocks, especially in intensive production systems. Manure can be liquid ('slurry' which is very low in dry matter) or solid containing straw and bedding materials as well. Animal manure is generally low in solids (<25%), with the exception of chicken droppings. Availability is generally below 100 tonnes for a farm per year, but it can be higher in case of large-scale farms.

2.2.2 Lignocellulosic Feedstocks

Lignocellulosic feedstocks include a wide range of materials consisting mostly of sugars (pentoses and hexoses) (60–75%); while the remainder (25%) is lignin plus a small fraction of proteins and lipids. Importantly, these sugars are present as polysaccharides, whether cellulose or hemicellulose. Cellulose is comprised of crystalline glucose, and is thus reluctant against enzymatic hydrolysis. It also excludes water and provides basic structural support. Hemicellulose consists of branched polymers of pentose sugars such as xylose and arabinose, acting as glue, holding together the cellulose fibres. Finally, lignin consists of complex aromatic polymers that help resist microbial degradation.

Lignocellulosic residues, or plant biomass with an elevated lignin content, come from a diversity of sources, the majority of which are agricultural or agroindustrial, and several examples are listed in Table 2.2. For instance, banana and palm oil agriculture generates large amounts of field residues after harvest (i.e., leftover banana trees, flowers, leaves), while processing activities like sugar and palm oil refinement produce large amounts of fibrous lignocellulosic biomass. Other lignocellulosic feedstocks include forestry residues, coffee pulp, and field residues such as corn stover. Major energy crops for biogas production are mostly maize silage and grass, which are harvested in summer in the Europe and could be fed into digesters throughout the year. Use of dedicated crops in other continents is limited.

As mentioned above, several tropical agricultural crops, such as bananas, sugarcane, and oil palm, leave significant residues in the field, which are lignocellulosic in nature. For instance, sugar cane field cropping produces significant amounts of leaves, as do agricultural activities in banana production. Availability of residues generally varies between 150 and 500 kg per tonne of raw harvested (food) material. Exceptions are straw/stover (high availability) and some minor residues. It should be noted that banana pseudo stems are high in weight but low in dry matter. As a rule, TS content in lignocellulosic materials can range from moderate (maize, grass) to high (stover, shells, press cake, bagasse). Most residues are high in cellulose and sometimes hemicellulose (together making up 40–60% of dry matter). Forest residues and palm kernel shells are high in lignin, which restricts their degradability (Table 2.2).

While the availability and energy content of these feedstocks mark them as favorable sources for valorisation, their complex structure, i.e., the interwoven network of crystalline cellulose, lignin, and associated hemicellulose, is highly resistant to hydrolysis into available sugars that are required for microbial utilization. Typical anaerobic digesters include specialist microorganisms capable of hydrolysing these materials, but the rate of digestion is reduced as a consequence, which in turn requires longer retention times and larger digester volumes.

Nevertheless, some techniques can be applied to reduce long retention times, such as different pre-treatment procedures which can greatly improve hydrolysis and utilization of these recalcitrant substrates. Pre-treatment techniques may be mechanical (e.g., milling), chemical (e.g., acid or alkali treatment), or thermal methods (e.g., steam explosion) in nature, but all in general try to expand the

Table 2.2 Availability and composition of different lignocellulosic residues

Residue	C:H:L	TS % FM	VS % TS	Availability (kg/tonne of raw product)	Yield (tonne/ha)	References
Silage maize	16:1:38	35	94	n.a.	30	Mumme et al. (2011)
Grass silage	31:29:10	50	92	n.a.	70	Cadavid-Rodríguez and Bolaños-Valencia (2016)
Wheat straw	33:22:19	98	93	1000	5.5	Niee Liew et al. (2012)
Corn stover	34:19:19	93	97	1000	10.0	Niee Liew et al. (2012) and Li et al. (2011)
Sugarcane Bagasse	42:22:18	94	97	280	22.4	Kumari and Das (2015)
Coffee pulp	31:11:23	55	91	550	2.8	Battista et al. (2016) and Syarief et al. (2012)
Coffee parchment	18:24:10	n/a	n/a	50	0.3	Battista et al. (2016) and Syarief et al. (2012)
Forestry residues	42:n/a:44	50	64	n/a	120	Teghamar et al. (2014) and Hoyne and Thomas (2001)
Oil palm fibre	30:21:24	76	78	130	2.6	Garcia-Nunez et al. (2016a, b)
Empty Fruit Bunches	34:21:22	64	80	220	4.4	Garcia-Nunez et al. (2016a, b) and Zhang et al. (2012)
Palm kernel shells	26:19:48	89	74	45	0.9	Garcia-Nunez et al. (2016a, b)
Banana leaves	31:8:18	n/a	n/a	156	6.2	Santa-Maria et al. (2013) and Nzila et al. (2015)
Banana flower stalks	45:8:18	n/a	n/a	468	18.7	Santa-Maria et al. (2013) and Nzila et al. (2015)
Banana pseudostems	56:8:18	5	4	1920	76.8	Santa-Maria et al. (2013) and Nzila et al. (2015)
Banana pseudostems (sun-dried)	56:8:18	92	83	104	76.8	Kalia et al. (2000)

C:H:L Cellulose:Hemicellulose:Lignin, *TS* Total Solids, *VS* Volatile solids, *FM* Fresh matter

lignocellulosic matrix, while solubilizing hemicellulose, or to a lesser extent cellulose and lignin. These approaches increase biogas production but have specific drawbacks, whether high energy costs, or the generation of inhibitory sub-products such as furfural or hydroxymethylfurfural (HMF). Novel pre-treatment methods are emerging, which focus on ionic liquid or supercritical CO_2 to solubilize and collect lignin, both increasing biogas production while also providing an additional revenue through lignin collection. Regardless of the pre-treatment technique used, this step is an essential consideration for improved biogas production from lignocellulosic feedstocks.

2.2.3 Co-digestion

Co-digestion is the simultaneous conversion of a mixture of different feedstocks, e.g., manure and plant biomass. In the past, AD was mostly referred to a single substrate/single output process but recently, co-digestion has become a standard technology in agricultural biogas production in many countries.

Four types of anaerobic digesters can be used to treat animal manure (Mathias 2014):

- Continuously Stirred Tank Reactors (CSTR);
- Upflow Anaerobic Sludge Blanket (UASB) reactors;
- Upflow Anaerobic Filter (UAF) digesters; and
- Baffled digesters.

Typically, digester type is to be selected depending on the characteristics of the major feedstock used, particularly TS. Feedstocks with high TS concentrations and slurry are mainly treated in CSTRs; while soluble organic wastes are mostly digested in anaerobic filters, fluidized bed reactors and upflow anaerobic sludge blanket (UASB) reactors (Mathias 2014).

Co-digestion is often carried out in wet single-step processes (e.g., CSTR). Substrates are diluted until dry solid content is between 8 and 15%. Wet systems are particularly useful when digestate is to be applied in the fields without separating the solid fraction (FNR 2010). Digestion of crops often requires long hydraulic retention times that may last up to months, and both mesophilic or thermophilic temperatures can be applied. Advantages of co-digestion include enhanced biogas yields and GHG reduction, homogenisation, high process stability, odour reduction, high nutrient recycling (nitrogen, phosphorus), large number of different substrates can be converted, possibility of integration into wastewater treatment in livestock production facilities, steady biogas production throughout the season, enhanced potential income from gate fees and waste treatment, and no 'indirect' effects such as land use change (Langeveld et al. 2016).

2.3 Biogas Yield and Digestate

2.3.1 Biogas Yield

While biogas yield is commonly expressed in cubic metre (m^3) of biogas per tonne of fresh or dry biomass, yields are often determined through biochemical methane potential (BMP) analysis, which is commonly used as a relative measure of the strength of sample per g of VS (mL biogas/kg of VS or, alternatively, m^3 biogas/tonne of VS). In another word, BMP normalizes available VS for a given substrate, and thus indicates the relative richness of a substrate for production of biogas. In brief, a given sample is added to a small volume of digester sludge, and incubated under mixing for 30 d at 35 °C, while the evolved gas is quantified and reported.

BMP analysis does not, however, account for the ratio of VS to neither TS nor FM, nor the recalcitrance of the substrate, which affects the required retention time for digestion. Thus, to calculate the volume of biogas per tonne FM from BMP values for a given substrate, VS and TS values are required. Therefore, due to the variable nature of all substrates, both VS and TS determination as well as BMP analysis are needed to accurately quantify the total available biogas from a given material for AD.

It should be noted that high variability in BMP values due to low sample size can reduce their accuracy, which has practical ramifications. This is especially true in case of manure, one of the most studied substrates for biogas potential, where differences in operational parameters such as diets, breeds, and management practices are likely to cause a high variability in BMP values. For instance, Labatut et al. (2011) found an average BMP value of 243 (±60) m^3 biogas/kg VS for 47 dairy manure samples, with values of the individual samples ranging from 127 to 329 m^3 biogas/kg VS. Thus, a large number of samples should be used to achieve a reasonably accurate assessment. Other sources of variability in BMP analysis include the type of inoculum used (Quintero et al. 2012). Thus, individual analysis of a given feedstock for biogas applications is highly beneficial, especially for larger operations.

As mentioned earlier, in practice, yields are often expressed as biogas or pure methane (CH_4), produced per unit of VS, TS, or FM. Thus, the TS and VS values provided in Tables 2.1 and 2.2 are crucial data that need to accompany biogas yield data for accurate conversion and comparison with other literature values. Table 2.3 provides yields reported for a large number of residues, showing that biogas yields can range from 105 to 700 m^3 of biogas per tonne of VS, depending on the composition of the solids.
Generalizing major feedstocks, three VS yield categories can be distinguished:

- **low**: <300 m^3 biogas/tonne VS (lignocellulose, cattle and pig manures)
- **modest**: 300–500 m^3 biogas/tonne VS (chicken manure, MSW, banana stalks)
- **high**: >500 m^3 biogas/tonne VS (palm oil mill effluent—OME, abattoir effluents, potato starch effluents).

Taking variations in percentage of TS and VS into account, biogas yields can be calculated per tonne of FM, which is important from a user viewpoint.
Feedstocks can be classified into five categories of FM biogas yield:

- **low**: <50 m^3 biogas per tonne FM (POME and coffee effluents, manure slurries, food residues)

Table 2.3 Biogas and methane yields of typical feedstocks

Liquid feedstocks	m^3 CH_4/ tonne VS	m^3 CH_4/ tonne TS	m^3 CH_4/m^3 effluent	References
Potato effluent	611	550	22	
POME	562	483	15	Langeveld and Quist-Wessel (2014)
Abattoir wastewater	700	560	84	Patinvoh et al. (2017)
Cattle slurry	234*	192*	21*	Calculated from FNR (2010)
Pig slurry	201*	181*	13*	Calculated from FNR (2010)
Solid feedstocks	m^3 CH_4/ tonne VS	m^3 CH_4/ tonne TS	m^3 CH_4/m^3 effluent	
Food residues	260	239	48	Yong et al. (2015)
Chicken manure	309*	252*	101*	Calculated from FNR (2010)
Cattle manure	236*	180*	45*	Calculated from FNR (2010)
MSW (biodegradable)	386*	348*	70*	Calculated from FNR (2010)
Lignocellulosic feedstocks	m^3 CH_4/ tonne VS	m^3 CH_4/ tonne TS	m^3 CH_4/m^3 effluent	
Bagasse	122	119	112	Kumar and Das (2015)
Pre-treated bagasse (NaOH)	177	172	162	Kumar and Das (2015)
Forest residues	214	137	103	Teghammar et al. (2014)
Pre-treated forest residues (NMMO)	266	170	128	Kabir et al. (2013)
Banana stalks	347	13	0.1	Li et al. (2016)
Banana stalks (sundried)	236	196	180	Kalia et al. (2000)
Coffee pulp	131	119	66	Ulsido and Li (2016)
Pre-treated coffee pulp (NaOH)	174	158	88	Battista et al. (2016)
Wheat straw	282	265	260	Yong et al. (2015)
Corn stover	296	288	268	Amin et al. (2017)
Maize silage	259	396*	139*	Li et al. (2015)
Grass silage	344–383	330*	180*	McEniry et al. (2014)

*Methane estimated as 60% of reported biogas yield values

- **modest**: 50–100 m^3 biogas per tonne FM (abattoir wastewater, municipal sewage sludge, coffee pulp)
- **high**: 100–200 m^3 biogas per tonne FM (chicken manure, lignocellulosic feedstocks).

It is exceedingly interesting to note the juxtaposition between biogas yields based on VS and FM. Specifically, it can be seen that while industrial effluents have the highest yields per VS, due to their dilute nature, FM biogas yields are exceedingly low. To overcome this, plant operators could reduce effluent dilution, giving a more concentrated stream that could increase operational biogas yields, or alternatively reduce digester sizes. It is also interesting to note that while banana stalks have one of the highest reported biogas production values per VS, biogas production is almost negligible per FM, due to high water content. However, basic pre-treatments (e.g., sun-drying) has been shown to dramatically increase biogas production from this substrate per FM.

While pre-treatment of various lignocellulosic substrates also increases yields, lignocellulosic substrates show the highest biogas yields per FM due to their high TS content. However, along with biogas yields, other parameters must also be considered. Specifically, recalcitrant lignocellulosic feedstocks show slower hydrolysis rates, requiring longer treatment times, and consequently larger reactor volumes to achieve comparable biogas production rates compared with more readily hydrolysable or soluble substrates. Physicochemical properties also need to be considered, as for instance, phase separation could occur with waste oil or other lipids. Moreover, waste grease, for example, can also disrupt methanogenesis at higher loading rates, likely as a result of acidification caused by the introduction of high-energy substrates (Zhu et al. 2011). Thus, these examples highlight the need for careful consideration of biogas yields, and the significance of providing excellent instructions on how to evaluate them.

2.3.2 *Fertilizer Production*

AD generates liquid and solid fertilizers which are superior to manure and compost in terms of nutrient availability (Nkoa 2013). In fact, AD process results in the mineralization of organically-bound nutrients, in particular nitrogen (N) and lowering the C/N ratio, which both in turn increase the short-term N delivery (Weiland 2010). In the case of phosphorus (P), some substrates such as manure can fulfil the P requirements of most crops after digestion, and the nitrogen requirement can similarly be fulfilled by up to 60–80% with the remaining nitrogen requirement provided with additional fertilizers (Holm-Nielsen 2009; Liedl et al. 2006). Such biofertilizers have been used in the cultivation of coffee and corn to reduce costs by 40% (Walter Borges de Oliveira et al. 2011).

Phosphorus is typically sequestered in the solid digestate, while nitrogen is typically present in the liquid phase (Liedl et al. 2006). Careful process planning is needed if digestate is to be used as fertilizer. Short hydraulic retention times used to

Table 2.4 Typical composition of cattle manure digestate

Parameter	Unit	Value range
Total Solids	%	1.5–45.7
Volatile Solids	%	38.6–75.4
Total N	% of DM	3.1–14
Total N	% of FM	0.12–1.5
Total NH_4^+	% Total N	35–81
Total P	% of DM	0.2–0.35
Total P	% of FM	0.04–0.26
Total K	% of DM	0.19–4.3
Total K	% of FM	0.12–1.15
pH	–	7.3–9.0

Source Adopted from Nkoa (2013)

increase efficiency may result in incomplete digestion, and as a result, the end-product digestate may have problems with odour emission, toxic organic compounds, pathogens, and phytotoxicity (Nkoa 2013). However, with thorough planning and analysis of biodigestate composition, it is very promising that these materials would represent an excellent source of fertilizer for agricultural use. An overview of main digestate characteristics is provided in Table 2.4.

It is important to note that there is a wide variation in the characteristics of anaerobic digestate, depending on the type of biomass inputs (feedstock) used, and the configuration of the digester. Spectroscopic techniques have recently demonstrated that anaerobic digestates inherit the chemical attributes of the feedstock from which they are produced (Nkoa 2013), and thus, biodigestate composition must be assessed on a case by case basis, evaluating combinations of feedstocks and digestion installations.

2.4 Biogas Chain Development

Literature on in situ performance of household or farm-scale digesters is limited. An extensive monitoring program in Germany, following the initiation of bioenergy supporting policies under the 'Energie Wende', however, provides a wealth of key indicators and other basic performance data for farm-digesters in this country (e.g., FNR 2009, 2010).

Table 2.5 provides an overview of typical digesters with capacities varying from 365 to 1100 kW (out of which 180–535 kWel). Most digesters in Germany are running on a combination of manure (pig, cattle) and crop materials (mostly maize silage). On a national level, over 40% of feedstocks in Germany consists of manure while 50% is maize and arable field residues. Reactor volumes vary between 1050 and 3800 m^3, out of which two thirds is used for the digestion process (the remainder is for gas storage). There is a large variation in retention time; systems with low shares of manure having an average retention time of over 120 d.

Table 2.5 Examples of co-digestion chains (Germany)

Chain	Feedstock (tonne/y)	Reactor volume (m^3)	Retention time (d)	CHP size (kW)
Small	Total: 7358 Manure: 73% Crop materials: 27%	Total: 1049 Nett: 903	46	180 kWel 185 kWth Total 365 Kw
Medium size	Total: 8419 Manure: 3% Crop materials: 97%	Total: 3225 Nett: 2000	123	500 kWel 600 kWth Total 1100 Kw
Medium size	Total: 10,403 Manure: 11% Crop materials: 90%	Total: 3800 Nett: 1730	146	535 kWel 551 kWth Total 1086 Kw
Medium size	Total: 23,009 Manure: 58% Crop materials: 42%	Total: 3180 Nett: 2650	43	526 kWel 566 kWth Total 1092 Kw

Source FNR (2009)

While the principle of AD has been applied for centuries in many parts of the world, there still is a lot of room for improvement in the design and operational management. This may refer to the impact of optimising feedstock loads, digester design and digester management. A large number of research projects have been implemented aiming to enhance biogas chain development in the Europe. Main emphasis is on feedstock analysis, planning of the production process, and economic sustainability. In the USA, a growing number of digesters are aiming to combine multiple feedstocks in the process. Many AD projects that originally started with manure, have incorporated other substrates in the process at a later point (Langeveld et al. 2016). The majority of biogas plants in Brazil process agricultural residues and MSW (Persson and Baxter 2015).

Specific biogas R&D programs have been implemented in the Asian countries like China and India. The Biogas Development and Training Centre is serving the east of India, implementing monitoring of biogas installations under the National Biogas Manure Management Programme (NBMMP). The Biogas Institute of the Ministry of Agriculture (BIOMA) in China, a part of the Chinese Academy of Agricultural Sciences (CAAS), focuses on issues like fundamental research on anaerobic microbiology and design of biogas projects. In Africa, dedicated programs like the Africa Biogas Partnership Programme (ABPP), a public- private partnership programme, aim to provide access-to-energy services through the installation of biogas digesters in partnership with local enterprises, NGOs, and governments. Apart from manure and crop residues, other potential feedstocks may include cassava, sugar cane and oil palm residues, as well as urban waste.

At the international level, the International Energy Agency (IEA) is providing a research and policy development framework for AD of a range of organic feedstocks including agricultural residues, energy crops, waste waters, MSW, and industrial organic wastes. Main interests of IEA Bioenergy Task 37 are biogas production for heat and power, biogas upgrading to biomethane, utilisation of biogas and biomethane for electricity grid balancing and production of high quality digestate that can be used as biofertilizer. Input to this task group is provided by members from Australia, Europe, Brazil, and Korea.

2.5 Conclusions

Methanogenic AD can convert dry as well as wet feedstocks from a range of sources into an important renewable energy carrier, i.e., biogas. This chapter has discussed potential use of a range of feedstocks including MSW, energy crops, field residues, and industrial residue streams for biogas production. Availability of wastes and residues is high and has been projected to increase marking them even more interesting as feedstocks for an emerging biogas industry. Feedstock composition is the major determining factor of its biogas potential and decomposition (hydrolysis) rate. Large variations are found in feedstocks with respect to share of high-energy compounds (sugars, starch, fats) and recalcitrant (lignocellulosic) materials. Slow hydrolysis rates have considerable impacts on digester design and chain performance as they will lead to low biogas yields and high retention times, and require larger (i.e., more expensive) reactors, pre-treatment techniques could to some extent address this challenge though. Digestion, especially methanogenesis, may also be disrupted by large amounts of high-energy substrates such as fats.

Operation of biogas chains shows large differences among different countries and regions, with feedstock use by-and-large depending on cost and availability of feedstocks as well as also on customs, level of experience, and background of the chain under consideration. Manure seems to be the most commonly chosen substrate while there is a large potential for digestion of urban waste and food processing residues. In many cases, only a small part of the potential has been currently realised. An increasing number of companies are offering evaluation of AD feedstocks in commercial packages; with the major focus currently placed on their biogas potential. In the future, an integrated analysis, including performance evaluation and optimization of feedstocks/digester combinations should be implemented to facilitate enhanced economic performance.

References

Amin F, Khalid H, Zhang H et al (2017) Pre-treatment methods of lignocellulosic biomass for anaerobic digestion. AMB Express 7:72

Amon Th, Amon B, Kryvoruchko V et al (2007a) Methane production through anaerobic digestion of various energy crops grown insustainable rotations. Biores Technol 98:3204–3212

Amon et al (2007b) Methane production through anaerobic digestion of various energy crops grown in sustainable crop rotations. Biores Technol 98:3204–3212

Battista F, Fino D, Mancini G (2016) Optimization of biogas production from coffee production waste. Biores Technol 200:884–890

Cadavid-Rodríguez LS, Bolaños-Valencia IV (2016) Grass from public green spaces an alternative source of renewable energy in tropical countries. Revista Ion 29:109–116
FAO (2013) Reviewed strategic framework. Paper discussed at the thirty-eight session, Rome, 15–22 June 2013
FNR (2009) Biogas-Messprogramm II. 61 Biogasanlagen im Vergleich. Gülzow, Fachagentur Nachwachsende Rohstoffe, p 168
FNR (2010) Leitfaden Biogas. Gülzow, Fachagentur Nachwachsende Rohstoffe, Von der Gewinnung zur Nutzung, p 272
Garcia-Nunez J, Rodriguez D, Fontanilla C et al (2016a) Evaluation of alternatives for the evolution of palm oil mills into biorefineries. Biomass Bioenerg 95:310–329
Garcia-Nunez J, Ramirez-Contreras B, Rodriguez D et al (2016b) Evolution of palm oil mills into bio-refineries: literature review on current and potential uses of residual biomass and effluents. Resour Conserv Recycl 110:99–114
Holm-Nielsen JB (2009) The future of anaerobic digestion and biogas utilization. Bioresour Technol 100:5478–5484
Hoyne S, Thomas A (2001) Forest residues: harvesting, storage and fuel value. COFORD. http://www.coford.ie/media/coford/content/publications/projectreports/residues.pdf
Hung YT, Lo HH, Awad A et al (2006) Potato wastewater treatment. In: Wang LK et al (eds) Waste treatment in the food processing industry. CRC Press, Boca Raton, FL, pp 193–254
Janke J, Nelles M, Leite A et al (2015) Biogas production from sugarcane waste: assessment on kinetic challenges for process designing. Intern J Mol Sci 16:20685–20703. https://doi.org/10.3390/ijms160920685
Kabir M, Castillo M, Taherzadeh M (2013) Effect of the N-Methylmorpholine-N-Oxide (NMMO) pre-treatment on anaerobic digestion of forest residues. BioResources 8:5409–5423
Kalia V, Sonakya V, Raizada N (2000) Anaerobic digestion of banana stem waste. Biores Technol 73:191–193
Kumari S, Das D (2015) Improvement of gaseous energy recovery from sugarcane bagasse by dark fermentation followed by biomethanation process. Biores Technol 194:354–363
Labatut R, Angenent T, Scott N (2011) Biochemical methane potential and biodegradability of complex organic substrates. Biores Technol 102:2255–2264
Langeveld JWA, Kalf R, Elbersen HW (2010) Bioenergy production chain development in the Netherlands: key factors for succes. Biofuels Bioprocess Biorefin 4:484–493
Langeveld JWA, Quist-Wessel PMF (2014) Variation in oil palm wastewater composition and its impact on potential biogas production. Report 1402. Wageningen, Biomass Research
Langeveld JWA, Guisson R, Stichnothe H (2016) Mobilising sustainable supply chains—biogas cases. In: Biogas production from municipal solid waste, oil palm residues and co-digestion. Paris, International Energy Agency
Li C, Liu G, Nges I, Deng L, Nistor M, Liu J (2016) Fresh banana pseudo-stems as a tropical lignocellulosic feedstock for methane production. Energy Sustain Soc 6:27
Li Y, Zhu J, Wan C, Park SY (2011) Solid-state anaerobic digestion of corn stover for biogas production. Am Soc Agric Biol Eng 54:1415–1421
Li X, Dang F, Zhang Y, Zou D, Yuan H (2015) Anaerobic digestion performance and mechanism of ammoniation pre-treatment of corn stover. BioResources 10:5777–5790
Liedl BE, Bombardiere J, Chatfield JM (2006) Fertilizer potential of liquid and solid effluent from thermophilic anaerobic digestion of poultry waste. Water Sci Technol 53:69
Liew L, Shi J, Li Y (2012) Methane production from solid-state anaerobic digestion of lignocellulosic biomass. Biomass Bioenergy 46:125–132
Mathias JFC (2014) Manure as a resource: livestock waste management from anaerobic digestion, opportunities and challenges in Brazil. Intern Food Agribusiness Manage Rev 17:87–110
Matulaitis R, Juškiene V, Juška R (2015) Measurement of methane production from pig and cattle manure in Lithuania. Zemdirbyste-Agriculture 102:103–110
McEniry J, Allen E, Murphy JD, O'Kiely P (2014) Grass for biogas production: the impact of silage fermentation characteristics on methane yield in two contrasting biomethane potential test systems. Renew Energy 63:524–530

Montenegro Campos CM, Calil Prado MA, Lopes Pereira E (2014) Kinetic parameters of biomass growth in a UASB reactor treating wastewater from coffee wet processing (WCWP). Rev Ambient Água 9

Mumme J, Eckervogt L, Pielert J et al (2011) Hydrothermal carbonization of anaerobically digested maize silage. Biores Technol 102:9255–9260

Nakada S, Saygin D, Gielen D (2014) Global bioenergy supply and demand projections: a working paper for REmap 2030. International Renewable Energy Agency, Abu Dhabi, United Arab Emirates

Nkoa R (2013) Agricultural benefits and environmental risks of soil fertilization with anaerobic digestates: a review. Agron Sustain Dev 34:473–492

Nzila C, Njuguna D, Madara D et al (2015) Characterization of agro-residues for biogas production and nutrient recovery in Kenya. J Emerg Trends Eng Appl Sci (JETEAS) 6:327–334

Patinvoh R, Osadolor O, Chandolias K, Horváth I, Taherzadeh M (2017) Innovative pre-treatment strategies for biogas production. Biores Technol 224:13–24

Persson T, Baxter D (eds) (2015) IEA bioenergy task 37—country reports summary 2014. IEA Bioenergy, Paris

Quintero M, Castro L, Ortiz C, Guzmán C, Escalante H (2012) Enhancement of starting up anaerobic digestion of lignocellulosic substrate: fique's bagasse as an example. Biores Technol 108:8–13

Santa-Maria M, Ruiz-Colorado A, Cruz G, Jeoh T (2013) Assessing the feasibility of biofuel production from lignocellulosic banana waste in rural agricultural communities in Peru and Colombia. Bioenergy Res 6:1000–1011

Sri Rahayu A, Karsiwulan D, Yuwono H et al (2015) Handbook POME to biogas. Project development in Indonesia, Washington, Winrock International

Syarief R, Novita E, Noor E, Mulato S (2012) Smallholder coffee processing design using wet technology based on clean production. J Appl Sci Environ Sanitation 7:93–102

Teghammar A, Forgács G, Horváth I, Taherzadeh M (2014) Techno-economic study of NMMO pre-treatment and biogas production from forest residues. Appl Energy 116:125–133

Ulsido MD, Li M (2016) Solid waste management practices in wet coffee processing industries of Gidabo watershed, Ethiopia. Waste Manage Res 34:638–645

Verheijen LAHM, Wiersema D, Hulshoff Pol LW, De Wit J (1996) Management of waste from animal product processing. Wageningen, International Agricultural Centre

Walter Borges de Oliveira SV, Leoneti AB, Magrini Caldo GM, Borges de Oliveira MM (2011) Generation of bioenergy and biofertilizer on a sustainable rural property. Biomass Bioenergy 35:2608–2618

Weiland P (2010) Biogas production: current state and perspectives. Appl Microbiol Biotechnol 85:849–60

Yong Z, Dong Y, Zhang X, Tan T (2015) Anaerobic co-digestion of food waste and straw for biogas production. Renew Energy 78:527–530

Zhang D, Ong Y, Li Z, Wu J (2012) Optimization of dilute acid-catalyzed hydrolysis of oil palm empty fruit bunch for high yield production of xylose. Chem Eng J 181–182:636–642

Zhu Z, Hsueh M, He Q (2011) Enhancing biomethanation of municipal waste sludge with grease trap waste as a co-substrate. Renew Energy 36:1802–1807

Chapter 3
Biogas Plants: Design and Fabrication

Luca Talia

List of Abbreviations

BoP	Balance of Plant
CAPEX	Cost of investment
CSTR	Continuous stirred-tank reactor
FIT	Feed-In Tariff
HRT	Hydraulic retention time
MSW	Municipal solid waste
OFMSW	Organic fraction municipal solid waste
OLR	Organic loading rate
PGY	Potential Gas Yield
Q_D	Quantity of dilution
TDS	Total dissolved solids
TS	Total solid content
TSS	Total suspended solids
UASB	Upflow anaerobic sludge blanket
VS	Volatile solids content

3.1 Plant Description

The structure of biogas plants is quite similar but the choices made during the design of the related details will be the key factors to lead a project to success (or not).

The typical configuration of a biogas plant consists of the followings areas:

1. Substrate management area (receipt, storage, transportation to feeding, etc.)
2. Feeding and/or pre-treatment area

L. Talia (✉)
SEBIGAS s.r.l., Olgiate Olona, Italy
e-mail: luca.talia@yahoo.it

M. Tabatabaei and H. Ghanavati (eds.), *Biogas*, Biofuel and Biorefinery Technologies 6, https://doi.org/10.1007/978-3-319-77335-3_3

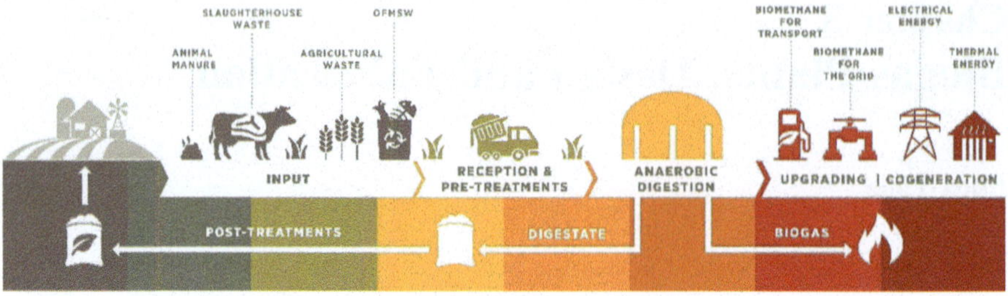

Fig. 3.1 Schematic presentation of the processes taking place in a biogas plant [SEBIGAS s.r.l.]

3. Anaerobic digestion area
4. Gas storage, treatment and usage
5. Digestate storage/usage/disposal

Materials flow through the different steps releasing their energy content in the form of biogas and are then extracted from the digesters, i.e., the digestate which could be disposed of or used as biofertilizer (Fig. 3.1).

Every step of the process needs to be carefully designed and evaluated, keeping in mind the final target of the project and it's constrains. There is a huge quantity of different possible combinations of equipment, processes, and technics that can be applied to optimise a given plant and to create a successful project.

The best design is well associated with possessing a sufficient deal of experience in every single area mentioned above as well as the equipment used in the plant and moreover, with having an in-depth understanding on different parameters affecting the process and consequently the overall functionality, performance, and return of investment.

3.2 Basic Design

For designing an anaerobic digestion plant, it is important to accurately establish the volume of the digestion (single or multiple digesters) and this in turn could impact the fluid dynamics and speed of reaction.

The calculation of the volume is generally performed after deciding on the other configurations and operation conditions of the plant such as process temperature, number of stages/phases, as well as type of process and technology.

The main technical parameters based on which an anaerobic digester could be classified include:

- Solids content
  The digestion process can be usually divided from wet to dry digestion accordingly
- Temperature of digestion
  Psychrophilic, mesophilic, thermophilic

- Technology
  From batch process to continuous process (e.g., plug-flow, CSTR, UASB, lagoon)
- Number of stages/phases
  Single stage or double stage/phase

To be able to select the correct application, every biogas plant design should start from the selection and analysis of the feedstock available and its final usage. In fact, if the project analysis is initiated by taking into account other factors rather than feedstock analysis, loss of time and resources would be expected.

3.2.1 Substrate Characteristics

The knowledge required on substrate should not be limited to the time of analyzing the project, but rather, the best scenario is to have a complete picture of the situation faced by acquiring the following data:

- Available quantity of feedstock per year; per day and receiving frequency
- Quality of the feedstock in terms of TS, VS, gas yield, N content (TKN), S content, etc. as well as their potential variations
- Suggested HRT, OLR, and temperature of digestion

This chapter is not dedicated to the analysis of the biological quality aspects of feedstock, but mainly on the effects that the parameters involved have during the design phase of a biogas plant project.

Total solid (TS) is a key parameter to be analysed as it could lead to completely different technology selection and design criteria. For the design of a plant, it is important to understand not only the initial TS of the substrate, but also its degradability and accordingly, the final TS after anaerobic digestion. The TS content inside digesters is strictly related to the final TS calculation, while the feeding capacity/technics are related to the initial TS.

To calculate the final TS after digestion, it is necessary to evaluate the mass balance for the substrate considered (Eq. 3.1), therefore:

$$TS' = \frac{TS - TS \times VS \times PGY \times \rho_{biogas}}{1 - TS \times VS \times PGY \times \rho_{biogas}} \tag{3.1}$$

where:

TS′	Final TS concentration of the substrate after anaerobic digestion [%]
TS	Initial TS concentration of the substrate [%]
VS	Volatile solids concentration referred to TS [$\%_{TS}$]
PGY	Potential gas yield [Nm^3_{biogas}/t_{VS}]
ρ_{biogas}	Specific weight of the biogas calculated as an approximation using the following formula (Eq. 3.2):

$$\rho_{biogas} = C_{CH_4}\, x\, \rho_{CH_4} + (1 - C_{CH_4})\, x\, \rho_{CO_2} \tag{3.2}$$

where:

C_{CH4} Percentage of CH_4 concentration in biogas
C_{CO_2} Percentage of CO_2 concentration in biogas
ρ_{CH_4} Specific weight of methane
ρ_{CO_2} Specific weight of carbon dioxide

As mentioned earlier, this value is an approximation because biogas contains other elements in small quantities (H_2S, H_2, O_2, N, etc.) as well that may slightly modify the final result.

The TS′ is basically lower than TS as a portion in form of VS is used by the microbial populations to produce biogas, and therefore the mass removed in form of biogas has to be removed through the mass balance calculations to determine the TS′.

TS′ concentrations below 2% would normally lead to the choice of a UASB or Lagoon system depending on the other biological parameters, while higher TS′ concentrations usually leads to the choice of CSTR technology (see paragraph 4.2.4.2) or even dry fermentation. In this chapter, the CSTR technology as the most common technology for industrial scale biogas plant is mainly analysed and discussed.

Hydraulic retention time (HRT) is the average time materials spend inside a digester and is calculated in day based on the substrate volume in input and digestion volume available (Eq. 3.3):

$$HRT = \frac{V_{net}}{Q_{sub}} \tag{3.3}$$

where:

V_{net} Available net digestion volume of the analysed system
Q_{sub} Volume of daily substrate input of the analysed system

Organic Loading Rate (OLR) represents the amount of volatile solids (kg) fed into the system analysed for every cubic meter of the available digestion volume every day. The OLR is therefore measured in $kg_{VS}/m^3/d$ as follows (Eq. 3.4):

$$OLR = \frac{Q_{VS}}{V_{net}} \tag{3.4}$$

where:

Q_{VS} Daily quantity of substrate volatile solids in the input of the analysed system

3.2.2 *Process Temperature*

Temperature of digestion is very important to finalise the design of a plant as a lot of components and equipment are affected by this parameter. In another word, the cost of the plant as well as the operation procedures may vary accordingly. In general, there are three different digestion temperatures:

Psychrophilic process is normally in use in lagoon systems while it is avoided in CSTR systems as it does not maximise the gas yield from the substrate and it is not easy to produce a constant quantity of gas out of the substrate either. The temperature range is usually considered lower than 25 °C, and therefore, there is no need for heating sources, however, the conversion efficiency of VS to biogas is lower than the other temperature regimes.

Mesophilic process is generally running smoothly and it does not require any particular attention for the selection of materials, it requires little thermal energy to keep the process temperature stable (especially compared with the thermophilic process). The temperature ranges between 37 and 42 °C.

Thermophilic process is "fast responding" to any modifications in the digestion parameters, which can lead to a more stressed operation of the plant and higher attention and effort would be needed to keep the same functionality. It should also be noted that the lifetime of some equipment (i.e. gasholder, mixers, pumps, etc.) could be negatively affected by high temperatures and hence, to keep the same reliability, the quality (and consequently costs) of the materials selected should be higher. On the other hand, this process usually allows the plants to operate at shorter HRTs at the same PGY rate due to the faster speed of reaction and therefore, the total digestion volume of the plant can be reduced.

In thermophilic processes, the speed of hydrolysis is normally 5 times higher the speed of mesophilic condition (Bouallagui et al. 2004; Converti et al. 1999). As a counter effect, the process is more affected by ammonia inhibition. Furthermore, the thermal energy need is sometime not affordable, especially with high moisture content substrates.

Overall and from the design and fabrication point of view, the main advantage of using a thermophilic process is the reduction of HRT. The temperature range is between 50 and 55 °C.

In general, different combinations of temperature regimes could also be used different in response to the specific case. For instance, a patented procedure by the Iowa State University (ISU) elaborated on the use of a thermophilic reactor to speed up the hydrolysis reaction in the first stage while a mesophilic reactor was used in the second stage to decrease the problems associated with ammonia inhibition and moreover, to have the possibility of using it as a back-up when encountering potential biological failures at the first stage (Han and Dague 1997).

The usage of intermediate temperatures between upper mesophilic limit (42 °C) and lower thermophilic limit (50 °C) could also be advantageous in terms of balancing the negative effects of inhibition caused by high temperatures and lower speed of reaction caused by lower temperature regimes.

It should be noted that rapid changes of temperature may damage and harm the microorganism, and therefore, the key issue during the design process regarding the temperature is not really to take necessary measures to ensure an exact temperature, but rather to prevent any fast temperature changes during the plant operation. In better words, if the temperature changes slowly in digesters, the existing microbial populations have the time required to adapt to the new situation without compromising the efficiency and the gas production.

3.2.3 Energy Outcome Target and Destination

The energy outcome target is basically established by the availability of substrate, the available size of the engine, the real need (or form) of energy production, and eventually by the limitations imposed by local authorities or feed-in tariff (FIT) policy.

As a general rule, the aforesaid feedstock parameters given at the time of the development of the project must be obviously taken into account, but precaution must be taken as they might easily change in time.

As the first step of calculation, it is possible to estimate the quantity of biogas that could be produced per year (Eq. 3.5):

$$Q_{biogas} = \sum_{i=1}^{n} Q_{sub_i} \cdot TS_i \cdot VS_i \cdot PGY_i \tag{3.5}$$

where:

N	Number of substrate used to feed the plant
Q_{biogas}	Quantity of biogas produced in a year [Nm^3/y]
Q_{sub}	Quantity of material available in a year [t/y]
TS	Initial TS concentration [%]
VS	Volatile solids concentration referred to TS [$\%_{TS}$]
PGY	Potential gas yield [Nm^3_{biogas}/t_{VS}]

Based on the expected percentage of methane content in biogas (based on the analysis of the substrate and its composition), it is possible to estimate methane production (Eq. 3.6):

$$Q_{CH_4} = Q_{biogas} \cdot C_{CH_4} \tag{3.6}$$

where:

Q_{CH_4}	Quantity of methane produced in one year [Nm^3/y]
C_{CH_4}	Concentration of methane expected in the biogas produced [%]

In case the biogas is converted into electricity and thermal power through a cogeneration system, the electrical power production can be approximated using the following formula (Eq. 3.7):

$$P_e = Q_{CH_4} \cdot 10 \cdot n_e \tag{3.7}$$

where:

P_e Electrical energy produced per year [kWh/y]
n_e Electrical efficiency of the cogeneration system selected [%]

3.2.4 *Digester Technology Classification*

Based on the characterisation of the substrate to be treated, the investment capital availability, the target outcome power, etc., it is possible to select the technology of digestion that would better suite the need of the project.

Here is a summarized list of the different reactors technology available.

3.2.4.1 Batch Process

Usually identified as sequencing batch reactor (SBR), in this type of application, the reactor is loaded one time and all digestate/percolate formed during digestion is recirculated until the end of the cycle (Fig. 3.2).

This type of application has of course the disadvantages of a non-constant production of gas due to the different speed of reaction during the retention time and the gas production peaks at around 50% of the retention time.

In order to stabilise the process, it is normally suggested to operate with multiple reactors in parallel, with staggered cycles.

This process is usually applied for dry fermentation and it is associated with low operational costs in spite of high energy consumption and maintenance costs.

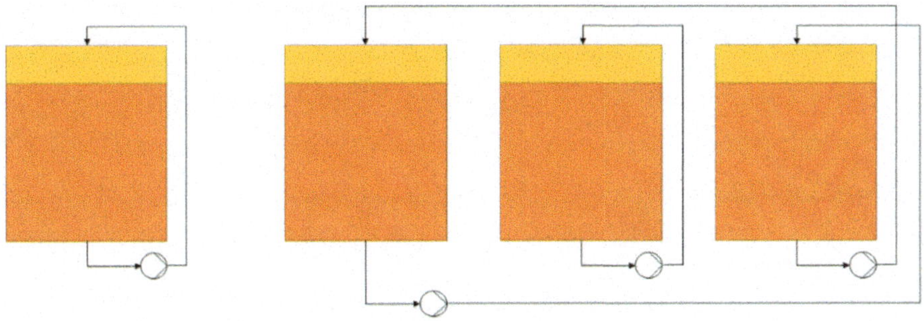

Fig. 3.2 Left: a single SBR reactor and Right: a multiple SBR reactor system

3.2.4.2 Continuous Process

In the continuous process, the substrate is fed constantly (continuously or in small batches with a defined interval of time), and as a direct consequence, the biogas production is almost stable.

The digesters can be vertical, horizontal, as well as in single or multiple stages. Depending on the type of mixing, they can also be further classified.

Plug flow

Thanks to the simplicity of the system and low investment costs, this system is normally adopted for farm liquid effluent. Concrete tanks, usually without mixing systems and possibly with internal baffles to differentiate the hydrolysis phase are used. In plug-flow systems where the solid content of the digestate would be lower than 10% TS, sedimentation of heavier parts and floatation of lighter parts might take place leading to heavy maintenance cost for emptying the digesters.

Continuous stirred tank reactor (CSTR)

This is the most common type of reactor operating under wet condition. This kind of reactor is suitable for the digestion of a high variety of substrates from agricultural waste to industrial waste or energy crops.

The reactor normally has a cylindrical shape with mixing system and it can be operated at different temperatures and OLRs (2–5 $kg_{VS}/m^3/d$, higher ORL have been used only with some special substrates and after continuous digestion tests) (Fig. 3.3).

The mixing technology used in these reactors is well known and it guarantees a high efficiency of VSs digestion compared with other systems. On the other hand, the investment costs for this kind of system is usually higher than the others.

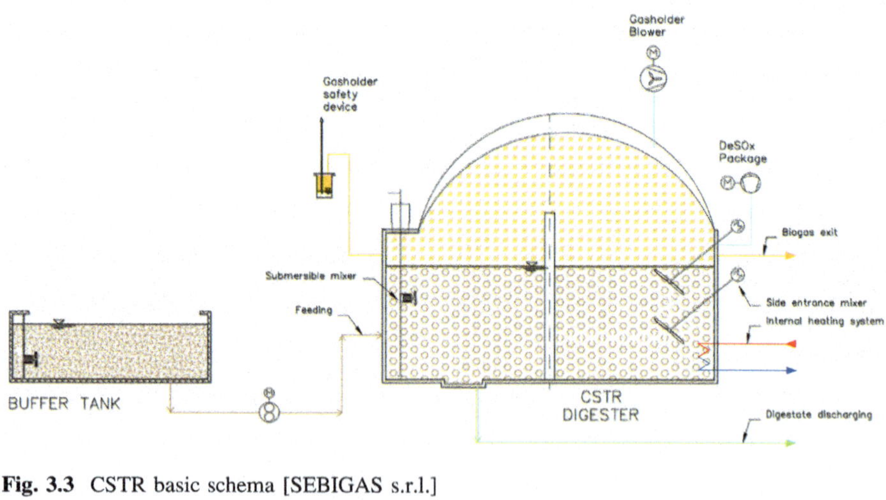

Fig. 3.3 CSTR basic schema [SEBIGAS s.r.l.]

High load reactor

Thanks to the retention of biomass and microbial population inside the reactor, this kind of process allows a higher efficiency and therefore, can operate at higher OLRs (8–20 kg_{VS}/m^3 d).

These systems are anyway not suitable for the digestion of substrates with high concentrations of particles i.e., total suspended solid (TSS) due to the tendency to accumulate the particles in the reactor.

The substrates usually treated are characterised by low level of TSS and high level of total dissolved solid (TDS) (such as industrial waste water, distillery waste, etc.).

The following types of reactor belong to this classification:

- Fixed or moving bed reactor
  The substrate and microbial population are attached to special supports (fixed or movable) installed in the reactor allowing the retention of the biomass while the substrate is recirculated to the supports to increase the contact time with the microbial biofilm.
- Upflow Anaerobic Sludge Blanket (UASB)
  In this system, a blanket of granular sludge is formed and kept in suspension in the tank. The combined action of the upward flow of the substrate and the gravity suspends the sludge blanket in the reactor (Fig. 3.4). This reactor is typically suited for wastewater digestion with low TSS.
- Membrane Reactor (MBR)
  This system uses a membrane to physically retain the biomass and microbial population inside the reactor while also separating the solid/liquid/gas phases. Thanks to its high separation capacity of and favourable performance in terms of achievable TSS, these reactors are particularly efficient for treating high pollutant content wastewaters.

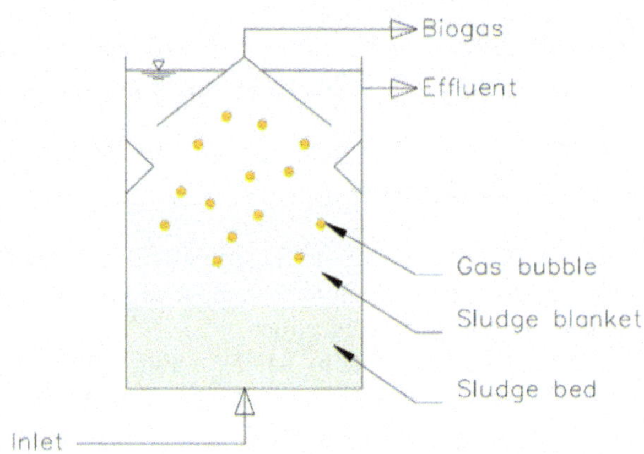

Fig. 3.4 A simplified schematic presentation of a UASB reactor

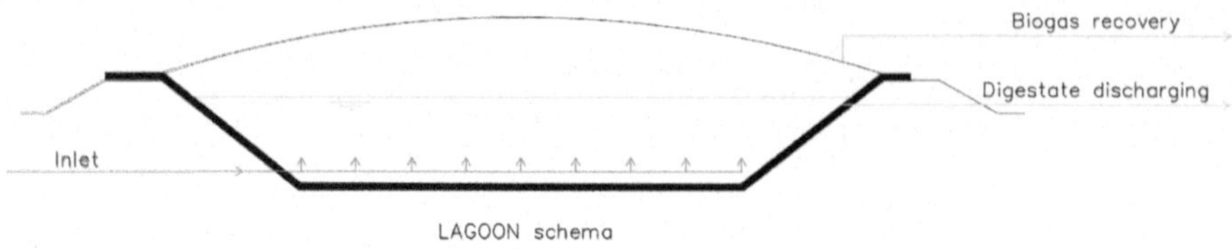

Fig. 3.5 Lagoon typical schema

Covered lagoon

This is the cheapest solution for anaerobic digestion process application (Fig. 3.5). It is a very simple and low investment cost application, but of course has some disadvantages like the high tendency for the formation of sedimented layers at the bottom of the system. This may necessitate emptying the system imposing high maintenance costs. Other disadvantages include huge area necessary, low efficiency due to non-controlled temperature of digestion, possible technical problems due to high volume of gas storage, leakage, etc.

The advantages of the system include possibility of designing a big volume system at a relatively low investment cost. This systems particularly of interest in case of very low TS/energy content substrates. Moreover, lagoons are easy to operate.

3.2.5 Solid Content Classification

Another type of classification for the digester technology is based on the solid content. Accordingly, the digestion process can be divided in two macro-group:

- Dry digestion
- Wet digestion

In wet digestion pumpable substrates are used while in dry digestion stackable substrates are used. In spite of that, there is no exact division line between these two types of digestion and it can just be assumed that wet digestion has an upper TS limit of around 10% while dry digestion has a lower TS limit of around 20%.

3.2.5.1 Dry Digestion

This process was initially used especially for the digestion of municipal solid waste (MSW) and organic fraction municipal solid waste (OFMSW), but it has also been used for energy crops digestion.

The advantages is that by having a high %TS inside a digester, the system could reach higher OLR values (even higher than 10 kgVS/m^3/d) which would consequently lead to smaller digestion volumes and lower investment costs. Moreover, there is no risk of sedimentation or floating layer formation as phase separation does not occur using a dry substrate.

To move dry materials, it is however necessary to have a strong system requiring much more maintenance and operation costs. It should also be highlighted that the absence of water implies a non-dilution of any possible inhibiting substances that may easily lead to a non-balanced system and acidogenesis phase prevalence.

3.2.5.2 Wet Digestion

Substrates like animal manure or wastewater/sludge do not need any additional liquid to reach the right TS% required by a wet system while for energy crops or other materials with TS% higher than 20–25%, it is mandatory to add water or to recirculate liquid digestate to keep the TS% at a right value in the digester allowing mixing and pumping of the substrate.

A typical configuration for wet digestion is CSTR digesters that need a good mixing technology to ensure an efficient contact between the microbial populations and the substrate.

The advantages are a smooth and economic operation giving stable production of biogas and simple usage of the plant.

The disadvantages include the possibility of the formation of sedimentation or floating layer (in case of non-optimised mixing design) which in extreme case may require emptying the digester. In some cases, it is also necessary to have a better pre-treatment system compared with dry digestion installations, but of course, the results would be higher efficiencies and higher specific gas production. An example is the OFMSW that needs a more accurate removal of plastics and sands before being introduced into a CSTR system, but the gas yield resulting from the process is higher than that of an average dry digestion system.

3.2.6 Process Design

After the selection of substrate to be used and the technology to be applied, it is possible to follow the main steps of biogas plant design as follows:

- Calculation of internal TS of the digester
- Recirculation or dilution
- Calculation of the volume necessary and selection of the type and dimension of the tanks

It is also important to keep in mind that every project has its own background and environment, and therefore, it is always important to study the local regulation regarding anaerobic digestion and in general, all construction regulations and permits that may affect the selection of the equipment.

3.2.6.1 Internal TS of Digester

The internal TS basically affects the mixing system (i.e., power and type of mixers or agitation system to be applied) and the possibility of sedimentation or phase separation.

In general, lower TS values are easier to mix, but phase separation, formation of floating layer and sedimentation are also more likely. On the contrary, higher TS values are more difficult to mix, but materials remain more homogeneous.

In some cases such as organic wastes, chicken manure, etc., it is essential to have a low TS at the entrance of the digester by allowing as much sedimentation as possible in the pre-treatment to avoid the introduction of large quantities of sands, shells and/or other undesired materials into the digester. In all other cases, the TS is a balance among TS′, and the recirculation and mixing system selected.

3.2.6.2 Recirculation or Dilution

To reach the selected target TS″ (TS inside the digester) the following steps should be followed (Fig. 3.6):

The quantity of added liquid to dilute (Q_D) at the beginning of the process (by recirculation or adding other liquids such as water, etc.) is calculated using the following formula (Eq. 3.8):

$$Q_D = Q' \cdot \frac{(TS' - TS'')}{(TS'' - TS_D)} \tag{3.8}$$

where:

Q_D Quantity of liquid added for dilution
TS_D Total solid concentration of the liquid added for dilution
Q' Quantity of substrate fed into digester
TS' Calculated TS concentration of the substrate after digestion
TS'' Target TS in digester

Therefore, the total quantity of liquid/solid fed into the digester (Q″) will be (Eq. 3.9):

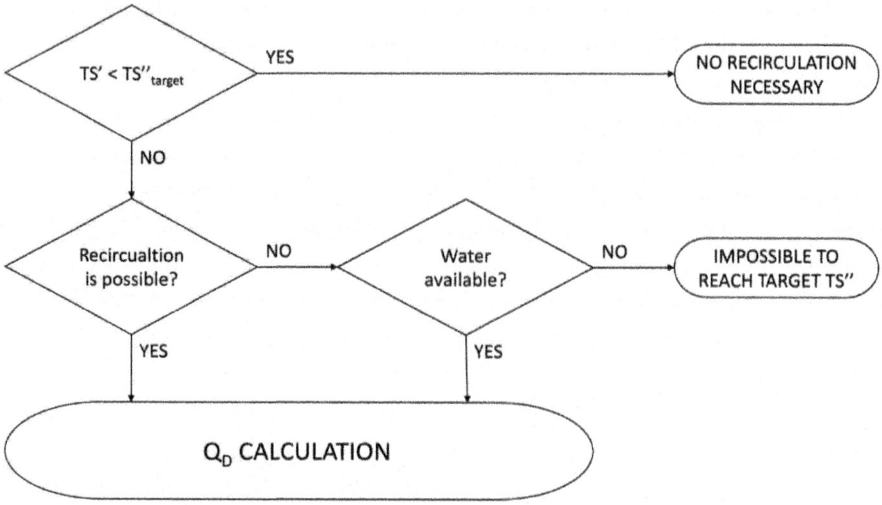

Fig. 3.6 Flow chart for the calculation of the recirculation or dilution

$$Q'' = Q' + Q_D \tag{3.9}$$

3.2.6.3 Calculation of Digestion Volume

The total digestion volume is defined by the HRT and OLR. Every substrate, depending on the digestion temperature and eventually on the pre-treatment applied, has its optimum retention time. Usually based on the findings of a batch test (that calculates the PGY), it is possible to predict the HRT of a given substrate.

OLR is estimated based on different biological parameters and represents the loading stress value of the digester, therefore, in general it is possible to force the OLR to higher values if the substrate is easy to digest by the microbial populations, while it should be kept at a lower level if the digestion process of recalcitrant substrates is intended.

Therefore, V_{HRT} [m^3] and V_{OLR} [m^3] can be computed as follows (Eqs. 3.10 and 3.11):

$$V_{HRT} = Q'' \cdot HRT \tag{3.10}$$

$$V_{OLR} = \frac{Q_{sub} \cdot 1000 \cdot TS \cdot VS}{OLR} \tag{3.11}$$

where:

V_{HRT} [m^3]	Net volume of digestion in accordance with target HRT
V_{OLR} [m^3]	Net volume of digestion in accordance with target OLR
Q'' [m^3/d]	Total daily volume fed into digester including eventual dilution
Q_{sub} [t/d]	Total quantity of substrate fed into digester daily
TS [%]	TS concentration of substrate
VS [%]	VS concentration of substrate
OLR [kg_{VS}/m^3d]	Target OLR for the considered substrate

The optimum total net digestion volume is the maximum value between the V_{HRT} and V_{OLR}.

In the design or construction phases of every and each biogas plant project, a big attention is focused on the investment costs. When the final selection on the digestion volume is made, the focus should not be only on maximising the biogas production or on the complete decomposition of the biomass, but rather efforts should be directed towards optimising the different parameters driving the Business Plan. It is sometimes possible to minimise the investment costs with a lower HRT and therefore, at the expense of lower specific biogas production and efficiency of the system, but this solution could be the only chance to have a feasible project.

From the environment point of view, the aim should always be to maximise the decomposition of the material to minimise the environment impacts of the digestate, but this also depends on the digestate usage.

3.2.6.4 Stages/Phases and Number of Digesters

The total volume necessary for the anaerobic digestion process can be divided in different tanks (or not) depending on the maximum dimension of 1 tank, the stirring capacity, the sedimentation expected, the number of stages desired, etc.

The complexity of the anaerobic digestion process lies in the fact that different phases exist throughout the process that may happen in sequence or simultaneously; each phase requiring its own optimal operating conditions such as temperature and pH.

Hydrolysis and acidogenesis are faster and they have higher efficiencies at low pH and high concentrations of substrate, while methanogenesis is inhibited by low pH values.

Inside a well-designed reactor, all four reactions can be taking place contributing to the overall anaerobic digestion process, there are for sure compromises though. Nevertheless, the concept of fully optimising every single step of digestion, necessitates separating different stages/phases of the process. Having more than one digester leads to other advantages as well such as:

- The calculated HRT is statistically more similar to the real one. The digester (e.g., CSTR) is continuously stirred and the materials inside should be approximately homogeneous, therefore, when a particle is extracted, its real retention time follows a normal distribution ("bell curve"), so there is a chance that a particle exits before or after the calculated HRT time.
  In case of an earlier exit, the digestion process could not be finished with a resulting lack of efficiency of the system and lower biogas production compared with the potential value, while if the particle exits later, there is almost no benefit.
  Based on this simple concept, it is clear that a double stage gives the possibility to mediate the normal distribution and reduce the standard deviation of the "bell curve".
- There is always a back-up. This means that in case of failure of one of the digesters, there is the possibility of using the other one to balance the process and the plant will not completely stop.

Single phase reactors have the all phases working at the same time in the same environment. This kind of design requires a lower OLRs (usually in the range of 1–4 kg_{VS}/m^3 d) in order to ensure a sufficiently high stability of the system.

Double phase reactors allow the optimisation of the four phases of digestion. In the first step, the best conditions for the hydrolysis and acidogenesis are usually achieved, i.e., a retention time of 1 to 5 d and OLR of >10 $kg_{VS}/m^3/d$ to increase the speed of the first 2 phases and avoid the start of the followings two.

The rest of the digestion takes place in the second step and the substrate is treated based on the design HRT.

Double stage reactors is a hybrid solution based on which the plant is designed with multiple digesters, but the system is not forced to have a clear separation of the digestion phases. In fact, all four phases happen together in different digesters, but in different percentages, so that the overall process is more stable and easier to operate even for non-skilled operators.

The choice of the configuration depends on the costs and benefits, and therefore it is always a good idea to discuss the different options and associated benefits during the preliminary design phase.

3.3 Plant Description and General Rules of Good Design

Once the biological and process parameters are established, it is then time to concentrate on the different areas forming a biogas plant. Biogas plants are usually non redundant systems as the investment costs target does not allow duplication of systems, equipment nor area of the plant. Due to this reason, it is more and more important to achieve an optimum design not only concerning the biological parameters such as HRT, OLR, and therefore net volume of digestion, but also it is decisive to have an overall harmonised design that avoid any problems in any areas.

Every step of the process from the receipt of material to the digestate usage and disposal could potentially be a bottle neck of the plant that may lead to stop of production.

A biogas plant can be divided into five main areas:

1. Substrate management area (receipt, storage, transportation to feeding, etc.)
2. Feeding and/or pre-treatment area
3. Anaerobic digestion area
4. Gas storage, treatment, and usage
5. Digestate storage/usage/disposal

The followings paragraphs aim at providing a general overview on the different areas explaining the possible choices and parameters that could influence the design.

3.3.1 Substrate Management Area

3.3.1.1 Substrate Receipt

It is important to keep into consideration the frequency, quantity, quality variation, and general variability of the material received to be prepared and ensure minimum possible energy loss prior to the biogas transformation.

Some substrates, like the energy crops in Europe, are usually received once per year, therefore, it is very important to be perfectly organised in the receipt, storage, and conservation technics choice so that the losses are minimised. In other countries, such as South East Asia or South America, there is a chance to have multiple harvests (2–6) per year for certain types of energy crops and this automatically influences the dimension of storage and the frequency of receipt. On the other side, energy crops collection in Europe is standardised and it is quite common to have very similar quality from one year to another, however, when harvesting more than once per year, the quality and characteristics of the materials are strongly influenced by the weather condition and seasonal changes and therefore, the variations in the substrate are less predictable, necessitating having a more flexible design.

A particular section could be dedicated to the wastes usually received 5 to 7 d a week (e.g., OFMSW or other industrial/food waste), but these wastes may produce reek and it is therefore suggested to receive them in a close building maintained slightly under pressure by using external blowers driving the air into odour control system (e.g., biofilters). The air recycle ratio changes in accordance with the local regulations, but as a general rules, it should be in the range of 2 to 4 air cycles/h (meaning that a building with an internal volume of 10,000 m^3 would need blowers with a capacity of 20–40,000 m^3/h to keep odour under control). It is usually suggested to keep separated the area where the operators stand or work continuously and the area where the operators visit just sometimes. In this way, it is possible to apply different air suction ratio to have 4 recycles/h in the working area and around 2 recycles/h for odour control.

Some other substrates like liquid or solid manure from animal farms do not even need any storage as they could be received on a daily base. In this case, the design should include a buffer area for the daily discharge and feeding into the plant.

3.3.1.2 Storage Area

The storage area is where the biomass to be used is stored for a certain period in accordance with the collection and receiving periods (which may vary from hours to 1 years).

The key point for a storage area are:

- To have enough volume to store the materials considering the flexibility and variation of characteristics of the materials to be used during the life of the plant.
- To keep the quality of the materials as much unchanged as possible to avoid loss of energy during storage period.

The storage structures in use are vertical silos or horizontal silos. Vertical silos are suitable for grains, cereals, or liquid products like oil or whey, while horizontal silos are more suitable for silage biomasses (grass silage, corn silage, energy crops silage, etc.). The horizontal silos could also be used to store agroindustry by-products or the other substrates to preserve their characteristics during the process.

Fig. 3.7 Examples of horizontal silo storage, **a** constructed on site and **b** pre-casted silo storage [Sebigas a division of Exergy S.p.A.]

The horizontal silo storage systems in biogas plant can be constructed in full from concrete or compacted ground or a combination of both.

The concrete solution is probably more reliable with the highest durability, but it is also costly to build. The wall can be constructed on site or pre-casted (Fig. 3.7a and b, respectively), the pre-casted concrete is generally more durable and holds more favourable acid resistant characteristics and is therefore, recommended if available on the market. The durability of walls constructed on-site depends on factors such as skills of the construction company, weather conditions, and concrete mix quality.

The fully compacted ground solution is associated with a high permeability of the percolate (liquid coming from silage/stocking procedure) into the ground, and should therefore be avoided.

It should be noted that a certain amount of percolate is generally generated through the silage procedure and it is a good practice to collect and eventually use it in the digester or anyway discharged/treated in accordance with local environmental rules and regulations.

The silobag is another possible solution for material storage. It is a kind of big plastic bag with a good resistance against severe weather conditions and that help to preserve the materials during storage. The costs of the machinery necessary to fill and empty the bag should be taken into consideration.

The size of the storage area basically depends on two parameters:

The total quantity of biomass to be used in digestion ($\mathbf{Q_y}$): is the total expected quantity of biomasses introduced to the anaerobic process to produce the target energy per year. $\mathbf{Q_y}$ can be calculated using the following equation (Eq. 3.12):

$$Q_y = \frac{Q_{\mathrm{CH_4}}}{PGY \cdot C_{\mathrm{CH_4}} \cdot VS \cdot TS \cdot (1 - \mu)} \tag{3.12}$$

where:

PGY Potential Gas Yield [$\mathrm{Nm^3_{biogas}/t_{VS}}$]

$Q_{\mathrm{CH_4}}$ Quantity of methane produced in one year [$\mathrm{Nm^3_{CH_4}/y}$]

$C_{\mathrm{CH_4}}$ Concentration of methane in the biogas [%$\mathrm{CH_4}$]

TS Initial TS concentration of the substrate [%]
VS Volatile solids concentration referred to TS [$\%_{TS}$]
μ Lost of VS during conservation process [%]

Duration of the availability (t_{sub}): is the duration in time the stored material is used in one year. Plants that use only energy crops with one harvest per year are forced to have a huge storage area to ensure sustainable material supply annually. On the contrary, plants that use multiple substrates and/or by-products with regular receipt are less subjected to the need for big storage areas.

Through these two parameters, it is possible to calculate the total volume of storage necessary (Eq. 3.13):

$$V_{storage} = \frac{Q_y}{\rho_{sub}} \cdot \frac{t_{sub}}{D_{365}} \cdot SF \quad (3.13)$$

where:

$V_{storage}$ Total volume of storage required [m^3]
Q_y Yearly quantity of substrate used [t/y]
ρ_{sub} Specific weight of the substrate silage [t/m^3]
t_{sub} Duration of the availability [days]
D_{365} Days of full production expected in one year [d/y]
SF Safety factor

The loss during the storage may vary considerably depending on different factors ranging between 5 and 7% (minimum and non-avoidable) for a good silage up to 30–40% in cases of misapplication of good silage procedures.

The specific weight of the substrate may also vary remarkably depending on the following non-exhaustive variables:

- Size of harvested material (usually 5–15 mm is the suggested length)
- Moisture content
- Press force during silage operation
- Height of the storage

Concerning the total volume of the storage, a minimum safety factor (suggested as 10–15%) should be taken into account to compensate for unexpected delays in the receipt of the subsequent batches of the raw materials, possible variations in the characteristics of the materials during storage time, etc.

All materials stored in horizontal silos must be covered by plastic lining to prevent oxygen entrance that may lead to oxidation and VS losses.

The silo storage may have different configurations depending on the available area or the construction technics (Figs. 3.8 and 3.9):

- Single or multiple
- Closed or open

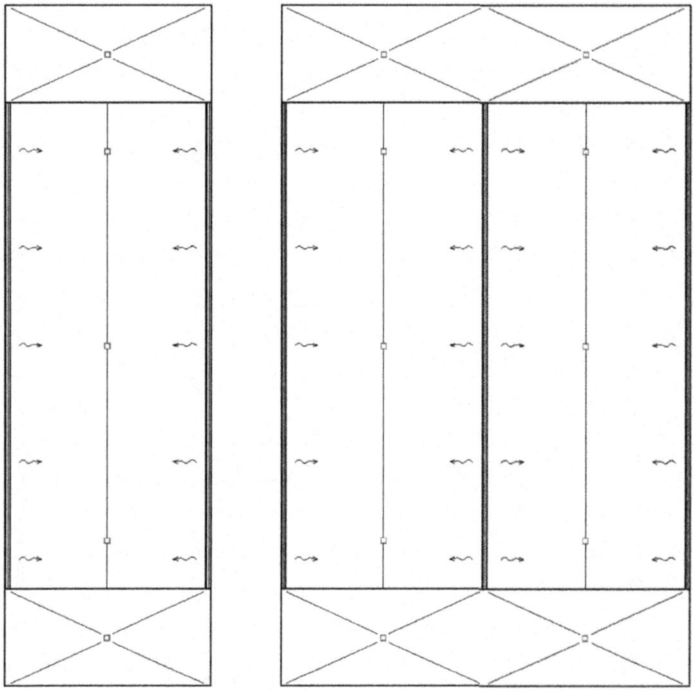

Fig. 3.8 Single silo storage (left) and multiple silo storage (right)

The width of each silo storage is designed to minimise the surface exposed to oxygen in the front. Moreover, the more the materials are pressed, the less the oxygen can penetrate through the surface.

As a general rule, it is suggested to have about 1 m of the front cut per day to minimise the exposure time of the front material and therefore, the loss of energy.

A silo storage with double side entrance allows a higher flexibility during plant operation as the "old" material that have been stored can be removed at first while the "new" materials can also be used thanks to the opposite access to the silo. This option allows a First in First out (FIFO) logic in the storage operation avoiding long time storage periods and consequent loss of energy. If a closed configuration is used, it is possible to increase the volume of the stocked material per square meter as the end wall allows the materials to be pressed without a slope, but it only allows a FILO strategy to be implemented.

A normal time limit for the silo storage is one year; two years is also possible but hardly suggestable.

Every silo storage releases percolate that could be posing pollution risks to soil and water resources. Therefore, it is strongly suggested to design a collection system and to send the percolate to the digestion process. It is also important not to underestimate the quantity of the percolate released in the early phase of the storage soon after silage operation.

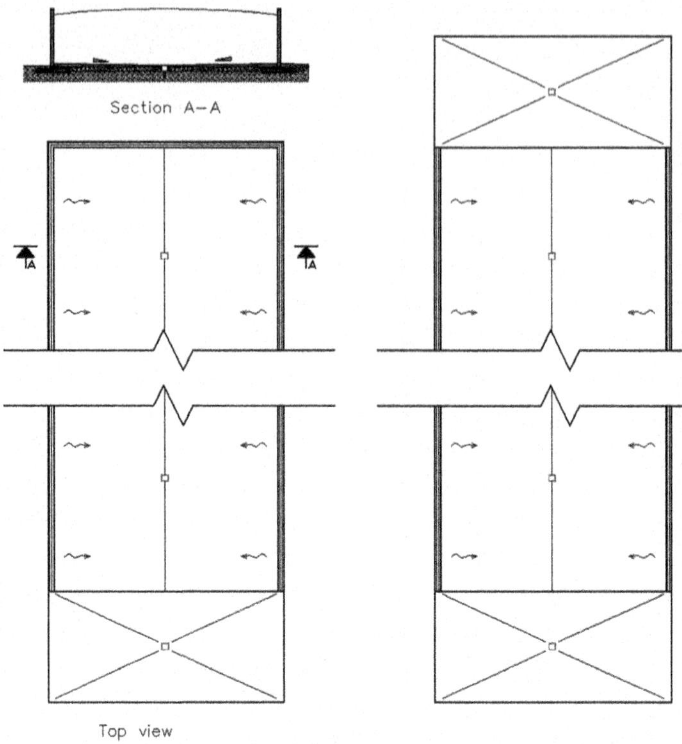

Fig. 3.9 A closed silo storage with one side entrance (left); an open silo storage with double side entrance (right)

The percolate usually has a very low pH, and therefore, it is important to take this information into consideration during the design of the civil construction to avoid any corrosion problems.

3.3.1.3 Transportation to the Feeding System

Wheel loaders are the most common transportation systems for the material from the storage area to the feeding system and has an average capacity of 2 tonne per trip. The distance between the silo storage and the feeding system is also a source of operational costs and should be considered.

Due to the obvious limited speed of the wheel loader, the longer the distance between the feeding system and farthest point of the storage, the more time-consuming the transportation and loading operation will be. The time needed by the operator to transfer the material from the storage to the feeding area should be calculated to prevent probable bottlenecks during plant operation.

Cranes are operated manually needing the presence of an operator during operation or are fully automatic transferring biomass from the storage area based on a predesigned route and algorithm. Cranes are a good solution when materials are received on a daily basis and are unloaded in a pit.

3.3.2 Feeding Equipment

The feeding system is the access door for the materials to the anaerobic digestion plant. There are various systems that can be applied to perform the same operation and of course there exist different prices and quality levels.

The feeding equipment is very important for the plant as there is no buffer nor spares in case of failures. It is preferable to design a biogas plant with at least two possible ways of feeding, a principal feeding system and a secondary system. The secondary feeding could also be a simpler and cheaper solution, but at least it would allow the operator to feed the plant in cases of maintenance of the principal system.

3.3.2.1 Liquid Feeding

Buffer tank

Also called receiving pit/tank, it is the tank used to collect and homogenise the pumpable substrate to be fed into the digesters. It is applicable as a primary feeding system for liquid manure, some vegetable waste, fruits, juice, etc.

In case of non-pumpable substrates (usually TS values over 10–12% depending on the substrate and the pump used), there is the possibility to dilute them in the buffer tank together with liquid fraction of digestate, water or digestate itself to have a pumpable, mixable, and homogeneous liquid. When the digestate is used, some gas is formed in the buffer tank which brings about some consequences:

- Gas forming and possible Atex classified area
- Loss of energy with linear relationship with time of contact between digestate and fresh material
- A higher quantity of recirculation has to be used to reach a favorable TS
- Not reducing the HRT as the digestate is just pumped outside for a while and pumped back.

While in case of using liquid fraction of digestate, the following consequences could be expected:

- Some bacteria are recirculated back to the fermentation and it may help the process
- It is effective on decreasing HRT

- Increasing inhibitory effects due to concentration of nitrogen, salts, and others contaminants
- Gas forming and possible Atex classified area.

In case of using water, the following points need to be considered:

- Higher cost of operation if water is not available for free
- Environmental impacts as the water used is mixed with the digestate and therefore, will then have to follow the same regulations
- Reduction of HRT (but less than using liquid fraction of digestate)
- Reduction of inhibitory effect.

The mixing system used in buffer tanks is very different from those used in digesters. The buffer tank usually receives materials in batch, in big quantities, and the scope is to homogenise the liquid and solids in a short time.

Near the charging point, a powerful mixer should be used to move away the biomass immediately after their introduction into the tank. The suction point of the pump from the buffer tank has to be kept clean to avoid any sedimentation blocking the suction pipe. Fast mixers are suitable for the application (see Sect. "3.3.3.4—Mechanical mixing system") and the installed power should be in the range of 30–70 W/m^3 depending on the characteristics of the materials to be homogenised.

The volume of the buffer tank should be calculated following the same considerations taken for the silo storage. It is important to consider the frequency of material received, the quantity of material, the total volume fed per day, the number of feeding cycle, and the availability of operator on site to follow the mixing and preparation activities.

A safety margin should also be considered when performing the volume calculations in provision of possible modifications in any of the mentioned criteria affecting the sizing of the tank. It is also suggested to have a minimum of 50 cm freeboard for open tanks and 20–30 cm freeboard for roof closed tanks.

Depending on the vehicle used to charge the tank, different constructions and dimension methods can be applied. In case the liquid is introduced with a slurry tank truck without discharging pumps, an underground solution would be suggested, while if there is the possibility to pump the liquid, an above-ground solution would be better allowing the installation of suction pumps to feed the digester at ground level, easing the empty procedure in case of maintenance.

When buffer tanks are used to mix and homogenise solid materials with the liquid fraction to create a pumpable solution, the level of the tank has to be calculated considering the maximum discharging height of the wheel loader/tipping trailer.

Direct feeding

The buffer tank can also be directly fed by pumps connected to the farm collection pit. In this case, it is recommended to include the function of the pumps and the instruments of the buffer tank in the PLC to check the level and to eventually stop pumps. Moreover, a remote control for start-stop could also be installed to ease the charging operation.

3.3.2.2 Solid Feeder

When solid biomass cannot be loaded in the buffer tank due to:

- Quantity
- TS (%)
- Missing recirculation
- Available space
- Operator availability
- Operator skill
- Atex area risk
- etc.

It is then possible to equip the plant with a dedicated solid feeder machine. There is a wide range of solid feeder types available on the market. These feeders can be classified according the following main characteristics:

1. Type of container and biomass transporting system
2. Wall and floor material
3. Type of injection system (injecting materials into the digester)

Other than the type of feeder, it is important to identify the volume necessary and the output capacity to fulfill the digester requirements in terms of daily volume/ weight and in terms of frequency of the cycles.

Type of container biomass transporting system

Walking floor is composed of pushing elements (sliding beams) arranged in parallel (Fig. 3.10). This system is suitable for transportation of light materials with low bulk weight. This technology have been used in the trucks' floor with automatic discharging system as well. A hydraulic unit normally activates the movement of the sliding beams at the required speed and the material is transported by friction.

Push floor has different carriers, depending on the task. The carriers slide on steel rails directly positioned at the bottom of the floor (Fig. 3.11). The racks of the

Fig. 3.10 Walking floor system [Huning Anlagenbau GmbH & Co.KG]

Fig. 3.11 Example of push floor rack installed in concrete bunker with lining protection from corrosion [Huning Anlagenbau GmbH & Co. KG]

push floor are moved forward and backward by hydraulic cylinders positioned on the head or on the tail of the container depending on the discharging system adopted.

The material is transported in the direction of the exit.

Push-off system consists of a shorts side wall actuated by a hydraulic cylinder, the wall slides along the floor in the direction of the exit of the system so that the material is physically pushed forward (Figs. 3.12 and 3.13). This system is a flexible solution that allows to transport different kinds of material, in particular heavy material or non-chopped material that might be difficult to transport forward with other systems.

Chain system is typically used for manure distribution on field trucks. This system is seldom used for feeding application in biogas plants due to the high maintenance costs.

Fig. 3.12 A typical solution of push-off system [Fliegl Agrartechnik GmbH]

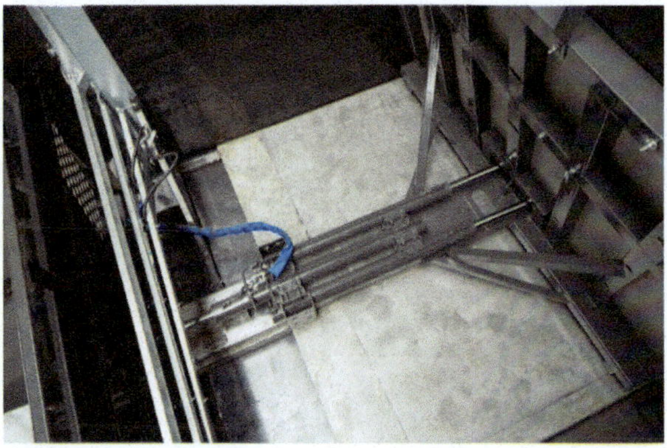

Fig. 3.13 The hydraulic cylinder and installation that moves the push-off wall [Fliegl Agrartechnik GmbH]

Horizontal screws is a method of transportation especially applied for big size biomass, often not chopped. A typical application is the slaughterhouse waste, animal carcass, etc.

Vertical mixing screws is based on the *Mixer-wagon* agricultural machinery normally used to mix the feeds for livestock and to distribute them along the feeding trough. The vertical mixing screws could be one or more (usually up to 3 maximum). This system has some disadvantages including have high electrical consumption (the screws need powerful motors), high maintenance (every screw is equipped with a gear reducer under the container or a bevel gear reducer just outside), and high wearing (the material is mixed continuously while transported towards the exit). One of the biggest advantages is that it is possible to load it with different layers of substrate while always having a homogeneous outcome material leading to a stable production with a well-balanced specific yield per ton of material. Given the high variations in the substrate used in biogas plants, if the material fed into the digesters is not homogenised, there might be fluctuations in gas production during the day (Fig. 3.14).

Fig. 3.14 The installation of a vertical mixing screws solid feeder (left) [Sebigas a division of Exergy S.p.A.] and a single vertical screw feeding system (right) [Trioliet B.V.]

Direct solid feeding is a very simple system used in cheap construction configuration for underground digesters ("round lagoon"). The system uses an opening on the side of the digester, under the gas holder, to create an opening wide enough to have the possibility to push the material inside the digester by means of a wheel loader. The system allows to have a very low construction costs, but may lead to higher operation costs as it is sometimes necessary to have an additional external mixing system operated by a tractor to mix the material at the entrance of the digester.

Wall and floor material

Different types of materials are used for the construction of solid feeders. The biomass container can be made of stainless steel, wood (Fig. 3.15), high density polyethylene (HDPE) (Fig. 3.17), carbon steel (with or without protection) (Fig. 3.16), concrete, etc.

It is important to plan the real lifetime usage of the plant to identify the correct materials suiting the duration the plant is supposed to be in use. The choice is made with reference to biomass characteristics:

- pH or corrosion action
- Wearing capability
- Liquid content and percolate

The costs is of course a key factor influencing the final choice.

Type of injection system (injecting materials into the digester)

After selection of the container, internal transportation system and their respective construction materials, it is time to analyse the advantages and disadvantages of the available solutions for the injection system—that is, the way to bring the material inside the digester.

Fig. 3.15 Wood wall solid feeder installed on Sebigas a division of Exergy S.p.A. plant

Fig. 3.16 Special polyethylene coating to protect the carbon steel frame [Trioliet B.V.]

Fig. 3.17 Feeding system with HDPE wall to prevent corrosion [Fliegl Agrartechnik GmbH]

The following principal injection systems are briefly described:

Screw conveyors are the most common injection system used in connection to solid feeders (Figs. 3.18 and 3.19). The conveyors can be classified based on screw diameter, pitch, type of pitch (constant or variable), thickness, construction material, internal guiding shaft, external protection, etc. These screws can be easily inclined and used in different situations. Being a closed system, it has the advantages of preventing powder losses and limiting unpleasant smell, yet it is costly and especially due to wearing of materials, it may need a high maintenance cost.

Belt conveyors are also quite common in European constructions. They allow the transportation of non-homogeneous materials and can not be easily blocked due to the absence of small gaps in the system. They also allow speed adjustment without facing rotation speed problems. The system also suffer from some disadvantages like high wearing and maintenance costs due to the high number of rollers and bearings, and limited inclination potential to avoid sliding back of the transported materials. The best application of this system is for waste transportation rather than energy crops.

Fig. 3.18 Combination of horizontal, vertical, and inclined screw conveyors to transport the material from a solid feeder to a digester

Fig. 3.19 Example of an inclined screw conveyor to transport the materials from a solid feeder to a digester [Sebigas a division of Exergy S.p.A.]

Solid-liquid pumps include a solid and liquid mixing buffer/box prior to the pumping body. Both lobe and screw pump are available for this solution. The advantages are a pre-mix of the substrate with the liquid (usually the digestate directly taken from the digesters), bearing a lower mixing electrical consumption to homogenise the materials, possibility of swift maintenance, and the possibility to install it far from the digester. It should be mentioned that the last item could also be unfavourable as it increases the cost of installation of the pipes and leads to pressure loss. The disadvantages of this system include, more complex piping system, as well as more complex software to control the liquid recirculation and liquid level inside the pump. Figures 3.20 and 3.21 presents example of solid-liquid pumps.

Fig. 3.20 Example of a solid-liquid pump located after the mixing box to send the materials into the digester [Pumpenfabrik Wangen GmbH]

Fig. 3.21 Example of a solid-liquid pump in combination with a grinder [Vogelsang Italia S.r.l.]

3.3.3 Anaerobic Digestion Area

This area of the plant is the core of the system and, even if apparently simple, includes small design details that may lead to successful or failed designs and operations.

Each digester has a volume ranging between 1.000 and 8.000 m^3 or more. Whenever it is necessary to empty such a volume, it takes a long time and the digestate has to be also temporarily stored in another tank. This operation may not be possible or may require additional costs for digestate disposal. For these reasons, it is important to have a reliable design incorporating different critical parts efficiently to maximise the availability of the plant allowing to reach the operation hours targeted in the original business plan.

3.3.3.1 Tank Construction

Digester tanks are made of reinforced concrete, carbon steel, stainless steel, or special coated steel. A digester, to keep an anaerobic process running inside, must be gas tight over liquid level and liquid tight under liquid level.

In any cases, it is important to carefully consider the underground condition and to perform a good geotechnical survey as well. This investigation is often underestimated (especially in some countries), yet there is a good chance to choose a more economical solution for the construction of the tank if the full data necessary for the geotechnical design and soil bearing capacity calculations are available.

In some cases, to bear the load of the digester and its liquid content, it is enough consolidate the ground or to stabilise it with stones and gravels, while in the majority of situations, it is necessary to implement more severe solutions such as piling, jet grouting, underground starting level, etc.

Reinforced concrete can be made on site by using a rebar beam cage or by using a special formwork without a reinforcing steel, allowing the structure to have a higher tensile strength (Fig. 3.22). There is also the possibility to install pre-casted panels and assemble them on site. Concrete constructions present considerable advantages in terms of flexibility in shape and dimension, possibility to be modified during construction in case of any modifications in the original design, capable of withstanding high loads, and being usually cheaper than other types of installation especially in large diameters. On the other side, highly skilled construction companies are required to ensure that the structure is liquid- and gas-tight. Strict tests are crucial carefully investigating the tanks as they must last for at least 15 or 20 years under severe environmental conditions. In case of wrong construction quality, it is important to take counter measures as soon as possible. During the operation, leaking of the digestate may affect the durability of the rebar embedded in the concrete due to action of liquid and oxygen, while in the top portion of the digester, possible gas leakage may lead to early corrosion due to acid gas (H_2S and moisture). Usage of high-grade concrete, right mix of additive, and professional

Fig. 3.22 Concrete casted digester during construction

Fig. 3.23 Glass fused tank under installation

planning of the pouring are mandatory for successful construction of reinforced concrete. The cement construction is cheaper compared with other installation, but it should be kept in mind that repairing costs could be exponentially higher as well.

Steel digesters are installed directly on the concrete basement, which is a key point for the tightness of the tank (Fig. 3.23). The parts of the tanks are usually connected with bolts followed by a cement pouring while a sealant on top seals the connection. The construction methodology can be very different: from welded metal sheet to bolted panels depending on corrosion and wearing effects. Usually V4A steel is suitable for H_2S contact, but with high concentration of oxygen and moisture, corrosion may become severe anyway. Coated panels (like glass fused or similar applications) are more resistant to the corrosion effect. Usually steel tank cannot be positioned underground as they have a very good resistance to the inside pressure (the pressure imposed by digestate on the wall), but have little resistance to the outside one (the pressure imposed by compacted ground on the outside wall of an empty tank). In case of underground solutions, the bottom rows of panels have to be oversized to resist to the pressure.

3.3.3.2 Tank Protection

The protection of the wall or roof of the digesters is very important to increase the durability of the structure during the lifetime of the project. Above the liquid level, concrete tanks can be coated with special PE membrane. The best option is to pour the membrane directly with the concrete so that no gas can penetrate between the protection membrane and wall causing corrosion (Fig. 3.24).

Special epoxy painting has been developed recently to protect the concrete and has been found promising especially in case of non-wearing material. In case of wearing material, there could be the need to open the digester after a certain time (years usually) to restore the original situation.

Fig. 3.24 PE membrane protection installed during pouring of the concrete to protect the surface above liquid level

3.3.3.3 Shape and Dimension

The selected net volume of digestion for each tank can be achieved through different shapes and dimensions. The main parameters are diameter, height, freeboard, thickness of the wall, bottom plate shape, and number of columns (if any).

Each parameter is driven by a mix of technical solutions and decisions. The diameter is related to the mixing system adopted and the substrate in the digestion. Lowering the ratio between the height and the diameter of digesters over 0.5–0.6 usually helps with the stratification and therefore, sedimentation and extraction from the bottom, but can lead to stratification of the digestate. Small diameter digesters are also easier to be mixed as the thrust of the mixer and distance of agitation can reach the middle of the tank. In flat digesters (the ratio between height and diameter in the range of 0.25–0.40), stratification (by energy crops, grass, etc.) could be avoided, but not all the surface is easily mixed and so it is probable to have areas with higher sedimentation.

A high digester also presents technical limitations for the installation of mixers, for example with heights over 10 m, the submersible mixer is not usually installed properly due to the length of the guiding shaft and the difficulty with lifting and lowering the mixer. Over 15 m, bottom lateral entrance mixers may face problems related to the sealing and tightness and special versions should be selected due to the pressure of the digestate. Vertical mixers installed at the centre of the tank should have longer shafts without guide at the bottom and therefore, they present the risk of high vibration and oscillation of the shaft.

The freeboard selection is mainly related to the position of the gas exit and the material used in digestion. The side exit of gas may reduce the possible freeboard as the bottom level of gas exit pipe should be at least 20–30 cm higher than the liquid level to prevent it from going into the gas line. In case when foam formation is expected, a higher safety factor should be applied and it is suggested to consider the bottom level of gas exit pipe at least 70 cm higher than the liquid surface.

The bottom plate can be flat or conical. The conical construction is not easy to build as it is necessary to utilise special tools during pouring. Conical shape allows the extraction of sedimentation through a pipe positioned at the centre of the cone. The flat surface is much easier to build and it can be cleaned by a submersible mixer with proper orientation and level adjustment. Usually conical-shaped digesters are used only for municipal solid waste or sometimes for poultry manure digestion.

In digesters with gas holders, it is usually necessary to have a central column that can be built in reinforced concrete, special wood or stainless steel. Reinforced concrete solution needs a special formwork, but it is a reliable solution while wood is a solution definitively affected by the quality of the available wood and is generally not recommended considering the installation environment. Stainless steel can be utilised, but as described, it is necessary to carefully control that no oxygen is present in the tank as it might affect the duration of the steel.

In case of digesters with roof, the presence of column depends on the diameter and design. It is normally enough to have one central column to bear the load of the roof, but also multiple column solutions can be adopted to decrease the thickness (and costs) of the roof as far as it provides the possibility to install an agitator at the centre of the tank.

3.3.3.4 Mechanical Mixing System

The mixing system is a key factor in CSTR technology as it results in the homogeneity of digestate inside the tank, but also accounts for the largest electrical self-consumption proportion. There is no exact formula to calculate the necessary characteristics of mixers. They can be classified based on speed, power, type, and installation.

- Speed: fast (shearing effect), slow (kneading effect)
- Power: expressed in kW
- Model: submersible motor, external motor
- Installation: vertical, horizontal, inclined, adjustable, fixed

The TS content and viscosity are the factors that drive the choice between the fast or slow mixing system, nevertheless, a combination of various systems is a common practice to enjoy the benefits of both systems. The agitator can operate continuously or intermittently with stirring interval that has to be set-up case by case based on the practice and experience during the operation of the plant. In the first period after start-up, it is usually a common practice to have longer and prolonged intervals of stirring times, while only after a certain period of stabilisation of the plant, it is possible to optimise the electrical self-consumption.

Submersible motor mixers are often used for waste water treatment applications and the design has been adapted for biogas applications. The motor can be

Fig. 3.25 Submersible mixer installed in concrete tank digester

either electrical or hydraulic, with gear reduction or with speed adjustment through motor winding design. The housing of the motor should be tight, as well as the electrical cable, with special attention to the gas side passage. The cooling of the motor is performed by the stirred liquid itself, therefore, in case of a wrong design, it is easy to reach a loop effect that could immediately lead to motor stop. If the mixer is too fast for the viscosity of the material and does not create enough flow through to cool down the motor housing, the motor temperature rises leading to over temperature stops. A slow or fast propeller can be installed to better adapt to the fluid characteristics. This kind of mixers are usually adjustable in level inside the tank which brings the big advantages of changing position, inclination and direction in accordance with the need of the plant. It is also easy to create a turbulent flow, break the scum (if formed) or floating layer as well as sediment material. Nevertheless, cavitation is possible, there are also a lot of moving parts and equipment inside the tank that may lead to difficult or frequent maintenance need, the motor may need be often extracted depending on the stirring time and interval. The extraction and replacement can be done through a special opening on the roof or gas holder, or a portion of the gas holder need to be opened. The guiding shaft is normally fixed at the wall of the digester so the mixer area of influence is the volume near the wall and the mixer effect cannot reach the centre of the tank (Fig. 3.25).

Different kinds of opening for maintenance are available on the market with different kinds of gas tightness systems. The best situation is to have an opening that could avoid gas exit during opening thanks to hydraulic sealing system (Fig. 3.26).

External motor agitators are characterised by a long through shaft that allows rotation transmission from the motor located outside to propeller located inside the digester (Fig. 3.27). Also in this configuration, slow speed or fast speed propeller can be installed to optimise the stirring of the fluid. In case of side entrance shaft

Fig. 3.26 Mixer dome for easy and safe maintenance of submersible mixer in gas tight tanks

Fig. 3.27 Inclined mixer with external motor during installation [Sebigas UAC Co., Ltd.]

and depending on the length, it is possible to install an additional support inside the tank to guide the shaft. In this case, special attention should be given to the choice of material and maintenance interval of the support, since it is necessary to empty the tank for maintenance purpose. The combination of external motor and slow speed propeller is quite common and gives the advantage of few moving parts inside the digester, possibility of continuous operation due to low electrical consumption, and prevention of scum formation. The area affected by the thrust and flow is much higher and allows a better homogenisation of the internal fluids.

In case of digesters with roof, it is possible to install a vertical shaft mixer positioned in the centre of the tank or anywhere near the centre of the tank (Fig. 3.28) (which is usually a badly mixed area) along with side entrance or submersible mixers. The flow direction is downwards below the mixer and upwards near the wall leading to bottom cleaning effect.

Fig. 3.28 Vertical shaft slow speed agitator installed at the centre of the digester roof [Stamo Maskin AB]

3.3.3.5 Access

During the design, it is normally suggested to keep an opening in the digester for future maintenance. In case of concrete or metal roof digesters, it is suggested to have a big opening to allow an easier passage of tools or machine in the digester for cleaning or emptying procedures. When a gas holder is installed on the roof, it is then enough to have a small passage to ease the entrance and exit of manpower during construction and installation phases.

3.3.3.6 Digestate—Sludge Transportation and Removal

Sludge and digestate are usually pumpable, and therefore, pumps driven by electrical motors are the most common solution for this purpose. Different parameters have to be considered for the pump selection:

- TS content
- Viscosity
- Need of specific flow not related to pressure drop

Fig. 3.29 Lobe pump in a special construction, directly connected to a grinder on the suction side [Vogelsang Italia S.r.l.]

- Installation position
- Maintenance need
- Pumps head
- Fluid temperature
- Running time
- Frequency of start and stop

Centrifugal pumps or positive displacement pumps like lobe (Fig. 3.29) or screw pumps are commonly used for digestate and sludge transportation.

Centrifugal pumps are suitable for low TS, low viscosity, variable flows in accordance with pressure loss and different temperatures. The advantages are the possibility to be installed either inside or outside the tank (submersible or external pumps), but this kind of pump is usually not the best choice for high head demanding.

Positive displacement pumps are volumetric pumps and the volume transported through each rotation and therefore, during each minute at a certain speed of rotation is always the same. Due to this reason, its application is particularly recommended when pumped volume needs be counted without instrument and of course with scant precision.

3.3.3.7 Heating System

As mentioned, the stability of the process temperature is a key factor to high efficiency. The environmental conditions affect both the entrance temperature of the biomass and the loss of temperature of the digestate in the tank.

To keep a stable process temperature, it is necessary to have a heating system controlled by the plant software. There are two possible ways to provide heat to the digestate: through internal pipes system or with external heat exchanger.

The internal pipe system is the cheapest solution both from investment and operational cost points of view. Two or more pipes are installed on the internal wall of the digester, in the bottom area. The hot water flow is forced to pass inside the circuit exchanging heat with the digestate around the pipes, the warmed digestate creates a convection flow keeping a homogeneous temperature in the digester. The application has to be installed in the presence of a good mixing system in order to prevent the formation of different temperature layers that may hinder the anaerobic digestion process.

In order of increase the efficiency, PVC, stainless steel rigid or stainless steel corrugated pipes can be used. The PVC pipes are cheaper, but need more rounds to have a good thermal exchange which could be an advantage in case of damage to one of the pipes. The stainless steel rigid or corrugated pipes are similar in terms of the corrosion and wearing points of view, but the corrugated pipe has of course a larger surface per linear length of the pipe which allows for a shorter line. Due to the corrugated surface, the corrugated pipes have a better heat conductivity and require a turbulent flow at lower speeds (consequently a lower flow is necessary from the recirculation pumps).

The external heat exchanger gives the absolute advantages of being cleanable and accessible for maintenance. The disadvantage is that it is necessary to recirculate the digestate in the heat exchanger while the hot water recirculation is also necessary. This increases the electrical self-consumption of the plant and it is therefore suggested in case of substrates that may lead to soiling effect or gluing effect on the surface of internal pipes or with high presence of sands that may have an early wearing effect.

A good solution to preserve heat and avoid temperature difference in the wall of the digester is to have insulated tanks. This is almost essential in thermophilic processes and highly suggested for mesophilic processes as well. The insulation can be done with different commercial material based on the application (Rockwool, high density closed cell extruded polystyrene, etc.).

The insulation should be protected from weather; corrugated steel plate can be installed for this purpose.

The thermal power needed for heat exchanger should be produced on site by the cogeneration system or by additional dedicated boilers.

3.3.4 Gas Handling

The biogas produced in the digester has to be used for the project purpose with a flow as stable as possible. To maximise the efficiency, reliability, and availability of the plant, the gas line should be designed to act as a buffering system to avoid stops.

3.3.4.1 Gas Storage

The biogas production also fluctuates in a certain range and to compensate for such fluctuations, a gas storage is usually planned.

In case of digesters with roof, the biogas is sent to an external or additional buffer storage. In case of gasholder mounted on top of the digester, the biogas produced is automatically stored.

The gasholder can be double or single membrane: single membrane means that there is only one membrane that store the gas dividing biogas from the atmosphere; while double membranes have an external membrane kept fully inflate by an air blower plus an internal membrane that can act like a real buffer for the gas storage passing from fully empty to fully inflated.

The material of the gas holder membrane can be either double-sided PVC-coated fiber fabric which are usually UV, microbial, abrasion, and biogas resistant; HDPE or ethylene propylene diene monomer (EPDM) rubber.

EPDM gas holders are elastic and therefore, characterised by a higher permeability; their duration is also affected by weather, especially by UV and therefore, are not considered as the best option. Breakage or leakage of this kind of membrane is frequent causing obvious loss of gas/money, but more importantly, leading to a high environmental impact that should be avoided especially in renewable energy production sites.

HDPE is a very common material, easy to weld and with a good permeability resistance. It is often used for lagoon constructions. It is cheaper, easier to weld on site and resistant over time. The limitation of HDPE is that it is not strong enough to be used at "high" pressures (not higher than 2-3 mbar usually).

PCV-coated fabric is the most common material for double membrane gas holder owing to its high resistance to permeability, strength, and duration over time (Fig. 3.30). PVC fabric is characterised by strength of the fabric and weight per square meter. The selection of the textile should be in accordance with the biogas pressure estimated in the design and presence of special shapes (balcony, etc.) that may increase the tension of the membrane in their proximity. The achievable pressure in PVC gas holder is in the range of 3–20 mbar. If there are no special needs, it is suggested to keep a lower pressure of the gas to prevent tension to the

Fig. 3.30 Example of PVC-coated fabric gasholder installed

textile and to increase the lifespan of the gas holder. Under the internal membrane, a net and belts system is installed to bear the load of the membrane when empty (belts) and to prevent the membrane to drop into the digestate (net). The net can also be used as a media for the growth of desulphurising bacteria.

The level of the internal membrane can be measured to provide a visual estimation or a signal to the PLC regarding the quantity of the gas stored in the membrane. There are many different systems by different commercial suppliers, from radar to water pressure based systems. If the level of the gas holder is one of the parameters used to control the engine power and start/stop, the measurement of the gas holder level should be quite precise and reliable.

The colour is also affecting the operation of the gas holder; a dark colour absorbs more UV and sunlight and therefore, its durability may be reduced. For the same reason, dark coloured membranes are subject to pressure variations in response to variations in weather conditions.

Finally, every storage system should be equipped with safety valves for over pressure and under pressure that release the gas in case of complete failure of normal operation.

3.3.4.2 Gas Usage

The biogas can be used for three main different purposes:

- Electrical and thermal power production (Cogeneration system)
- Thermal power production (Boiler)
- Biomethane production (Upgrading)

The most commonly installed system is the Cogeneration as it is a well-known system already used for electrical power production from natural gas and with a good after sales service worldwide. There are different suppliers for this kind of engine (Fig. 3.31). In the selection, it is important to check H_2S resistance, $\%CH_4$ accepted, efficiency, cost of maintenance including overall maintenance (usually around 60,000 h of operation), and availability of spare parts and fast emergency service.

The boiler is usually a simpler system and it is normally important to check the resistance to H_2S.

The upgrading systems are different from each other and a specific analysis considering the required outcome, the input variability and characteristics of biogas, the cost of operation and maintenance, and long terms reliability of the supplier should be performed.

Fig. 3.31 MTU biogas engine installed in power-house solution [MSM Energy Solutions Co., Ltd.]

3.3.4.3 Gas Treatment

Every biogas using system (upgrading, cogeneration unit or boiler) has a minimum required quality of biogas in terms of pressure, CH_4 content, H_2S content, moisture, O_2, and others contaminants. There are different systems to meet biogas quality requirements: water scrubber, activated carbon, biological treatment, etc.

Every situation requires a dedicated study and selection of the system (if required). There are a lot of commercial products that can meet the requirements of every project and different systems can be combined to achieve a better result. For biogas engine, the minimum requirement is usually a chiller to remove the moisture (with dew point in accordance to the minimum ambient temperature) and a gas blower to increase the pressure.

3.3.5 Digestate Area

At the end of the process, the digestate should be extracted by the last digester. The quantity of digestate can be calculated through the following mass balance equation (Eq. 3.14).

$$Q_{dig} = \Sigma Q_{subi} - Q_{subi} \cdot TS_i \cdot VS_i \cdot PGY_i \cdot \rho_{biogas} \tag{3.14}$$

where:

Q_{dig}	Daily quantity of digestate produced [t/d]
Q_{sub}	Daily quantity of substrate in input [t/d]
TS	Initial TS concentration of the substrate [%]
VS	Volatile solids concentration [$\%_{TS}$]
PGY	Potential Gas Yield [Nm^3_{biogas}/t_{VS}]
ρ_{biogas}	Specific weight of the biogas [t/Nm^3]

3.3.5.1 Solid/Liquid Separation

The digestate coming out from the digesters may still have a high TS concentration. Under certain circumstances, it could be economically and technologically viable to install a solid/liquid separation system. The horizontal or vertical screw press is the most commonly used. The digestate is pumped into the separation system and it allows the liquid passing through the screen while the solid goes out from the front or top of the machine.

The efficiency of this type of equipment highly depends on the quality of the material: viscosity, size of particle, content of fibres, TS concentration, etc.

A solid/liquid separation system usually allows to remove 1–3% of the digestate TS from the liquid fraction while producing a solid fraction with 20–25% TS concentration.

Based on the TS assumption, it is possible to estimate the quantity of liquid and solid fractions.

$$Qs = \mathrm{Qdig}\frac{TS' - TSL}{TSS - TSL} \tag{3.15}$$

$$QL = Qdig - Qs \tag{3.16}$$

where:

Q_S Quantity of solid fraction
Q_L Quantity of liquid fraction
TS_S TS concentration in the solid fraction
TS_L TS concentration in the liquid fraction

Belt-type filter presses, centrifuges and worm separation are other solid/liquid separation systems used in biogas plants.

The separation of the digestate would be advantageous leading to:

- Reduced liquid quantity and final storage volume
- Have a portion of stackable product
- Reduced floating layer and solidification of surface in the final tank

3.3.5.2 Digestate Storage

The digestate produced is usually stored in final tanks with cylindrical or rectangular shapes. These tanks can be equipped with mixers so that the liquid can be homogenised before discharging. The agitator can be permanently installed, removable or tractor-tow driven.

Following the local regulations about odour control and nitrogen loss, the final storage tanks can be covered with gas-tight membranes or simply with odour control membranes.

The choice of gas tight cover is subjected to discussion in relation to the feedstock used, HRT, OLR, and resulting efficiency of the designed process.

A high efficiency plant usually releases a digestate with less than 2–3% remaining biogas potential, therefore, under non-anaerobic conditions and lower temperature of the final storage environment, it is difficult to have big loss in atmosphere. The gas tight solution is usually non-viable.

The dimension of the storage is defined primarily by the frequency of emptying the tank. This interval may vary from a couple of days to months (usually 180 d in Europe). The rainfall expected during such period of time should be added to the storage volume to prevent early full state of the tank.

3.3.5.3 Digestate Treatment

According to the local regulations, the digestate derived from certain feedstock cannot be spread on the field (example: slaughterhouse waste, household waste, OFMSW, restaurant waste, some industrial waste, etc.). Under such circumstances, an additional WWT should be added to remove N, salt, TS content and the other contaminants until reaching the quality required for sewerage or river discharge.

References

Bouallagui H, Haouari O, Touhami Y, Cheikh BR, Marouani L, Hamdi M (2004) Effect of temperature on the performance of an anaerobic tubular reactor treating fruit and vegetable waste. Process Biochem 39:2143–2148

Converti A, Del Borghi A, Zilli M, Arni S, Del Borghi M (1999) Anaerobic digestion of the vegetable fraction of municipal refuses: mesophilic versus thermophilic conditions. Bioprocess Eng 21(4):371–376

Han Y, Dague RR (1997) Laboratory studies on the temperature-phased anaerobic digestion of domestic wastewater sludge. Wat Env Res 69(6):1139–1143

Chapter 4
Biogas Production Systems

Elena Valijanian, Meisam Tabatabaei, Mortaza Aghbashlo, Alawi Sulaiman and Yusuf Chisti

4.1 Basics of Anaerobic Digestion

Anaerobic digestion converts organic waste to biogas, a gas mixture containing methane and other components. This microbial process is carried out in the absence of oxygen. The waste feedstock and microbial culture are mixed in a bioreactor, or anaerobic digester, to achieve the desired conversion. Digester feedstock may be wet (water-rich), or dry (solids-rich). A wet digestion process typically uses a pumpable aqueous slurry containing $\leq 15\%$ (w/w) dry solids as feedstock. In contrast, the dry feedstocks are sufficiently dry to be "stackable". A digester may be operated continuously, that is with continuous input and removal of material. Alternatively, a batch operation may be used in which the digester is fed and harvested periodically.

E. Valijanian · M. Tabatabaei
Biofuel Research Team (BRTeam) / Agricultural
Biotechnology Research Institute of Iran (ABRII), Agricultural Research, Education and Extension Organization (AREEO), Karaj, Iran

M. Aghbashlo
Department of Mechanical Engineering of Agricultural Machinery, Faculty of Agricultural Engineering and Technology, College of Agriculture and Natural Resources, University of Tehran, Karaj, Iran

A. Sulaiman (✉)
Faculty of Plantation and Agrotechnology, Universiti Teknologi MARA, 40450 Shah Alam, Selangor, Malaysia
e-mail: asuitm@yahoo.com

Y. Chisti (✉)
School of Engineering, Massey University, Palmerston North, New Zealand
e-mail: Y.Chisti@massey.ac.nz

M. Tabatabaei and H. Ghanavati (eds.), *Biogas*, Biofuel and Biorefinery Technologies 6, https://doi.org/10.1007/978-3-319-77335-3_4

Anaerobic digestion involves multiple steps with different groups of microorganisms contributing to the different steps. Initially, the complex biopolymers (carbohydrates, proteins) and other large molecules (fats) are broken down to simpler molecules (sugars, fatty acids, amino acids) in a hydrolysis step. The products of the hydrolysis are used by bacteria known as acidogens in an acidogenesis step to produce volatile fatty acids (VFA) and alcohols. Other byproducts include sulfide, carbon dioxide and ammonia. The next step is acetogenesis during which VFAs and alcohols are converted by acetogenic bacteria (or acetogens) to hydrogen and carbon dioxide. The final step is methanogenesis, or the conversion of hydrogen, acetic acid (one of the VFAs), carbon dioxide and water to methane and carbon dioxide. The bacteria involved are known as methanogens. Methanogenesis is the slowest step, or the rate limiting step, of the entire anaerobic digestion process. Methanogens are pH-sensitive and inhibited by acidic pH. Too rapid a buildup of VFAs can lower the reactor pH sufficiently to inactivate the methanogens and stop biogas generation. This "souring" of the digester must be avoided for stable operation. Other toxic substances and inhibitors in the feed may adversely impact anaerobic digestion.

Depending on the digester, either a single or multiple types of feedstocks may be processed. Different feedstocks may have to be blended to obtain a composition suitable for biogas production. Feedstock may be subjected to various pretreatments (e.g. size reduction) to facilitate subsequent digestion. The digestion process may be mesophilic or thermophilic. Mesophilic digester typically operate in the temperature range of 30–42 °C. A higher temperature range (50–60 °C) is generally used for thermophilic digestion. Mesophilic and thermophilic digestion rely on different groups of microorganisms. If the waste being digested is from human (e.g. municipal wastewater) and animal sources (e.g. manure, slaughterhouse waste), the use of thermophilic digestion is preferable as it is more effective in killing pathogens. Dewatered solid residue, or digestate, of a thermophilic digester may be applied to land directly as compost for crop production with no risk of spreading parasites and diseases. Thermophilic digestion is rapid, but more expensive and harder to control.

A digester is one component of the process for producing biogas from waste. This chapter is focused on the different kinds of bioreactors, or digesters, used in anaerobic digestions. The effectiveness of digestion is influenced by prior steps. Prior to digestion, there may be other feedstock preparation and blending steps. Similarly, following digestion, the crude biogas may have to be cleaned, or upgraded, usually for onsite combustion to produce heat or electricity. In addition, the liquid effluent from the digester needs to be suitably managed (Tricase and Lombardi 2012). Typically, the solids are dewatered and used as compost and soil conditioners. The liquid effluent may be used for irrigation. Irrespective of the digestion system, the biogas product typically consists of methane (50–80% by volume), carbon dioxide (20–50% by volume) and trace levels of nitrogen, hydrogen sulfide and water (Tricase and Lombardi 2012).

A great variety of organic feedstocks may be potentially anaerobically digested, but typical feedstocks consist of animal manure, municipal waste and agroindustrial

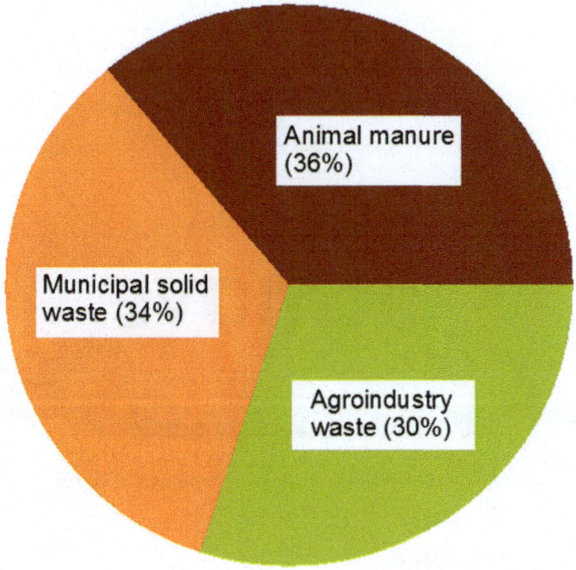

Fig. 4.1 Main organic feedstocks used in biogas production worldwide

waste (Fig. 4.1). Organic sludge of aerobic wastewater treatment processes and animal manure are among the most widely used feedstocks (Chaiprasert 2011; Horváth et al. 2016).

The crude biogas produced by an anaerobic digester typically needs to be upgraded for further use. This generally requires removal of some hydrogen sulfide (H_2S) and, in some cases, removal of carbon dioxide. Biogas used in steam boilers and internal combustion engines for electricity generation must not contain more than 200 ppm of the corrosive hydrogen sulfide, although carbon dioxide is acceptable in many applications. Some H_2S reduction processes use biological desulfurization of the gas to reduce the hydrogen sulfide content. The various technologies for gas cleaning have been reviewed in the literature (Ryckebosch et al. 2011; Sun et al. 2015).

As the demand for gas fluctuates, on site storage is typically necessary (Al Seadi et al. 2008). Low pressure (<140 mbar) storage is typical. For this purpose, a floating top holder integrated with the digester may be used. Other commonly used options are gas-tight polymer membrane (e.g. polyester fabric covered with polyethylene or chlorosulfonated polyethylene) domes and spheres (Fig. 4.2) located on top of the digester or separate from it. Membranes vary in thickness from 0.5 to 2.5 mm. Such membranes are weatherproof, resistant to hydrogen sulfide and ultraviolet light, and withstand the local temperatures (Voicu et al. 2015). A vent with a biogas flare is provided in case excess gas needs to be released (Hijazi et al. 2016).

Fig. 4.2 A biogas storage dome made of polymer membranes. *Courtesy* Atmove Biomethane Solutions, Vienna, Austria

4.2 Anaerobic Digesters

Important considerations in selection and operation of anaerobic digesters are the following: installation and operational costs; the necessary residence time, or hydraulic retention times (HRT) of the waste in the digester; the amount of waste to be processed; the organic loading, or strength of the waste, and dry solids content; the nature of the waste, in particular the relative ratios of nitrogen and carbon; whether a wet digestion (dry matter content $\leq 15\%$), dry digestion (dry matter content >15%), or a combination of the two is to be used (Tricase and Lombardi 2012); the mixing requirements and the method of achieving mixing (Budzianowski 2016); and heating requirements depending on whether the process is mesophilic or thermophilic (Tricase and Lombardi 2012).

All process choices have associated advantages and drawbacks. For example, the digestate of a dry system is easier to dewater whereas a wet process produces a lot of liquid effluent to be disposed. A dry digestion process typically requires a larger inoculum, longer retention time and is prone to be unstable compared to a wet digestion process (Budzianowski 2016).

Anaerobic digestion may consist of a single-stage operation, or a two-stage process. Single stage operation is less efficient, but most commonly used (Ke et al. 2005) because of its simplicity. In a single-stage digester, all the different reactions (i.e. hydrolysis, acidogenesis, acetogenesis, methanogenesis) occur within a single vessel operated at a given set of conditions that are not ideal for any of the multiple reactions that occur. Single-stage digesters typically require longer hydraulic retention times on account of the long doubling time of the methanogens. Therefore, a conventional single-stage digester is typically larger and consequently takes more energy to mix and heat (Lehtomäki et al. 2008; Budzianowski 2016) compared to a two-stage digester.

A two-stage digester effectively separates the methanogenesis steps from the rest of the treatment process. The first reaction chamber is optimized to maximize hydrolysis and production of volatile fatty acids whereas the second chamber is optimized for methane production. Therefore, a two-stage digestion is more efficient overall compared to a single-stage process (Nair et al. 2005; Ahamed et al. 2015).

4.2.1 *Conventional Anaerobic Bioreactors*

4.2.1.1 Anaerobic Sequencing Batch Reactor

Anaerobic sequencing batch reactor (ASBR) is one of the simplest systems used for producing biogas. ASBRs are used mostly for treating wastewaters of diverse sources including effluents of food processing facilities, slaughterhouses, animal farms and pharmaceutical industries (Zaiat et al. 2001). The reactor consists of a single vessel. At the end of a batch operation, the vessel is partly drained and then filled to the initial volume with fresh organics-loaded wastewater (Fig. 4.3). This fill-and-draw operation is repeated at intervals (Mao et al. 2015). The residual sludge from a previous batch becomes the inoculum for the next batch. This design is suitable if the volume to be processed is relatively small. The operation can be adjusted for different strengths of waste (Zupančič and Jemec 2010). An ASBR operation is shown in Fig. 4.3.

Both mesophilic and thermophilic ASBRs are used. The processing capacity depends on the number of feeding-retention-emptying cycles that can be accommodated in a given period (Dague 1993). ASBRs tend to be simple and easy to operate with little attention (Dutta et al. 2014). An activated sludge aerated

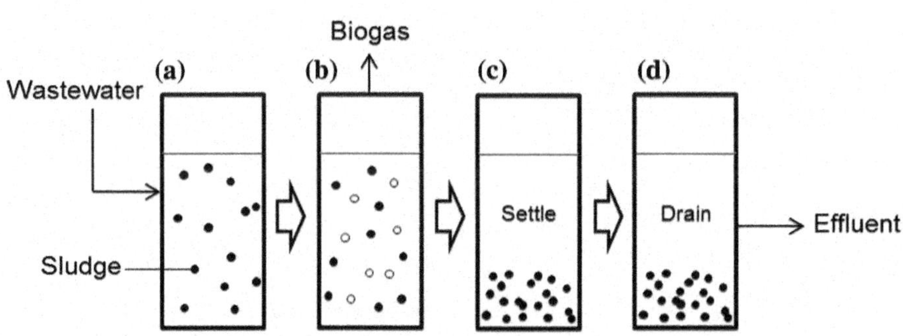

Fig. 4.3 An anaerobic sequencing batch reactor operation: **a** an inoculum-containing digester is filled with wastewater; **b** digestion occurs during the retention time of the batch; **c** the sludge is allowed to settle by gravity; **d** the treated water is decanted while the sludge is left behind as inoculum for the next batch

operation is readily combined with the ASBR. For example, once the vessel is filled with fresh wastewater, it may be aerated to oxidize the waste in an activated sludge type of operation. At some point, the aerated operation would switch to anaerobic operation to produce biogas. Once gas production declines to a low level, the reactor is partly emptied and the entire process sequence is repeated.

Improved mixing during anaerobic operation enhances performance (Maurina et al. 2014). Ideally, prior to emptying the digester, a quiescent period is used to settle some of the microbial biosolids for preferential retention in the digester so that a rich inoculum is available to the next batch (Fig. 4.3). In practice this can be difficult because gas bubbles entrapped within microbial solids make sedimentation difficult.

4.2.1.2 Continuous Stirred Tank Reactor

Continuous flow stirred tank reactors (CSTRs) are perhaps the most commonly used reactor configurations in production of biogas. They are attractive because of the simplicity of their design compared to many other types of biogas digesters. Typically, CSTRs are used to process slurries with total solids content of 5–10% (Browne et al. 2013). Slurries of animal manure and organic industrial wastes are treated using CSTRs. The reactor consists of a rectangular or cylindrical tank with one or more mechanical stirrers (Fig. 4.4) (Mao et al. 2015). CSTRs are generally low-cost and easy to operate. Mechanical mixing assures good contact between the microorganisms and the waste material. Mixing may be continuous, or intermittent (Rico et al. 2011). Sufficient mixing is essential to prevent accumulation of large amounts of VFAs that would result in souring, or acidification, and a consequent inhibition of biogas production (Liao et al. 2006; Ozgun et al. 2013). Mixing also affects the formation, structure and metabolic efficacy of microbial flocs (Jiang et al. 2016). Excessive mixing can lead to reduced generation of biogas from VFAs, their accumulation and reactor souring.

As a drawback, CSTRs have long retention times (Carrillo-Reyes et al. 2016) and may be more energy intensive than some of the other types of reactors. Performance of CSTRs is improved by recycling microbial solids, or enhancing retention of the active biomass. A gravity sedimentation tank located at the exit of

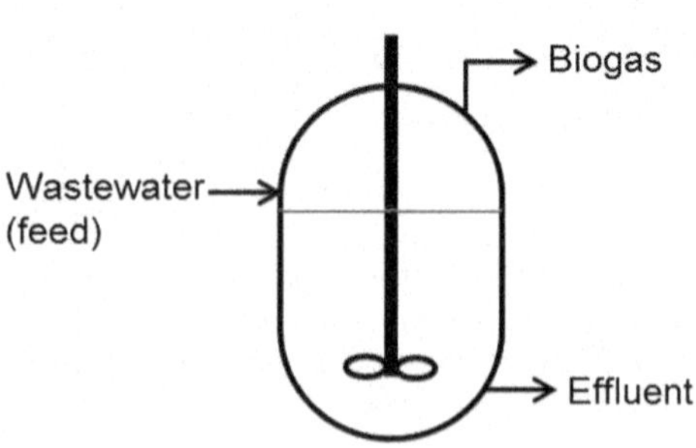

Fig. 4.4 A continuous stirred tank type of anaerobic digester

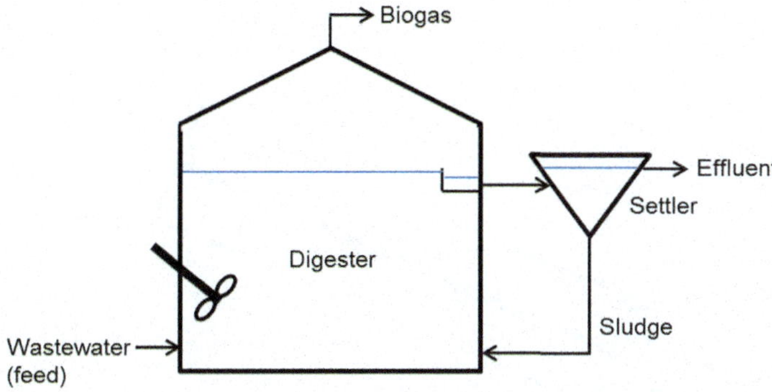

Fig. 4.5 A continuous stirred tank digester with biomass recycle via a gravity sedimentation tank

the CSTR may be used for biomass recycle (Fig. 4.5). Alternatively, membranes-based retention systems (Wei et al. 2014) and various methods of immobilizing the biomass, on inert suspended particles, for example, may be used. Presence of an elevated concentration of the active biomass in the reactor improves substrate conversion and shortens the required retention time (Wu et al. 2008).

4.2.1.3 Anaerobic Plug-flow Reactor

Anaerobic plug-flow reactors (APFRs) are typically long rectangular channels, with the flow entering one end and leaving at the distant end. There is relatively little mixing in the direction of flow. The tanks, or channels, are generally placed above ground. Both mesophilic and thermophilic operations are used (Kim et al. 2003). APFRs are commercially used for treating diverse kinds of organic wastes including slurries of animal manure, distillery wastewater, and the organic fraction of municipal solid waste (Rajeshwari et al. 2000; Sharma et al. 2000). A plug flow configuration in principle can provide a gradation of local environmental conditions to favor different aspects of anaerobic digestions in different parts of the reactor. For example, hydrolysis may be predominant in the entry zone of the reactor whereas methanogenesis may be the dominant activity near the exit. Compared to a single-stage CSTR, plug flow reactors are generally more efficient in converting the substrate to biogas and are more stable to operate (Mao et al. 2015). Microbial sludge builds up along the length of the rector due to growth and a high overall sludge content explains both their better efficiency and stability (Mao et al. 2015). The APFRs are relatively simple to build and maintain (Lansing et al. 2008). Plug flow reactors with agitators have been described (Karellas et al. 2010) (Fig. 4.6). Agitation is used to improve local mixing, while minimizing mixing in the direction of low.

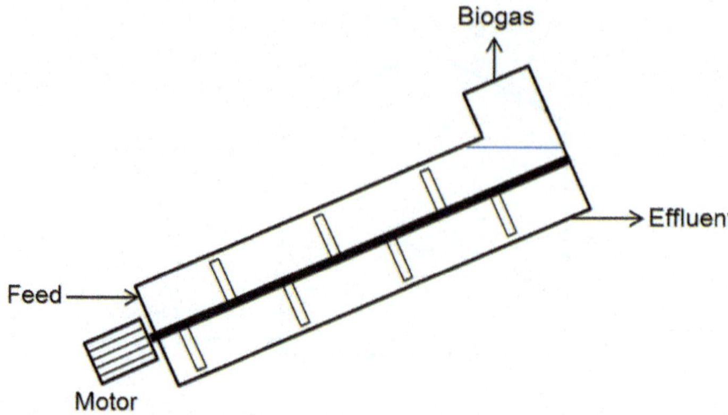

Fig. 4.6 A plug flow digester with mechanical mixing. Modified from Sharma et al. (2000)

4.2.2 Sludge Retention Reactors

Conversion of waste to biogas is brought about by the active microbial sludge in the reactor. Therefore, increased retention of the active sludge increases the rate of waste conversion and shortens the hydraulic retention time of the effluent. Increased sludge retention also enhances stability of the reactor. Enhancing sludge retention is the basis of the design of the various kinds of sludge retention reactors that have been developed. These are discussed in this section.

4.2.2.1 Anaerobic Contact Reactor

Anaerobic contact reactor (ACR) is essentially a fully mixed mechanically stirred tank with recycle of sludge (Fig. 4.5). The effluent from the tank flows into some kind of a solid-liquid separator (e.g. gravity sedimentation tank, lamella clarifier, sludge flotation device) and the recovered solids are returned to the anaerobic reactor. Because of a high concentration of active microbial biomass, ACRs are capable of treating high-strength waste with a high concentration of digestible solids. Wastewaters of food processing industry and pulp and paper mills are some examples (Capela et al. 2009; Şentürk et al. 2010; Şentürk et al. 2013). Hydraulic retention times are short (Aslanzadeh et al. 2013) and fluctuations in organic loading are well tolerated. The ACRs are relatively less susceptible to souring and other inhibitors (Rajeshwari et al. 2000). They are considered superior to some of the other types of digesters such as the up-flow anaerobic sludge blanket reactor. An organics loading rate of up to 8 kg chemical oxygen demand (COD) m^{-3} d^{-1} may be handled with COD removal efficiencies of 78–95% (Lo and Liao 1986).

Design of the sludge recovery and recycle system can have important consequences on operation of the reactor. For example, the use of high-shear pumps in returning the recovered sludge could disrupt microbial flocs or sludge granules and this may negatively impact biogas production (Brockmann and Seyfried 1997; Ozgun et al. 2013). Stirred digesters coupled to some kind of a membrane-based cell retention have proved highly effective in biogas production (Wei et al. 2014; Chen et al. 2016).

4.2.2.2 Up-Flow Anaerobic Sludge Blanket Reactor

An up-flow anaerobic sludge blanket (UASB) reactor consists of a rectangular or cylindrical unmixed tank fed with the waste stream near the bottom. Its main distinguishing feature is a bed of dense granular sludge confined mostly to the lower zone of the reactor where the wastewater enters (Mao et al. 2015; Reungsang et al. 2016) (Fig. 4.7). The granular sludge is kept suspended by the up flow of wastewater and the rising bubbles of biogas (Jiang et al. 2014). The top of the reactor may be expanded (see Sect. 4.2.2.3), or otherwise modified, to facilitate retention of the granular sludge by the action of gravity (Fig. 4.8).

The performance of the reactor depends critically on development and retention of the granular sludge (Horváth et al. 2016). Granular sludge is formed through self-immobilization of microorganisms (Schmidt and Ahring 1996), typically in an environment that is relatively quiescent. A high sludge load improves efficiency resulting in short hydraulic retention times and high permissible organics loading rates (Ahmad et al. 2011). Start-up periods can be long because of slow development of granular sludge (Schmidt and Ahring 1996). Variations in hydraulic loadings can be accommodated within narrow limits, as the sludge granules need to remain suspended without being washed out (Liu and Tay 2004). Both the quality of the sludge granules formed and the rate of their formation depend on the type of waste being treated (Bhatti et al. 1995). Wastewater containing a high proportion of

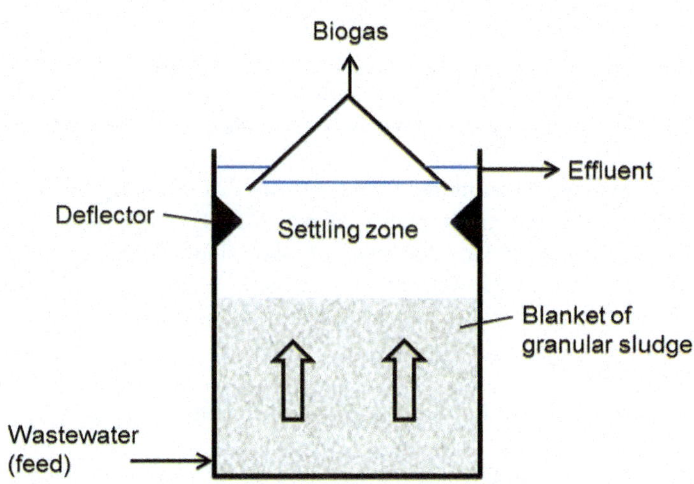

Fig. 4.7 An up-flow anaerobic sludge blanket reactor

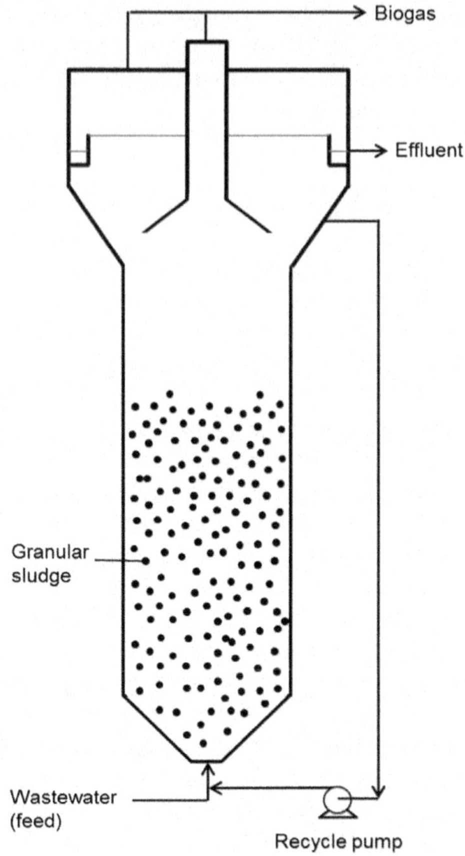

Fig. 4.8 An expanded bed granular sludge blanket reactor

suspended organic matter is not effectively treated in UASB reactors (Chernicharo et al. 2009). This limitation may be overcome by using a two-stage digestion system: an initial stirred-tank reactor to achieve hydrolysis and acid formation followed by a UASB for methanogenesis (Aslanzadeh et al. 2013; Mao et al. 2015; Horváth et al. 2016).

UASB reactors with membrane-based retention of sludge granules have been used, but the membranes also retain nongranulated sludge fines leading to a degradation of the granular character of the sludge (Ozgun et al. 2013). A continuous washout of the fine sludge is actually a necessary selection pressure for stable maintenance of the granular sludge morphology in a UASB.

4.2.2.3 Expanded Granular Sludge Blanket Reactor

The expanded granular sludge blanket (EGSB; Fig. 4.8) reactor is a variation of the UASB digester. Compared to the conventional UASB, it permits improved mixing and mass transfer between the sludge granules and the surrounding liquid (Mao et al. 2015). Compared to the UASB, the zone containing the suspended granular sludge is taller and narrower, but the top region of the reactor has an expanded cross-section (Fig. 4.8). A relatively high up-flow velocity (Kato et al. 1994) of the wastewater can be used as the up-flow is slowed down in the expanded upper zone sufficiently so that the sludge granules settle by gravity into the narrower part of the column. If the available flow of wastewater to be treated is insufficient to obtain the required high up-flow velocity in the narrower column of the reactor, some of the treated effluent may be recycled to increase the flow rate (Fig. 4.8).

EGSB systems are effective for treating medium and low-strength wastewaters containing soluble organics and more complex compounds such as lipids (Lettinga et al. 1997). They are effective at temperatures as low as 10 °C (Chu et al., 2005). They have a higher throughput compared to UASB systems (Rana et al. 2017) and a smaller footprint (Van Lier et al. 2015). Their scale-up based on data obtained in the laboratory is apparently not straightforward because of the scale-dependent differences in microbial ecology and microcosm physiology (Connelly et al. 2017).

ESGBs with membranes to retain sludge granules have been used (Chen et al. 2016), but actually defeat the purpose of developing a granular sludge morphology that inevitably degrades if the fines are not allowed to washout.

4.2.2.4 Up-Flow Anaerobic Solid-State Reactor

The up-flow anaerobic solid-state reactor (UASS) is conceptually new (Mumme et al. 2010). It is intended for treating ground biomass solids such as corn silage and barley straw. The solids are fed to the bottom of a tall column (Fig. 4.9) and rise because of their natural buoyancy relative to the liquid (Pohl et al. 2013). A confining sieve at the top of the column arrests the rise of solids and as more solids are fed a sort of a floating bed of solids builds up in the column (Fig. 4.9). Microorganism-containing liquor from a separate digester (an anaerobic filter) is pumped to the bottom of the column of floating feed solids, rises through the bed and is recycled to the digester (Fig. 4.9). The biomass in the bed is slowly digested to biogas. Utility, scalability, operability and stability of UASS systems are barely known.

4.2.2.5 Anaerobic Baffled Reactor

Anaerobic baffled reactors (ABR) are various kinds of reactor in which an elongated vessel has been partitioned by a number of baffles to produce fully or partly separated reactor units arranged in series (Mao et al. 2015). The baffles direct the flow to effectively generate a plug flow system. A five-compartment anaerobic granular bed baffled reactor is shown in Fig. 4.10 (Zwain et al. 2017).

The different compartments of an ABR may use entirely different principles to treat the waste for improved production of biogas. For example, Ran et al. (2014)

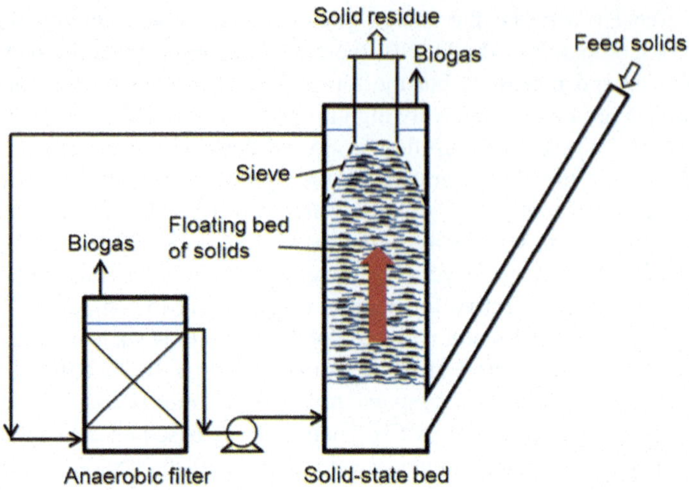

Fig. 4.9 A up-flow anaerobic solid-state reactor with liquor recirculation via anaerobic filters. Adapted from Mumme et al. (2010)

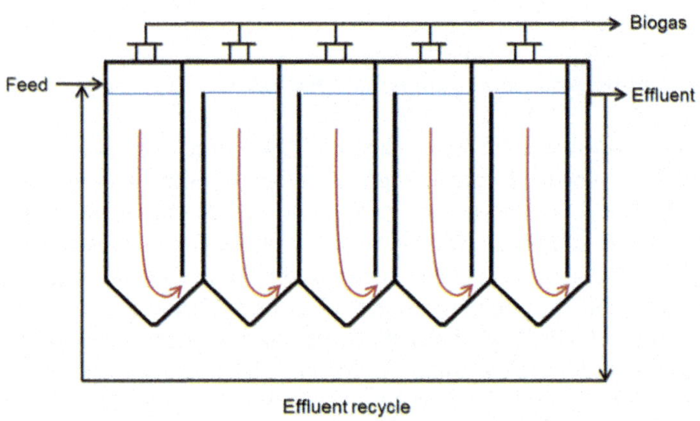

Fig. 4.10 A five-compartment granular bed baffled reactor. Based on Zwain et al. (2017)

evaluated a four-compartment ABR for simultaneous production of biogas and hydrogen from organic waste. The compartments were of an equal size. The first compartment was designed for hydrogen production by dark fermentation. The remaining three compartments served as microbial electrolysis cells for methane production. The ABR was operated with a hydraulic retention time of 24 h and the COD of the wastewater fed was in the range of 3500–4000 mg/L (Ran et al. 2014). The gas from the dark fermentation compartment contained 20.7% methane by volume. The methane levels in the gas from the compartments 2, 3 and 4 were 98.0, 93.6 and 70.1% (Ran et al. 2014). The total COD removal was 98%. Other similar electrically-enhanced biogas production systems have been described (Moreno et al. 2016; Khan et al. 2017), but their energetics and practical utility remain uncertain.

4.2.2.6 Internal Circulation Reactor

An internal circulation (IC) reactor is similar to an up-flow anaerobic sludge blanket reactor, but is generally much taller. A typical internal circulation reactor is shown in Fig. 4.11. The wastewater is distributed at the bottom of the reactor in a zone that contains most of the granular microbial sludge. In this zone the wastewater is intimately contacted with the sludge and much of the biogas production occurs. The intermediate gas-liquid-solid separator located above the granular sludge zone (Fig. 4.11) confines most of the granular sludge to the lower part of the reactor and separates the rising biogas bubbles from the liquid. The gas from the intermediate separator is collected in up-flow pipes, or risers, and rises rapidly by a gas lift action (Chisti 1998) to the top of the reactor. The rising gas transports liquid along with it. At the top, above the second gas-liquid separator, the biogas is separated from the liquid and leaves the reactor. The gas-free liquid being denser than the gas-liquid dispersion in the reactor, is transported by gravity to the bottom of the reactor using a down flow pipe, or downcomer (Fig. 4.11). The reactor zone located above the intermediate gas-liquid-solid separator is a polishing zone that contains some finer granular sludge and here the residual organics in the water generate more biogas. The gas bubble rise to the upper gas-liquid-solid separator where the gas is collected and rises to the top of the reactor through risers (Fig. 4.11). The gas-lift action causes recirculation of liquid within the reactor between the top and bottom zones. Internal circulation reactors are used commercially. They have performed well in COD removal and production of biogas (Tauseef et al. 2013; Mao et al. 2015; Wang et al. 2017). Organics loading rates can be as high as 35 kg COD/m^3 d (Mutombo 2004). These reactors are used to treat wastewater from breweries, the pulp and paper industry, distilleries, fermentation processes, and the petrochemical processes (Mutombo 2004; Mao et al. 2015).

4.2.2.7 Anaerobic Fluidized Bed Reactor

Anaerobic fluidized bed reactors (AFBR) are conceptually similar to the expanded granular sludge blanket reactor (Sect. 4.2.2.3), but instead of granular sludge they use relatively heavy small inert particles (e.g. fine sand or alumina) supporting a self-immobilized microbial biofilm (Zhang et al. 2008). The particles are maintained in suspension by a constant up-flow of the wastewater (Fig. 4.12) (Mao et al. 2015). Good mixing of the suspended solids and a high relative velocity between them and the liquid result in good mass transfer of organics to the biofilm. As a consequence of a high biomass loading and good biodegradation activity,

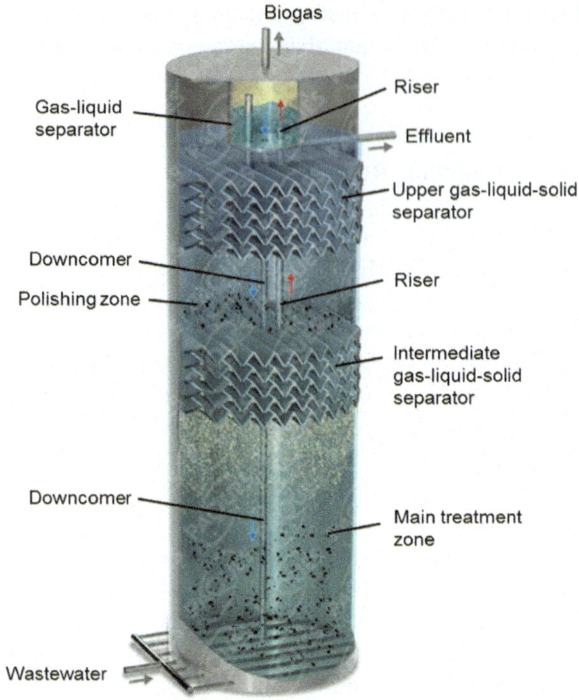

Fig. 4.11 A BIOPAQ® IC anaerobic internal circulation digester. *Courtesy* Paques BV, the Netherlands

the reactors are able to handle a high organics load and better tolerate inhibitory chemicals (Karadag et al. 2015).

4.2.3 Anaerobic Membrane Reactors

Membrane bioreactors (Fig. 4.13) use water-permeable microporous polymeric or ceramic membranes mainly to retain active biomass in an anaerobic wastewater treatment unit (Skouteris et al. 2012; Visvanathan and Abeynayaka 2012; Ozgun et al. 2013; Horváth et al. 2016). Biomass is retained on the upstream side of the membrane in contact with the waste material. Treated water free of suspended solids permeates through the membrane to the other side. Retention of active microbial biomass greatly enhances COD removal and biogas generation performance of membrane bioreactors compared to systems having freely suspended microbial cells without any retention mechanisms. For example, a membrane bioreactor with a hydraulic retention time of 1 day achieved a biogas production rate of 1–2 L/day compared to 0.05–0.12 L/day achieved with a comparable reactor containing freely suspended cells (Youngsukkasem et al. 2013).

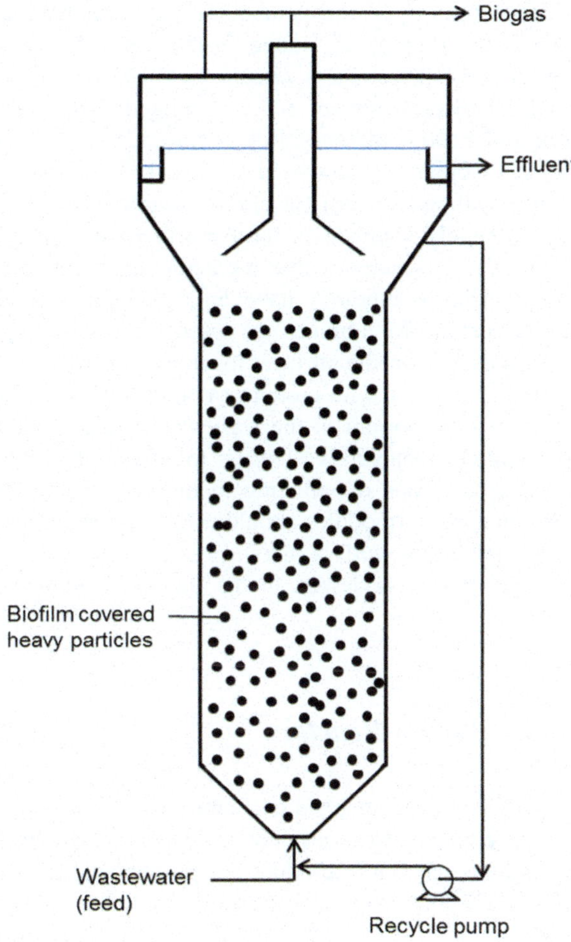

Fig. 4.12 An anaerobic fluidized bed reactor

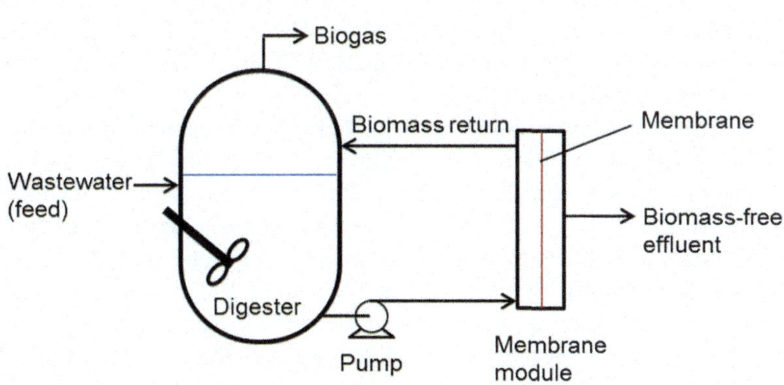

Fig. 4.13 A digester with an externally installed membrane module to prevent loss of biomass

Much higher concentrations of active biomass may be retained in a membrane bioreactor compared to a granular sludge UASB system, for example. This increases the efficiency of the reactors (Kanai et al. 2010) and a high organics loading can be used (Umaiyakunjaram and Shanmugam 2016) on account of the high concentration of the active biomass. Retention times can be short. Membranes also ease separation of biogas (Bakonyi et al. 2014). The flow rate through the reactor can be varied without concern for biomass washout.

Membranes are generally supplied in the form of easily replaceable modules. Membrane modules may be placed within a reactor vessel, or in an external circulation loop (Fig. 4.13). Membranes have an associated capital expense and require periodic replacement. Membranes are susceptible to fouling, or adsorption of material within pores and on the surface, resulting in a reduced rate of permeation. Fouling is often associated with proteins (Youngsukkasem et al. 2013), lipids and surfactants that may be present in the wastewater. Adding granular activated carbon and other solid adsorbents to wastewater to remove the fouling compounds has been suggested as a means of reducing membrane fouling (Skouteris et al. 2015). Prior to use, membrane modules should always be experimentally tested in the laboratory for operability with a given wastewater. Anaerobic membrane bioreactors have been further reviewed in the literature (Skouteris et al. 2012; Ozgun et al. 2013).

4.2.4 Anaerobic Biofilm Reactors

Biofilms are microbial consortia attached to a support material. The support surface is often inert and may be fixed or suspended. Anaerobic microbial biofilms can effectively digest organic material to produce biogas (Karadag et al. 2015). A large mass of immobilized biofilm and mass-transfer promoting movement of liquid around the film allow biofilm reactors to handle high organics loading and tolerate well any fluctuations in hydraulic or organics loads (Karadag et al. 2015). Once the biofilm has developed, startup periods are short compared to the other conventional anaerobic treatment systems (Patel et al. 1995). The nature of the support material influences the development of the biofilm (Liu et al. 2017) and its strength of attachment, or mechanical stability.

Inert solids packed in a column as in a trickle bed bioreactor are examples of fixed supports (Fig. 4.14). Fixed bed biofilm reactors are also known as anaerobic filters (Lemmer and Krümpel 2017). The wastewater enters the bottom and rises up through the packed bed. The packing is generally selected to have a large amount of interstitial space to reduce resistance to flow of wastewater and the biogas. For example plastic packing with a specific surface area of 100–200 m^2 per m^3 may be used. A biofilm develops on the surfaces of the static packing and provides the activity necessary for converting the dissolved organics to biogas. Anaerobic filters tend to be compact. The may be used as stand-alone units, or in series with one of the other types of digesters to further polish the effluent (Bodkhe 2008; Rajinikanth

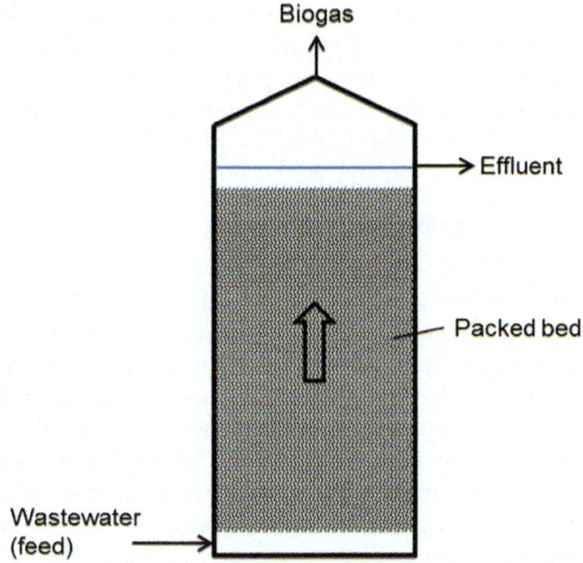

Fig. 4.14 An up-flow anaerobic filter

et al. 2009; Lemmer and Krümpel 2017). Clogging of the filter can be a problem (Bodkhe 2008). Excess biofilm sloughs off and leaves with the effluent. Moderate organic loads (e.g. 5–10 kg COD/m^3) are generally handled best in anaerobic filters.

Biofilm may be developed also on inert particles suspended in a liquid (Wang et al. 2009), as in a fluidized bed system (see Sect. 4.2.2.7). Bioreactors of this type are known as anaerobic moving biofilm reactors. An example is the anaerobic fluidized bed reactor (Fig. 4.12). Moving biofilm reactors allow good contact between the biofilm and wastewater, but retention mechanism may be needed to prevent washout of the biofilm supporting particles.

Various hybrid anaerobic biofilm reactors have also been developed by combing the biofilm systems with other types of reactors (Büyükkamaci and Filibeli 2002; Najafpour et al. 2006).

4.2.5 High-Rate Reactors

High-rate reactors are all those configurations that have somehow been modified to enhance the rate of degradation of organics, reduce retention time and increase organics loading and generation of biogas. Enhanced degradation performance may be achieved by one or more of the following approaches: better retention of the biomass to maintain a high concentration of active microorganisms in the reactor; enhanced mixing to improve mass transfer between the wastewater and microbial solids; improved temperature control; and compartmentalization to provide optimal conditions for hydrolysis-acidogenesis and the rate-limiting methanogenesis reactions in different compartments (Grobicki and Stuckey 1991; Dahiya and Joseph 2015; Horváth et al. 2016). Digesters mimicking the conditions in the various zones of the digestive tracts of ruminants are being developed (Zhang et al. 2014).

4.3 Concluding Remarks

Many different configurations and operational schemes have been developed for anaerobic digesters for use in different applications, as reviewed in this chapter. The aims generally are to reduce washout of active biomass, shorten the start-up period, minimize operational instabilities and attempt to better accommodate the inevitable variations in feed composition. Cost of installation, operation and maintenance are other factors that substantially affect the economics of biogas production. Single-stage digesters are most commonly used on account of their simplicity, but two-stage digesters are more efficient overall. No particular digester type can be recommended as being universally suitable. The choice in a given scenario must consider many factors including the following: the nature and strength of the waste stream; the expense of construction and operation; the availability and skills level of the local workforce; local climatic conditions, infrastructural support and cost of energy; and prospects for disposal of the digestate and the effluent. Biogas production by anaerobic digestion is a useful method of recovering energy from organic waste while greatly mitigating the environmental impact of the waste.

References

Ahamed A, Chen C-L, Rajagopal R, Wu D, Mao Y, Ho I, Lim J, Wang J-Y (2015) Multi-phased anaerobic baffled reactor treating food waste. Biores Technol 182:239–244

Ahmad A, Ghufran R, Wahid ZA (2011) Role of calcium oxide in sludge granulation and methanogenesis for the treatment of palm oil mill effluent using UASB reactor. J Hazard Mater 198:40–48

Al Seadi T, Rutz D, Prassl H, Köttner M, Finsterwalder T, Volk S, Janssen R (2008) *Biogas Handbook*. University of Southern Denmark Esbjerg, Esbjerg

Aslanzadeh S, Rajendran K, Jeihanipour A, Taherzadeh MJ (2013) The effect of effluent recirculation in a semi-continuous two-stage anaerobic digestion system. Energies 6:2966–2981

Bakonyi P, Nemestóthy N, Simon V, Bélafi-Bakó K (2014) Fermentative hydrogen production in anaerobic membrane bioreactors: a review. Biores Technol 156:357–363

Bhatti ZI, Furkukawa K, Fujita M (1995) Comparative composition and characteristics of methanogenic granular sludges treating industrial wastes under different conditions. J Ferment Bioeng 79:273–280

Bodkhe S (2008) Development of an improved anaerobic filter for municipal wastewater treatment. Biores Technol 99:222–226

Brockmann M, Seyfried CF (1997) Sludge activity under the conditions of crossflow microfiltration. Water Sci Technol 35:173–181

Browne JD, Allen E, Murphy JD (2013) Improving hydrolysis of food waste in a leach bed reactor. Waste Manag 33:2470–2477

Budzianowski WM (2016) A review of potential innovations for production, conditioning and utilization of biogas with multiple-criteria assessment. Renew Sustain Energy Rev 54:1148–1171

Büyükkamaci N, Filibeli A (2002) Concentrated wastewater treatment studies using an anaerobic hybrid reactor. Process Biochem 38:771–775

Capela I, Bilé MJ, Silva F, Nadais H, Prates A, Arroja L (2009) Hydrodynamic behaviour of a full-scale anaerobic contact reactor using residence time distribution technique. J Chem Technol Biotechnol 84:716–724

Carrillo-Reyes J, Cortés-Carmona MA, Bárcenas-Rùiz CD, Razo-Flores E (2016) Cell wash-out enrichment increases the stability and performance of biohydrogen producing packed-bed reactors and the community transition along the operation time. Renew Energy 97:266–273

Chaiprasert P (2011) Biogas production from agricultural wastes in Thailand. J Sustain Energy Environ Spec Issue 63–65

Chen C, Guo W, Ngo HH, Lee D-J, Tung K-L, Jin P, Wang J, Wu Y (2016) Challenges in biogas production from anaerobic membrane bioreactors. Renew Energy 98:120–134

Chernicharo CDL, Almeida P, Lobato LCS, Couto T, Borges J, Lacerda Y (2009) Experience with the design and start up of two full-scale UASB plants in Brazil: enhancements and drawbacks. Water Sci Technol 60:507–515

Chisti Y (1998) Pneumatically agitated bioreactors in industrial and environmental bioprocessing: hydrodynamics, hydraulics and transport phenomena. Appl Mech Rev 51:33–112

Chu L-B, Yang F-L, Zhang X-W (2005) Anaerobic treatment of domestic wastewater in a membrane-coupled expended granular sludge bed (EGSB) reactor under moderate to low temperature. Process Biochem 40:1063–1070

Connelly S, Shin SG, Dillon RJ, Ijaz UZ, Quince C, Sloan WT, Collins G (2017) Bioreactor scalability: laboratory-scale bioreactor design influences performance, ecology, and community physiology in expanded granular sludge bed bioreactors. Front Microbiol 8:664

Dague RR (1993) Anaerobic sequencing batch reactor. US Patent 5,185,079

Dahiya S, Joseph J (2015) High rate biomethanation technology for solid waste management and rapid biogas production: an emphasis on reactor design parameters. Biores Technol 188:73–78

Dutta K, Tsai C-Y, Chen W-H, Lin J-G (2014) Effect of carriers on the performance of anaerobic sequencing batch biofilm reactor treating synthetic municipal wastewater. Int Biodeterior Biodegradation 95:84–88

Grobicki A, Stuckey D (1991) Performance of the anaerobic baffled reactor under steady-state and shock loading conditions. Biotechnol Bioeng 37:344–355

Hijazi O, Munro S, Zerhusen B, Effenberger M (2016) Review of life cycle assessment for biogas production in Europe. Renew Sustain Energy Rev 54:1291–1300

Horváth IS, Tabatabaei M, Karimi K, Kumar R (2016) Recent updates on biogas production—a review. Biofuel Res J 3:394–402

Jiang J, Wu J, Zhang J, Poncin S, Li HZ (2014) Multiscale hydrodynamic investigation to intensify the biogas production in upflow anaerobic reactors. Biores Technol 155:1–7

Jiang J, Wu J, Poncin S, Li HZ (2016) Effect of hydrodynamic shear on biogas production and granule characteristics in a continuous stirred tank reactor. Process Biochem 51:345–351

Kanai M, Ferre V, Wakahara S, Yamamoto T, Moro M (2010) A novel combination of methane fermentation and MBR—Kubota Submerged Anaerobic Membrane Bioreactor process. Desalination 250:964–967

Karadag D, Köroğlu OE, Ozkaya B, Cakmakci M (2015) A review on anaerobic biofilm reactors for the treatment of dairy industry wastewater. Process Biochem 50:262–271

Karellas S, Boukis I, Kontopoulos G (2010) Development of an investment decision tool for biogas production from agricultural waste. Renew Sustain Energy Rev 14:1273–1282

Kato MT, Field JA, Versteeg P, Lettinga G (1994) Feasibility of expanded granular sludge bed reactors for the anaerobic treatment of low-strength soluble wastewaters. Biotechnol Bioeng 44:469–479

Khan MZ, Nizami AS, Rehan M, Ouda OKM, Sultana S, Ismail IM, Shahzad K (2017) Microbial electrolysis cells for hydrogen production and urban wastewater treatment: A case study of Saudi Arabia. Appl Energy 185:410–420

Ke S, Shi Z, Fang HH (2005) Applications of two-phase anaerobic degradation in industrial wastewater treatment. Int J Environ Pollut 23:65–80

Kim M, Gomec CY, Ahn Y, Speece R (2003) Hydrolysis and acidogenesis of particulate organic material in mesophilic and thermophilic anaerobic digestion. Environ Technol 24:1183–1190

Lansing S, Víquez J, Martínez H, Botero R, Martin J (2008) Quantifying electricity generation and waste transformations in a low-cost, plug-flow anaerobic digestion system. Ecol Eng 34:332–348

Lehtomäki A, Huttunen S, Lehtinen T, Rintala J (2008) Anaerobic digestion of grass silage in batch leach bed processes for methane production. Biores Technol 99:3267–3278

Lemmer A, Krümpel J (2017) Demand-driven biogas production in anaerobic filters. Appl Energy 185:885–894

Lettinga G, Field J, van Lier J, Zeeman G, Pol LH (1997) Advanced anaerobic wastewater treatment in the near future. Water Sci Technol 35:5–12

Liao B-Q, Kraemer JT, Bagley DM (2006) Anaerobic membrane bioreactors: applications and research directions. Critical Rev Environ Sci Technol 36:489–530

Liu Y, Tay J-H (2004) State of the art of biogranulation technology for wastewater treatment. Biotechnol Adv 22:533–563

Liu Y, Zhu Y, Jia H, Yong X, Zhang L, Zhou J, Cao Z, Kruse A, Wei P (2017) Effects of different biofilm carriers on biogas production during anaerobic digestion of corn straw. Biores Technol 244:445–451

Lo K, Liao P (1986) Digestion of cheese whey with anaerobic rotating biological contact reactors. Biomass 10:243–252

Mao C, Feng Y, Wang X, Ren G (2015) Review on research achievements of biogas from anaerobic digestion. Renew Sustain Energy Rev 45:540–555

Maurina G, Rosa L, Beal L, Baldasso C, Gimenez J, Torres A, Sousa M (2014) Effect of internal recirculation velocity in an anaerobic sequencing batch reactor (ASBR). Braz J Chem Eng 31:895–903

Moreno R, San-Martín MI, Escapa A, Morán A (2016) Domestic wastewater treatment in parallel with methane production in a microbial electrolysis cell. Renew Energy 93:442–448

Mumme J, Linke B, Tölle R (2010) Novel upflow anaerobic solid-state (UASS) reactor. Biores Technol 101:592–599

Mutombo D T (2004) Internal circulation reactor: pushing the limits of anaerobic industrial effluents treatment technologies. In: Proceedings of the 2004 Water Institute of Southern Africa (WISA) Biennial Conference, 2004, pp 608–616

Nair S, Kuang Y, Pullammanappallil P (2005) Enhanced degradation of waste grass clippings in one and two stage anaerobic systems. Environ Technol 26:1003–1012

Najafpour G, Zinatizadeh A, Mohamed A, Isa MH, Nasrollahzadeh H (2006) High-rate anaerobic digestion of palm oil mill effluent in an upflow anaerobic sludge-fixed film bioreactor. Process Biochem 41:370–379

Ozgun H, Dereli RK, Ersahin ME, Kinaci C, Spanjers H, van Lier JB (2013) A review of anaerobic membrane bioreactors for municipal wastewater treatment: integration options, limitations and expectations. Sep Purif Technol 118:89–104

Patel P, Desai M, Madamwar D (1995) Biomethanation of cheese whey using anaerobic upflow fixed film reactor. J Ferment Bioeng 79:398–399

Pohl M, Heeg K, Mumme J (2013) Anaerobic digestion of wheat straw–Performance of continuous solid-state digestion. Biores Technol 146:408–415

Rajeshwari K, Balakrishnan M, Kansal A, Lata K, Kishore V (2000) State-of-the-art of anaerobic digestion technology for industrial wastewater treatment. Renew Sustain Energy Rev 4:135–156

Rajinikanth R, Ganesh R, Escudie R, Mehrotra I, Kumar P, Thanikal JV, Torrijos M (2009) High rate anaerobic filter with floating supports for the treatment of effluents from small-scale agro-food industries. Desalin Water Treat 4:183–190

Ran Z, Gefu Z, Kumar JA, Chaoxiang L, Xu H, Lin L (2014) Hydrogen and methane production in a bio-electrochemical system assisted anaerobic baffled reactor. Int J Hydrogen Energy 39:13498–13504

Rana S, Singh L, Wahid Z, Liu H (2017) A recent overview of palm oil mill effluent management via bioreactor configurations. Curr Pollut Rep. https://doi.org/10.1007/s40726-017-0068-2

Reungsang A, Sittijunda S, Sreela-Or C (2016) Methane production from acidic effluent discharged after the hydrogen fermentation of sugarcane juice using batch fermentation and UASB reactor. Renew Energy 86:1224–1231

Rico C, Rico JL, Muñoz N, Gómez B, Tejero I (2011) Effect of mixing on biogas production during mesophilic anaerobic digestion of screened dairy manure in a pilot plant. Eng Life Sci 11:476–481

Ryckebosch E, Drouillon M, Vervaeren H (2011) Techniques for transformation of biogas to biomethane. Biomass Bioenerg 35:1633–1645

Schmidt JE, Ahring BK (1996) Granular sludge formation in upflow anaerobic sludge blanket (UASB) reactors. Biotechnol Bioeng 49:229–246

Şentürk E, İnce M, Engin GO (2010) Kinetic evaluation and performance of a mesophilic anaerobic contact reactor treating medium-strength food-processing wastewater. Biores Technol 101:3970–3977

Şentürk E, İnce M, Engin GO (2013) The effect of transient loading on the performance of a mesophilic anaerobic contact reactor at constant feed strength. J Biotechnol 164:232–237

Sharma V, Testa C, Lastella G, Cornacchia G, Comparato M (2000) Inclined-plug-flow type reactor for anaerobic digestion of semi-solid waste. Appl Energy 65:173–185

Skouteris G, Hermosilla D, López P, Negro C, Blanco Á (2012) Anaerobic membrane bioreactors for wastewater treatment: a review. Chem Eng J 198–199:138–148

Skouteris G, Saroj D, Melidis P, Hai FI, Ouki S (2015) The effect of activated carbon addition on membrane bioreactor processes for wastewater treatment and reclamation–a critical review. Biores Technol 185:399–410

Sun Q, Li H, Yan J, Liu L, Yu Z, Yu X (2015) Selection of appropriate biogas upgrading technology-a review of biogas cleaning, upgrading and utilization. Renew Sustain Energy Rev 51:521–532

Tauseef S, Abbasi T, Abbasi S (2013) Energy recovery from wastewaters with high-rate anaerobic digesters. Renew Sustain Energy Rev 19:704–741

Tricase C, Lombardi M (2012) Environmental analysis of biogas production systems. Biofuels 3:749–760

Umaiyakunjaram R, Shanmugam P (2016) Study on submerged anaerobic membrane bioreactor (SAMBR) treating high suspended solids raw tannery wastewater for biogas production. Biores Technol 216:785–792

Van Lier J, van der Zee F, Frijters C, Ersahin M (2015) Celebrating 40 years anaerobic sludge bed reactors for industrial wastewater treatment. Rev Environ Sci Bio/Technol 14:681–702

Visvanathan C, Abeynayaka A (2012) Developments and future potentials of anaerobic membrane bioreactors (AnMBRs). Membrane Water Treat 3:1–23

Voicu G, Dincă M, Paraschiv G, Moiceanu G (2015) A review regarding the biogas production through anaerobic digestion of organic waste. Adv Eng Forum 13:185–193

Wang S, Rao NC, Qiu R, Moletta R (2009) Performance and kinetic evaluation of anaerobic moving bed biofilm reactor for treating milk permeate from dairy industry. Biores Technol 100:5641–5647

Wang T, Huang Z, Ruan W, Zhao M, Shao Y, Miao H (2017) Insights into sludge granulation during anaerobic treatment of high-strength leachate via a full-scale IC reactor with external circulation system. J Environ Sci. https://doi.org/10.1016/j.jes.2017.06.024

Wei C-H, Harb M, Amy G, Hong P-Y, Leiknes T (2014) Sustainable organic loading rate and energy recovery potential of mesophilic anaerobic membrane bioreactor for municipal wastewater treatment. Biores Technol 166:326–334
Wu S-Y, Hung C-H, Lin C-Y, Lin P-J, Lee K-S, Lin C-N, Chang F-Y, Chang J-S (2008) HRT-dependent hydrogen production and bacterial community structure of mixed anaerobic microflora in suspended, granular and immobilized sludge systems using glucose as the carbon substrate. Int J Hydrogen Energy 33:1542–1549
Youngsukkasem S, Barghi H, Rakshit SK, Taherzadeh MJ (2013) Rapid biogas production by compact multi-layer membrane bioreactor: efficiency of synthetic polymeric membranes. Energies 6:6211–6224
Zaiat M, Rodrigues JAD, Ratusznei SM, De Camargo EFM, Borzani W (2001) Anaerobic sequencing batch reactors for wastewater treatment: a developing technology. Appl Microbiol Biotechnol 55:29–35
Zhang Z-P, Show K-Y, Tay J-H, Liang DT, Lee D-J (2008) Biohydrogen production with anaerobic fluidized bed reactors—a comparison of biofilm-based and granule-based systems. Int J Hydrogen Energy 33:1559–1564
Zhang M, Zhang G, Zhang P, Fan S, Jin S, Wu D, Fang W (2014) Anaerobic digestion of corn stovers for methane production in a novel bionic reactor. Biores Technol 166:606–609
Zupančič GD, Jemec A (2010) Anaerobic digestion of tannery waste: semi-continuous and anaerobic sequencing batch reactor processes. Biores Technol 101:26–33
Zwain HM, Aziz HA, Ng WJ, Dahlan I (2017) Performance and microbial community analysis in a modified anaerobic inclining-baffled reactor treating recycled paper mill effluent. Environ Sci Pollut Res 24:13012–13024

Chapter 5
Biogas Production: Mechanical and Thermal Pre-treatment Technologies

Ing. Dieter Jürgen Korz

5.1 Introduction

Huge amounts of organic waste are generated in urban, industrial, and agricultural areas worldwide. Unfortunately, organic wastes are still usually disposed of in landfills instead of being utilized for energy production and nutrients recycling. On the other hand, disposal of these materials in landfills causes very significant methane emissions and methane is one of the most climate-damaging greenhouse gases. There is no more doubt that any process/event leading to negative environmental impacts such as global warming and climate change must be antagonized for the sake of the future. The problem of global warming must be resolved globally, and waste treatment based on recycling and energy recovery could be an important solution to this crisis.

Solid and liquid organic wastes are generated in many different areas and thus, have very diverse compositions. Depending on the applied waste management strategies, the organic fraction of municipal solid waste (MSW) could be in a range of 20–70%. In many countries source separated collection systems for organic waste are already in place or being implemented. In the case of source separated organic wastes, the amount of biodegradable organic fraction is high but there always exist undesired contamination with plastic, glass, metals, stones, sand, etc. It has been proven that from source separated organic wastes, high-quality organic fertilizers can be produced, and this is a very efficient way of nutrients recycling.

Organic wastes are also produced within the commercial and industrial areas mainly in the food and beverage processing industry, in restaurants, canteens, and supermarkets. These wastes also include a high level of contamination by packaging materials which must be considered in the subsequent waste treatment processes.

Ing.Dieter Jürgen Korz (✉)
Proweps Envirotec GmbH, Rißstraße 19, 88400 Biberach, Germany
e-mail: dieter.korz@proweps-envirotec.com

M. Tabatabaei and H. Ghanavati (eds.), *Biogas*, Biofuel and Biorefinery Technologies 6, https://doi.org/10.1007/978-3-319-77335-3_5

A well proven technology to treat organic wastes in an eco-friendly manner is anaerobic digestion (AD) for biogas production. The key process step in biogas production plants is the pre-treatment of the organic wastes because of their inhomogeneity and fluctuating compositions. It is vitally important to separate undesired waste fractions such as plastic, glass, stones, and metals to ensure the greatest possible extent a trouble free AD process as well as the production of a high-quality fertilizer. Furthermore, the biodegradable fraction of the organic wastes should be disintegrated very efficiently to obtain a homogenized feedstock and ensure the highest possible conversion of the organic materials into biogas in anaerobic digesters.

Besides mechanical treatment of organic wastes which must always be the first processing step, further treatments have also been developed and are used mainly to increase biogas production. Such technologies are based on biological or thermal treatment of the biodegradable organic fraction before AD.

Pre-treatment systems for organic waste must comply with some important requirements such as:

- Flexibility to treat different types of organic waste
- Substrate homogenization
- Efficient removal of contaminants
- Energy efficiency
- Wear resistance
- High biogas production in anaerobic digesters.

The composition of organic waste originated from different areas (municipal, industrial, commercial) differs significantly. Many reasons cause such changes in composition as waste collection strategies, seasonal fluctuations, different packaging material, etc.

The waste composition is the most important criteria for the selection of a suitable pre-treatment technology. Furthermore, for the selection of the most suitable pre-treatment technology, it is important to know which type of AD, i.e., wet or dry digestion systems is to be used for treating the organic waste.

Wet anaerobic systems are operated at a lower solid concentration and use pre-treatment systems to remove the undesirable contaminants before the AD process. The digestate after AD can be used directly as high-quality fertilizer and no further digestate treatment (post-composting, compost refining, etc.) is normally necessary. Due to the efficient pre-treatment of organic waste, the biogas production is high. Wet AD systems are the preferred technologies for treating wet organic waste such as food leftovers, packaged food, and food waste. Although organic fraction from MSW can be treated in wet AD systems.

Dry anaerobic systems are operated at higher solid concentrations and use simpler pre-treatment systems before the AD process. As the efficiency of contaminant separation is not sufficient to use the digestate directly as high-quality fertilizer, additional digestate treatment (i.e., post-composting, compost refining) is usually required if the input material is contaminated. Dry anaerobic systems are mainly used if the organic waste includes a high percentage of garden waste but also to treat the organic fraction of MSW after a mechanical separation process.

This chapter describes the different pre-treatment technologies used in industrial AD plants for treating different kinds of organic waste. Only systems that are already in operation in industrial scale plants over a long period and have been proven to be suitable for organic waste processing are considered and discussed herein.

5.2 Organic Waste

As mentioned earlier, organic waste is produced in different areas and its composition varies depending on its origin and the applied waste management strategies. A good understanding of waste composition is very important to decide on the best suitable, most efficient, and most economic treatment technology. This section presents the different types of organic waste that is produced in different areas.

Biowaste is collected by municipalities in cities where food waste is not separated from garden waste and both waste fractions are usually collected in a common biowaste bin. This is a typical collection system used is some countries such as Germany. Therefore, the biowaste includes a high amount of garden waste especially during the growing season from spring to autumn. In winter, the amount of garden waste in biowaste is low and that means that less organic waste is produced and the organic waste has a higher percentage of food waste and moisture. Due to these variations in garden and food waste, the moisture content also varies. In winter, due to the higher amount of food waste, the moisture content of biowaste is much higher than in other seasons.

It is also important for the design of a treatment plant to consider the fact that garden waste also includes huge amounts of sand and stones but this waste also includes plastic and other non-organic contaminations which must be considered in the plant design.

Many municipalities have separate collection systems for food and for garden waste. This makes a lot of sense because food waste with its high specific energy yield content can be processed efficiently in AD plants to produce biogas. While garden waste due to its high amount of cellulose and lingo-cellulose is more suitable for aerobic composting. The specific biogas yield of garden waste is much lower compared with food waste and is therefore, not advisable to treat it in AD plants.

Food waste has a high moisture content and is also always contaminated especially with plastics as well as other contaminants as metals, glass, stones etc. Therefore, an efficient pre-treatment before AD is crucial.

On some occasions, municipal food waste is collected in paper bags with very low levels of contamination. However, the pre-treatment stage for such type of food waste is also important because the paper together with the organic part of the food waste should still be separated from some contaminations prior to treatment in anaerobic digesters. Paper bags should not be separated as contaminants.

Food waste from restaurants and kitchen typical have a high moisture content (75–90%) as well as significant amounts of contamination (packaging material, metals from cutlery, broken plates, cups, and glass). For instance, food waste from restaurants in China contain a lot of chop sticks besides other contaminants which should be considered. Another type of food waste is the biomass generated through vegetable processing in kitchens. Low level of contamination and a low moisture content are the main characteristics of this waste. The pre-treatment of such kind of organic waste involves at least an efficient disintegration and homogenization before the AD process while the possibility of the presence of contaminants in the bins should not be overlooked.

Packaged food waste including expired food is another stream of food-related wastes. Such types of organic waste contain mainly plastic and other packaging material as contaminants. Due to the heterogeneity of organic waste the pre-treatment process before AD is extremely important.

5.3 Pre-treatment Technologies

Different pre-treatment technologies are available for organic waste processing prior to an AD process. The present section explains the most widely used and most experienced pre-treatment technologies.

5.3.1 Mechanical Pre-treatment Systems

5.3.1.1 Separation Hammer Mills

Separation hammer mills are mainly used to treat packaged food waste and food leftovers from restaurants. The principal function of the separation mills is the crushing of organic waste using hammers installed on an electrically-driven shaft. The separation hammer mills are equipped with a screen and a separation system to remove the packaging material which is mainly the plastic fraction. Figure 5.1 shows a separation hammer mill with charging and discharging systems.

The plastic fraction has a low density and is separated by means of an air suction out of the separation mill into a screw separator that removes and transfers the plastics into a container. The separation mill is equipped with a screen with a defined size of the screen wholes. Crushed biodegradable organic material as well as crushed contaminants (plastic and glass) with particle sizes smaller than the screen whole size are separated. This crushed organic material is then used for further processing in anaerobic systems. Usually process or fresh water is added to assure an efficient separation of contaminants from the organic material.

Several separation hammer mill systems from different suppliers are available on the market e.g.:

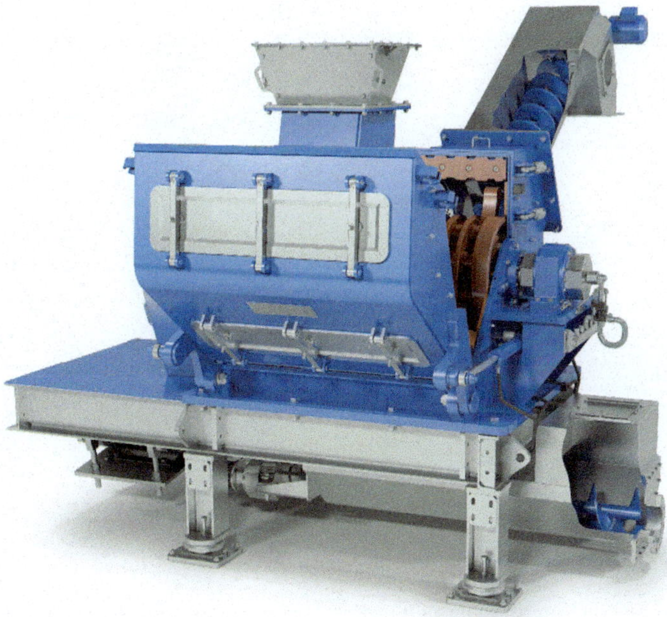

Fig. 5.1 A typical separation hammer mill. *Courtesy* Haarslev Industries A/S (datasheet_Haarslev2.0)

- Wackerbauer
- Haarslev
- Hybag
- Atritor

It needs to be stressed that the separation mills do not separate inert material as glass or sand. As such contaminants are usually included in the organic waste, it is normally necessary to install additional equipment to separate inert materials especially if wet AD systems are used for organic processing.

An example of a separation mill combined with a hydrocyclone to remove inerts is the co-digestion plant in Innsbruck (Austria). Figures 5.2a and b show the separation mill and the hydrocylone system in the sewage water treatment plant in Innsbruck, respectively. The processed biowaste after removal of the plastic and packaging contamination is then further treated in the hydrocylone system.

At high flow velocities of an organic waste liquid suspension, heavy material is separated. The heavy fraction separation efficiency of hydrocyclones is good when the solid content of the organic slurry is below 10%. As biowaste usually has a higher solid content, a dilution with fresh or process water is required. As mentioned earlier, water is usually added directly into the separation mill to adjust the required solid content.

(a) (b)

Fig. 5.2 **a** Separation hammer mill and **b** hydrocylone system installed for separation of inerts in the sewage treatment plant in Innsbruck, Austria. *Courtesy* DI Reinhard Oberguggenberger (KA Betriebsinfo)

5.3.1.2 Impact Reactor Combined with a Reject Separation System

This pre-treatment technology combines a waste disintegration based on the impact technology with a separation system to remove light and heavy contaminants. The process flow of the two-step pre-treatment technology is shown in Fig. 5.3.

In the impact reactor equipped with a special solid rotor, the organic waste is disintegrated selectively. Plastic bags are opened but not crushed in small pieces and the coarse plastic pieces can then easily and efficiently be removed in the reject separation system by the screening process. The impact technology disintegrates the biodegradable organic waste fraction very efficiently into small particles. This is important to ensure a high organic conversion into biogas in the digester.

The disintegration process does not require any water addition and the adjustment of the moisture content for AD is managed in the reject separation system. Thus, electricity consumption in the impact reactor is minimized. Figure 5.4 shows an impact reactor installed in an AD plant.

The second and very important part of the pre-treatment is the separation of biodegradable organic fraction from light and heavy contaminants performed using reject separation systems (Fig. 5.5).

In the reject separation system, light fraction (plastics, textiles, wood pieces etc.) is first separated by screening processes. The solid content of the waste slurry is first adjusted by adding internal process water according to the digestion requirement and to ensure an optimal separation of biodegradable organic material from

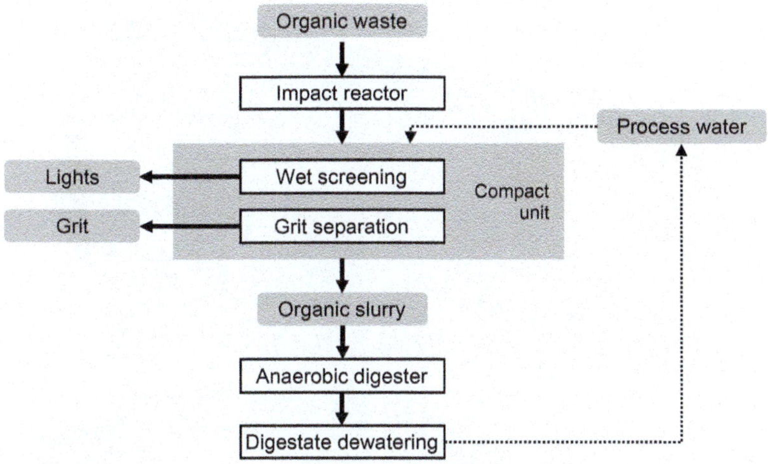

Fig. 5.3 The flowchart of the impact reactor combined with reject separation system pre-treatment

Fig. 5.4 Impact reactor

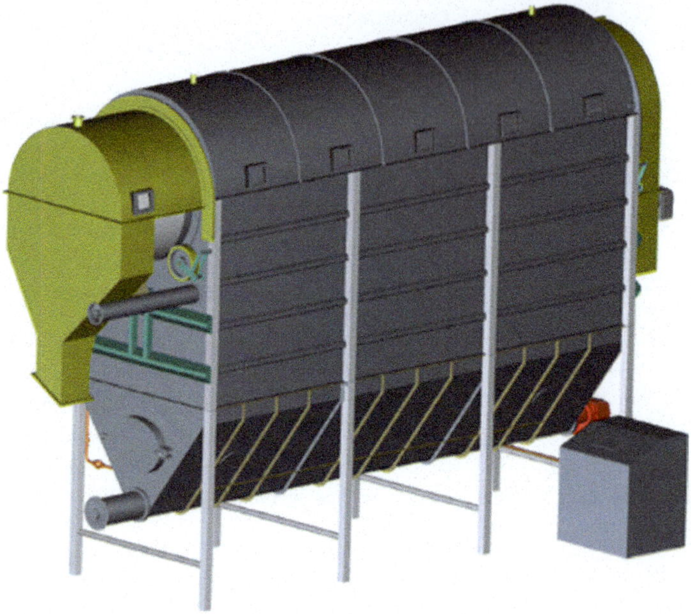

Fig. 5.5 Reject separation system

contaminants. Disintegrated organic material as well as small inert material (glass, sand, and stones) are subsequently passed through the screen into an aerated grid separation system. The aeration of the grid separation system reduces the viscosity of the organic slurry and therefore, the separation efficiency will be improved. The grid separation system is also equipped with a top skimmer and very small and light particles as polystyrene pieces are separated. The separation of fine plastic particles is of great importance when the organic fraction separated from municipal solid waste is to be treated. The system produces a clean organic slurry that can be used without further treatment in an anaerobic digester. The supplier of this pre-treatment technology is Dieffenbacher GmbH.

5.3.1.3 Pulper Technologies

Pre-treatment technologies based on pulper technology are used prior to wet AD systems. The system consists of several components and prepares a clean, homogenous, pumpable organic fraction from organic waste that will be then treated in the digester.

More specifically, using the pulper technology, biowaste is first treated in a crusher to open plastic bags and reduce the particle size of the organic waste. The pre-crushed organic waste is then mixed with process water in the pulper to produce a homogenous waste suspension with a solid content of approx. 10%. Pulpers are equipped with central mixers to ensure efficient mixing of organic waste with process water as well as to further reduce the particle size of the organic material. At the bottom of the pulper, a grit separation chamber is installed where heavy particles such as glass, stones, metals, bones are collected and can then be easily removed. This is an important function of the process to prevent these heavy pieces from being pumped into the AD system. The organic slurry will then pass through a screen that separates waste components which are not useful in AD process including plastic, textiles, and wood also known as the light fraction. The cleaned organic slurry will the subsequently pumped into a buffer tank. As the organic slurry may still contain inert materials (such as sand, glass splitter, and stones), it is important to make sure all these materials are removed as they would cause wear in the plant but also would lead to sedimentation in the digester. Therefore, a hydrocylone system is used to remove this fine inert material before the organic slurry is introduced into anaerobic digesters.

Various systems using the pulper technology are available. They differ in the systems used to separate the contaminants from the biodegradable organic fraction. For instance, Fig. 5.6 shows the flow chart of a pulper technology combined with a hydrocylone system used to process organic waste, i.e., BTA-process (biowaste, food waste) and Fig. 5.7 presents a BTA pre-treatment installation in Austria.

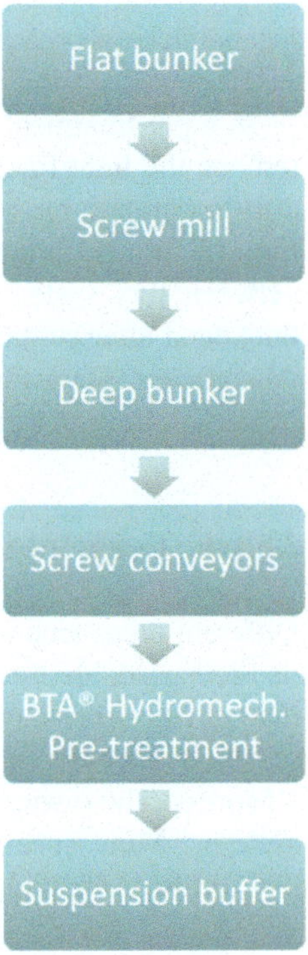

Fig. 5.6 Pulper system combined with hydrocylone (BTA-process). *Courtesy* BTA-International Co. (Kübler et al. 2015)

Fig. 5.7 BTA-plant Zell am See, Austria *Courtesy* BTA-International Co. (Kübler et al. 2015)

In a pre-treatment system developed by Lohse GmbH, a screen drum is used to remove the light fraction from the organic waste slurry while the inert materials are also separated by a hydrocylone system. Ecogy pulper technology shown in Fig. 5.8 is also another technology available on the market. The key difference between this system and the other pulper systems is that the organic slurry is thickened after contaminants removal to increase the slurry solid content before the AD process. Thus, the hydraulic loading and consequently, the energy consumption of the whole biogas production plant will be reduced.

5.3.1.4 Organics Separation with Press System

Different press technologies are used for pre-treatment of organic waste to separate its biodegradable fraction from contaminants. The systems that are available on the market use either the screw press technology or piston press technology. One of the most advanced systems is the hydraulic press developed many years ago by VMpress. The owner of this technology is Anaergia offering complete pre-treatment systems. More specifically, the organic waste is treated in a cylindrical chamber with a piston that applies a high pressure on the organic waste. The fractioning of the organic waste is carried out by particle size based on a screen system that is part

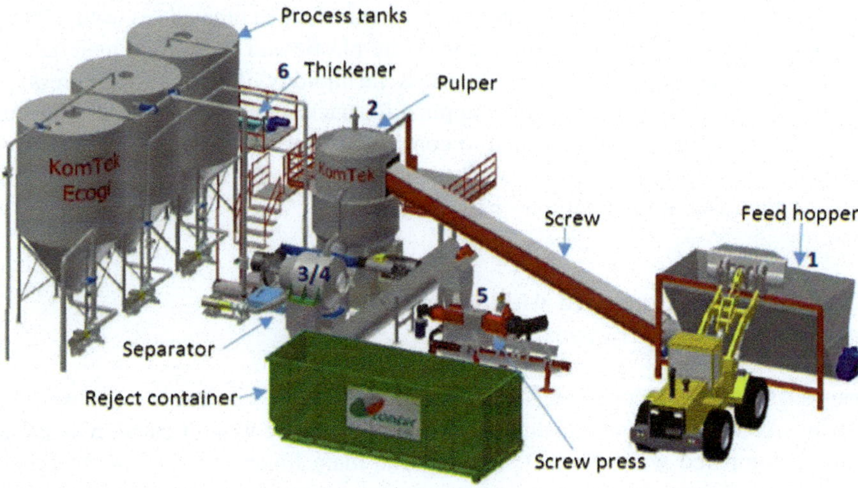

Fig. 5.8 Ecogy pulper system. *Courtesy* Gemidan Ecogi A/S (http://ecogi.dk/en/technology/)

of the press chamber. Biodegradable fraction of the organic waste as well as small light and heavy waste particles (the light fraction with higher calorific values, e.g., plastics, textiles, and cardboard) are separated from bigger waste particles. As the organic waste is being treated in the press without water addition, the organic fraction separated has the original solid content. The removal of small contaminants (plastic and grid) from the organic fraction before AD requires additional processing to produce a cleaner organic slurry. Anaergia offers a dynamic cyclone system to remove small light pieces. An adjustment of the solid content of the organic fraction for these additional processing steps is then required. Moreover, this technology requires additional separation systems to remove fine inert material from the wet organic fraction separated in the press.

Other suppliers of hydraulically operated presses include:

- Finsterwalder Umwelttechnik GmbH
- Putzmeister Solid Pumps GmbH.

The separation process of organics from packaging material with a hydraulically operated press is identical in all the systems but the applied pressure is different. The pressure varies from 30 to 250 bar. Systems that are operated at higher pressure values have a higher organics separation efficiency but on the other, require a higher investment cost. As disposal costs for rejects from organics processing are high, the target of the treatment should always be to minimize organic losses and to produce rejects streams with low water contents.

Screw presses such as those supplied by Bellmer Kufferath are also used to separate the organic fraction from packaging material after the waste is crushed with a shredder. The crushed organic waste is then mechanically pressed with a screw. The pressure-built-up is reached with a conical screw with substantial flight heights.

The liquid fraction is separated with a screw inside the machine and is used directly for AD. The solid fraction contains mainly the plastic and packaging material.

Another supplier of such type of press technology (known as Biopress) is Doppstadt. Their systems are mainly applied for packaged food and food leftovers. As mentioned earlier, it is important to consider additional processing steps before AD if the waste includes heavy contaminants such as glass, stones, and sand as some of such heavy material may pass the screen.

5.3.1.5 Crusher Combined with Screening (Pre-treatment for Dry AD Systems)

Dry AD systems are operated at higher solid contents compared with wet AD systems. Accordingly, the moisture content of the organic waste treated in digesters is lower compared with wet AD systems. The main function of the pre-treatment system is a rough depackaging (opening and removing of plastic bags), size reduction, and separation of oversized material that is not suitable for AD. Organic waste is first disintegrated by crushers followed by a classification in different particle sizes using drum or star screen separators. Figure 5.9 shows a crusher with two shafts.

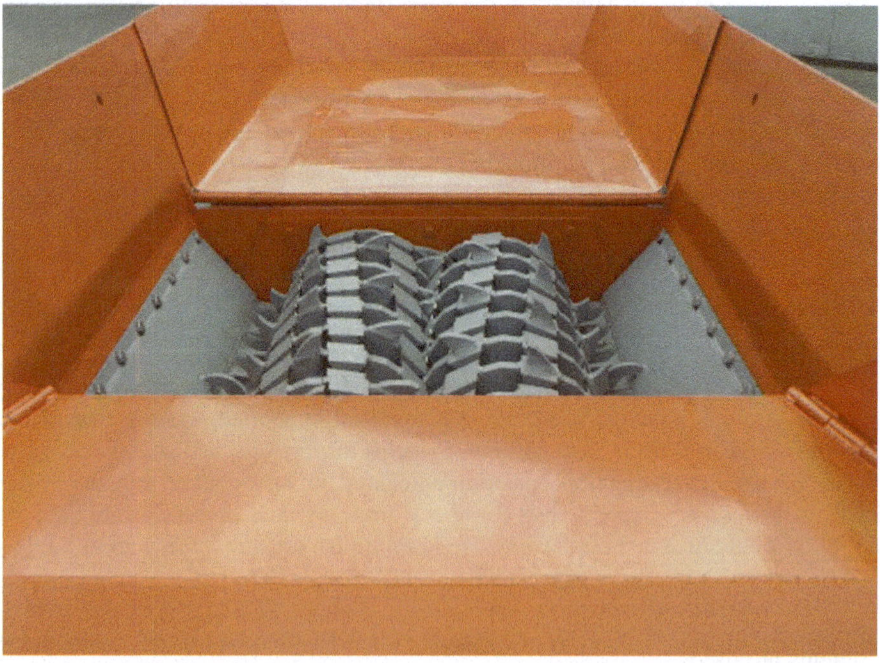

Fig. 5.9 Crusher used for managing biowaste. *Courtesy* ARJES (http://www.arjes.de/de/produkte/vorzerkleinerer/vz-850/)

The screening process usually separates the organic waste in fractions <60–80 and >60–80 mm. In order to minimize losses of biodegradable organic material in the oversized fraction, the material is usually crushed and screened again. Fe-metals can be separated from the organic waste fraction <60–80 mm before the material is fed into the digesters. Before feeding the crushed organic waste into the digester the moisture content is adjusted according to the process specific requirements by adding internal process water separated in a digestate dewatering process. It must be considered that the organic fraction still includes contaminants as glass, plastic, and stones which must be removed after AD. Efficient separation of such contaminants requires a moisture content of ideally <40% and that means that an additional composting process after AD is usually necessary to reduce the moisture content of the digestate to the required level. After the composting process, contaminants are removed by screening (removal of heavy material as glass and stones) and wind sifting (plastic removal) to produce a compost with the required quality according to national rules and regulations.

5.3.2 Thermal Pre-treatment Systems

5.3.2.1 Thermal Hydrolysis

Following mechanical treatment which is important to separate contaminants and to produce a homogenous biomass, the processed organic waste can be further treated at high temperature to not only increase the biogas production but also to pasteurize the organic waste. The sanitation of organic waste, i.e., to eliminate pathogenic microbes such as *Salmonella*, is an important part of waste processing. AD plants must comply with the national regulations related to the pasteurization requirements for both digestate and compost products.

High temperature values should improve the microbial degradation of organic material and consequently a higher biogas production in the digesters is expected. In another word, the thermal treatment of organic biomass destroys persistent structures of the cells and complex molecules and therefore, an optimized microbial degradation in the digesters is achieved.

The heating of organic waste to high temperature values before AD is not very common as the technology is very expensive and the expected increase in energy production in the digester does not usually justify the additional investment. Furthermore, an efficient pre-treatment of organic waste especially food waste is usually sufficient to obtain a high conversion rate into biogas and the additional increase in biogas production because of thermal treatment is not considerable. Thermal treatment leads to higher organic conversion efficiencies when substrates such as sewage sludge or agricultural residues (e.g., straw) with high lignocellulosic contents are used.

Among thermal hydrolysis technologies available the CAMBI thermal hydrolysis process is used in organic waste treatment plants for biowaste. In this system, following mechanical treatment and before AD, the organic waste is heated up to 165–170 °C and is treated at 6–7 bar for 30 min at this temperature. For heating the organic waste slurry to the required processing temperature, steam is required. After thermal treatment, the temperature of the organic slurry is reduced to the AD temperature. The system is equipped with a heat recovery to minimize the consumption of thermal energy.

The thermal treatment at such high temperatures is advantageous only if organic waste includes animal byproducts of class 1 or 2 according the European animal by-products regulation. If animal by-products class 3 (food waste) is to be treated in AD plants, the pasteurization of the organic waste before AD at 70 °C for 1 h is sufficient and accepted according the EU requirements unless stated otherwise by national regulations.

If a thermal treatment at high temperatures is required, a specific economical evaluation is required. It should also be noted that since renewable energies in many countries are no longer supported by subsidies and must survive considering the market conditions, the additional energy production in response to thermal treatment may not justify the respective high investment cost for such a system.

5.3.2.2 Pasteurization of Organic Waste

National and international regulations usually require a pasteurization of organic waste. The system usually used to address this need is a thermal treatment at a temperature >70 °C with a minimum hydraulic retention time of 1 h. In the EU, it is also required to reduce the particle size of the organic waste for pasteurisation below 12 mm.

Different pasteurization systems working either in batch- or continuous modes are available on the market. Figure 5.10 shows a continuous food waste pasteurization system.

To achieve a continuous operation, at least three tanks are used while they alternatively undergo the above-mentioned pasteurization conditions. Pasteurised organic slurry is then cooled down to the digester process temperature. With a heat recovery system, extra heat through cooling down the slurry after pasteurisation, is used to heat up the unpasteurized organic slurry. Based on a continuous operation, a high heat recovery efficiency is ensured.

If the pre-treatment and pasteurisation of organic waste is carried out before the AD, then the digestate can be used directly as high-quality fertilizer and additional processing (post-composting, compost refining) is not required.

Other common methods for pasteurisation of organic waste are:

- Thermophilic digester operation at temperatures of 55–58 °C
- Aerobic post-composting of digestate.

Fig. 5.10 Continuous food waste pasteurization system

Post-composting of digestate requires the addition of structural material (garden waste) because it is necessary to adjust the characteristics (moisture content, air porosity) of the organic material for an efficient composting process to ensure sufficient heat development for pasteurisation.

5.4 Conclusions

Organic waste is very heterogeneous, while its moisture content as well as level of contamination vary significantly. Thus, pre-treatment before AD is a key process.

Different pre-treatment technologies have been developed and are successfully installed in many AD plants all over the world. The pre-treatment of organic waste is the key process step in biogas production plants ensuring:

- Efficient separation of contaminants
- High availability of AD plants
- High biogas yields
- Production of high quality fertilizers.

In any AD application treating organic waste, a mechanical pre-treatment is installed. Different pre-treatment technologies are available to reduce the size of the organic waste and to separate the plastic and packaging material from the biodegradable fraction of the waste. As organic waste almost always includes contaminations such as glass, metals, stones, and sand; additional systems are also required to deal with such heavy contaminants of the waste. Thereby, plants are generally highly flexible to treat all kinds of organic waste without any quality restrictions.

Pre-treatment systems should be most efficient in terms of organics disintegration and in terms of the separation of the contaminants. It is very important to avoid the severe crushing of the contaminants as this makes their separation before or after AD very difficult. A selective and efficient organics disintegration results in higher specific biogas yields in the digesters.

Thermal treatment of organic waste at high temperature and pressure values is more common when treating sewage sludge or agricultural biomass, e.g., straw. Such types of organic biomass are difficult to degrade anaerobically due to their rigid structure. On the contrary, food waste only requires a mechanical processing and can be efficiently converted in AD systems to biogas. Therefore, the additional biogas production that can be obtained normally does not justify the high investments for thermal treatment systems for such type of organic waste. The pasteurization of organic waste does not require such high temperatures as a treatment at 70 °C for 1 h is sufficient and accepted according to international regulations.

One of the key targets of AD besides energy production is the recycling of valuable nutrients back into the natural cycle. But this is only possible if the fertilizer is of high quality and has lowest possible level of contaminations. Strict and continuously monitored fertilizer quality requirements have already been established in many countries around the world. Following the requirements plays a key role for future development of AD plants. Therefore, an efficient pre-treatment of organic waste also ensures the production of high quality fertilizers and consequently, additional, expensive digestate processing after AD can be avoided.

Efficient separation of contaminants and preparation of a homogenous organic waste slurry also prevents process interruptions mainly caused by clogging, sedimentation, floating, wear, and ensures high plant availability. As one of the main revenues of waste treatment plants is the gate fee for treating waste, a high plant availability is of greatest significance for the plant economics. Moreover, minimizing costs associated with replacing worn-out parts is also important for the plant economics.

High biogas yields in anaerobic digesters are achieved if the biodegradable organic material is well crushed in the pre-treatment and a large surface are for microbial degradation is achieved.

Finally, it should be stressed out that pre-treatment technologies should be flexible as organic waste composition due to different reasons is unsteady. Seasonal fluctuations in waste composition such as changes in the general waste management strategies must be considered and more importantly, the pre-treatment technology used should be able to accommodate such variations. This also concerns the type and amount of contaminants as well as the moisture content of the waste. Plant operators are also advised to consider using different types of organic waste available on the market to enhance the plant economics.

References

Haarslev company information: waste food depacker (datasheet_Haarslev2.0)

KA Betriebsinfo: Behandlung von Bioabfällen mit anschließender Co-Vergärung, pp 2108–2112

Kübler H, Muck O, Patzies M, Schulte S, BTA International GmbH L. Winter, Biogas ZEMKA GmbH (2015) High substrate flexibility and intelligent biogas valorization concept: the biogas plant ZEMKA in Austria. Presentation BIOGAS 2015, 24. Internationale Jahrestagung und Fachmesse, p 12, 29 Jan 2015

Gemidan Ecogi A/S Ecogi—Technical Description (available on homepage http://ecogi.dk/en/technology/)

http://www.arjes.de/de/produkte/vorzerkleinerer/vz-850/

Chapter 6
Prominent Parameters in Biogas Production Systems

Meisam Tabatabaei, Elena Valijanian, Mortaza Aghbashlo, Hossein Ghanavati, Alawi Sulaiman and Minato Wakisaka

6.1 Introduction

Anaerobic digestion (AD) process has been investigated comprehensively over the last decades for both waste/wastewater treatment and renewable energy production in both the industrial and agricultural sectors (Lindmark et al. 2014). The growth and activity of anaerobic microorganisms, i.e., the beating heart of the AD, and consequently the efficiency of the process are significantly impacted by some prominent parameters and therefore, it is crucial to ensure that these parameters are as optimized as much as possible. These parameters include constant temperature values favoring microbial growth, pH-value, sufficient nutrient supply (substrate composition and C/

M. Tabatabaei · E. Valijanian
Biofuel Research Team (BRTeam) / Agricultural Biotechnology Research Institute of Iran (ABRII), Agricultural Research, Education and Extension Organization (AREEO), Karaj, Iran

M. Aghbashlo
Department of Mechanical Engineering of Agricultural Machinery, Faculty of Agricultural Engineering and Technology, College of Agriculture and Natural Resources, University of Tehran, Karaj, Iran

A. Sulaiman (✉)
Faculty of Plantation and Agrotechnology, Universiti Teknologi MARA, 40450 Shah Alam, Selangor, Malaysia
e-mail: asuitm@yahoo.com

M. Wakisaka (✉)
Department of Biological Functions and Engineering, Graduate School of Life Science and Systems Engineering, Kyushu Institute of Technology, 2-4 Hibikino, Wakamatsu-Ku, Kitakyushu, Fukuoka 808-0196, Japan
e-mail: wakisaka@life.kyutech.ac.jp

H. Ghanavati
Agricultural Biotechnology Research Institute of Iran (ABRII), Agricultural Research, Education and Extension Organization (AREEO), Karaj, Iran

M. Tabatabaei and H. Ghanavati (eds.), *Biogas*, Biofuel and Biorefinery Technologies 6, https://doi.org/10.1007/978-3-319-77335-3_6

N ratio), mixing intensity, retention time as well as presence and amount of inhibitors (e.g., ammonia and heavy metals) (Al Seadi et al. 2008). Among the various groups of anaerobic microorganisms involved in the AD process, methanogenic archaea handling the very last stage of biogas production (BP), are very sensitive to any changes in their environmental conditions (Voicu et al. 2015). Since these microorganisms are in fact the rate-limiting factor of the whole BP process, it is essential to carefully monitor the environmental conditions. In light of that, maintaining these parameters within their appropriate ranges as far as microbial growth and activity are concerned is key to achieve long-term stable AD operation (Zhang et al. 2014). This chapter is aimed at reviewing and discussing the aforementioned parameters.

6.2 Temperature

Temperature is undoubtedly a crucial parameter that could profoundly impact the AD process (Chen et al. 2016). AD process are performed under psychrophilic (15 ± 1 °C), mesophilic (37 °C), and thermophilic (55–70 °C) conditions. Among them, thermophilic condition has been reported to possess the fastest reaction rates and highest load bearing capacity, consequently, leading to the highest productivity (Mao et al. 2015). However and despite of the favorable advantages associated with thermophilic AD, the existing high temperature is believed to intensify ammonia toxicity which could eventually result in unstable digestion process and in severe cases its complete inhibition (Weiland 2010). Table 6.1 tabulates the typical retention times required under different thermal conditions (Al Seadi et al. 2008).

It has been claimed that BP under psychrophilic and mesophilic conditions using anaerobic membrane bioreactors (AnMBRs) led to comparable methane production rates, significant methane loss in the permeate under psychrophilic conditions was observed though (Martinez-Sosa et al. 2011; Trzcinski and Stuckey 2010; Smith et al. 2013). Moreover, low temperatures corresponded to a slightly higher fouling rate possibly caused by VFA accumulation and the release of protein-dominated extracellular polymeric substances (EPS) (Gao et al. 2014a).

In addition to the AD process, temperature has also been reported to be effective on microbial population balance and BP through thermal pre-treatment. In a study, Ennouri et al. (2016) employed thermal pre-treatment prior to the AD of waste activated sludge (WAS). They argued that sludge thermal pre-treatment at 120 °C led to the highest biogas yields of 0.42 L/g volatile solid (VS) removed and

Table 6.1 Thermal conditions and their typical retention times

Thermal stage	Process temperatures (°C)	Minimum retention time (days)
Psychrophilic	<20	70–80
Mesophilic	30–42	30–40
Thermophilic	43–55	15–20

*With permission from Al Seadi et al. (2008) Copyright© 2017

0.37 L/g VS removed for urban and industrial sludge samples, respectively. They attributed the positive impacts of thermal pre-treatment to the enhanced growth of hydrogen-using methanogens (HUMs), and the consequent rapid consumption of the generated H_2 leading to enhanced acetogenesis (Ennouri et al. 2016).

It should be noted that variations in temperature could also adversely affect the microorganisms involved in the AD process and consequently BP. In line with that, (Bowen et al. 2014) showed that under-optimal temperature values led to reduced substrate utilization and VFA production rates. On the contrary, increases in operating temperatures are generally accompanied with less negative effects probably owing to the higher ability of microbial community to adapt to such temperature conditions. This was practically confirmed by (Kundu et al. 2014a) who investigated the recovery of a mesophilic anaerobic reactor after sharply increasing the operating temperature to 65 °C. They reported that the temperature shock widely affected the acetoclastic methanogenic populations, which consequently led to a prompt performance deterioration of the AD system. Nevertheless, they also observed a gradual recovery of the system and attributed the improvements in BP to the emergence of thermophilic methanogenic populations under the new temperature regime (Kundu et al. 2014a). Similar observations were made by (Westerholm et al. 2017) who looked into thermophilic-to-mesophilic temperature adaptation.

Overall, it should be noted that in a stable AD system, there is a balance among the different microbial populations contributing to the process and therefore, there is no accumulation of intermediates, e.g., the end products of a specific population. Sudden changes such as abrupt variations in temperature, could disturb this balance jeopardizing the whole microbial structure, FA concentration and consequently BP process (Kundu et al. 2014a; Kwietniewska and Tys 2014). When deciding on the temperature choice of an AD system, various factors should be taken into account. In better words, there is trade-off between the pros and cons of different temperature regimes, e.g., mesophilic and thermophilic conditions. For instance, increased process temperatures are in principle associated with increased microbial metabolic rates, higher degradation rates of organic materials and COD removal as well as with increased BP potentials. However, higher concentrations of FA and their potential accumulation in case of any process deficiencies as well as higher susceptibility to ammonia inhibition are also more probable under thermophilic conditions compared with mesophilic ones (Chen et al. 2008). Hence and as mentioned earlier, for large-scale BP operations, different factors including ambient conditions, type of wastes used, as well as the cost and expenses needed to maintain thermophilic conditions should be investigated to ensure the economic viability of the choice of temperature regime for AD systems (Khan et al. 2016). For instance, Arikan et al. (2015) concluded that for a 400 m^3 digester in Maryland with an ambient temperature of 13 °C, energy requirements could be decreased to 70% if the operating temperature would be reduced from 35 to 28 °C.

6.3 pH

Microbial growth in an anaerobic digester and consequently BP are largely dependent on the pH (Yang et al. 2015). Optimal growth of different microbial groups involved in the AD process take place under different pH values, e.g., acidogens (5.0–6.0) and methanogens (6.5–8.0) (Kundu et al. 2017). This along with the other existing differences such as different doubling time (growth rate) has been the driving factor for separating acidogenesis and methanogenesis processes through the application of two-stage reactor configurations (Chaikasem et al. 2015). Since the operational pH could directly and significantly affect the whole BP process, by optimizing each stage in separate reactors, various advantages could be achieved including reduced VFA accumulation, higher process stability, as well as higher buffering capacity against loading rate and toxicity (Maspolim et al. 2015). It should also be noted that drastic declines of pH could lead to irreversible inhibition of the BP, i.e., the process cannot be recovered even if the pH of the digester is restored to optimal values (Dareioti et al. 2017). Nevertheless, phases operation or in another word, conducting different stages of AD in separate reactors imposes a significantly higher capital and maintenance costs.

If single-phase operation is intended, the favourable pH range to partially meet the requirements of different microbial groups involved in the AD process would be 6.8–7.4. Fang and Liu (2002) revealed that the relative abundance of microbial species substantially increased from 6 at pH 4.0–14 at pH 7.0. In general, maintaining pH in a single-phased anaerobic digester is challenging as due to the high activity of hydrolytic enzymes, high VFA concentrations are present and if by any means their uptake is disturbed, pH could easily decline to values at which the whole BP operation would be at stake (Jiang et al. 2013). In a study, Hernández and Rodríguez (2013) looked into the impact of low pH values (i.e., 5.0, 5.5, and 6.0) using an anaerobic batch reactor and claimed that at pH values lower than 6.0, methane production stood at >1% while hydrogen production was recorded at high levels showing the prevalence of acid-producing bacteria in the reactor. These findings are in line with those of Nakasaki et al. (2015) who also recorded a significantly lower archaeal cell density at low pH values versus bacteria.

Yang et al. (2015) argued that methane yield was increased by more than 7 times when pH was controlled in comparison with the conditions of pH uncontrolled group. This reveals the importance of different pH control strategies. Among these strategies is the application of controlled organic loading rate (OLR) in order to prevent the accumulation of VFAs, also known as acidification. However, there is always a trade-off between the benefits achieved by employing high OLRs and the cost associated to maintain the pH at optimum range for methanogens (Mao et al. 2015). Another strategy is the extraction of propionic acid, e.g., by solvent extraction (Wang et al. 2009). Nevertheless, economic feasibility of this method at large-scale industrial operation is in question. Control of pH through the addition of sodium hydroxide has also been largely investigated (Chandra et al. 2012). However, the applicability of this method when dealing with industrial scale digesters is in question. It should also be mentioned that excess sodium ion could also play the role of an inhibitor of the methanogenesis process, for instance at concentrations above 3500 mg/L under mesophilic conditions (Chen et al. 2008). Another technology in which no additives would be required is the pH-dependent recycling of effluent from an anaerobic filter into the acidification reactor (Lindner et al. 2015).

Sharp pH variations could also adversely affect some reactor designs and configurations such as AnMBRs. More specifically, serious membrane fouling has been reported as a result of sludge flocs breakage and the accumulation of fine particles in the bulk sludge caused by pH shocks (Gao et al. 2010; Zhao et al. 2015). In light of that, phased AnMBRs, in which acidogenesis and methanogenesis processes are carried out in separate reactors, are highly recommended (Lin et al. 2013).

6.4 Volatile Fatty Acids and Alkalinity

Volatile fatty acids (VFAs) are important intermediates generated during the hydrolysis and acidogenesis stages of the AD process which also hinder the process if accumulated beyond certain limits. VFAs consist of organic acids such as acetic, propionic, butyric and valeric acids (Haugen 2014) and are used as substrate for the acetogenesis and methanogenesis stages to produce the final product of the process, i.e., methane. VFAs concentration is a good indicator for the evaluation of the stability of the digestion process. In general, abrupt changes in the environmental conditions in the digester such as substrate composition, inhibitors overload and/or temperature instability could result in the inhibition of the microbial population involved in the process leading to increased concentrations of VFAs to over 1500–2000 mg/L, jeopardizing the BP process (Falk 2012; Labatut and Gooch 2014). In another word, accumulation of these acids in particular propionic acid could significantly reduce the pH and subsequently, inhibit the methanogenesis (Jain et al. 2015; Drosg 2013). It has been highlighted in the published literature that fast biodegradation of organic macromolecules like proteins, fats, and carbohydrates in food and agro-industrial wastes is the main cause of such VFA accumulations and the resultant reactor imbalances (Jain et al. 2015). It should be mentioned that acetic acid is the final precursor to methane and therefore, its moderate accumulation is normal. Overall, the ratio of acetic acid to propionic acid as well as butyric or valeric acid concentration are acceptable indicator for determination of digester stability, or conversely, process instability (Drosg 2013; Jain et al. 2015). Argyropoulos (2013) reported that methane production was inhibited by more than 50% at acetate, butyrate and propionate concentrations above 13, 15, and 3.5 g/L, respectively. Different approaches and methods have been presented for VFAs measurement, such as steam distillation, photometric, colorimetric, chromatographic, and titration methods as well as infrared spectrometry. While the measurement of individual VFA is carried out by high pressure liquid chromatography (HPLC) or gas chromatography (GC) analysis (Aceves-Lara et al. 2012; Drosg 2013).

In addition to VFAs level, alkalinity or buffering capacity (measured as mg of $CaCO_3$/L) is also a typical fast indicator for monitoring the digestion process. In the AD process, there exist two buffering systems contributing to pH stability. One buffering system concerns the equilibrium between dissolved carbon dioxide and CO_3^{2-} (pKa = 6.35). At the pH value of 4, this equilibrium is shifted in favor of free carbon dioxide, while at the pH value of 13, all carbon dioxide is bound as carbonate in the system. For monitoring the process, a rise of the carbon dioxide percentage in the biogas can be an indicator of process disturbance. The other buffer system concerns the equilibrium between the ammonia and ammonium (pKa = 9.25), hydrogen sulfide ($H_2S/HS^-/S^{2-}$, pKa 7.1 and 13.3), and hydrogen phosphate ($H_3PO_4/H_2PO_4^-/HPO_4^{2-}/PO_4^{3-}$, pKa 2.1, 7.2 and 12.3), responsible for preventing weak acidification conditions. Therefore, substrates containing bicarbonate buffering capacity and high ammonia content generally maintain the pH stable around weak alkaline condition, providing the digester with tolerance against high VFAs values preventing pH drops (Boe and Angelidaki 2006; Falk 2012; Lahav and Morgan 2004). In spite of the presence of these buffering systems, high organic loads of easy degradable carbohydrates or introduction of toxic substances could still disturb pH stability, causing VFAs accumulation (Falk 2012). If the buffering capability of the system is low, both alkalinity and VFA measurements are useful indicators to monitor the system whereas under highly buffered conditions, only VFAs measurement is reliable for checking probable process imbalances (Esteves et al. 2013). One of the best method for measuring carbonate alkalinity is titrating a diluted sample with 0.1 N hydrochloric acid until a pH of 4.3 is reached (Effenberger 2008).

The VFA to total alkalinity ratio (VFA/TAC) indicates the quantity of volatile organic acids to the buffer capacity of carbonate (total alkaline carbonate) in a digester (Deublein and Steinhauser 2011). In anaerobic digesters working under appropriate conditions, this ratio is normally in the range of 0.1–0.35 (Esteves et al. 2013). In fact, this ratio amplifies changes related to acidification, because both parameters are usually correlated, i.e., when VFAs increases, in most cases, alkalinity decreases (Clemens 2012). Overall, in order to prevent the inhibition of methane production by VFAs accumulation during the AD process, co-digesting with other feedstocks or using a two-stage digestion system have been proven to be effective (Jain et al. 2015).

6.5 Total and Volatile Solids

Different substrates have a vast variation in terms of their moisture and solid contents and it is important to analyze these parameters for effectively determine a stable organic loading rate and to consequently achieve a stable and continuous gas production in the AD process (Falk 2012). Total solid (TS) and VS concentrations of the substrates under digestion provide useful insights about the biogas yield that could be potentially produced. Furthermore, mechanical parts such as pipes, cutters, pumps, and mixers can only work effectively with a certain TS concentration. The TS of the substrate fed into wet and dry AD systems is generally kept approximately in the range of 10–15% and 25–40%, respectively (Li 2015). However, it should be noted that the standard range of TS in wet biogas plants with agricultural substrate is between 6 and 10% (Wolf 2013). The TS of the waste could also influence the activity of the microorganisms involved in the AD process (Krishania et al. 2013). Overall, the TS of organic matter substrate and TS during the process are two main factors for selecting the type of system and reactor design. For instance, considering a similar substrate, when TS is between 7 and 9%, floating dome is considered as the most suitable design while, TS values of 4–15% and 15–30% could be efficiently digested using CSTR and dry type anaerobic digesters, respectively (Krishania et al. 2013). Determination of the TS is carried out by drying the sample in a gravity convection oven at 105 °C overnight (>8 h) following measuring weight reduction.

It should be highlighted that TS represents the total organic and inorganic materials in a sample while VS represents the organic compounds content. The degree of degradation of a given substrate could be determined by comparing the VS values of the digester's influent and effluent (Wolf 2013). Determination of VS is accomplished by burning the 105 °C-dried sample in a muffle furnace at 550 °C for 2 h, followed by measuring weight reduction (Labatut and Gooch 2014). Accordingly, TS and VS could be calculated by using the following equations (Eqs. 6.1 and 6.2):

$$\mathrm{TS}(\%) = \frac{\text{Dry weight of sample}}{\text{Wet weight of sample}} \times 100 \tag{6.1}$$

$$\mathrm{VS}(\%) = \frac{\text{Dry weight of sample} - \text{Burned weight of sample}}{\text{Dry weight of sample}} \times 100 \tag{6.2}$$

6.6 Organic Loading Rate (OLR)

OLR is among the most important parameters affecting both microbial populations during AD as well as reactor performance and BP (Kundu et al. 2017). OLR is the amount of organic feed, introduced daily to a digester (Drosg 2013; Schnurer and Jarvis 2010). More specifically, OLR is the quantity of VS fed per working volume of a digester per day and is expressed as kg VS/m^3 digester/d (Esteves et al. 2013). OLR is calculated using the following equation (Eq. 6.3):

$$\mathrm{OLR}\left(\text{Kg VS}/\text{m}^3\ \text{digester}/\text{d}\right) = \frac{\text{Total daily VS of substrate (kg)}}{\text{Total active volume of digester }(\text{m}^3)} \tag{6.3}$$

It should be noted that OLR should be adjusted depending on the type of substrate (Esteves et al. 2013). The critical issue which may be probably faced when increasing this parameter is the possibility of acidification by overloading a digester with organic materials leading to decrease or stop of methane production (Drosg 2013; Moriarty 2013). Accordingly, during the start-up of an anaerobic digester, the OLR should be increased slowly in order to ensure an efficient adaptation of the microorganisms involved in the AD process. An appropriate OLR in mesophilic CSTR digesters ranges between 3 and 5 kg VS/m^3/d depending on the type of substrate (Drosg 2013), while the microflora is generally inhibited at OLR values exceeding 6.4 kg VS/m^3/d (Moriarty 2013). To achieve a successful start-up, it is advisable to start at an OLR as low as 0.5 kg VS/m^3/d; and provided that BP initiates reaching a constant values after for 4 d, the OLR can then be increased by 0.5 kg VS/m^3/d every two weeks till reaching the target range at which BP is expected to slop up ("Operational guidelines for Plönninge biogas plant," n.d.; Wellinger et al. 2013).

Table 6.2 presents optimum OLR and pH range for methane production using different type of substrates. As mentioned earlier, in principle, OLR reflects the daily amount of VS introduced into a digester under the continuous operation mode (Mao et al. 2015). Therefore, BP is expected to increase by increasing OLR. However, this should not exceed the capacity of the various groups of the microbial population handling AD. Accordingly, most biogas plants are deliberately operated at under-optimal organic material loading or in better word, under safer condition, with an aim to minimize process errors and therefore, they are bound to ignore existing potentials. It is significant to note that by doubling the OLR, the plant capacity could be theoretically be doubled as well without needing to build any

Table 6.2 Optimum OLR and pH range for methane production using different type of substrates*

Substrate	Reactor type	pH range	OLR	References
Sugar beet cossettes Pig manure	Semi-continuous stirred tank reactor	7.4–7.8	11.2 g VS/ $L_{reactor}$ d	Aboudi et al. (2015)
High COD wastewater	AnMBR	>7.4	11.81 kg COD kg/ VSS/d	Yu et al. (2016)
Dairy waste	Two stage induced bed reactor	6.8–7.5	32.9 g-COD/ L/d	Zhong et al. (2015)
Food waste	Thermophilic and mesophilic digester with recirculation	7.6–8.1	18.5 g VS/d	Zamanzadeh et al. (2016)
Vegetable waste	Completely stirred tank reactor (acidogenesis) fixed-bed biofilm (methanogenesis)	5.1 ± 0.1 (acidogenic reactor) 7.6 ± 0.1 (methanogenic reactor)	3.0 g VS/L/d	Zuo et al. (2015)

*With permission from Khan et al. (2016). Copyright© 2017

additional digesters (Falk 2012). Above-optimal OLRs or in another words, introduction of a large volume of organic materials into a digester on a daily basis may result in unfavourable changes in the digester's environment which could eventually inhibit the BP. In fact, extremely high OLRs encourage high hydrolysis and acidogenesis activities which in turn lead to high VFA production (acetic acid, propionic acid, butyric acid, and valeric acid (Zhang et al. 2014). Under the circumstances when methanogenesis activity could not cope with this high VFA concentration (i.e., VFA uptake be methanogens for methane generation), VFA accumulation, decreased pH, and consequently irreversible acidification of the digester would be likely (Palacio-Barco et al. 2010; Zhang et al. 2013a). Thereafter, the maximum endurable OLR needs to be predicted accurately for different cases (Mao et al. 2015). It should also be noted that the application of thermophilic conditions and effluent recirculation has been reportedly suggested to address the overloading inhibition caused by high OLRs (Rincón et al. 2008).

Low OLRs on the other hand could result in starvation of the microbial populations, low overall performance of the digester, and consequently low BP rate. Moreover, from the engineering point of view, targeting under-optimal OLRs when designing digesters will be accompanied by larger reactor investment and operational costs (Jiang et al. 2012).

OLR could also strongly affect the dynamics of bacterial communities. Rincón et al. (2008) revealed that at low OLRs, *Firmicutes* were the predominant bacteria while at high OLRs, *Gammaproteobacteria*, *Actinobacteria*, *Bacteroidetes* and *Deferribacteres* were the major bacteria present in the digester. Kundu et al. (2013) also observed significant changes in dynamics of archaeal community (methanogens) tin response to variations in organic loads owing to the different level of tolerance of different archaea to organic loads shock.

Temporal condition of the reactor could also affect its resistance to organic loads shocks. In a study on the effect of higher OLRs under mesophilic and thermophilic conditions during AD of food wastes in stirred tank reactors, Guo et al. (2014) concluded that the reactor under mesophilic condition showed a comparatively better performance under the organic loads shock owing to the higher richness and evenness of the microbial community.

As mentioned earlier, higher OLRs is preferred from the engineering point of view since less reactor volume would be required to treat a certain volume of waste stream. The above-mentioned limitations faced by employing high OLRs could be addressed through using two-stage anaerobic processes as possible inhibition of methanogenesis by acidification could be efficiently avoided (Khan et al. 2016).

6.7 Hydraulic and Solid Retention Times

Retention time is in fact the time required to complete the digestion of organic materials introduced into a digester and is directly related to substrate composition and microbial growth as well as a number of process parameters such as temperature and OLR (Ekama and Wentzel 2008). The main factors which should be considered when adjusting retention time, are the composition of the substrate and the digestion temperature. For example, in the case of cellulose-rich plant matters, the microbial population needs more time for degradation; and so retention time should be prolonged to ensure of an efficient hydrolysis (Schnurer and Jarvis 2010).

Hydraulic retention time (HRT) and solid retention time (SRT) are two important factors in the design and process control of AD systems. HRT refers to the average duration the feed material stays inside the digester and is calculated as a ratio of the active volume of the digester and the daily volume of the feed material (Eq. 6.4) (Drosg 2013).

$$\mathrm{HRT(d)} = \frac{\text{Total active volume of digester }(\mathrm{m}^3)}{\text{Total daily feed materials }(\mathrm{m}^3)} \tag{6.4}$$

Low HRT values could lead to the wash-out of the active biomass including methanogens whereas high HRT values could result into low biogas productivity of the digester. Thus, it is important to make sure that HRT is adjusted at an appropriate values for a given substrate fed into the digester (Drosg 2013; Wolf 2013).

SRT is defined as the length of the residence time solids stay in a digester. SRT is a critical operational parameter because it not only affects the process efficiency, but also controls the biological characteristics, functionality, as well as stability of the digester (Eq. 6.5) (Schnurer and Jarvis 2010).

$$\mathrm{SRT(d)} = \frac{\text{Dry matter in the whole digester (kg)}}{\text{Dry matter in the digester feed (kg/d)}} \tag{6.5}$$

In many cases, HRT and SRT are equal, but in a digestion tank in which part of the residues are returned to the process, the SRT value exceed that of the HRT (Schnurer and Jarvis 2010). If SRT is prolonged beyond optimal values, it could result in low BP while shorter SRT values could lead to insufficient degradation of VS and consequently decreased BP (Argyropoulos 2013). Accordingly, some digester designs have been developed to increase SRT during the AD process (i.e., SRTs much higher than HRTs) to prevent biomass loss and to increase BP efficiency. These kinds of reactors include the anaerobic fluidized bed reactors (AFBR) in which microorganisms are attached to carrier materials, (Kumar et al. 2008), as well as up-flow anaerobic sludge blanket (UASB), and expanded granular sludge bed (EGSB) reactors in which microorganisms are accumulated and aggregated in clumps. In such digesters, despite the high inflow of substrate, the microorganisms can be retained in the digestion tank effectively (Schnurer and Jarvis 2010). High-rate systems are generally run at HRTs less than 5 d and are used for wastewater treatment. However, in the CSTR systems, the biomass is suspended in the liquid phase and will be removed together with the slurry; therefore, the SRT is advisable to be equal to the HRT (usually as long as 10–20 d) in these systems.

This would avoid biomass wash-out and is critical to run the digester efficiently (Boe and Angelidaki 2006).

HRT is in fact not only associated with the economic aspects of biogas plants but could also exert determining impacts of the microbial community involved in the AD process and consequently biogas yield. More specifically and from the economical point of view, shorter HRTs are associated with lower volume reactors, i.e., less capital and maintenance costs (Stuckey 2012). From the microbial point of view, short HRTs are in favor of microbial groups of high growth (doubling) rate and low substrate affinity (Kundu et al. 2017) and could increase the risk of biomass wash-out which could in turn lead to detoriarated AD and biogas yields (Kwietniewska and Tys 2014). It has been reported that the adverse impacts of short HRTs on microbial population are more intensified in stirred tank reactors (STR) in comparison with other reactor configurations such as AnMBRs and upflow anaerobic sludge blanket (UASB) (Kundu et al. 2017). This could be explained by the fact that there are literally no specific mechanisms in STRs to maintain biomass in response to extremely short HRTs, unlike the other reactor configurations where membrane and granules serve as such, respectively.

In general, HRT value could range from a few hours such as 2 h as (Kim et al. 2010) to 30 d as reported by Jeong et al. (2010). Nevertheless, the optimal value of HRT should be determined case by case by taking into account different parameters including as feed characteristics, system hydraulics, sludge properties, reactor design/configuration, etc. (Chen et al. 2016). For instance, in a study using an integrated anaerobic fluidized-bed MBR, (Gao et al. 2014b) decreased HRT from 8 to 6 h and found that methane productivity increased in response to the change applied. They attributed this finding to the consequent increase in OLR. On the contrary when they further decreased HRT, metahne productivity decreased probably due to VFAs accumulation (Gao et al. 2014b). In a different study, Linke et al. (2013) compared the operation of 24 full-scale biogas plants in Germany under different temperature regimes. They stated that at less than 20 °C, longer HRTs were required to reach a certain level of degradation compared with operation above 35 °C. Therefore, it could be concluded that HRT should be determined based the operating parameters such as temperature and OLR in a particular digester design/configuration. As an example, Table 6.3 tabulates the effects of different operational parameters on BP in AnMBRs including possible suggestions for optimizing BP (Chen et al. 2016).

6.8 Toxic Compounds

The presence of toxic compounds could severely jeopardize AD process leading to substantial accumulation of volatile fatty acids (VFAs) and consequently reductions in BP or sometimes complete inhibition of the process. These toxic substances could be introduced into the process with feedstock such as oil, grease, phenols, paracetamol, caffeine, ibuprofen, triclosan, heavy metals, and volatile aromatics

Table 6.3 The effects of operational factors on BP in AnMBRs and possible suggestions for optimizing BP*

Factors	The effects on BP process	Possible suggestions for optimized BP from AnMBRs
Temperature	*Thermophilic*: – Faster reaction rates/higher-load bearing capacity/higher biogas productivity – Possible acidification/inhibition of BP – Decreased stability and increased toxicity/poor methanogensis/higher net energy input and larger investments – Difficulty in anaerobic biomass immobilization/poor sludge settling characteristics/reduced methanogenic activities/poor effluent quality – Less cooling required/improved process economics – Sludge decay with non-adapted mesophilic sludge/serious membrane fouling – Reduced sludge viscosity/a higher flux/process efficiency/lower shear rates/lower energy requirement – A lower permeate viscosity/ increased membrane permeability by decreasing TMP – More compact cake layer/higher cake layer resistance/server fouling issues/very low long-term flux/ process inefficiency *Mesophilic*: – Better process stability, higher biomass richness, better permeate quality but possible low methane yields and poor biodegradability and nutrient imbalance *Psychrophilic*: – Enhanced methane solubility/loss of methane in effluent/lower methane recovery – TSS and soluble COD accumulation and a higher viscosity/increased filtration resistance/increased fouling and operational cost – Enhanced membrane removal and compensation for the decreased SMA and bulk sludge removal	– Two phase AnMBRs with thermophilic hydrolysis/ acidogenesis and mesophilic methanogenesis – Avoidance of drastic temperature changes

(continued)

Table 6.3 (continued)

Factors	The effects on BP process	Possible suggestions for optimized BP from AnMBRs
	– Energy requirement for operating the system is lower – Reduced reaction and hydrolysis rates/reduced methanogenic activity *Temperature changes:* – Temperature decrease/decreases in the VFA production rate, the ammonia concentration, the substrate utilization rate and the metabolic rate of the microorganisms/increased start-up times/decreasing CH_4 and H_2 yields – Temperature increase/increase in pH, hydrolysis of organic particulates/ increase in methane potential – Temperature increase/free ammonia concentration/methanogenic inhibition – Temperature fluctuation/stress on biomass/increase membrane fouling and operational cost	
pH	– Extremely low pH value/ acidification and VFA accumulation/ reduced methane yield – Extremely high pH value/increased ammonia toxicity and VFA inhibition/reduced methane yield – pH shocks/dispersion of sludge flocs/the accumulation of colloids, solutes or biopolymers in the bulk sludge suspension/deteriorated membrane performance and BP potential	– Two phase AnMBR with optimized conditions for both acidogenic and methanogenic reactor to bring biogas yield optimization – Minimize pH shock loading by neutralizing the feed with chemicals such as sodium biocarbonate
HRT	– Optimum HRT exists, which ensures the maximum methane productivity – HRT lower than the optimal value/ VFA accumulation/reduced methane yield/server fouling – HRT above the optimal value/ insufficient utilization of biogas digester component/reduced methane production	– Avoid operation at too high or too low a HRT – Operate AnMBRs for maximum BP at optimal HRT
SRT	– Long SRT/enhance dominancy of methanogenesis/enhanced methane yield – Long SRT/reduced dissolved methane/higher methane recovery – Long SRT/reduced sludge disposal and cost	– Long SRT is generally recommended for AnMBRs operation – Additional care is required for fouling mitigation at long SRT

(continued)

Table 6.3 (continued)

Factors	The effects on BP process	Possible suggestions for optimized BP from AnMBRs
	– Long SRT/reduced sludge particle size and release of SMP/membrane fouling – Long SRT/cake formation and consolidation/increased fouling cost – Long SRT/accumulation of inorganic solids/inorganic fouling	
OLR	– Increased OLR/higher metabolic activity of methanogens/increase biogas yield and methane content in the biogas to certain extent – High OLR/VFA accumulation/ irreversible acidification/risk of a deteriorated biogas yield – High OLR or organic shock loading/ release of tight EPS/SMP and accumulation of fine particles/serious membrane fouling	– Operating AnMBR at sustainable OLR to maximize the methane yield – Thermophilic systems and effluent recirculation can help relieve systems from the overloading issues

*With permission from Chen et al. (2016). Copyright© 2017

(Haak et al. 2016; Yang et al. 2016) or could be generated during AD process such as ammonia, VFAs, and hydrogen sulfide (Kwietniewska and Tys 2014). The level of resistance of the anaerobic microbial communities to toxic compounds varies depending on different environmental conditions inside a digester (Al Seadi et al. 2008). For instance, ammonia, a product of protein and N-rich compounds biodegradation, could be toxic at certain concentrations, it also plays an important role in microbial growth under optimal concentrations though (Yenigün and Demirel 2013; Whelan et al. 2010). Inhibiting ammonia concentrations for mesophilic AD range from 2.8 to 8 g/kg and from 2.5 to 4 g/kg for thermophilic processes (Poggi-Varaldo et al. 1997; Angelidaki and Ahring 1993; Li 2015).

The toxicity level of some inhibitors varies depending on the pH value of the digestion environment as well. For instance, both ammonia and hydrogen sulfide are toxic only when they are in their non-ionized forms (Lay et al. 1997), and the key factor governing their conversion from non-ionized to ionized forms is pH. Accordingly, ammonia is considered toxic at pH values exceeding pH 7, while hydrogen sulfide shows its inhibitory effects at pH values below 7 (Li 2015). Feedstocks such as slaughterhouse wastewater as well as pig or poultry manures are rich in nitrogen and could lead to a high level of ammonia production when they undergo AD. Between the two forms of ammonia, free ammonia, has been shown to possess higher inhibitory effect on methanogenesis (Westerholm et al. 2011b).

The susceptibility of different methanogenic archaea to increasing ammonia concentrations is also different. For instance, Niu et al. (2013) observed that at a total ammonia nitrogen (TAN) as high as 8000 mg/L, hydrogenotrophic

Fig. 6.1 Total ammonia percentages present in solution at different temperatures (20, 35 and 55 °C) and varying pH values. With permission from Zhang et al. (2014) Copyright© 2017

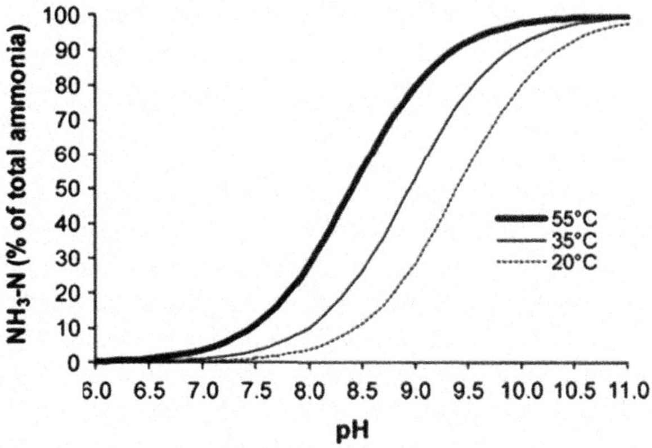

Methanothermobacter thermautotrophicus was predominant in the digester while *Methanosarcina* was not detected. Accordingly, they argued that acetoclastic methanogens are more sensitive to ammonia inhibition. Moreover, temperature could also impact the severity of ammonia inhibition (Rajagopal et al. 2013). More specifically, ammonia inhibition could be strengthened under thermophilic conditions compared with mesophilic conditions (Kayhanian 1999). This is ascribed to the fact that an increase in temperature not only generally increases the metabolic rate of the microorganisms involved in the AD process but also results in increased concentrations of free ammonia nitrogen (FAN) (Fig. 6.1). The following equation shows the relationship between FAN concentration and both pH and temperature (Eq. 6.6) (Rajagopal et al. 2013).

$$\text{Ammonia}\left(\frac{\text{mg}}{\text{L}}\right) = \text{Ammonium}\left(\frac{\text{mg}}{\text{L}}\right) \Big/ \left(1 + 10^{-pH} \Big/ 10^{-\left(0.09018 + \frac{2729.92}{\text{T(Kelvin)}}\right)} - 1\right) \tag{6.6}$$

Various strategies have been laid forth in order to minimize ammonia inhibition such as dilution of substrate, air-stripping, application of materials with ion exchange capacity or carbon fibre, etc. (Westerholm et al. 2016). As mentioned earlier, acetoclastic methanogens are more sensitive to ammonia inhibition compared with hydrogenotrophic methanogens. High ammonia concentrations could gradually lead to a transition from acetoclastic methanogens to hydrogenotrophic methanogens through the development of syntrophic acetate oxidation (SAO) bacteria. In fact, SAO bacteria are relatively highly ammonia-tolerant and compete for the acetate present under a wide range of operating conditions oxidizing it into hydrogen, carbon dioxide, and formate. Subsequently, their products could be consumed by hydrogenotrophic methanogens for methane generation (Westerholm et al. 2016). Examples of SAO bacteria are the thermophilic *Thermacetogenium phaeum* and *Pseudothermotoga lettingae*, the thermotolerant *Tepidanaerobacter acetatoxydans* and the mesophilic *Clostridium ultunense* and *Syntrophaceticus schinkii* (Westerholm et al. 2016). In light of that, bioaugmentation with these syntrophic co-cultures has also been proposed as a strategy to improve adaptability to high ammonia concentration under both mesophilic and thermophilic conditions. Table 6.4 tabulates a number of studies in which the digesters under investigation were dominated by SAO populations. Analyses of the recently published genomes of *C. ultunense*, *T. phaeum*, *S. schinkii* and *T. acetatoxydans* could have provided insights into the functional genes that can be related to the SAO process, and could assist in better understanding of these populations and their functions within the AD community (Westerholm et al. 2016).

Research findings have also revealed that through acclimatization, microbial tolerance to inhibitory compounds could be increased. In line with that, Fotidis et al. (2013) investigated the impacts of increasing concentrations of acetate and ammonia on methanogenesis under acclimatized and non-acclimatized conditions. Their results suggested an increased tolerance to increasing concentrations of the inhibitory compounds investigated when acclimatization was carried out first.

Long chain fatty acids (LCFAs) at concentrations as low as 50 mg/L could also play an inhibitory role to both acetogens and methanogens during the AD process (Dasa et al. 2016). Inhibition of some of these microorganisms could directly adversely affect the degradation of LCFAs themselves too. In better words, acetoclastic methanogens are in charge of the degradation of these compounds during the AD process via the acetate yielding mechanism of β-oxidation and therefore, their inhibition would lead to further accumulation of LCFAs and consequently more intensified inhibition (Koster and Cramer 1987). This was also showed earlier by Hanaki et al. (1981) who argued that during fats digestion under anaerobic conditions, the first step, i.e., hydrolysis of fats into glycerol and fatty acids proceeds at a high rate while the second step, i.e., the degradation of LCFAs is generally hindered owing to the inhibitory nature of these compounds to a large proportion of the bacteria involved in the AD process (Koster and Cramer 1987). Liu et al. (2011) claimed that addition of calcium could result in decreased LCFA-driven inhibition during AD through the formation of calcium stearate.

6.9 C/N Ratio

The availability of nutrients is of critical importance for microbial growth inside digesters and C/N ratio is indicative of the combination of organic matters. In fact, an optimized C/N ratio could assist with preventing serious problems during the AD process such as ammonia inhibition while ensuring a sufficient availability of nitrogen required for microbial biomass production. In better words, to achieve the highest degradation rate of organic materials and consequently highest BP potential, C/N ratio should be constantly managed and monitored (Mao et al. 2015). It should also be noted that both above- and under-optimal C/N ratios are associated with adverse effects on methane production rate, and this highlights the significance of substrate selection. On the other hand, AD is a complex process in which different groups of microorganisms are involved further emphasizing the importance of substrate composition (Klimiuk et al. 2015). Regueiro et al. (2014) showed that

Table 6.4 Operating conditions and molecular investigations of anaerobic digesters (laboratory- or industrial-scale) and batch/enrichment cultures dominated by syntrophic acetate oxidation (SAO)*

Biological system	Ammonia g NH_3-N/L (g NH_4^+-N/L)	Operating parameters/ experimental set-up	Microbial community investigation	References
Mesophilic				
LS-CF*	n/a*	37 °C, pH 7 Acet: 0.01 g/L Dilution rate: 0.025/day	Quantitative RT-PCR of mcrA transcripts	Shigematsu et al. (2004)
IS-CF*	4–5.6 g N/L	37–38 °C, pH n/a VFA: 1.8–2.7 g/L HRT: 20–25 days	FISH analyses of methanogens	Karakashev et al. (2006)
LS-CF	0.6–1.0 (5.5–6.9)	37 °C, pH 7.9–8.0 VFA: 18–30 g/L HRT: 30 days OLR: 3 g VS/(L day)	qPCR analyses of methanogens and characterised SAOB, T-RFLP and clone library analyses of acetogenic communities (fhs gene), illumina amplicon sequencing of bacterial 16S rRNA genes	Westerholm et al. (2011a) and Müller et al. (2016)
Batch	n/a	37 °C, pH 7.2–7.4 HRT: 30 days OLR 1.5 g COD/(L day)	MAR-FISH with ^{14}C-acetate, RNA-SIP with $^{13}C6$-glucose and $^{13}C3$-propionate to identify and quantify acetate-utilising communities	Ito et al. (2011)
LS-CF	0.07–0.5 (1.5–11)	37 °C, pH 6.5–7.8 Acet: <0.1–10, prop: <0.1–10 HRT: 26–57 days OLR: 0.8–3.6 g VS/(L day)	qPCR analyses of methanogens and characterised SAOB	Westerholm et al. (2012)
LS-CF with/without TE*	0.3–0.5 (3.6)	37 °C, pH 7.9–8.1 Acet: 0.6–3.5 g/L, prop: 0.1–2.2 g/L HRT: 30 days	qPCR analyses of methanogens and characterised SAOB	Karlsson et al. (2012)
Batch	n/a	38 °C, pH 8.1 Acetate: 0.7 g/L		Polag et al. (2013)
IS-CF	0.2–0.5 (3.3–4.9)	36–40 °C, pH 7.6–8.0 VFA: 3–13 g/L	qPCR analyses of methanogens and characterised SAOB	Sun et al. (2014)
IS-CF	0.3–0.4 (2.9–4.6)	37–38 °C, pH 7.9 VFA: 0.6–0.8 g/L	FISH analyses of methanogens	Fotidis et al. (2014)

(continued)

Table 6.4 (continued)

Biological system	Ammonia g NH_3-N/L (g NH_4^+-N/L)	Operating parameters/ experimental set-up	Microbial community investigation	References
LS-CF	0.2–0.3 (4.2–5.2)	35 °C, pH 7.5 VFA: 1 g/L OLR 2.2 g VS/(L day) 16–25% SAO contribution	Shotgun sequencing, DNA-SIP with ^{13}C-acetate and FISH-NanoSIMS analyses of bacterial communities	Werner et al. (2014)
LS-CF with/ without TE	0.4–1.5 (5.4–5.8)	37–42 °C, pH 7.9–8.1 Acet: <0.1–3.4 g/L, prop: <0.1–6.3 g/L HRT: 30 days sOLR: 2.3 g VS/(L day)	qPCR analyses of methanogens and characterised SAOB, T-RFLP and/or clone library analyses of bacterial (16S rRNA), acetogenic (*fhs* gene) and methanogenic (*mcrA* gene) communities	Westerholm et al. (2015)
IS-CF	0.4–1.5 (2.4–4.2)	37–40 °C, pH 7.7–8.2 Acet: 0.3–0.5 g/L, prop: 0.005–0.02 g/L HRT: 21–32 days	Illumina amplicon sequencing of bacterial and archaeal 16S rRNA genes	Luo et al. (2016)
Thermophilic				
Acetate enrichment	n/a	60 °C, pH 6.5–6.8 Acet: 3 g/L		Zinder and Koch (1984)
Acetate chemostat		60 °C, pH 6.5–6.8 Acet: 0.6 g/L	Microscopic examinations	Petersen and Ahring (1991)
IS-CF	2.2–2.6 g N/L	52–55 °C, pH n/a VFA: 0.2–0.8 g/L HRT: 15–25 days	FISH analyses of methanogens	Karakashev et al. (2006)
Batch	n/a	55 °C, pH–7		Qu et al. (2009)
Batch	n/a	55 °C, pH 6.8–7.8 Acet: 6 g/L BM		Hao et al. (2010)
LS-CF	n/a	55 °C, pH 7.2 Acet: 0.1 g/L, prop: 0.07 g/L HRT: 4 days OLR 6.25 gCODcr/(L day)	Clone libraries and sequencing of bacterial and archaeal 16S rRNA genes	Sasaki et al. (2011)
Batch	n/a	55 °C, initial pH 5.5 Acet: 6 g/L BM	qPCR analyses of methanogenic (16S rRNA) and acetogenic (*acsB* and *fhs* genes) communities	Hao et al. (2012)

(continued)

Table 6.4 (continued)

Biological system	Ammonia g NH_3-N/L (g NH_4^+-N/L)	Operating parameters/ experimental set-up	Microbial community investigation	References
Batch	n/a	55 °C, pH > 7.5 Acet: 9–12 g/L BM	ARISA of archaeal and bacterial communities; qPCR analyses of acetogens (*acsB* and *fhs* genes)	Hao et al. (2013)
LS-CF	0.7–1.0	55 °C, pH 6.7–7.0 Acet: 0.07–0.30 g/L COD, prop: 0.01–0.14 g/L COD HRT: 2–4 days	FISH analyses and 16S rRNA gene pyrosequencing of bacterial and methanogenic communities	Ho et al. (2013)
Batch	0.06 (1)	53 °C, pH 7.3 Acet: 1 g/L BM	FISH analyses of archaeal community	Fotidis et al. (2013)
Batch	0.09–2.7 (6–7)	55 °C, pH 6.6–8.2 Acet: 15 g/L	qPCR analyses of methanogens	Lü et al. (2013)
IS-CF	0.2–0.8 (2.0–3.2)	48–55 °C, pH 7.7–8.1 VFA: 1.9–3.8 g/L HRT: 20–101 days OLR: 2.5–3.5 g VS/(L day)	qPCR analyses of methanogens and characterised SAOB	Sun et al. (2014)
IS-CF	0.5 (2.0–2.4)	52–55 °C, pH 7.9–8.0 VFA: 0.9–1.8 g/L	FISH analyses of methanogens	Fotidis et al. (2014)
LS-CF	0.7–1.0 g NH_4^+/L	55–65 °C, pH 6.7–7.1 Acet: 0.06–2.1 g COD/L, prop: 0.04–0.6 g COD/L HRT: 2–4 days	454 pyrosequencing of bacterial and archaeal 16S rRNA genes	Ho et al. (2014)
Batch	TAN 1.8 g/L	52 °C, pH 7.7 Acetate: 0.2–6 g/L	Proteome analyses	Mulat et al. (2014)
IS-CF	2.2–3.4 (0.7–1.5)	50–53 °C, pH 8.0–8.4 Acet: 0.05–1.6 g/L, prop: 0.01–0.2 g/L HRT: 3–15 days	Illumina amplicon sequencing of bacterial and archaeal 16S rRNA genes	Luo et al. (2016)

*With permission from Westerholm et al. (2016). Copyright© 2017

LS-CF laboratory-scale (semi) continuously fed digesters; *IS-CF* industrial-scale continuously fed digesters; *TE* addition of trace element mixture including iron; *HRT* hydraulic retention time; *n/a* not available substrate composition could exert a solid effect on both the activity and structure of the microbial communities during AD. They indicated that under stable performance of reactors, achieved by providing an optimal C/N ratio, *Methanosaeta* is the most dominant methanogenic archaea under mesophilic condition (Regueiro et al. 2014).

In general, at an optimal C/N ration, by increasing the proportion of biodegradable organic fraction or volatile solids of a substrate, BP will be boosted (Schlegel et al. 2008; Divya et al. 2015). For instance, animal manure in particular of cow origin because of its low C/N ratio could serve as an ideal candidate for anaerobic co-digestion with carbon rich substrates such as crops residues. Carbon-rich substrates such as crop silage are also needed and widely used when anaerobically digesting other feedstocks such as food processing by-products (Linke et al. 2013). In general, the optimal C/N ratio for AD ranges between 20 and 30, while 25 as most widely practiced value (Yen and Brune 2007; Zhang et al. 2013b).

Apart from C and N, other macroelements such as phosphorus and sulphur are also equally vital to ensure maximal growth and activity of the microorganisms involved in AD. The optimal ratio of these macronutrients, i.e., carbon, nitrogen, phosphor, and sulphur (C:N:P:S) should be reportedly 600:15:5:1 (Oleszek et al. 2014). Trace or micro elements such as iron, nickel, cobalt, selenium, etc. are also as important for the microbial growth and BP as the macroelements and should therefore be taken into serious consideration.

6.10 Mixing and Shear Stress

Mixing and the consequent hydrodynamic shear is an important parameter affecting mass transfer in anaerobic digesters. Moreover, it also affect the formation, structure, and metabolism of the microbial populations involved in the AD process (Jiang et al. 2016). In fact, mixing is essential for homogenizing the influent and guarantees an efficient contact between the substrates and the microbial biomass. Regardless of the type of mixing applied, it has been shown that both the mixing mode and mixing intensity are directly influential on digester's performance and biogas yield (Lindmark et al. 2014). Mixing is also important in terms of removal of metabolic end-products and uniform distribution of heat (temperature) throughout the digester (Deublein and Steinhauser 2008).

In spite of the importance of mixing, it should be noted that above-optimal mixing intensities, i.e., high shear stress, could pose serious threats to the survival of the microorganisms (Deublein and Steinhauser 2008). Among different methanogens, it has been reported that Methanosaetaceae, the main bridging agent in the aggregates, are the most sensitive group to increased shear (Kundu et al. 2014b). On the other hand, it also should be noted that beside inefficient contact between microbes and feed, the absence of proper mixing is accompanied by other adverse phenomena as well, such as scum formation and uneven distribution of heat throughout the digester (Divya et al. 2015). Therefore, the impacts of various mixing regimes and the resultant shear stress on microbial population and BP need to be thoroughly investigated to ensure efficient contact between the microbes and the feed while avoiding any harmful effects on the microbial biomass (Lindmark et al. 2014). For instance, Lebranchu et al. (2017) employed combined experimental and computational fluid dynamics simulations in order to investigate the impacts of shear stress and impeller design (i.e., double helical ribbon and classical Rushton turbine) on BP. Their findings indicated that the double helical ribbon led to 50% higher methane production rate in comparison with classical Rushton turbine owing to a significantly faster dispersion of substrate.

In a study, Kundu et al. (2014b) gradually incremented shear force in a hybrid anaerobic digester by increasing the up-flow velocity from 4 to 10 m/h. They observed that up-flow velocities as high as up to 6 m/h resulted in improved reactor performance and biogas yield, but velocities >6 m/h sharply reduced the overall performance of the digester. These observations could be attributed to the granule breakage and the wash-out of the active biomass (i.e., bacteria and methanogens) (Hoffmann et al. 2008). Similar results were obtained in a different study by Jiang et al. (2016) who also claimed that methane content decreased continuously in response to increasing shear rate.

6.11 Conclusions

As far as commercial BP is concerned, maximum energy recovery with limited input in terms of energy, time, labour, etc. is the major target of a successful AD process. In line with that, preserving the process against any fluctuations in operating parameters is of crucial importance in order to decrease the risks of process failure. Variations in these parameters (i.e., temperature, pH, OLR, HRT, toxic compounds, etc.) can harm the microbial communities involved in the BP process which consequently lead to poor overall performance of the system. Hence, in-depth understanding of prominent parameters in BP and characterization of any detrimental variations in these parameters are undoubtedly necessary. The present chapter has concisely elaborated on these parameters including their roles in the process while also presenting some findings reported in the existing literature.

References

Aboudi K, Álvarez-Gallego CJ, Romero-García LI (2015) Semi-continuous anaerobic co-digestion of sugar beet byproduct and pig manure: effect of the organic loading rate (OLR) on process performance. Biores Technol 194:283–290

Aceves-Lara CA, Latrille E, Conte T, Steyer JP (2012) Online estimation of VFA, alkalinity and bicarbonate concentrations by electrical conductivity measurement during anaerobic fermentation. Water Sci Technol 65:1281–1289

Al Seadi T, Rutz D, Prassl H, Köttner M, Finsterwalder T, Volk S, Janssen R (2008) Biogas handbook. University of Southern Denmark Esbjerg. ISBN 978-87-992962-0-0
Angelidaki I, Ahring BK (1993) Thermophilic anaerobic digestion of livestock waste: the effect of ammonia. Appl Microbiol biotechnol 38:560–564
Argyropoulos A (2013) Soft sensor development and process control of anaerobic digestion
Arikan OA, Mulbry W, Lansing S (2015) Effect of temperature on methane production from field-scale anaerobic digesters treating dairy manure. Waste Manag 43:108–113
Boe K, Angelidaki I (2006) Online monitoring and control of the biogas process. Technical University of Denmark Danmarks Tekniske Universitet, Department of Systems Biology Institut for Systembiologi
Bowen EJ, Dolfing J, Davenport RJ, Read FL, Curtis TP (2014) Low-temperature limitation of bioreactor sludge in anaerobic treatment of domestic wastewater. Water Sci Technol 69:1004–1013
Chaikasem S, Jacob P, Visvanathan C (2015) Performance improvement in a two-stage thermophilic anaerobic membrane bioreactor using PVA-gel as biocarrier. Desalin Water Treat 53:2839–2849
Chandra R, Takeuchi H, Hasegawa T, Kumar R (2012) Improving biodegradability and biogas production of wheat straw substrates using sodium hydroxide and hydrothermal pretreatments. Energy 43:273–282
Chen C, Guo W, Ngo HH, Lee DJ, Tung KL, Jin P, Wang J, Wu Y (2016) Challenges in biogas production from anaerobic membrane bioreactors. Renew Energy 98:120–134
Chen Y, Cheng JJ, Creamer KS (2008) Inhibition of anaerobic digestion process: a review. Biores Technol 99:4044–4064
Clemens J (2012) How to optimize the biogas process according to process control monitoring data in biogas plants. Saksa
Dareioti MA, Dokianakis SN, Zafiri C, Kornaros M (2017) Enhancement of biogas production by using a two-stage process for the anaerobic digestion of cheese whey. Linnaeus Eco-Tech 130–140
Dasa KT, Westman SY, Millati R, Cahyanto MN, Taherzadeh MJ, Niklasson C (2016) Inhibitory effect of long-chain fatty acids on biogas production and the protective effect of membrane bioreactor. BioMed Res Intern 2016
Deublein D, Steinhauser A (2008) Biogas from waste and renewable resource. Wiley-VCH Verlag GmbH & Co. KgaA, Weinhem
Deublein D, Steinhauser A (2011) Biogas from waste and renewable resources: an introduction. Wiley
Divya D, Gopinath L, Christy PM (2015) A review on current aspects and diverse prospects for enhancing biogas production in sustainable means. Renew Sustain Energy Rev 42:690–699
Drosg B (2013) Process monitoring in biogas plants. IEA Bioenergy Task
Effenberger M (2008) Influence of temperature and feeding mode on digestion and sanitation efficiency during multiple-stage anaerobic treatment of liquid dairy cattle manure. Cuvillier Verlag
Ekama G, Wentzel M (2008) Organic material removal. In: Henze M, van Loosdrecht MCM, Ekama GA, Brdjanovic D (eds) Biological wastewater treatment: principles, modelling and design. Published by IWA Publishing, London, UK, pp 53–86
Ennouri H, Miladi B, Diaz SZ, Güelfo LAF, Solera R, Hamdi M, Bouallagui H (2016) Effect of thermal pretreatment on the biogas production and microbial communities balance during anaerobic digestion of urban and industrial waste activated sludge. Biores Technol 214:184–191
Esteves S, Miltner M, Puchas K (2013) Monitoring review and guide: for the optimization of anaerobic digestion and biomethane plants
Falk HM (2012) Monitoring the anaerobic digestion process. Jacobs University Bremen
Fang HH, Liu H (2002) Effect of pH on hydrogen production from glucose by a mixed culture. Biores Technol 82:87–93

Fotidis I, Karakashev D, Angelidaki I (2014) The dominant acetate degradation pathway/methanogenic composition in full-scale anaerobic digesters operating under different ammonia levels. Int J Environ Sci Technol 11:2087–2094

Fotidis IA, Karakashev D, Kotsopoulos TA, Martzopoulos GG, Angelidaki I (2013) Effect of ammonium and acetate on methanogenic pathway and methanogenic community composition. FEMS Microbiol Ecol 83:38–48

Gao DW, Hu Q, Yao C, Ren NQ (2014a) Treatment of domestic wastewater by an integrated anaerobic fluidized-bed membrane bioreactor under moderate to low temperature conditions. Biores Technol 159:193–198

Gao DW, Hu Q, Yao C, Ren NQ, Wu W-M (2014b) Integrated anaerobic fluidized-bed membrane bioreactor for domestic wastewater treatment. Chem Eng J 240:362–368

Gao WJ, Lin H, Leung K, Liao B (2010) Influence of elevated pH shocks on the performance of a submerged anaerobic membrane bioreactor. Process Biochem 45:1279–1287

Guo X, Wang C, Sun F, Zhu W, Wu W (2014) A comparison of microbial characteristics between the thermophilic and mesophilic anaerobic digesters exposed to elevated food waste loadings. Biores Technol 152:420–428

Haak L, Roy R, Pagilla K (2016) Toxicity and biogas production potential of refinery waste sludge for anaerobic digestion. Chemosphere 144:1170–1176

Hanaki K, Matsuo T, Nagase M (1981) Mechanism of inhibition caused by long-chain fatty acids in anaerobic digestion process. Biotechnol Bioeng 23:1591–1610

Hao LP, Lü F, He PJ, Li L, Shao LM (2010) Predominant contribution of syntrophic acetate oxidation to thermophilic methane formation at high acetate concentrations. Environ Sci Technol 45:508–513

Hao LP, Lü F, Li L, Shao LM, He PJ (2012) Shift of pathways during initiation of thermophilic methanogenesis at different initial pH. Biores Technol 126:418–424

Hao L, Lü F, Li L, Wu Q, Shao L, He P (2013) Self-adaption of methane-producing communities to pH disturbance at different acetate concentrations by shifting pathways and population interaction. Biores Technol 140:319–327

Haugen F (2014) Optimal design, operation and control of an anaerobic digestion reactor. Ph.D. Dissertation, Høgskolen i Telemark

Hernández M, Rodríguez M (2013) Hydrogen production by anaerobic digestion of pig manure: effect of operating conditions. Renew Energy 53:187–192

Ho D, Jensen P, Batstone D (2014) Effects of temperature and hydraulic retention time on acetotrophic pathways and performance in high-rate sludge digestion. Environ Sci Technol 48:6468–6476

Ho DP, Jensen PD, Batstone DJ (2013) Methanosarcinaceae and acetate-oxidizing pathways dominate in high-rate thermophilic anaerobic digestion of waste-activated sludge. Appl Environ Microbiol 79:6491–6500

Hoffmann RA, Garcia ML, Veskivar M, Karim K, Al-Dahhan MH, Angenent LT (2008) Effect of shear on performance and microbial ecology of continuously stirred anaerobic digesters treating animal manure. Biotechnol Bioeng 100:38–48

Ito T, Yoshiguchi K, Ariesyady HD, Okabe S (2011) Identification of a novel acetate-utilizing bacterium belonging to Synergistes group 4 in anaerobic digester sludge. ISME J 5:1844–1856

Jain S, Jain S, Wolf IT, Lee J, Tong YW (2015) A comprehensive review on operating parameters and different pretreatment methodologies for anaerobic digestion of municipal solid waste. Renew Sustain Energy Rev 52:142–154

Jeong E, Kim HW, Nam JY, Shin HS (2010) Enhancement of bioenergy production and effluent quality by integrating optimized acidification with submerged anaerobic membrane bioreactor. Biores Technol 101:S7–S12

Jiang J, Wu J, Poncin S, Li HZ (2016) Effect of hydrodynamic shear on biogas production and granule characteristics in a continuous stirred tank reactor. Process Biochem 51:345–351

Jiang J, Zhang Y, Li K, Wang Q, Gong C, Li M (2013) Volatile fatty acids production from food waste: effects of pH, temperature, and organic loading rate. Biores Technol 143:525–530

Jiang Y, Heaven S, Banks CJ (2012) Strategies for stable anaerobic digestion of vegetable waste. Renew Energy 44:206–214

Karakashev D, Batstone DJ, Trably E, Angelidaki I (2006) Acetate oxidation is the dominant methanogenic pathway from acetate in the absence of Methanosaetaceae. Appl Environ Microbiol 72:5138–5141

Karlsson A, Einarsson P, Schnürer A, Sundberg C, Ejlertsson J, Svensson BH (2012) Impact of trace element addition on degradation efficiency of volatile fatty acids, oleic acid and phenyl acetate and on microbial populations in a biogas digester. J Biosci Bioeng 114:446–452

Kayhanian M (1999) Ammonia inhibition in high-solids biogasification: an overview and practical solutions. Environ Technol 20:355–365

Khan M, Ngo HH, Guo W, Liu Y, Nghiem LD, Hai FI, Deng L, Wang J, Wu Y (2016) Optimization of process parameters for production of volatile fatty acid, biohydrogen and methane from anaerobic digestion. Biores Technol 219:738–748

Kim J, Kim K, Ye H, Lee E, Shin C, McCarty PL, Bae J (2010) Anaerobic fluidized bed membrane bioreactor for wastewater treatment. Environ Sci Technol 45:576–581

Klimiuk E, Gusiatin ZM, Pokój T, Rynkowska S (2015) ADM1-based modeling of anaerobic codigestion of maize silage and cattle manure—a feedstock characterisation for model implementation (part I)/Modelowanie kofermentacji kiszonki kukurydzy i obornika bydlęcego za pomocą ADM1–charakterystyka wsadu surowcowego (część I). Archives Environ Protect 41:11–19

Koster IW, Cramer A (1987) Inhibition of methanogenesis from acetate in granular sludge by long-chain fatty acids. Appl Environ Microbiol 53:403–409

Krishania M, Kumar V, Vijay VK, Malik A (2013) Analysis of different techniques used for improvement of biomethanation process: a review. Fuel 106:1–9

Kundu K, Bergmann I, Klocke M, Sharma S, Sreekrishnan T (2014a) Impact of abrupt temperature increase on the performance of an anaerobic hybrid bioreactor and its intrinsic microbial community. Biores Technol 168:72–79

Kundu K, Bergmann I, Klocke M, Sharma S, Sreekrishnan TR (2014b) Influence of hydrodynamic shear on performance and microbial community structure of a hybrid anaerobic reactor. J Chem Technol Biotechnol 89:462–470

Kundu K, Sharma S, Sreekrishnan T (2013) Changes in microbial communities in a hybrid anaerobic reactor with organic loading rate and temperature. Biores Technol 129:538–547

Kundu K, Sharma S, Sreekrishnan T (2017) Influence of process parameters on anaerobic digestion microbiome in bioenergy production: towards an improved understanding. BioEnergy Research 10:288–303

Kumar A, Yadav AK, Sreekrishnan TR, Satya S, Kaushik CP (2008) Treatment of low strength industrial cluster wastewater by anaerobic hybrid reactor. Biores technol 99:3123–3129

Kwietniewska E, Tys J (2014) Process characteristics, inhibition factors and methane yields of anaerobic digestion process, with particular focus on microalgal biomass fermentation. Renew Sustain Energy Rev 34:491–500

Labatut RA, Gooch CA (2014) Monitoring of anaerobic digestion process to optimize performance and prevent system failure

Lahav O, Morgan B (2004) Titration methodologies for monitoring of anaerobic digestion in developing countries—a review. J Chem Technol Biotechnol 79:1331–1341

Lay JJ, Li YY, Noike T (1997) Influences of pH and moisture content on the methane production in high-solids sludge digestion. Water Res 31:1518–1524

Lebranchu A, Delaunay S, Marchal P, Blanchard F, Pacaud S, Fick M, Olmos E (2017) Impact of shear stress and impeller design on the production of biogas in anaerobic digesters. Bioresour Technol

Li YF, Nelson MC, Chen PH, Graf J, Li Y, Yu Z (2015) Comparison of the microbial communities in solid-state anaerobic digestion (SS-AD) reactors operated at mesophilic and thermophilic temperatures. Appl microbiol biotechnol 99:969–980

Lin H, Peng W, Zhang M, Chen J, Hong H, Zhang Y (2013) A review on anaerobic membrane bioreactors: applications, membrane fouling and future perspectives. Desalination 314:169–188

Lindmark J, Thorin E, Fdhila RB, Dahlquist E (2014) Effects of mixing on the result of anaerobic digestion. Renew Sustain Energy Rev 40:1030–1047

Lindner J, Zielonka S, Oechsner H, Lemmer A (2015) Effect of different pH-values on process parameters in two-phase anaerobic digestion of high-solid substrates. Environ Technol 36:198–207

Linke B, Muha I, Wittum G, Plogsties V (2013) Mesophilic anaerobic co-digestion of cow manure and biogas crops in full scale German biogas plants: a model for calculating the effect of hydraulic retention time and VS crop proportion in the mixture on methane yield from digester and from digestate storage at different temperatures. Biores Technol 130:689–695

Liu Y, Chen X, Zhu B, Yuan H, Zhou Q, Xia Y, Li X (2011) Formation and function of calcium stearate in anaerobic digestion of food waste. Chin J Environ Eng 5:2844–2848

Lü F, Hao L, Guan D, Qi Y, Shao L, He P (2013) Synergetic stress of acids and ammonium on the shift in the methanogenic pathways during thermophilic anaerobic digestion of organics. Water Res 47:2297–2306

Luo G, Fotidis IA, Angelidaki I (2016) Comparative analysis of taxonomic, functional, and metabolic patterns of microbiomes from 14 full-scale biogas reactors by metagenomic sequencing and radioisotopic analysis. Biotechnol Biofuels 9:51

Mao C, Feng Y, Wang X, Ren G (2015) Review on research achievements of biogas from anaerobic digestion. Renew Sustain Energy Rev 45:540–555

Martinez-Sosa D, Helmreich B, Netter T, Paris S, Bischof F, Horn H (2011) Anaerobic submerged membrane bioreactor (AnSMBR) for municipal wastewater treatment under mesophilic and psychrophilic temperature conditions. Biores Technol 102:10377–10385

Maspolim Y, Zhou Y, Guo C, Xiao K, Ng WJ (2015) Comparison of single-stage and two-phase anaerobic sludge digestion systems—performance and microbial community dynamics. Chemosphere 140:54–62

Moriarty K (2013) Feasibility study of anaerobic digestion of food waste in St. NREL, Bernard, Louisiana

Mulat DG, Ward AJ, Adamsen APS, Voigt NV, Nielsen JL, Feilberg A (2014) Quantifying contribution of synthrophic acetate oxidation to methane production in thermophilic anaerobic reactors by membrane inlet mass spectrometry. Environ Sci Technol 48:2505–2511

Müller B, Sun L, Westerholm M, Schnürer A (2016) Bacterial community composition and fhs profiles of low- and high-ammonia biogas digesters reveal novel syntrophic acetate-oxidising bacteria. Biotechnol Biofuels 9:48

Nakasaki K, Kwon SH, Takemoto Y (2015) An interesting correlation between methane production rates and archaea cell density during anaerobic digestion with increasing organic loading. Biomass Bioenerg 78:17–24

Niu Q, Qiao W, Qiang H, Li Y-Y (2013) Microbial community shifts and biogas conversion computation during steady, inhibited and recovered stages of thermophilic methane fermentation on chicken manure with a wide variation of ammonia. Biores Technol 146:223–233

Oleszek M, Król A, Tys J, Matyka M, Kulik M (2014) Comparison of biogas production from wild and cultivated varieties of reed canary grass. Biores Technol 156:303–306

Palacio-Barco E, Robert-Peillard F, Boudenne JL, Coulomb B (2010) On-line analysis of volatile fatty acids in anaerobic treatment processes. Anal Chim Acta 668:74–79

Petersen SP, Ahring BK (1991) Acetate oxidation in a thermophilic anaerobic sewage-sludge digestor: the importance of non-aceticlastic methanogenesis from acetate. FEMS Microbiol Lett 86:149–152

Poggi-Varaldo HM, Rodriguez-Vazquez R, Fernandez-Villagomez G, Esparza-Garcia F (1997) Inhibition of mesophilic solid-substrate anaerobic digestion by ammonia nitrogen. Appl Microbiol Biotechnol 47:284–291

Polag D, Heuwinkel H, Laukenmann S, Greule M, Keppler F (2013) Evidence of anaerobic syntrophic acetate oxidation in biogas batch reactors by analysis of 13C carbon isotopes. Isot Environ Health Stud 49:365–377

Qu X, Vavilin V, Mazéas L, Lemunier M, Duquennoi C, He PJ, Bouchez T (2009) Anaerobic biodegradation of cellulosic material: Batch experiments and modelling based on isotopic data and focusing on aceticlastic and non-aceticlastic methanogenesis. Waste Manag 29:1828–1837

Rajagopal R, Massé DI, Singh G (2013) A critical review on inhibition of anaerobic digestion process by excess ammonia. Biores Technol 143:632–641

Regueiro L, Veiga P, Figueroa M, Lema JM, Carballa M (2014) Influence of transitional states on the microbial ecology of anaerobic digesters treating solid wastes. Appl Microbiol Biotechnol 98:2015–2027

Rincón B, Borja R, González J, Portillo M, Sáiz-Jiménez C (2008) Influence of organic loading rate and hydraulic retention time on the performance, stability and microbial communities of one-stage anaerobic digestion of two-phase olive mill solid residue. Biochem Eng J 40:253–261

Sasaki D, Hori T, Haruta S, Ueno Y, Ishii M, Igarashi Y (2011) Methanogenic pathway and community structure in a thermophilic anaerobic digestion process of organic solid waste. J Biosci Bioeng 111:41–46

Schlegel M, Kanswohl N, Rossel D, Sakalauskas A (2008) Essential technical parameters for effective biogas production. Agronomy Res 6:341–348

Schnurer A, Jarvis A (2010) Microbiological handbook for biogas plants. Swedish Waste Manage 2009:1–74

Shigematsu T, Tang Y, Kobayashi T, Kawaguchi H, Morimura S, Kida K (2004) Effect of dilution rate on metabolic pathway shift between aceticlastic and nonaceticlastic methanogenesis in chemostat cultivation. Appl Environ Microbiol 70:4048–4052

Smith AL, Skerlos SJ, Raskin L (2013) Psychrophilic anaerobic membrane bioreactor treatment of domestic wastewater. Water Res 47:1655–1665

Stuckey DC (2012) Recent developments in anaerobic membrane reactors. Biores Technol 122:137–148

Sun L, Müller B, Westerholm M, Schnürer A (2014) Syntrophic acetate oxidation in industrial CSTR biogas digesters. J Biotechnol 171:39–44

Trzcinski AP, Stuckey DC (2010) Treatment of municipal solid waste leachate using a submerged anaerobic membrane bioreactor at mesophilic and psychrophilic temperatures: analysis of recalcitrants in the permeate using GC-MS. Water Res 44:671–680

Voicu G, Dincă M, Paraschiv G, Moiceanu G (2015) A review regarding the biogas production through anaerobic digestion of organic waste. Adv Eng Forum

Wang K, Chang Z, Ma Y, Lei C, Wang J, Zhu T, Liu H, Zuo Y, Li X (2009) Study on solvent extraction of propionic acid from simulated discharged water in vitamin B 12 production by anaerobic fermentation. Biores Technol 100:2878–2882

Weiland P (2010) Biogas production: current state and perspectives. Appl Microbiol Biotechnol 85:849–860

Wellinger A, Murphy JD, Baxter D (2013) The biogas handbook: science, production and applications. Elsevier

Werner JJ, Garcia ML, Perkins SD, Yarasheski KE, Smith SR, Muegge BD, Stadermann FJ, Derito CM, Floss C, Madsen EL (2014) Microbial community dynamics and stability during an ammonia-induced shift to syntrophic acetate oxidation. Appl Environ Microbiol 80:3375–3383

Westerholm M, Dolfing J, Sherry A, Gray ND, Head IM, Schnürer A (2011a) Quantification of syntrophic acetate-oxidizing microbial communities in biogas processes. Environ Microbiol Rep 3:500–505

Westerholm M, Müller B, Arthurson V, Schnürer A (2011b) Changes in the acetogenic population in a mesophilic anaerobic digester in response to increasing ammonia concentration. Microbes Environ 26:347–353

Westerholm M, Isaksson S, Sun L, Schnürer A (2017) Microbial community ability to adapt to altered temperature conditions influences operating stability in anaerobic digestion. Energy Procedia 105:895–900

Westerholm M, Levén L, Schnürer A (2012) Bioaugmentation of syntrophic acetate-oxidizing culture in biogas reactors exposed to increasing levels of ammonia. Appl Environ Microbiol 78:7619–7625

Westerholm M, Moestedt J, Schnürer A (2016) Biogas production through syntrophic acetate oxidation and deliberate operating strategies for improved digester performance. Appl Energy 179:124–135

Westerholm M, Müller B, Isaksson S, Schnürer A (2015) Trace element and temperature effects on microbial communities and links to biogas digester performance at high ammonia levels. Biotechnol Biofuels 8:154

Whelan M, Everitt T, Villa R (2010) A mass transfer model of ammonia volatilisation from anaerobic digestate. Waste Manag 30:1808–1812

Wolf C (2013) Simulation, optimization and instrumentation of agricultural biogas plants. Doctoral Dissertation, Department of Electronic Engineering National University of Ireland, Maynooth

Yang L, Huang Y, Zhao M, Huang Z, Miao H, Xu Z, Ruan W (2015) Enhancing biogas generation performance from food wastes by high-solids thermophilic anaerobic digestion: effect of pH adjustment. Int Biodeterior Biodegradation 105:153–159

Yang S, Hai FI, Price WE, McDonald J, Khan SJ, Nghiem LD (2016) Occurrence of trace organic contaminants in wastewater sludge and their removals by anaerobic digestion. Biores Technol 210:153–159

Yen HW, Brune DE (2007) Anaerobic co-digestion of algal sludge and waste paper to produce methane. Bioresour Technol 98:130–134

Yenigün O, Demirel B (2013) Ammonia inhibition in anaerobic digestion: a review. Process Biochem 48:901–911

Yu D, Liu J, Sui Q, Wei Y (2016) Biogas-pH automation control strategy for optimizing organic loading rate of anaerobic membrane bioreactor treating high COD wastewater. Biores Technol 203:62–70

Zamanzadeh M, Hagen LH, Svensson K, Linjordet R, Horn SJ (2016) Anaerobic digestion of food waste—effect of recirculation and temperature on performance and microbiology. Water Res 96:246–254

Zhang C, Su H, Baeyens J, Tan T (2014) Reviewing the anaerobic digestion of food waste for biogas production. Renew Sustain Energy Rev 38:383–392

Zhang C, Xiao G, Peng L, Su H, Tan T (2013a) The anaerobic co-digestion of food waste and cattle manure. Biores Technol 129:170–176

Zhang T, Liu L, Song Z, Ren G, Feng Y, Han X, Yang G (2013b) Biogas production by co-digestion of goat manure with three crop residues. PLoS One 8:e66845

Zhao J, Yang Q, Li X, Wang D, An H, Xie T, Xu Q, Deng Y, Zeng G (2015) Effect of initial pH on short chain fatty acid production during the anaerobic fermentation of membrane bioreactor sludge enhanced by alkyl polyglcoside. Int Biodeterior Biodegradation 104:283–289

Zhong J, Stevens DK, Hansen CL (2015) Optimization of anaerobic hydrogen and methane production from dairy processing waste using a two-stage digestion in induced bed reactors (IBR). Int J Hydrogen Energy 40:15470–15476

Zinder SH, Koch M (1984) Non-aceticlastic methanogenesis from acetate: acetate oxidation by a thermophilic syntrophic coculture. Arch Microbiol 138:263–272

Zuo Z, Wu S, Qi X, Dong R (2015) Performance enhancement of leaf vegetable waste in two-stage anaerobic systems under high organic loading rate: role of recirculation and hydraulic retention time. Appl Energy 147:279–286

Chapter 7
Biogas Production: Microbiological Aspects

Gerianne Robles, Ramkumar B. Nair, Sabine Kleinsteuber, Marcell Nikolausz and Ilona Sárvári Horváth

List of Abbreviations

AD	Anaerobic digestion
COD	Chemical oxygen demand
CSTR	Continuous stirred-tank reactor
DGGE	Denaturing gradient gel electrophoresis
FTHFS	Formyltetrahydrofolate synthetase
HRT	Hydraulic retention time
IA/TA	Intermediate alkalinity to total alkalinity
MSW	Municipal solid waste
OTU	Operational taxonomic unit
OLR	Organic loading rate
PCR	Polymerase chain reaction
RT-PCR	Reverse transcription polymerase chain reaction
SSU	Small-subunit ribosomal RNA
SRT	Solid retention time
TRFLP	Terminal restriction fragment length polymorphism
TAN	Total ammonia nitrogen
USEPA	United States Environmental Protection Agency

G. Robles · R. B. Nair · I. Sárvári Horváth (✉)
Swedish Centre for Resource Recovery, Unversity of Borås, 501 90 Borås, Sweden
e-mail: ilona.horvath@hb.se

G. Robles
School of Engineering Royal Melbourne Institute of Technology,
124 La Trobe Street, Melbourne, VIC 3000, Australia

S. Kleinsteuber · M. Nikolausz
Department of Environmental Microbiology, Helmholtz Centre
for Environmental Research—UFZ, Permoserstr 15, 04318 Leipzig, Germany

Present Address:
R. B. Nair
Mycorena AB, Holtermansgatan 1, 411 29 Gothenburg, Sweden

M. Tabatabaei and H. Ghanavati (eds.), *Biogas*, Biofuel and Biorefinery
Technologies 6, https://doi.org/10.1007/978-3-319-77335-3_7

UASB	Upflow anaerobic sludge blanket
VS/TS	Volatile solids/total solids
VFA	Volatile fatty acids

7.1 Introduction

While the world population keeps increasing, the generation of waste is also multiplying in similitude with it, posing serious threat to both health and the environment. With increasing concerns over climate change and high energy consumption, there is an increased demand for the renewable fuel alternatives such as bioethanol and biogas (Wang et al. 2013; Wan et al. 2011). Energy crops, being an easily accessible biomass for bioenergy, are popular substrates for industrial-scale production of bioethanol and biogas; however they are now facing the food versus fuel dispute (Salehian and Karimi 2013; Nair et al. 2017). Therefore, alternative feedstocks must be utilized for a sustainable bioenergy generation. In light of these issues, the idea of processing low-value organic waste materials into more value-added final end products is becoming more and more attractive with ongoing advances in resource recovery technologies.

The process of anaerobic digestion (AD) of organic waste is well established, with a large proportion of associated research and development studies based in '*bioenergy-developed*' countries such as ones in Europe and the USA. The aim of this process is the production of a methane rich biogas through biological decomposition of organic matter, in an oxygen-free environment. The produced biogas can be used for power and heat production, or can be upgraded and used as vehicle fuel in the transport sector. AD can be considered as a low-cost environmental friendly waste management process, since it reduces the emission of greenhouse gases (GHGs); meanwhile it reduces and stabilizes the wastes. One of the major benefits of AD is its versatility to handle a wide range of organic substrates. So far, mainly household wastes, food waste, sewage sludge, agricultural residues, manure and energy crops are being used. In addition, the by-product of AD, the '*digestate residue*', can be further utilized as a fertilizer on the agricultural land.

In general, AD of organic material requires the combined activity of several different groups of microorganisms with different metabolic capabilities (Gerardi 2003). The conversion of organic material to methane involves four main steps: *Hydrolysis*, *Acidogenesis*, *Acetogenesis* and *Methanogenesis*. AD can serve as a process to produce either high value but low volume products (intermediary products within the anaerobic degradation chain) or high volume but low value products (end products, like biomethane and biofertilizer), or both in a biorefinery system. Many types of microorganisms and biochemical pathways are involved and consequently a large number of intermediate bioproducts are formed (Agler et al. 2011). This also led to a current expanded application of this process for the production of value added chemicals through mixed culture biotechnologies.

In this chapter the parameters affecting the performance of AD processes will be discussed, together with a brief description of the current AD technologies. Since the degradation of organic material requires a synchronized action of different groups of microorganisms, this chapter will also present recent developments in molecular biology techniques providing us a valuable tool for improved understanding of this complex microbial system.

7.2 Organic-Waste Recycling: Comparison of Treatment Processes

A number of different treatment processes currently exist for the management of organic waste; some more technologically advanced than the others, and some more established in certain countries where the policy and legislation pushes for certain environmental targets. The four alternative processes explored in this section are the systems which are currently applied worldwide, i.e. Landfilling, Aerobic Composting, Incineration and Anaerobic Digestion. A comparison of the advantages and disadvantages of each process is outlined in Table 7.1.

Landfilling involves the collection and dumping of waste into designated areas of land. Landfilling of organic waste is the most undesirable management method due to its high land requirement as well as a number of environmental concerns, including the leachate production, threatening groundwater health, as well as the potential breakdown of organics into harmful greenhouse gases such as methane and carbon dioxide that are released into the atmosphere (Khalid et al. 2011; Cecchi 2011). However, fugitive gas emissions can be avoided with appropriate landfill management in place. Nevertheless, apart from being a convenient disposal option, landfilling provides less valuable products than the other alternative management options.

Composting is a highly implemented management process applied to organic waste, diverting organics from landfill and converting it to a stabilized product with increased value in the agronomic industry (Tognetti et al. 2011). Within the US, the USEPA reports approximately 31 million tonnes of garden waste (commonly called yard trimmings) are generated annually, and approximately 60% was recycled for composting in 2013 (Ge et al. 2016). However, if aerobic conditions are not managed perfectly this technology can also lead to methane emissions.

Incineration (also known as combustion or oxidation) is regarded as a thermal treatment process converting waste materials into energy carriers, which can then be used to generate electricity and heat along with by-products, like ash (inorganics) and flue gas. One concern is the large amounts of gaseous pollutants such as dioxins and heavy particulate metals which are produced during the process. However, certain feedstock, such as food waste, tend to have a high moisture content, making it undesirable for conserving energy via incineration processes. On the other hand, the higher water content makes it a perfect feedstock for anaerobic digestion (Owens and Chynoweth 1993).

Table 7.1 Comparison of treatment processes for the management of organic waste

Management technologies	Advantages	Disadvantages
Landfilling	• Convenient disposal option	• Has high negative impacts on the environment with accumulation of wastes linked to groundwater leaching and fugitive gas emissions
Aerobic composting	• Has a low capital cost and produces stabilized product with high agronomic value	• Compost does not have high commercial value compared to other by-products of alternative management technologies • Can lead to unwanted methane emissions if oxic conditions are not perfectly managed • Energy consuming process (consumes 30–35 kWh per ton of input waste) (Hartmann and Ahring 2006)
Incineration	• Generates heat and steam that drives turbine to produce electricity	• Cannot be used to treat wet organic wastes • Produces large amounts of ash and flue gas which may contain pollutants such as particulate matter, heavy metals, dioxins, furans, sulfur dioxide, and hydrochloric acid • Chlorinated compounds found in organic fraction of municipal solid waste can contribute to the formation of hydrogen chloride and products of incomplete combustion (Hartmann and Ahring 2006) • Bottom ash must be disposed-off as a 'hazardous' waste
Anaerobic digestion	• Produces high quality biogas (an alternative energy source) and digestate (organic fertilizer) • Energy producing process, (produces 100–150 kWh per tonne of input waste) (Hartmann and Ahring 2006)	• Has a high capital cost

Anaerobic digestion optimizes the natural process taking place in any oxygen free environments, such as lake sediments, marshes, landfills, for the increased recovery of high quality biogas. It is performed under anoxic conditions (in absence of oxygen) where different groups of microorganisms break down complex organic materials. The chosen feedstock is widely varied and can include energy crops, agricultural residues, municipal solid waste (MSW), organic waste and wastewater from industries. The final product of anaerobic digestion is biogas, typically with mainly 50–75% methane and 25–50% carbon dioxide content, as well as a high quality digestate, which can be applied as a soil fertilizer or conditioner (Soccol 2011; Ahring 2003). The produced biogas can then be utilized to produce electricity, or can be upgraded to methane and used as vehicle fuel.

7.3 Biochemistry of Anaerobic Digestion

The biochemical decomposition process can be seen as a synergistic process carried out by various array of microorganisms, which can however, generally be divided into four main stages; *Hydrolysis*, *Acidogenesis*, *Acetogenesis*, and *Methanogenesis* (Fig. 7.1) (Liu et al. 2012).

In the very first step, complex organics undergo *hydrolysis* by the action of extracellular enzymes, produced by hydrolytic bacteria, resulting in simple soluble molecules viz., amino acids from proteins, long-chain fatty acids from lipids, and simple sugars from complex carbohydrates (Li et al. 2011). Hydrolysis can occur relatively fast if the substrates are easily degradable and enough physical contact between the enzymes and the substrate is provided (Taherzadeh and Karimi 2008). However, substrates with more recalcitrant structure require longer time, and the degradation is usually not complete (Deublein and Steinhauser 2008; Nair et al. 2015). Hence, for substrates with complex structure, such as lignocelluloses, which

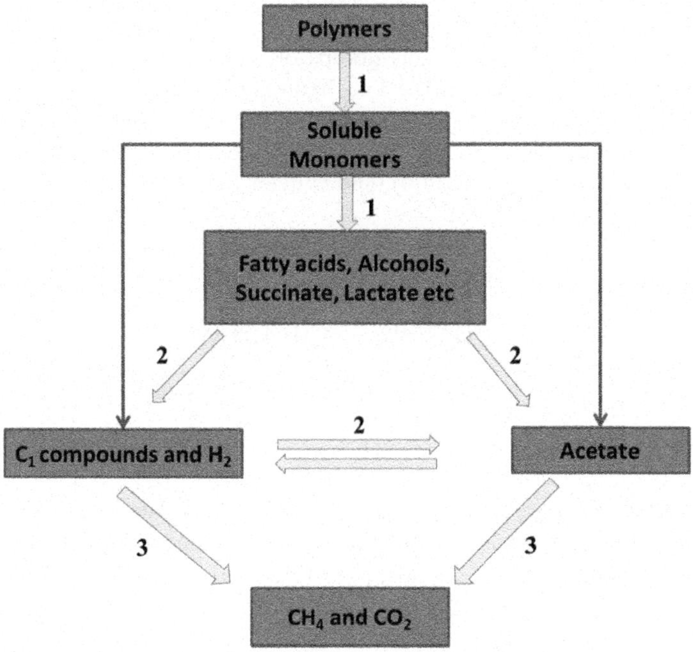

Fig. 7.1 Methane production process involving steps: 1—Hydrolysis and Acidogenesis; 2—Acetogenesis and Syntrophy; 3—Methanogenesis. Adapted from Schnürer (2016)

are barely accessible to the enzymes, the hydrolysis step is often considered as the rate-limiting step (Taherzadeh and Karimi 2008; Vavilin et al. 1996).

The smaller molecules produced will then be further converted to short chain volatile fatty acids (VFAs), alcohols, carbon dioxide and hydrogen in the *acidogenesis* step. The partial pressure of the hydrogen regulates the expected products in this step. In general, the most favorable pathway of primary fermentative bacteria is the production of acetate via pyruvate with production of hydrogen. Hence, in a well-balanced process, with low partial pressure of hydrogen, the main products are acetate, carbon dioxide, and hydrogen. However, if the environmental conditions are not optimal, more intermediates, such as other volatile fatty acids and alcohols, are formed at high partial pressure of hydrogen. These products are more reduced than the products generated under optimal conditions, hence need to be further modified before they can be converted into biogas (Schnürer and Jarvis 2009; Schink 1997).

In the next stage, the *acetogenesis* step, obligatory hydrogen producing bacteria convert VFAs longer than two carbon atoms and alcohols longer than one carbon atom, further to acetate, hydrogen, and carbon dioxide (Schink 1997; Bryant 1979). At standard conditions, the reactions accomplished by acetogenic microorganisms are not exergonic. Here, low partial pressures of hydrogen (lower than 10^{-5} bar) are needed for the reactions to be energetically feasible. The syntrophic association occurring between the hydrogen-producing bacteria and methane-producing *archaea* in the next *methanogenic* step can preserve the partial pressure of hydrogen within the range suitable for the growth of the acetogenic microorganisms (Schink 1997).

In the final stage of AD, various groups of methanogens consume the acetate (acetoclastic methanogens, Eq. 7.1), hydrogen and carbon dioxide (hydrogenotrophic methanogens, Eq. 7.2) and methyl alcohol and methyl amines (methylotrophic methanogens, Eq. 7.3, or H_2-dependent methylotrophic pathway of Methanomassillicoccales, Eq 7.4), and convert these intermediate products into methane.

$$4CH_3COOH \rightarrow 4CH_4 + 4CO_2 \quad (7.1)$$

$$CO_2 + 4H_2 \rightarrow CH_4 + 2H_2O \quad (7.2)$$

$$4CH_3OH \rightarrow 3CH_4 + CO_2 + 2H_2O \quad (7.3)$$

$$CH_3OH + H_2 \rightarrow CH_4 + H_2O \quad (7.4)$$

The theoretical or maximum methane yield of a substrate can be calculated (Eq. 7.5) from the elemental composition of the substrate $C_cH_hO_xN_nS_s$, there 22.4 used as the molar volume of any ideal gas (Yang et al. 2004).

$$Y_{CH_4} = \frac{22.4\left(\frac{c}{2} + \frac{h}{8} - \frac{x}{4} - \frac{3n}{8} - \frac{s}{4}\right)}{12c + h + 16x + 14n + 16s} \tag{7.5}$$

The methane yield obtained in practice; however, rarely achieves more than 60% of the calculated theoretical yield since the *substrate* can contain other compounds which are resistant to the degradation, such as lignin, or compounds with slower degradation rate, such as cellulose, hemicellulose or proteins (Yang et al. 2004).

Biogas produced is often used on-site at the biogas plants or fed into the public gas grid after upgrading. It can also be applied as a vehicle fuel; however such applications requires the biogas to be processed and upgraded to high quality methane (Weiland 2003). Furthermore, the digested residue, which exits the digester, is a nutrient rich and highly stabilized fertilizer. Thus anaerobic digestion implements high end recycling technology through the generation of biogas and digestate residue, with limited odour issues in contrast to the aerobic composting.

7.4 Process Performance: General Scheme and Factors Affecting

Commercial feasibility of AD is heavily reliant on its process stability as well as its ability to handle a largely diverse heterogeneous feedstock (Kayhanian 1999). Different factors including pH, temperature, mixing rate, organic loading rate and retention time, micro- and macro-nutrient availability play a crucial role for the performance of the digestion process. Table 7.2 summarizes the process parameters identified as being the most vital for digester performance and that are indicators for the digester lifetime (Crolla 2012). Therefore, to preserve high process efficiency, these parameters should be effectively controlled and kept within the optimum range (Ward et al. 2008). However, the feedstock structure and characteristics also have a significant impact on the performance of the digestion.

Table 7.2 Conditions required for a stabilized anaerobic digester. Adapted from Crolla et al. (2012)

Parameter	Optimal range
Alkalinity	1500–4000 mg $CaCO_3$/L
pH	6.8–7.2
VS/TS (volatile solids/total solids)	>45%
TAN (total ammonia nitrogen)	<1500 mg/L
C:N:P (carbon-nitrogen-phosphorous ratio)	100–120:5:1
C:N (carbon-nitrogen ratio)	20–30
IA/TA (intermediate alkalinity to total alkalinity)	0.1–0.2 (<0.4)

7.4.1 Organic Loading Rate (OLR) and Hydraulic or Solid Retention Time (HRT or SRT)

The organic loading rate (OLR) is the amount of volatile solids that are loaded to a unit of volume of the digester under a unit of time. Providing different feedstock will lead to a better nutritional balance resulting in an improved microbial diversity, with a stable and robust system capable of resisting and tolerating certain fluctuations in the process conditions as a consequence.

In case of systems treating solid waste, OLR is expressed as $kg_{VS}\ m^{-3}\ d^{-1}$. Hence the organic loading rate may be calculated using the following equation (Eq. 7.6), where OLR = organic loading rate (kg VS substrate/m^3 digester/day), S = fresh substrate added daily (kg/day), VS = VS content of substrate (%VS of substrate), V = volume of bioreactor (m^3)

$$OLR = \frac{S.VS}{V.100} \tag{7.6}$$

Accordingly, for liquid feedstock OLR is measured based on the amount of chemical oxygen demand (COD) added, thus expressed as $kg_{COD}\ m^{-3}\ d^{-1}$ (Vandevivere et al. 2003). In general, lower OLRs are applied during the start-up period of a process, while a balanced and well-functioning process can handle higher OLRs. The biological performance of the AD system is very sensitive to the amounts of loading as well as to waste composition (Zuo et al. 2013; Sharma et al. 1999). Since the hydrolysis and acidogenic fermentation steps usually perform faster, overloading the digester normally leads to the accumulation of VFAs or other inhibitors, which may finally terminate the methane production (Bouallagui et al. 2004; Mata-Alvarez et al. 2000). On the other hand, under-loaded systems, i.e. running the digester at low OLRs, are not economically feasible, since the capacity of the digester is not entirely utilized.

Another important parameter that controls the rate of bioconversion is the retention time. The retention time is usually expressed as hydraulic retention time (HRT) representing the estimated time while the liquid sludge is present in the digester (Eq. 7.7):

$$HRT = \frac{V}{Q} \tag{7.7}$$

where V is the reactor volume (m^3) and Q is the flow rate of fresh substrate (m^3/day).

The retention time can also be expressed as solid retention time (SRT) considering the time that microorganisms/solids spend in the digester (Appels et al. 2008). HRT and SRT are equal in many cases when continuous stirred tank reactors (CSTR) are employed, but for process configurations in which part of the residues are recirculated back to the process, SRT gets longer than HRT. SRT can be also prolonged compared to HRT in high-rate processes; such as upflow anaerobic sludge blanket (UASB) reactor (Bal and Dhagat 2001), fluidized bed reactors (Moharram et al. 2016) and anaerobic expanded bed reactors (Kato et al. 1994), where the microorganisms form granules or are attached to certain carrier material, whereby retained in the system for longer residence time. Longer SRT also enables the viable biomass to be adapted to the inhibiting substances such as ammonia, sulfides, and others that might otherwise be toxic at high concentrations (Schnürer and Jarvis 2010). Shorter retention times are normally favorable to increase the efficiency of the process and reduce the system costs (Chandra et al. 2012). However, there must always be a balance between OLR and HRT in order to optimize digestion efficiency. Therefore, at higher OLRs, retention times should be sufficiently long, providing the microorganisms with enough time to degrade the substrate (Demirer and Chen 2005). In industrial sewage sludge, where the feedstock has a low total solid content, recirculation of the thickened sludge including the biomass would allow longer retention time for the microorganisms to degrade the organic matter (Schnürer and Jarvis 2010).

7.4.2 Operating Temperature

There are three main operating temperatures for the AD process; *psychrophilic* (optimum at 10 °C), *mesophilic* (optimum at 37 °C), or *thermophilic* (optimum above 50 °C) (Kabir et al. 2015a). Temperature is a critical factor in AD process as it affects the activity of the microorganisms in the digester. Hydrolytic and acidogenic bacteria are not much sensitive to temperature changes, generally due to functional redundancy as well as higher diversity of bacterial communities. On the other hand, the acetogenic and methanogenic processes are significantly influenced by changes in the temperature. The generation time for methanogenic archaea ranges from about 3 days at 35 °C to 50 days at 10 °C.

Lower temperatures are known to result in slower microbial growth and substrate utilization, resulting in a decreased biogas production; however high temperatures can also be associated with decreased biogas yield due to the increased concentration (only the ammonia dissolved in the liquid inhibits AD) of free ammonia, which may inhibit the gas production (Khalid et al. 2011). Moreover, at higher temperature the digester might suffer from VFA accumulation easier, affecting the overall digestion performance. When operating temperature of around 35–37 °C is considered to be appropriate, any change from mesophilic to thermophilic will slow down the biogas production rate; but it will increase again as the necessary shift in microbial populations occur (Khalid et al. 2011). The optimal growth temperature for most methanogenic archaea is between 30 and 40 °C, with a few genera growing best between 50 and 60 °C (Table 7.3). Therefore, the changes in the operational temperature range also result in changes in dominant species.

In general, mesophilic processes can be seen as more established due to their robustness and stability, but tend to have a slower start-up phase. Due to a greater diversity of the microorganisms at this range of temperature, the process is more

Table 7.3 Optimal growth temperature for methanogenic archaea. Adapted from Gerardi (2003)

Genus	Temperature range (°C)
Methanobacterium	37–45
Methanobrevibacter	37–40
Methanosphaera	35–40
Methanothermus	83–88
Methanococcus	35–40
	65–91
Methanocorpusculum	30–40
Methanoculleus	35–40
Methanogenium	20–40
Methanoplanus	30–40
Methanospirillum	35–40
Methanococcoides	30–35
Methanohalobium	50–55
Methanohalophilus	35–45
Methanolobus	35–40
Methanosarcina	30–40
	50–55
Methanotrix	35–50

robust and balanced, thus handling fluctuations better (Zhao and Kugel 1996; Levén et al. 2007). On the other hand, thermophilic conditions result in a more rapid conversion of organic acids (Ge et al. 2016). The benefits of a higher temperature (thermophilic) system also mean higher bioconversion rates with shorter retention times and smaller reactor volumes resulting in an end 'digestate' product typically pathogen free and of higher quality (Li et al. 2011).

7.4.3 *Volatile Fatty Acids, pH and Alkalinity*

The concentration of volatile fatty acids is determined by their production and consumption rates, the loading rate and the characteristics of feedstock (Karthikeyan and Visvanathan 2012). VFAs can accumulate in a system when methanogens cannot keep up with the rate of breakdown in the earlier degradation stages, causing a drop in the pH and alkalinity, which in turn will inhibit methanogens, and finally resulting in system failure (Yang et al. 2015). Hence accumulation of VFAs can occur if the digester is overloaded, especially when the substrate is easily digestible. Generally it is estimated that to have a stable process the concentration of volatile fatty acids, particularly acetic acid, should be below 2 g/L (Jain and Mattiasson 1998). Methane is not the only product of AD, as carbon dioxide is also produced and through the degradation of urea, proteins and amino acids ammonia is also released. Carbon dioxide and ammonia result in the production of carbonic acid, bicarbonate, and ammonium ions (Eqs. 7.8 and 7.9):

$$CO_2 + H_2O \leftrightarrow H_2CO_3 \leftrightarrow H^+ + HCO_3^- \leftrightarrow 2H^+ + CO_3^{2-} \tag{7.8}$$

$$NH_3 + H^+ \leftrightarrow NH_4^+ \tag{7.9}$$

Buffering capacity or alkalinity is referred to the equilibrium between carbon dioxide and bicarbonate, with ammonia as the major cation, which cause a significant resistance to pH changes. In order to preserve optimum pH in the digester, it is vital to have a high and stable alkalinity. The major buffering capacity in anaerobic digesters is connected to the presence of bicarbonate (HCO_3^-), with a pKa value of 6.3, and the main acids are VFAs, with an aggregate pKa around 4.8 (Boe and Angelidaki 2006). From bicarbonate and ammonium ions ammonium bicarbonate is formed, which then can react with VFAs forming salts (Eq. 7.10):

$$NH_4HCO_3 + RCOOH \rightarrow RCOONH_4 + H^+ + HCO_3^- \tag{7.10}$$

The higher the bicarbonate concentration in the digester medium, the greater the alkalinity and hence the resistance to changes in pH (Alvarez et al. 2006). However, a sudden change in pH can occur, for instance, if the system is overloaded when the feed rate is increased significantly. Since the methanogens grow slower than the fermentative bacteria, VFA accumulation will result in a pH drop. In a research study, Khanal (2008) states that the VFA to alkalinity ratio is an indication of imbalance in the system, where this value is ≤ 0.4 in a healthy system and a ratio of ≥ 0.8 will result in process failure. In general, the optimal pH for methanogenesis is between 6.8 and 7.2, while for hydrolysis a pH of 5.5 and for acidogenesis a pH of 6.5 are considered more suitable (Khalid et al. 2011).

7.4.4 Nutritional Requirements

As for any biological processes, where microorganisms are involved, both macro- and micronutrients should be provided for the microbial community in a well-balanced manner to be able to achieve a stable and efficient biogas production process. It is reported that in an ideal AD system the nutrients should be found in excess in the digester as even small shortage of any of them may inhibit the process (Chan 2003). Therefore, in the case of feedstock nutrient deficiencies, supplementary nutrients must be added to stimulate the digestion process. Fundamental macronutrients such as carbon (C), nitrogen (N), phosphorus (P) and sulfur (S) are necessary for microbial growth. Organic carbon is the primary source of energy and the basic building block of the cell material. Nitrogen and phosphorus are used in the protein and nucleic acid synthesis. For optimal gas production the C/N ratio has been suggested to be set between 15:1 and 25:1 (Esposito et al. 2012). The required amount of phosphorus is 6–7 times less than that of nitrogen. Other essential macronutrients are potassium (used in cellular transport and cation balancing) and sulfur (required in numerous enzymes). Among micronutrients iron (Fe), nickel (Ni), cobalt (Co), molybdenum (Mo), tungsten (W) are the most important ones, since they are necessary co-factors for the unique enzyme systems of anaerobic microorganisms (Zandvoort et al. 2006); thus the presence of several trace metals is obligatory for methane production (Oleszkiewicz and Sharma 1990). Nickel constitutes, for example, the active centre of acetate formation enzymes and H_2-consuming hydrogenases (nickel-containing coenzymes F_{420} and F_{430} are important hydrogen-carriers). Cobalt plays an important role in the transfer of methyl-groups by the coenzyme cobalamin, so it is essential for all methanogenic pathways. However, the correct concentrations and proportions of the required trace metals should be determined individually in each case because that depends on the composition of the microbial community and the substrate, as well as the bioavailability of the micronutrients (Jagadabhi 2011).

7.4.4.1 C/N Ratio and Ammonia Inhibition

Nitrogen is necessary for the growth of the microorganisms. On the one hand, nitrogen deficiency can result in insufficient consumption of the carbon source, resulting in a reduced microbial growth and finally leading to a decrease in the biogas production (Resch et al. 2011). Furthermore, free ammonia or ammonium ions are produced by the breakdown of nitrogenous matter in the digester, commonly present in the form of proteins and urea (Chandra et al. 2012). Around 60–80% of total nitrogen (from proteins and other organic compounds) are converted to ammonia-N during AD (Karthikeyan and Visvanathan 2012). An estimation of the amount of ammonia that is generated during the degradation of a substrate can be calculated using the following equation (Tchobanoglous et al. 1993) based on stoichiometric relationships (Eq. 7.11):

$$C_aH_bO_cN_d + \frac{4_a - b - 2_c + 3_d}{4}H_2O \rightarrow \frac{4_a + b - 2c - 3d}{8}CH_4 + \frac{4_a - b + 2c + 3d}{8}CO_2 + dNH_3 \tag{7.11}$$

A healthy system will have a concentration of around 200 mg/L to support anaerobic growth of the bacteria. Concentrations of ammonia-N over 1500 mg/L will cause moderate inhibition, while concentrations greater than 3000 mg/L will cause 100% inhibition of methanogenesis (Gerardi 2003). However, in some cases this value can be exceeded by systematic adaptation of the microbial community to higher nitrogen levels (Dai et al. 2016). The free ammonia is a main source of inhibition as it can diffuse into the cell causing proton imbalance (Klass 1984). The chemical equilibrium between ammonium and free ammonia depends on the temperature and pH. As the temperature or the pH increases, this equilibrium would shift toward free NH_3 resulting in inhibition (Klass 1984; Chen et al. 2008).

Nevertheless, the effects of ammonia inhibition can be moderated by dilution with water in extreme ammonia overloads, or altering feedstock to adjust C/N ratio in slighter overloads (Kayhanian 1999).

7.5 Process Configurations: Substrates and Operations

The biodegradability of different organic waste fractions is highly defined by their characteristics and chemical composition, i.e. the amount of lipids, proteins, and carbohydrates those including cellulose, hemicelluloses, as well as lignin (Hartmann and Ahring 2006). Optimally, the substrates need to have a composition that meets the nutritional requirements of the microorganisms involved in the degradation process.

As it was discussed earlier, the expected theoretical methane potential can be calculated (Eq. 3.5 in Chap. 3). Applying general chemical formula for carbohydrates ($C_6H_{10}O_5$), proteins ($C_5H_7O_2N$), and lipids ($C_{57}H_{104}O_6$) will give 0.42, 0.50, and 1.01 m^3 CH_4 kg^{-1} volatile solids, respectively (Møller et al. 2004a, b). Although the methane yield of proteins and lipids is higher than that of the other components, they often cause inhibition in the digestion process due to the accumulation of ammonia or fatty acids. Moreover, studies have reported that the rate of hydrolysis for fat-rich components is slower compared to less fatty wastes; since the lipids can be adsorbed onto the surface of other components interrupting the hydrolysis process by reducing the accessibility for the hydrolytic enzymes (Neves et al. 2008). Similarly, the composition of the waste also determines the C/N ratio present in the waste. A high C/N ratio can cause nitrogen deficiency, while the degradation of waste with a very low C/N ratio can results in ammonia inhibition followed by process failure (Hartmann and Ahring 2006).

Apart from the composition, the degradability of a substrate may also depend on the particle size, the collection method, weather and seasonal fluctuations, and also the cultural habits of the community that the waste is collected from. Regarding the particle size, a smaller particle size offers a larger accessible surface area for enzymatic attacks, hence an enhanced biodegradability (Hartmann and Ahring 2006; Hajji and Rhachi 2013). Table 7.4 summarizes the methane potential reported in the literature for some organic waste fractions.

7.5.1 Pretreatment for Enhanced Yield

Pretreatments are usually applied for two different reasons: (1) making the material mechanically easier to handle or (2) altering the structure of the material to make it more digestible. Regarding the first option, waste fractions that are composed of a wide variety of components of different sizes and shapes (like municipal solid waste), must undergo pretreatment to obtain a reasonably uniform final product,

Table 7.4 The methane potential of organic waste fractions. Adapted from Kabir et al. (2015b)

Waste types	Methane yield (m^3 CH_4 kg^{-1} VS)	Reference
Industrial and commercial waste		
Expired food	0.47–1.10	Braun et al. (2003)
Sludge from distilleries	0.40–0.47	-do-
Potato waste	0.69–0.89	-do-
Molasses	0.31	Angelidaki and Ellegaard (2003)
Edible-oil sludge	1.10	Braun et al. (2003)
Municipal and industrial		
Household waste	0.40–0.50	Angelidaki and Ellegaard (2003)
Garden waste	0.10–0.20	-do-
Paper	0.08–0.37	Owens and Chynoweth (1993)
Market waste	0.90	Braun et al. (2003)
Municipal solid waste	0.20–0.22	Chynoweth et al. (1993)
Banana peel	0.27–0.32	Gunaseelan (2004)
Citrus waste	0.43–0.73	-do-
Vegetable waste	0.19–0.40	-do-
Animal and slaughterhouse waste		
Animal fat	1.00	Braun et al. (2003)
Stomach and gut contents	0.4–0.46	Ahring et al. (1992)
Rumen content	0.35	Braun et al. (2003)
Blood	0.65	-do-
Agricultural waste		
Cow manure	0.15–0.30	Angelidaki and Ellegaard (2003)
Swine manure	0.30–0.51	Møller et al. (2004), Ahring et al. (1992)
Poultry manure	0.30	Ahring et al. (1992)
Straw and other plant residues	0.15–0.36	Møller et al. (2004), Angelidaki and Ellegaard (2003)
Green plant, crops, grain	0.18–0.28	Angelidaki and Ellegaard (2003)
Sugarcane	0.23–0.30	Chynoweth et al. (1993)
Sorghum	0.26–0.39	-do-

which has a greater density than the original form. Inert materials such as sand, clay, glass, or floating materials like plastics also need to be removed. Shredders, crushers, or millers are often used to carry out the size reduction process.

In the case of waste fractions with complex recalcitrant structure, the application of suitable pretreatment methods aiming to enhance the accessibility of the enzymes to the biomass is needed (Taherzadeh and Karimi 2008; Nair et al. 2016). A proper pretreatment can increase the methane yield by improving the hydrolysis rate and different pretreatment methods such as mechanical, chemical, biological, or physicochemical methods, are reported in the literature. Pretreatment processes leading to significant improvements in methane yields for lignocelluloses, such as woody biomass and straw (Kabir et al. 2015c; Teghammar et al. 2012; Aslanzadeh et al. 2014), paper residues (Teghammar et al. 2010), citrus waste (Forgács et al. 2012) as well as keratin rich waste fractions (Kabir et al. 2013; Forgács et al. 2013) are reported in various research studies.

7.5.2 Co-digestion for Improved Process Efficiency

Co-digestion is the simultaneous digestion of more than one substrate with complementary characteristics and has become popular as the digestion of several materials can give higher methane yields than those expected when single materials are treated individually (Weiland 2003; Pagés-Díaz et al. 2011, 2015). Some of the reasons linked with the enhancement are related to the combinations of substrates that result in a positive interaction within the system, influencing C/N ratio and reactor stability, balancing buffer capacity, supplementing nutrients, or reducing negative effects of toxic or inhibitory compounds.

7.5.3 Operational Methods and Reactor Designs

The design of any anaerobic digester has to address three fundamental requirements such as: capable to continuously handle a high organic loading rate; to have a short hydraulic retention time in order to have smaller reactor volume; and to produce a high volume of high quality biogas (Ward et al. 2008). There are several different types of digesters, which are used in the industry including batch, continuous one-stage system, or continuous two-stage/multi-stage systems. Additional configurations, such as the anaerobic sequencing batch reactor (ASBR), upflow anaerobic sludge blanket (UASB) reactor, tubular reactor, plug-flow systems and anaerobic filters also exist (Bouallagui et al. 2005). These reactors can be compared based on the biological and technical performance and characteristics. Table 7.5 summarizes the advantages and disadvantages of a number of different types of digesters:

Batch reactors are quick, require inexpensive equipment, and are the simplest to operate since they are fed with feedstock and left for a longer period before being emptied (Khalid et al. 2011). The methane production is generally the highest at the beginning and decreases toward the end of the process as the substrate is being utilized.

Continuous systems are fed continuously, while the digestate residue is discharged at the same rate, allowing a steady state to occur, leading to a constant gas production rate. However, this type of operation is only possible for substrates,

Table 7.5 Comparison of digester types. Adapted from Korres et al. (2013)

Reactor configurations	Disadvantages	Advantages
One-stage system	• Longer retention time • Potential failure due to foam and scum formation	• Less technical therefore simple design
Two-stage system	• Complex design • Higher capital and maintenance cost • Solid portion of feedstock to be removed in second stage	• Increase in biomass digestion due to recirculation • Constant feed flow rate to second methanogenic stage • A more robust system therefore less susceptible to failure
Dry digestion	• Waste handling is more complex and therefore more costly • Feed material must be structure with high solid content • More difficult mixing	• Higher solids loading and biomass retention • Feed is more controlled • Simpler pretreatment
Wet digestion	• Formation of scum when crops are digested • High water and energy requirement • Short-circuiting may occur • Sensitive to shock loads	• Higher water volume results in dilution of inhibitors
Batch operation	• Larger tank volume • Overall lower biogas yield	• No required pumping or mixing • Low capital cost • Low process and mechanical requirements
Continuous operation	• Higher potention for acidification and VFA accumulation to occur	• Simple design and operation • Low capital cost • Higher biomass retention
High-rate systems	• Long start-up phase • Channeling can occur at low feed rate	• More control over feeding • Low capital cost

which can be pumped for continuous feeding. Otherwise, a semi-continuous process is applied with a discrete amount of feed several times a day. The two commonly used reactor types are: continuous stirred-tank reactors (CSTR, using mechanical agitation or effluent or biogas recirculation for mixing), and plug-flow reactors (PFR, where the content of the reactor is pushed along a horizontal reactor). PFRs are usually used in dry digestion processes treating substrates with high solid content (Patinvoh et al. 2017), while CSTRs are applied in wet digestion systems.

The choice of wet or dry digestion technology depends on the total solid content (TS) of the material treated. Wet systems are designed for processing dilute organic slurry with a TS content of maximum 10–15%. Substrates having TS higher than 15% are usually diluted with fresh or recirculated process water, or will be co-digested with co-substrates of lower TS content. Wet anaerobic digestion systems have been successfully applied to treat various ranges of low solid materials, including sewage sludge and food industrial effluents.

On the other hand, in solid-state fermentation processes, also called dry digestion, the substrates used have high solid content (25–40% TS), hence a fundamentally different technical approach regarding the waste handling and treatment is needed (Verma 2002). Due to the high viscosity in the dry digestion systems, heat and nutrient transfer is not as efficient as it is in wet processes, hence mixing is very important to prevent local overloading and acidification (Luning et al. 2003; Wellinger et al. 1993). However, conventional mechanical mixers are not appropriate for solid-state processes; instead, recirculation of the waste or re-injection of the produced biogas is usually used in these kinds of reactors to solve the mixing problems (Luning et al. 2003).

Furthermore, the CSTR design is often applied in single stage systems, there the reactor operates, favoring both acidogenic and methanogenic microorganisms. These types of systems are simple to operate and have lower capital and operating costs, making them attractive for a wide range of applications during the last decades (Vandevivere et al. 2003; Kelleher 2007).

However, the conversion of organic matter to biogas is performed through a sequence of biochemical reactions, which do not necessarily have the same optimal environmental conditions. Two- and multi-stage systems have therefore been developed providing optimal conditions for the different groups of microorganisms involved in the degradation process, which in turn leads to higher reaction rates and consequently, a higher biogas yield (Ghosh et al. 2000). In two-stage reactors, hydrolysis/acidification and acetogenesis/methanogenesis are separated. Therefore, the first phase can operate at lower pH, which is more favorable for the growth of hydrolytic and acidogenic microorganisms; whereas the second step is optimized to favor the growth of methane forming microorganisms (Ince 1998). The rate limiting factor in the second stage is typically the rate of microbial growth (Chaudhary 2008) since methane producing archaea have longer generation times, and therefore longer biomass retention times are needed in this second phase, which in turn enhances the biogas yield (Verma 2002). These kinds of digesters usually have a more stable performance than single-stage digesters, since they do not suffer from the process disturbances caused by for example the changes in the pH or by ammonia accumulation (Chaudhary 2008; Joshua et al. 2012). An even better phase separation option can be provided in multi-stage reactors, which can offer process control and optimization for each conversion step, leading to increased methane production (Griffin 2012).

7.6 Assessment of the Microbial Community Structure

Although biogas production via anaerobic digestion of organic matter has a long tradition, the whole process was considered as a black box system by process engineers for a long while and the detailed understanding of the microbiome behind the process has been ignored. It was mainly due to the fact that classical microbiological techniques based on cultivation and pure culture studies provided limited information about the diversity, community composition and physiological function of the major key players. Obtaining and handling anaerobic pure cultures have been further challenges in microbiology. The introduction of molecular techniques offered detailed insights into the diversity and dynamics of microbial communities involved in the biogas process in a similar manner as they caused a paradigm shift in microbial ecology studies of natural systems. Investigating the microbiology of the biogas process is still a challenge due to the enormous diversity, but also offers the beauty of exploring novel taxa, new pathways, interesting syntrophic interactions, and also new concepts regarding evolutionary mechanisms. The following chapter describes the exploration of the microbiome in anaerobic digesters by the application of molecular biology techniques.

7.6.1 Classical Molecular Biology Techniques

The standard approach of cultivation-independent analysis of complex microbial communities usually involves the isolation of nucleic acids from environmental samples, followed by polymerase chain reaction (PCR) amplification of a phylogenetic marker gene, which is in most cases the 16S rRNA gene. This gene encodes the small subunit (SSU) ribosomal RNA and is highly conserved among Bacteria and Archaea. It contains highly conserved regions as ideal targets for primer development, as well as variable regions allowing phylogenetic classification. Although the design of perfectly matching universal primers remains an unsolved problem and preferential amplification cannot be ruled out (Sipos et al. 2007), the frequently used domain-specific primers have been thoroughly tested (Klindworth et al. 2013) and can give a broad overview of the microbial diversity of the biogas process. PCR followed by cloning and sequencing provides a first estimation of the diversity of Bacteria and Archaea in biogas reactors. In a pioneer study, Godon and co-workers investigated the microbial community structure of a fluidized-bed reactor fed with vinasses by cloning and sequencing of SSU rRNA gene PCR products (Godon et al. 1997). The predominant bacterial operational taxonomic units (OTUs) were affiliated to low-GC Gram-positive bacteria, *Cytophaga-Flexibacter-Bacteroides* group, and Proteobacteria. The most dominant Archaea OTUs were very similar to the already known methanogenic species *Methanosarcina barkeri*, *Methanosarcina frisius*, and *Methanobacterium formicicum*. As another example, McHugh et al. (2006) investigated the microbial community dynamics during the treatment of sucrose and VFA-based artificial wastewater under mesophilic and psychrophilic conditions by the cloning and sequencing approach. A distinct dominance shift from acetoclastic to hydrogenotrophic methanogens in response to a decrease in the operation temperature was observed in both reactors. The majority of the clones were closely related to the *Bacillus–Lactobacillus–Streptococcus* group in the sucrose-fed reactor, while sequences affiliated to *Bacteroides* and *Deltaproteobacteria* were dominant in the VFA-fed reactor. A major drawback of the cloning and sequencing approach is that due to the time and cost demand the sample throughput is very low and the method is not suitable for reactor monitoring, only for snapshot analysis (Talbot et al. 2008).

Molecular fingerprinting methods have been developed for the fast comparison of PCR products of numerous samples by providing profiles or patterns describing the diversity of amplified DNA sequences. These techniques have enabled a fast comparison of numerous samples and thus the study of temporal and spatial shifts in microbial community structure. Studies of biogas reactors were performed by denaturing gradient gel electrophoresis (DGGE) (Liu et al. 2009), single-strand conformation polymorphism (SSCP) analysis (Leclerc et al. 2004) and terminal restriction fragment length polymorphism (T-RFLP) profiling (Ziganshin et al. 2011), to mention a few examples. However, adding taxonomic information to the community patterns is complicated (Nikolausz et al. 2005) and usually requires the establishment of supporting clone libraries. A detailed review on the application of classical 16S rRNA gene-based methods for the investigation of anaerobic bioreactors can be found elsewhere (Talbot et al. 2008).

The analysis of functional genes instead of rRNA genes is an alternative approach, which allows the targeted investigation of distinct functional guilds. The *mcrA* gene encoding the alpha-subunit of methyl coenzyme M reductase is the most widely used molecular marker for the assessment of the biogas process, targeting the methanogens (Lueders et al. 2001; Luton et al. 2002). An advantage of the *mcrA* gene approach is that it strictly targets methanogens whereas Archaea domain-specific 16S rRNA gene primers also amplify DNA from non-methanogenic Archaea. A further advantage is that transcripts of functional genes represent the actual metabolic activity of microorganisms due to the short half-life of mRNA. Therefore, RNA-based assessment of the methanogenic community provides more reliable information on the active members of a community (Munk et al. 2012; Lv et al. 2014; Nikolausz et al. 2013; Wintsche et al. 2016).

Amplicons of *mcrA* genes or their transcripts can be further analysed by the cloning and sequencing approach. Nettmann et al. (2008) investigated the methanogenic community structure of an agricultural biogas plant supplied with cattle manure and maize silage under mesophilic conditions. *Methanomicrobiales*, *Methanosarcinales* and *Methanobacteriales* were the major orders and hydrogenotrophic methanogenesis was the presumed major pathway of methane formation. Rastogi et al. (2008) compared the methanogenic communities during summer and winter seasons in a biogas reactor running on cattle dung without temperature control. The summer clone library was more diverse with sequences affiliated to *Methanomicrobiales* (41.7%), *Methanosarcinales* (30%), and *Methanobacteriales* (19%), while the winter clone library was dominated by *Methanomicrobiales* (98.6% of all clones).

Due to the degeneracy of the genetic code (amino acids with multiple codons) the primer design is challenging and requires the involvement of degenerate positions. PCR amplification of a single template with such primers may result in multiple bands in DGGE and SSCP patterns after gel separation, making the adaptation of molecular fingerprinting techniques for the functional genes to be more complicated. Thus, only few attempts have been made for methanogenic communities in biogas reactors (e.g. PCR-SSCP Munk et al. 2010). Since terminal restriction fragment (T-RF) length is not influenced by the degenerate primer positions, T-RFLP fingerprinting targeting the *mcrA* gene was successfully applied by many studies, which analysed the methanogenic communities in bioreactors digesting various substrates including energy crops and agricultural wastes (Ziganshin et al. 2016a; Leite et al. 2015; Mulat et al. 2015; Lucas et al. 2015; Popp 2015; Zhang et al. 2014) to mention few examples. Due to the relatively low diversity of methanogens it was possible to develop a T-RFLP approach based on an improved primer set (Steinberg and Regan 2008) and a database facilitating the fast identification of methanogens, thus avoiding the need of cloning and sequencing (Bühligen et al. 2016).

Microorganisms involved in reductive acetogenesis or syntrophic acetate oxidation employing the Wood-Ljungdahl pathway can be targeted by the *fhs* gene encoding the formyltetrahydrofolate synthetase (FTHFS). The development of suitable primers is challenging because of specificity issues (Westerholm et al. 2011; Gagen et al. 2010). The *fhs* gene diversity in anaerobic digesters has been less frequently studied compared to the *mcrA* gene, but few investigations already revealed the diversity in a mesophilic laboratory-scale biogas reactor (Westerholm et al. 2011) and gene abundance by qPCR in natural and engineered environments (Xu et al. 2009) as well as homoacetogenic activity during acidification in a thermophilic anaerobic digester (Akuzawa et al. 2011).

Genes encoding [FeFe] hydrogenases (*hydA*) can also be used as specific biomarkers of some groups of H_2-producing bacteria (Vignais et al. 2001; Huang et al. 2010). The activities of hydrogen-producing bacteria together with the active methanogens (RNA-based *mcrA* gene approach) were investigated in a study of various reactor systems during the reduction of the hydraulic retention time (Ziganshin et al. 2016b).

7.6.2 Next Generation Amplicon Sequencing

The major drawbacks of the classical PCR-based techniques are that they have either high throughput but low taxonomic resolution (molecular fingerprinting techniques) or vice versa (cloning and sequencing). The application of the next generation high-throughput sequencing technologies addressed this challenge by providing an unparalleled resolution of the diversity (sequencing depth) at reduced cost and required time for performing the assays. The increased need for sequencing was mainly driven by the Human Genome Project. Nevertheless, the biogas microbiology research also benefitted from the revolution of DNA sequencing. A detailed overview on the technical details and description of the various sequencing platforms are beyond the scope of this book chapter. Descriptions of the technology developments can be found in other related reviews (Goodwin et al. 2016; Rothberg and Leamon 2008).

Next generation amplicon sequencing was used for the detailed description of the microbial communities of laboratory-scale (Wintsche et al. 2016; Leite et al. 2016; Ziganshin et al. 2013; Poirier et al. 2016; Sträuber et al. 2016; Town et al. 2014; Popp 2017; Sun et al. 2015) and full-scale anaerobic digesters (Lucas et al. 2015; Werner et al. 2011; Sundberg et al. 2013; Lee et al. 2012; Maus et al. 2017; Sun et al. 2016; Abendroth et al. 2015; De Vrieze et al. 2015; Li et al. 2015). These investigations revealed a much higher level of diversity compared to the classical molecular techniques (Werner et al. 2011; Lee et al. 2012) and the increased sequencing depth provided new insights into the contribution of low abundant community members during inoculation, enrichment or start-up of the biogas process (Leite et al. 2016; Ozbayram et al. 2017; Goux et al. 2016). The major bacterial taxa found by amplicon sequencing in various biogas systems are summarized in Table 7.6.

The microbial composition of biogas reactors is mainly shaped by the substrate composition and the process temperature (Ziganshin et al. 2013; Luo et al. 2015), the hydraulic retention time (Regueiro et al. 2015) and the ammonia level (Ziganshin et al. 2013; Sun et al. 2015; Sundberg et al. 2013; De Vrieze et al. 2015). The bacterial community is generally more diverse than the archaeal community. Due to the lower diversity of the methanogenic communities a high read number during amplicon sequencing is not required to identify the major key players of the community. Only few studies focused on the *mcrA* gene for amplicon sequencing and high throughput description of the methanogens (e.g. Popp 2017; Ellis et al. 2012; Wilkins et al. 2015). However, the method can be useful to analyse the role of less abundant methanogens, which cannot be detected by T-RFLP. Such methanogens below one percent relative abundance may appear and become even dominant when the reactor conditions change (Leite et al. 2016). Agneessens and co-workers found an increase of the relative abundance of a Methanobacteriales affiliated OTU from below 1 to 6.1% after H_2 addition to a system digesting sludge and straw by the application of *mcrA* gene amplicon sequencing (Agneessens et al. 2017).

Table 7.6 Examples for the application of next generation amplicon sequencing to reveal the phylogenetic diversity of full-scale biogas reactor communities

Reactor type/ temperature	Substrate	Sequencing platform	Most abundant bacterial taxa	Reference
UASB[a] reactors (variable)	Brewery wastewater	454 pyrosequencing	Bacteroidetes, Syntrophobacterales, Desulfuromonadales, Spirochaetes	Werner et al. (2011)
One and two-stage reactors (mesophilic and thermophilic)	Municipal wastewater	454 pyrosequencing	Proteobacteria, Bacteroidetes, Firmicutes, Spirochaetes, Chloroflexi	Lee et al. (2012)
Mesophilic CSTR[b] systems	Energy crops (97% maize silage)	454 pyrosequencing	Firmicutes, Bacteroidetes, Cloacimonetes	Lucas et al. (2015)
CSTRs[b] and liquid pump/ wet fermenters (mesophilic and thermophilic)	Energy crops and manure	Illumina MiSeq	Firmicutes, Bacteroidetes, Spirochaetes, Tenericutes	Maus et al. (2017)
Various mesophilic and thermophilic CSTR[b] systems	Sewage sludge (SS), co-digestion of various wastes from slaughterhouses, restaurants, households	454 pyrosequencing	Actinobacteria, Proteobacteria, Chloroflexi, Spirochetes (SS); Firmicutes (co-digestion)	Sundberg et al. (2013)
Various waste treatment systems (mesophilic and thermophilic)	Various waste streams	Illumina MiSeq	Firmicutes, Bacteroidetes, Proteobacteria	De Vrieze et al. (2015)
Small CSTR[b] farm reactor	Rye and maize silages, hay, straw, green grass, solid meat- and dairy-cattle manure	Illumina MiSeq	Proteobacteria, Bacteroidetes, Firmicutes, Tenericutes	Goux et al. (2016)

[a]Upflow anaerobic sludge blanket
[b]Continuous stirred tank reactor

7.6.3 Omics Approach

The advances in sequencing technologies made it possible to analyse not just single genes, but to target all genes or gene transcripts even in complex microbial communities. Metagenomics is a complex investigation approach including high throughput sequencing and bioinformatics tools to characterize the genetic content of complex microbial communities (Thomas et al. 2012). In a similar way metatranscriptomics is defined as the approach to characterize the gene transcripts of a complex community by deep sequencing the reverse transcribed RNA isolated from a complex sample. The above discussed (6.2) single-gene (16S rRNA gene or *mcrA* gene) amplicon sequencing approach is also frequently described as metagenomics in the literature, which is a wrong interpretation and should be avoided, because the holistic element of "omics" is missing in this approach (Prosser 2015). The omics approach goes beyond the description of community structure by providing information about the potential function (DNA) or its expression (RNA) as it analyses not just single taxonomic or functional markers but sets of genes or genomic fragments with the potential to explore metabolic pathways and novel or unexpected functions. Another advantage of the omics approach is that it does not require *a prior* knowledge for primer design, and completely unknown genes might be explored. The relative abundance data of genes or taxa are less distorted due to the lack of PCR-associated biases. However, PCR-based amplicon sequencing offers a higher coverage of rare taxa, considering the same sequencing depth (Lebuhn et al. 2014). A detailed review providing practical advice on sample processing, sequencing technology, assembly, binning, annotation, experimental design, statistical analysis, data storage, and data sharing can be found elsewhere (Thomas et al. 2012).

Although the ultimate aim of metagenomics is the complete coverage of all genes and construction of population genomes (genome-centric metagenomics), this was not possible for a while in case of very complex communities, and the reconstruction of complete genomes could be achieved only for low to medium diversity samples (Tyson et al. 2004). Another problem is that such assembled individual genomes are most probably chimeras of genetic information derived from closely related microorganisms. An alternative strategy is to directly analyse the unassembled sequence data (gene-centric metagenomics) by comparing reads directly to protein databases without linking phylogenetic information to function (Jaenicke et al. 2011; Wirth et al. 2012; Li et al. 2013). The first metagenome studies of biogas systems were performed with samples from a full-scale agricultural biogas plant (Krause et al. 2008; Schlüter et al. 2008). These initial results suggested main contributions of *Methanoculleus* to hydrogenotrophic methanogenesis and the role of Clostridia in hydrolysis of cellulosic biomass.

Wirth and co-workers investigated a laboratory-scale CSTR system fed with maize silage and pig manure slurry under mesophilic conditions. A metagenomics approach by using extremely parallel SOLiD™ short-read DNA sequencing was employed to resolve the functional and taxonomic complexity of the reactor microbiome. Firmicutes was the predominant bacterial phylum with Clostridia (36%) and Bacilli (11%) as dominant classes, while the major methanogenic order was the Methanomicrobiales and the most abundant species was *Methanoculleus marisnigri*. Functional genes revealed the importance of hydrogen metabolism by identification of genes involved in both production and consumption (Wirth et al. 2012). Solli et al. (2014) studied the start-up of four parallel laboratory-scale CSTRs co-digesting fish waste and cow manure. The most abundant phyla in all reactors were Firmicutes followed by Bacteroidetes and the candidate phylum Cloacimonetes (WWE1). The Cloacimonetes increased in relative abundance in all reactors compared to the original inoculum. The predominant methanogen was affiliated to the hydrogenotrophic genus *Methanoculleus*. The acetoclastic genera *Methanosarcina* and *Methanosaeta* were present but their abundance was significantly lower. Genes encoding enzymes for methane formation from both CO_2/H_2 and acetate were present in the reactors. A high number of reads were annotated as genes involved in amino acid and carbohydrate metabolism, which is consistent with the finding that many species found in biogas reactors are involved in protein and carbohydrate digestion. Further examples for the application of metagenomics of the microbiomes of full-scale biogas plants are shown in Table 7.7.

Table 7.7 Examples for the application of whole genome sequencing (metagenomics) to reveal the phylogenetic diversity and major metabolic pathways of full-scale biogas reactor communities

Reactor type/ temperature	Substrate	Sequencing platform	Most abundant bacterial taxa	Reference
CSTR (mesophilic)	Maize silage, green rye, chicken manure	454 pyrosequencing	Clostridiales (phylum Firmicutes), Bacteroidales (phylum Bacteroidetes), Methanomicrobiales	Krause et al. (2008), Schlüter et al. (2008), Kröber et al. (2009), Jaenicke et al. (2011)
Wastewater treatment plants (mesophilic)	High-strength wastewater and sludge	Illumina HiSeq	Clostridia, Proteobacteria	Cai et al. (2016)
Three connected identical cylindrical digesters (thermophilic)	Maize silage, barley, cattle manure, pig manure	Illumina MiSeq	Firmicutes, Synergistetes, Thermotogae	Maus et al. (2016)
Pilot-scale CSTRs (mesophilic to thermophilic)	Sugar beet pressed pulp	Ion Torrent	Bacteroidetes, Firmicutes, Proteobacteria	Tukacs-Hajos et al. (2014)
Agricultural biogas plant (mesophilic)	Maize silage, cow manure, chicken manure	454 pyrosequencing	Firmicutes, Bacteroidetes, Spirochaetes	Güllert et al. (2016)

A recent study aimed to characterize the core microorganisms in manure-based CSTRs and successfully assembled 157 new genomes from the extended data set (Treu et al. 2016a). In a similar way, Vanwonterghem et al. managed to assemble 101 population genomes from a metagenome dataset obtained from a laboratory-scale biogas reactor study (Vanwonterghem et al. 2016). The metabolic potential analysis of such assembled populations revealed metabolic networks with a high level of functional redundancy as well as niche specialization. The study of Frank et al. (2016) is a good example to go beyond a simple description of the community and its metabolic potential. They investigated a commercial, ammonia-tolerant biogas reactor fed with slaughterhouse and municipal waste and reported the discovery and dominance of a novel uncultured phylotype (unFirm_1). They managed to reconstruct the respective genome, and a quantitative metaproteome analysis implied a function in syntrophic acetate oxidation. Although other cultured syntrophic acetate oxidizing bacteria were also identified in the reactor, their limited proteomic representation suggested that unFirm_1 plays an important role in converting acetate to methane in syntrophic interaction with hydrogenotrophic methanogens.

Metagenome datasets were also analysed for the presence of putative pathogenic microorganisms, such as *Clostridium botulinum* and *Escherichia coli*, to address controversial discussion about the potential role of biogas plants in spreading of pathogens (Eikmeyer et al. 2013). Only very few sequences were predicted to originate from pathogenic clostridial species, and mapping of metagenome reads revealed that only species that are more or less related to pathogenic ones were present and known virulence determinants could hardly be detected. Another interesting opportunity for the utilization of the extended metagenome data is to discover novel lignocellulose hydrolyses with potential application for the enhancement of lignocellulose-rich biomass degradation in AD systems (Pandit et al. 2016; Yang et al. 2016; Yan et al. 2013; Xia et al. 2013). Such gene mining might be extended by cloning and biochemical characterization of novel genes assembled from metagenomic data as it was demonstrated recently (Wang et al. 2015).

Metatranscriptome analyses have the advantage that they provide information on the metabolically active community members. Methodological challenges are associated with RNA isolation, overrepresentation of ribosomal RNA and the short half-life of mRNA. The first metatranscriptome analysis of a biogas reactor was performed by Zakrzewski et al. (2012). Firmicutes was the dominant active bacterial phylum followed by Bacteroidetes, Actinobacteria and Synergistetes. Abundant transcripts were identified to encode enzymes involved in substrate hydrolysis, acidogenesis, acetate formation and methanogenesis. Transcripts for enzymes functioning in methanogenesis were more abundant than it was deduced from the 16S rRNA sequence tags. This result emphasised that key enzymes of the methanogenesis are highly expressed, and despite the low relative abundance of methanogens compared to bacteria, they can be highly active as terminal key players of the biogas process.

A combined application of metagenomics and metatranscriptomics with increased sequencing depth on samples from an agricultural production-scale biogas plant was recently demonstrated (Bremges et al. 2015). Sequencing at least one order of magnitude deeper than previous studies enabled the mapping of transcripts to reconstructed genome sequences and hence, the identification of active metabolic pathways in target organisms. As an example the reconstruction of most genes involved in the methane metabolism was demonstrated, although not all key gene transcripts were found by metatranscriptomics. Further studies with the combined application of metagenomics and metatranscriptomics investigated the anaerobic digestion of *Spirulina* microalga at extreme alkaline conditions (Nolla-Ardevol et al. 2015), and the effect of long chain fatty acids (oleate) addition on the microbial community as well as on the transcriptional responses (Treu et al. 2016b). A deep metagenome and metatranscriptome analysis was conducted to reveal the differences in lignocellulose digestion strategies among the microbiomes of an industrial biogas fermenter, a cow rumen, and elephant gut communities (Güllert et al. 2016). The relatively lower hydrolysis rates in the biogas plant compared to the animal digestive systems was partially explained by the lower concentration of cellulolytic glycoside hydrolase (GH) genes (as number of cellulolytic GH genes per Mbp sequence data). Moreover, highly transcribed cellulases in the biogas plant samples were four times more often affiliated with the phylum Firmicutes compared to the Bacteroidetes, while an equal distribution was observed in the elephant feces sample.

The metaproteome of microbial communities in anaerobic digesters can be investigated by protein extraction and fractionation followed by chromatographic separation and tandem mass spectrometric analysis. However, to exploit the potential of the metaproteomics approach, a more comprehensive biogas-related database is needed, because current databases are full with hypothetical proteins with unknown functions. To understand the function and role of these proteins, currently a supportive metagenome information as scaffold is needed as it was demonstrated by the study of Hanreich et al. (2012). Similarly, Ortseifen and co-workers found out during an integrated metagenome and metaproteome analysis of a biogas plant that public databases yielded insufficient identification rates compared to a corresponding metagenome database from the same sample (Ortseifen et al. 2016). The application of metaproteomics for biogas plant samples was recently reviewed including an overview of the workflow and potential pitfalls (Heyer et al. 2015).

A recent study used multiple meta-omic approaches (including quantitative metaproteomics) to characterize an industrial biogas reactor treating food waste at high temperature and elevated free ammonia levels. Metaproteomics data suggested acclimatization and activity of a *Methanosaeta* phylotype even at high ammonia levels. Moreover, a metabolic scenario was drafted whereby multiple uncultured syntrophic acetate-oxidizing bacteria are capable of syntrophically oxidizing acetate as well as longer-chain fatty acids (via the β-oxidation and Wood-Ljundahl pathways) to hydrogen and carbon dioxide. A recent large-scale metaproteomics study investigated 35 different industrial biogas plants (Heyer et al. 2016), the same ones previously assessed by 16S rRNA gene amplicon sequencing by

De Vrieze et al. (2015). Similar microbial key players were identified and additionally the main biomass degradation pathways were elucidated. An important finding was that high ammonia levels correlated with hydrogenotrophic methanogens and bacterial one-carbon metabolism.

These meta-omics studies also unveiled that most of the microorganisms were still unexplored and only limited functional information could be derived due to missing reference genome information (Treu et al. 2016a). The next-generation sequencing technology is still advancing rapidly and a substantial cost reduction per read can be expected in future, which will further accelerate the application of omics approaches in the field of biogas microbiology research (Vanwonterghem et al. 2014). Future studies should go beyond snapshot analyses and need to support complex experiments carefully designed to answer specific ecological questions (Prosser 2015). Examples for recent laboratory-scale reactor studies are the investigation of the effect of bioaugmentation in case of biogas production from protein-rich substrates (Kovacs et al. 2015), comparison of optimal and acidified straw digesting systems (Pore et al. 2016), assessment of the effect of long chain fatty acids pulses (Kougias et al. 2016), and study of the effect of alkaline sludge pretreatment on the microbiome (Wong et al. 2013).

7.6.4 Importance of Cultivation to Aid and Benefit from Molecular Biology Techniques

Linking physiological function to molecular datasets by comparing sequences to closely-related cultured relatives has many shortcomings. Even closely related microorganisms might have completely different functions and sequencing of 16S rRNA genes can generally achieve only genus level classification, which further reduces the predictability of the metabolic function. Moreover, most of the sequences are only related to unknown species. The gap between the number of cultivated and well-described microorganisms and the putative microorganisms described only by DNA sequences is widening in an alarming rate. As a consequence, there is a similar gap between the characterised and hypothetical proteins identified only by metagenomics (Hugenholtz and Tyson 2008). A good example is the candidate phylum Cloacimonetes (WWE1), which is often an abundant taxon in biogas systems (Lucas et al. 2015; Ozbayram et al. 2017; Solli et al. 2014) but it has no cultivated member so far.

These observations highlight the need for obtaining more key players of the biogas process in pure culture by applying novel and much more sophisticated cultivation methods. A very elegant approach was applied recently by Maus et al. (2016) to study a thermophilic biogas plant by including metagenome and metatranscriptome analyses complemented by the cultivation of hydrolytic and acid-producing bacteria as well as methanogenic archaea. The value of new pure cultures genomes for the interpretation of metagenome and metatranscriptome data was also demonstrated. As an example, the up to now only poorly recognized role of *Defluviitoga tunisiensis* (also obtained in pure culture by Maus et al. (2016)) in thermophilic anaerobic digestion of agricultural residues was underlined.

Assembled genome information from metagenome data can also help in the development of isolation and cultivation strategies as it was demonstrated by the study of Pope et al. (2011). A dominant bacterial species (WG-1) affiliated to the family Succinivibrionaceae and implicated in lower methane emissions from starch-containing diets was isolated from the wallaby gut microbiota. The successful cultivation strategy to obtain an axenic culture was devised from the partial reconstruction of the bacterium's metabolism from binned metagenome data. A similar strategy could be applied for the targeted isolation of abundant but so far not cultivated microorganisms of the complex microbiome in anaerobic digesters.

7.7 Concluding Remarks

AD can be regarded as a dual-purpose technology as it stabilizes the solid biomass waste and also converts the complex organic material of such waste into biogas, thus having the ability to convert a waste management issue into a profit option. It is regarded as one of the most widely implemented biotechnology solutions in the management of organic waste as it exhibits both economical and energy recovery benefits (Weiland 2003). This process takes advantage of waste material as feedstock which are available all-round the year at low cost, producing high-quality end products (Tognetti et al. 2011). One metric ton of organic solid waste, if not treated, has the potential to breakdown and emit 50–110 m^3 of carbon dioxide CO_2 and 90–140 m^3 of methane into the atmosphere, equivalent to around 400 kWh of power (Cecchi 2011; Vietitez and Ghosh 1997). Hence, the application of AD technology is of outmost importance to attain the environmental goals that our society has set regarding sustainable development. Knowledge on process technology and process microbiology has expanded rapidly during the recent years. This is mainly due to the ongoing revolution of the molecular biological techniques and associated data analysis methods for the assessment of structures, metagenomes and metatranscriptomes of microbial communities in biogas plants. Although the continuously growing databases produced by these metaomics approaches and their correlations to the reactor parameters revealed many interesting aspects of the biogas process, which cannot be considered as a black box process anymore, the complex biogas microbiome is so complex, that its effective management is still a challenge for the future.

References

Abendroth C et al (2015) Eubacteria and archaea communities in seven mesophile anaerobic digester plants in Germany. Biotechnol Biofuels 8:87

Agler MT et al (2011) Waste to bioproduct conversion with undefined mixed cultures: the carboxylate platform. Trends Biotechnol 29(2):70–78

Agneessens LM et al (2017) In-situ biogas upgrading with pulse H2 additions: the relevance of methanogen adaption and inorganic carbon level. Bioresour Technol 233:256–263

Ahring BK et al (2003) Biomethanization of the organic fraction of municipal solid wastes. In: Mata-Alvarez J (ed) IWA Publishing, London, UK

Ahring BK, Angelidaki I, Johansen K (1992) Anaerobic treatment of manure together with industrial waste. Water Sci Technol 25(7):311–318

Akuzawa M et al (2011) Distinctive responses of metabolically active microbiota to acidification in a thermophilic anaerobic digester. Microb Ecol 61(3):595–605

Alvarez R, Villca S, Liden G (2006) Biogas production from llama and cow manure at high altitude. Biomass Bioenerg 30(1):66–75

Angelidaki I, Ellegaard L (2003) Codigestion of manure and organic wastes in centralized biogas plants. Appl Biochem Biotechnol 109(1):95–105

Appels L et al (2008) Principles and potential of the anaerobic digestion of waste-activated sludge. Prog Energy Combust Sci 34(6):755–781

Aslanzadeh S et al (2014) Biogas production from N-methylmorpholine-N-oxide (NMMO) pretreated forest residues. Appl Biochem Biotechnol 172(6):2998–3008

Bal AS, Dhagat NN (2001) Upflow anaerobic sludge blanket reactor–a review. Indian J Environ Health 43(2):1–82

Boe K, Angelidaki I (2006) Online monitoring and control of the biogas process. Technical University of DenmarkDanmarks Tekniske Universitet, Department of Systems BiologyInstitut for Systembiologi

Bouallagui H et al (2004) Two-phases anaerobic digestion of fruit and vegetable wastes: bioreactors performance. Biochem Eng J 21(2):193–197

Bouallagui H et al (2005) Bioreactor performance in anaerobic digestion of fruit and vegetable wastes. Process Biochem 40(3):989–995

Braun R, Brachtl E, Grasmug M (2003) Codigestion of proteinaceous industrial waste. Appl Biochem Biotechnol 109(1–3):139–153

Bremges A et al (2015) Deeply sequenced metagenome and metatranscriptome of a biogas-producing microbial community from an agricultural production-scale biogas plant. Gigascience 4:33

Bryant MP (1979) Microbial methane production—theoretical aspects. J Anim Sci 48(1):193–201

Bühligen F et al (2016) A T-RFLP database for the rapid profiling of methanogenic communities in anaerobic digesters. Anaerobe 39:114–116

Cai M et al (2016) Metagenomic reconstruction of key anaerobic digestion pathways in municipal sludge and industrial wastewater biogas-producing systems. Front Microbiol 7:778

Cecchi F et al (2011) Anaerobic digestion of the organic fraction of municipal solid waste for methane production: research and industrial application, in comprehensive biotechnology. In: Moo-Young M (ed) Elsevier, Waterloo, Canada, pp 463–472

Chan ECS (2003) Microbial nutrition and basic metabolism. In Handbook of water and wastewater microbiology. Academic Press, London, pp 3–33

Chandra R, Takeuchi H, Hasegawa T (2012) Methane production from lignocellulosic agricultural crop wastes: a review in context to second generation of biofuel production. Renew Sustain EnergyRev 16(3):1462–1476

Chaudhary BK (2008) Dry continuous anaerobic digestion of municipal solid waste in thermophilic conditions. Master of Engineering thesis: School of Environment, Resources and Development; Asian Institute of Technology. http://faculty.ait.ac.th/visu/public/uploads/Data/AIT-Thesis/Master%20Thesis%20final/Binod.pdf

Chen Y, Cheng JJ, Creamer KS (2008) Inhibition of anaerobic digestion process: a review. Biores Technol 99(10):4044–4064
Chynoweth DP et al (1993) Biochemical methane potential of biomass and waste feedstocks. Biomass Bioenergy 5(1):95–111
Crolla A et al (2012) Advantages and limitations with using various substrates in manure biogas plants. In: 4th annual canadian farm and food biogas conference and exhibition. London Convention Center, London, Ontario
Dai X et al (2016) Metabolic adaptation of microbial communities to ammonium stress in a high solid anaerobic digester with dewatered sludge. Sci Rep 6:28193
De Vrieze J et al (2015) Ammonia and temperature determine potential clustering in the anaerobic digestion microbiome. Water Res 75:312–323
Demirer GN, Chen S (2005) Two-phase anaerobic digestion of unscreened dairy manure. Process Biochem 40(11):3542–3549
Deublein D, Steinhauser A (2008) Biogas from waste and renewable resources: an introduction. In: Deublein D, Steinhauser A (eds) Wiley-VCH Verlag GmbH & Co. KGaA, Mörlenbach, Germany
Eikmeyer FG et al (2013) Detailed analysis of metagenome datasets obtained from biogas-producing microbial communities residing in biogas reactors does not indicate the presence of putative pathogenic microorganisms. Biotechnol Biofuels 6(1):49
Ellis JT et al (2012) Characterization of a methanogenic community within an algal fed anaerobic digester. ISRN Microbiol 2012:753892
Esposito G et al (2012) Anaerobic co-digestion of organic wastes. Rev Environ Sci Bio/Technol 11(4):325–341
Forgács G et al (2012) Methane production from citrus wastes: process development and cost estimation. J Chem Technol Biotechnol 87(2):250–255
Forgács G et al (2013) Pretreatment of chicken feather waste for improved biogas production. Appl Biochem Biotechnol 169(7):2016–2028
Frank JA et al (2016) Novel syntrophic populations dominate an ammonia-tolerant methanogenic microbiome. MSystems 1(5)
Gagen EJ et al (2010) Functional gene analysis suggests different acetogen populations in the bovine rumen and tammar wallaby forestomach. Appl Environ Microbiol 76(23):7785–7795
Ge X, Xu F, Li Y (2016) Solid-state anaerobic digestion of lignocellulosic biomass: recent progress and perspectives. Biores Technol 205:239–249
Gerardi MH (2003) The microbiology of anaerobic digesters. In: Gerardi MH (ed) Wastewater microbiology series. Wiley Incorporated, Hoboken, NJ
Ghosh S et al (2000) Pilot-scale gasification of municipal solid wastes by high-rate and two-phase anaerobic digestion (TPAD). Water Sci Technol 41(3):101–110
Godon JJ et al (1997) Molecular microbial diversity of an anaerobic digestor as determined by small-subunit rDNA sequence analysis. Appl Environ Microbiol 63(7):2802–2813
Goodwin S, McPherson JD, McCombie WR (2016) Coming of age: ten years of next-generation sequencing technologies. Nat Rev Genet 17(6):333–351
Goux X et al (2016) Start-up phase of an anaerobic full-scale farm reactor—appearance of mesophilic anaerobic conditions and establishment of the methanogenic microbial community. Bioresour Technol 212:217–226
Griffin LP (2012) Anaerobic digestion of organic wastes: the impact of operating conditions on hydrolysis efficiency and microbial community composition. Master of Science thesis, Department of Civil and Environmental Engineering; Colorado State University
Güllert S et al (2016) Deep metagenome and metatranscriptome analyses of microbial communities affiliated with an industrial biogas fermenter, a cow rumen, and elephant feces reveal major differences in carbohydrate hydrolysis strategies. Biotechnol Biofuels 9:121
Gunaseelan VN (2004) Biochemical methane potential of fruits and vegetable solid waste feedstocks. Biomass Bioenergy 26(4):389–399
Hajji A, Rhachi M (2013) The influence of particle size on the performance of anaerobic digestion of municipal solid waste. Energy Procedia 36:515–520

Hanreich A et al (2012) Metaproteome analysis to determine the metabolically active part of a thermophilic microbial community producing biogas from agricultural biomass. Can J Microbiol 58(7):917–922

Hartmann H, Ahring B (2006) Strategies for the anaerobic digestion of the organic fraction of municipal solid waste: an overview. Water Sci Technol 53:7–22

Heyer R et al (2015) Metaproteomics of complex microbial communities in biogas plants. Microb Biotechnol 8(5):749–763

Heyer R et al (2016) Proteotyping of biogas plant microbiomes separates biogas plants according to process temperature and reactor type. Biotechnol Biofuels 9:155

Huang Y et al (2010) Succession of the bacterial community and dynamics of hydrogen producers in a hydrogen-producing bioreactor. Appl Environ Microbiol 76(10):3387–3390

Hugenholtz P, Tyson GW (2008) Microbiology: metagenomics. Nature 455(7212):481–483

Ince O (1998) Performance of a two-phase anaerobic digestion system when treating dairy wastewater. Water Res 32(9):2707–2713

Jaenicke S et al (2011) Comparative and joint analysis of two metagenomic datasets from a biogas fermenter obtained by 454-pyrosequencing. PLoS ONE 6(1):e14519

Jagadabhi PS (2011) Methods to enhance hydrolysis during one and two-stage anaerobic digestion of energy crops and crop residues. http://urn.fi/URN:ISBN:978-951-39-4448-3

Jain SR, Mattiasson B (1998) Acclimatization of methanogenic consortia for low pH biomethanation process. Biotech Lett 20(8):771–775

Joshua LR et al (2012) Anaerobic digestion technologies for the treatment of municipal solid waste. Int J Environ Waste Manag 9(1–2):100–122

Kabir MM, Forgács G, Horváth IS (2013) Enhanced methane production from wool textile residues by thermal and enzymatic pretreatment. Process Biochem 48(4):575–580

Kabir MM, Forgács G, Sárvári Horváth I (2015a) Biogas from lignocellulosic materials. In: Karimi K (ed) Lignocellulose-based bioproducts. Springer International Publishing, Cham, pp 207–251

Kabir M et al. (2015b) Biogas from wastes: processes and applications. In: Resource recovery to approach zero municipal waste. CRC Press, pp 107–140

Kabir MM et al (2015c) Experimental and economical evaluation of bioconversion of forest residues to biogas using organosolv pretreatment. Bioresour Technol 178:201–208

Karthikeyan OP, Visvanathan C (2012) Bio-energy recovery from high-solid organic substrates by dry anaerobic bio-conversion processes: a review. Rev Environ Sci Biotechnol 12(3):257–284

Kato MT et al (1994) Feasibility of expanded granular sludge bed reactors for the anaerobic treatment of low-strength soluble wastewaters. Biotechnol Bioeng 44(4):469–479

Kayhanian M (1999) Ammonia inhibition in high-solids biogasification: an overview on practical solutions. Environ Technol 20(4):11

Kelleher M (2007) Anaerobic digestion outlook For MSW streams. Biocycle 48:51

Khalid A et al (2011) The anaerobic digestion of solid organic waste. Waste Manag 31(8):1737–1744

Khanal SK (2008) Anaerobic biotechnology for bioenergy production: principles and applications. Blackwell Publishing Company

Klass DL (1984) Methane from anaerobic fermentation. Science 223(4640):1021–1028

Klindworth A et al (2013) Evaluation of general 16S ribosomal RNA gene PCR primers for classical and next-generation sequencing-based diversity studies. Nucleic Acids Res 41(1):e1

Korres N et al (2013) Bioenergy production by anaerobic digestion using agricultural biomass and organic wastes. Routledge Taylor and Francis Group

Kougias PG et al (2016) Dynamic functional characterization and phylogenetic changes due to long chain fatty acids pulses in biogas reactors. Sci Rep 6:28810

Kovacs E et al (2015) Augmented biogas production from protein-rich substrates and associated metagenomic changes. Bioresour Technol 178:254–261

Krause L et al (2008) Taxonomic composition and gene content of a methane-producing microbial community isolated from a biogas reactor. J Biotechnol 136(1–2):91–101

Kröber M et al (2009) Phylogenetic characterization of a biogas plant microbial community integrating clone library 16S-rDNA sequences and metagenome sequence data obtained by 454-pyrosequencing. J Biotechnol 142(1):38–49
Lebuhn M et al (2014) Towards molecular biomarkers for biogas production from lignocellulose-rich substrates. Anaerobe 29:10–21
Leclerc M, Delgenes JP, Godon JJ (2004) Diversity of the archaeal community in 44 anaerobic digesters as determined by single strand conformation polymorphism analysis and 16S rDNA sequencing. Environ Microbiol 6(8):809–819
Lee SH et al (2012) Monitoring bacterial community structure and variability in time scale in full-scale anaerobic digesters. J Environ Monit 14(7):1893–1905
Leite AF et al (2015) Improved monitoring of semi-continuous anaerobic digestion of sugarcane waste: effects of increasing organic loading rate on methanogenic community dynamics. Int J Mol Sci 16(10):23210–23226
Leite AF et al (2016) Lessons learned from the microbial ecology resulting from different inoculation strategies for biogas production from waste products of the bioethanol/sugar industry. Biotechnol Biofuels 9:144
Levén L, Eriksson AR, Schnürer A (2007) Effect of process temperature on bacterial and archaeal communities in two methanogenic bioreactors treating organic household waste. FEMS Microbiol Ecol 59(3):683–693
Li Y, Park SY, Zhu J (2011) Solid-state anaerobic digestion for methane production from organic waste. Renew Sustain Energy Rev 15(1):821–826
Li A et al (2013) A pyrosequencing-based metagenomic study of methane-producing microbial community in solid-state biogas reactor. Biotechnol Biofuels 6(1):3
Li J et al (2015) Substrate type and free ammonia determine bacterial community structure in full-scale mesophilic anaerobic digesters treating cattle or swine manure. Front Microbiol 6:1337
Liu FH et al (2009) The structure of the bacterial and archaeal community in a biogas digester as revealed by denaturing gradient gel electrophoresis and 16S rDNA sequencing analysis. J Appl Microbiol 106(3):952–966
Liu X et al (2012) Pilot-scale anaerobic co-digestion of municipal biomass waste: focusing on biogas production and GHG reduction. Renew Energy 44:463–468
Lucas R et al. (2015) Long-term monitoring reveals stable and remarkably similar microbial communities in parallel full-scale biogas reactors digesting energy crops. FEMS Microbiol Ecol 91(3)
Lueders T et al (2001) Molecular analyses of methyl-coenzyme M reductase alpha-subunit (mcrA) genes in rice field soil and enrichment cultures reveal the methanogenic phenotype of a novel archaeal lineage. Environ Microbiol 3(3):194–204
Luning L, van Zundert EH, Brinkmann AJ (2003) Comparison of dry and wet digestion for solid waste. Water Sci Technol 48(4):15–20
Luo G et al (2015) New steady-state microbial community compositions and process performances in biogas reactors induced by temperature disturbances. Biotechnol Biofuels 8:3
Luton PE et al (2002) The mcrA gene as an alternative to 16S rRNA in the phylogenetic analysis of methanogen populations in landfill. Microbiol Sgm 148:3521–3530
Lv Z et al (2014) Influences of the substrate feeding regime on methanogenic activity in biogas reactors approached by molecular and stable isotope methods. Anaerobe 29:91–99
Mata-Alvarez J, Mace S, Llabres P (2000) Anaerobic digestion of organic solid wastes. An overview of research achievements and perspectives. Biores Technol 74(1):3–16
Maus I et al (2016) Unraveling the microbiome of a thermophilic biogas plant by metagenome and metatranscriptome analysis complemented by characterization of bacterial and archaeal isolates. Biotechnol Biofuels 9:171
Maus I et al (2017) Biphasic study to characterize agricultural biogas plants by high-throughput 16S rRNA gene amplicon sequencing and microscopic analysis. J Microbiol Biotechnol 27(2):321–334

McHugh S, Collins G, O'Flaherty V (2006) Long-term, high-rate anaerobic biological treatment of whey wastewaters at psychrophilic temperatures. Bioresour Technol 97(14):1669–1678
Moharram MA, Abdelhalim HS, Rozaik EH (2016) Anaerobic up flow fluidized bed reactor performance as a primary treatment unit in domestic wastewater treatment. HBRC J 12(1):99–105
Møller HB, Sommer SG, Ahring BK (2004a) Biological degradation and greenhouse gas emissions during pre-storage of liquid animal manure. J Environ Qual 33(1):27–36
Møller HB, Sommer SG, Ahring BK (2004b) Methane productivity of manure, straw and solid fractions of manure. Biomass Bioenerg 26(5):485–495
Mulat DG et al (2015) Changing feeding regimes to demonstrate flexible biogas production: effects on process performance, microbial community structure, and methanogenesis pathways. Appl Environ Microbiol 82(2):438–449
Munk B et al (2010) Population dynamics of methanogens during acidification of biogas fermenters fed with maize silage. Eng Life Sci 10(8):496–508
Munk B et al (2012) A metabolic quotient for methanogenic archaea. Water Sci Technol 66 (11):2311–2317
Nair RB et al (2015) Dilute phosphoric acid pretreatment of wheat bran for enzymatic hydrolysis and subsequent ethanol production by edible fungi *Neurospora* intermedia. Ind Crops Prod 69:314–323
Nair RB et al. (2016) Optimizing dilute phosphoric acid pretreatment of wheat straw in the laboratory and in a demonstration plant for ethanol and edible fungal biomass production using Neurospora intermedia. J Chem Technol Biotechnol n/a–n/a
Nair RB, Lennartsson PR, Taherzadeh MJ (2017) Bioethanol production from agricultural and municipal wastes. In: Current developments in biotechnology and bioengineering. Elsevier, pp 157–190
Nettmann E et al (2008) Archaea diversity within a commercial biogas plant utilizing herbal biomass determined by 16S rDNA and mcrA analysis. J Appl Microbiol 105(6):1835–1850
Neves L et al (2008) Influence of composition on the biomethanation potential of restaurant waste at mesophilic temperatures. Waste Manag 28(6):965–972
Nikolausz M et al (2005) Observation of bias associated with re-amplification of DNA isolated from denaturing gradient gels. FEMS Microbiol Lett 244(2):385–390
Nikolausz M et al (2013) Evaluation of stable isotope fingerprinting techniques for the assessment of the predominant methanogenic pathways in anaerobic digesters. Appl Microbiol Biotechnol 97(5):2251–2262
Nolla-Ardevol V, Strous M, Tegetmeyer HE (2015) Anaerobic digestion of the microalga Spirulina at extreme alkaline conditions: biogas production, metagenome, and metatranscriptome. Front Microbiol 6:597
Oleszkiewicz JA, Sharma VK (1990) Stimulation and inhibition of anaerobic processes by heavy metals—a review. Biol Wastes 31(1):45–67
Ortseifen V et al (2016) An integrated metagenome and -proteome analysis of the microbial community residing in a biogas production plant. J Biotechnol 231:268–279
Owens JM, Chynoweth DP (1993) Biochemical methane potential of municipal solid waste (MSW) components. Water Sci Technol 27(2):14
Ozbayram EG et al (2017) Effect of bioaugmentation by cellulolytic bacteria enriched from sheep rumen on methane production from wheat straw. Anaerobe 46:122–130
Pagés-Díaz J et al (2011) Co-digestion of different waste mixtures from agro-industrial activities: kinetic evaluation and synergetic effects. Biores Technol 102(23):10834–10840
Pagés-Díaz J et al (2015) Semi-continuous co-digestion of solid cattle slaughterhouse wastes with other waste streams: interactions within the mixtures and methanogenic community structure. Chem Eng J 273:28–36
Pandit PD et al (2016) Mining of hemicellulose and lignin degrading genes from differentially enriched methane producing microbial community. Bioresour Technol 216:923–930
Patinvoh RJ et al (2017) Dry fermentation of manure with straw in continuous plug flow reactor: reactor development and process stability at different loading rates. Biores Technol 224: 197–205

Poirier S et al (2016) Anaerobic digestion of biowaste under extreme ammonia concentration: identification of key microbial phylotypes. Bioresour Technol 207:92–101
Pope PB et al (2011) Isolation of Succinivibrionaceae implicated in low methane emissions from *Tammar* wallabies. Science 333(6042):646–648
Popp D et al (2015) Biogas production from coumarin-rich plants–inhibition by coumarin and recovery by adaptation of the bacterial community. FEMS Microbiol Ecol 91(9):fiv103
Popp D et al (2017) Inhibitory effect of coumarin on syntrophic fatty acid oxidizing and methanogenic cultures and biogas reactor microbiomes. Appl Environ Microbiol
Pore SD et al (2016) Metagenome changes in the biogas producing community during anaerobic digestion of rice straw. Bioresour Technol 213:50–53
Prosser JI (2015) Dispersing misconceptions and identifying opportunities for the use of 'omics' in soil microbial ecology. Nat Rev Microbiol 13(7):439–446
Rastogi G et al (2008) Investigation of methanogen population structure in biogas reactor by molecular characterization of methyl-coenzyme M reductase A (mcrA) genes. Biores Technol 99(13):5317–5326
Regueiro L, Lema JM, Carballa M (2015) Key microbial communities steering the functioning of anaerobic digesters during hydraulic and organic overloading shocks. Bioresour Technol 197:208–216
Resch C et al (2011) Enhancement options for the utilisation of nitrogen rich animal by-products in anaerobic digestion. Biores Technol 102(3):2503–2510
Rothberg JM, Leamon JH (2008) The development and impact of 454 sequencing. Nat Biotechnol 26(10):1117–1124
Salehian P, Karimi K (2013) Alkali pretreatment for improvement of biogas and ethanol production from different waste parts of pine tree. Ind Eng Chem Res 52(2):972–978
Schink B (1997) Energetics of syntrophic cooperation in methanogenic degradation. Microbiol Mol Biol Rev 61(2):262–280
Schlüter A et al (2008) The metagenome of a biogas-producing microbial community of a production-scale biogas plant fermenter analysed by the 454-pyrosequencing technology. J Biotechnol 136(1–2):77–90
Schnürer A (2016) Biogas production: microbiology and technology. Adv Biochem Eng Biotechnol 156:195–234
Schnürer A, Jarvis Å (2009) Mikrobiologisk handbok för biogas anläggningar. Rapport U2009 3
Schnürer A, Jarvis Å (2010) Microbiological handbook for biogas plants. Swedish Waste Management U2009:03. Swedish Gas Centre Report 207 Volume 1–138
Sharma A, Unni BG, Singh HD (1999) A novel fed-batch digestion system for biomethanation of plant biomasses. J Biosci Bioeng 87(5):678–682
Sipos R et al (2007) Effect of primer mismatch, annealing temperature and PCR cycle number on 16S rRNA gene-targetting bacterial community analysis. FEMS Microbiol Ecol 60(2):341–350
Soccol RC et al (2011) Lignocellulosic bioethanol: current status and future perspectives. In: Biofuels. Academic Press, Amsterdam, pp 101–122
Solli L et al (2014) A metagenomic study of the microbial communities in four parallel biogas reactors. Biotechnol Biofuels 7(1):146
Steinberg LM, Regan JM (2008) Phylogenetic comparison of the methanogenic communities from an acidic, oligotrophic fen and an anaerobic digester treating municipal wastewater sludge. Appl Environ Microbiol 74(21):6663–6671
Sträuber H, Lucas R, Kleinsteuber S (2016) Metabolic and microbial community dynamics during the anaerobic digestion of maize silage in a two-phase process. Appl Microbiol Biotechnol 100 (1):479–491
Sun L et al (2015) Characterization of microbial community structure during continuous anaerobic digestion of straw and cow manure. Microb Biotechnol 8(5):815–827
Sun L et al (2016) The microbial community structure in industrial biogas plants influences the degradation rate of straw and cellulose in batch tests. Biotechnol Biofuels 9:128
Sundberg C et al (2013) 454 pyrosequencing analyses of bacterial and archaeal richness in 21 full-scale biogas digesters. FEMS Microbiol Ecol 85(3):612–626

Taherzadeh MJ, Karimi K (2008) Pretreatment of lignocellulosic wastes to improve ethanol and biogas production: a review. Int J Mol Sci 9(9):1621–1651

Talbot G et al (2008) Evaluation of molecular methods used for establishing the interactions and functions of microorganisms in anaerobic bioreactors. Water Res 42(3):513–537

Tchobanoglous G, Theisen H, Vigil S (1993) Integrated solid waste management: engineering principles and management issues. McGraw-Hill Companies, Incorporated

Teghammar A et al (2010) Pretreatment of paper tube residuals for improved biogas production. Biores Technol 101(4):1206–1212

Teghammar A et al (2012) Enhanced biogas production from rice straw, triticale straw and softwood spruce by NMMO pretreatment. Biomass Bioenergy 36:116–120

Thomas T, Gilbert J, Meyer F (2012) Metagenomics—a guide from sampling to data analysis. Microb Inform Exp 2(1):3

Tognetti C, Mazzarino MJ, Laos F (2011) Comprehensive quality assessment of municipal organic waste composts produced by different preparation methods. Waste Manag 31(6):1146–1152

Town J et al (2014) Microbial community composition is consistent across anaerobic digesters processing wheat-based fuel ethanol waste streams. Bioresour Technol 157:127–133

Treu L et al (2016a) Deeper insight into the structure of the anaerobic digestion microbial community; the biogas microbiome database is expanded with 157 new genomes. Bioresour Technol 216:260–266

Treu L et al (2016b) Untangling the effect of fatty acid addition at species level revealed different transcriptional responses of the biogas microbial community members. Environ Sci Technol 50 (11):6079–6090

Tukacs-Hajos A et al (2014) Monitoring of thermophilic adaptation of mesophilic anaerobe fermentation of sugar beet pressed pulp. Bioresour Technol 166:288–294

Tyson GW et al (2004) Community structure and metabolism through reconstruction of microbial genomes from the environment. Nature 428(6978):37–43

Vandevivere P, Baere LD, Verstraete W (2003) Types of anaerobic digester for solid wastes. In: Mata-Alvarez J (ed) Biomethanization of the organic fraction of municipal solid wastes. IWA Publishing, pp 111–140

Vanwonterghem I et al (2014) Linking microbial community structure, interactions and function in anaerobic digesters using new molecular techniques. Curr Opin Biotechnol 27:55–64

Vanwonterghem I et al (2016) Genome-centric resolution of microbial diversity, metabolism and interactions in anaerobic digestion. Environ Microbiol 18(9):3144–3158

Vavilin VA, Rytov SV, Lokshina LY (1996) A description of hydrolysis kinetics in anaerobic degradation of particulate organic matter. Biores Technol 56(2):229–237

Verma S (2002) Anaerobic digestion of biodegradable organics in municipal solid wastes. M.Sc. Dissertation, Department of Earth and Environmental Engineering; School of Engineering and Applied Science, Columbia University

Vietitez ER, Ghosh S (1997) Biogasification of solid wastes by two-phase anaerobic fermentation. Biomass Bioenergy 16:11

Vignais PM, Billoud B, Meyer J (2001) Classification and phylogeny of hydrogenases. FEMS Microbiol Rev 25(4):455–501

Wan C, Zhou Y, Li Y (2011) Liquid hot water and alkaline pretreatment of soybean straw for improving cellulose digestibility. Bioresour Technol 102(10):6254–6259

Wang W, Ji S, Lee I (2013) Fast and efficient nanoshear hybrid alkaline pretreatment of corn stover for biofuel and materials production. Biomass Bioenerg 51:35–42

Wang M et al (2015) Synergistic function of four novel thermostable glycoside hydrolases from a long-term enriched thermophilic methanogenic digester. Front Microbiol 6:509

Ward AJ et al (2008) Optimisation of the anaerobic digestion of agricultural resources. Bioresour Technol 99(17):7928–7940

Weiland P (2003) Production and energetic use of biogas from energy crops and wastes in Germany. Appl Biochem Biotechnol 109:1–2

Wellinger A, Wyder K, Metzler AE (1993) Kompogas—a new system for the anaerobic treatment of source separated waste. Water Sci Technol 27(2):153–158

Werner JJ et al (2011) Bacterial community structures are unique and resilient in full-scale bioenergy systems. Proc Natl Acad Sci USA 108(10):4158–4163
Westerholm M et al (2011) Changes in the acetogenic population in a mesophilic anaerobic digester in response to increasing ammonia concentration. Microbes Environ 26(4):347–353
Wilkins D et al (2015) Pyrosequencing of mcrA and archaeal 16S rRNA genes reveals diversity and substrate preferences of methanogen communities in anaerobic digesters. Appl Environ Microbiol 81(2):604–613
Wintsche B et al (2016) Trace elements induce predominance among methanogenic activity in anaerobic digestion. Front Microbiol 7:2034
Wirth R et al (2012) Characterization of a biogas-producing microbial community by short-read next generation DNA sequencing. Biotechnol Biofuels 5:41
Wong MT et al (2013) Towards a metagenomic understanding on enhanced biomethane production from waste activated sludge after pH 10 pretreatment. Biotechnol Biofuels 6(1):38
Xia Y et al (2013) Mining of novel thermo-stable cellulolytic genes from a thermophilic cellulose-degrading consortium by metagenomics. PLoS ONE 8(1):e53779
Xu K et al (2009) Real-time PCR assays targeting formyltetrahydrofolate synthetase gene to enumerate acetogens in natural and engineered environments. Anaerobe 15(5):204–213
Yan X et al (2013) Discovery of (hemi-) cellulase genes in a metagenomic library from a biogas digester using 454 pyrosequencing. Appl Microbiol Biotechnol 97(18):8173–8182
Yang Y et al (2004) Performance of a fixed-bed reactor packed with carbon felt during anaerobic digestion of cellulose. Biores Technol 94(2):197–201
Yang L et al (2015) Challenges and strategies for solid-state anaerobic digestion of lignocellulosic biomass. Renew Sustain Energy Rev 44:824–834
Yang C et al (2016) Discovery of new cellulases from the metagenome by a metagenomics-guided strategy. Biotechnol Biofuels 9:138
Zakrzewski M et al (2012) Profiling of the metabolically active community from a production-scale biogas plant by means of high-throughput metatranscriptome sequencing. J Biotechnol 158(4):248–258
Zandvoort MH et al (2006) Trace metals in anaerobic granular sludge reactors: bioavailability and dosing strategies. Eng Life Sci 6(3):293–301
Zhang C, Yuan Q, Lu Y (2014) Inhibitory effects of ammonia on methanogen mcrA transcripts in anaerobic digester sludge. FEMS Microbiol Ecol 87(2):368–377
Zhao Q, Kugel G (1996) Thermophilic/mesophilic digestion of sewage sludge and organic wastes. J Environ Sci Health Part A 31(9):2211–2231
Ziganshin AM et al (2011) Bacteria and archaea involved in anaerobic digestion of distillers grains with solubles. Appl Microbiol Biotechnol 89(6):2039–2052
Ziganshin AM et al (2013) Microbial community structure and dynamics during anaerobic digestion of various agricultural waste materials. Appl Microbiol Biotechnol 97(11):5161–5174
Ziganshin AM et al (2016a) Comparative analysis of methanogenic communities in different laboratory-scale anaerobic digesters. Archaea 2016:3401272
Ziganshin AM et al (2016b) Reduction of the hydraulic retention time at constant high organic loading rate to reach the microbial limits of anaerobic digestion in various reactor systems. Bioresour Technol 217:62–71
Zuo Z et al (2013) Effects of organic loading rate and effluent recirculation on the performance of two-stage anaerobic digestion of vegetable waste. Biores Technol 146:556–561

Chapter 8
Biogas Production Systems: Operation, Process Control, and Troubleshooting

Hossein Ghanavati

8.1 Introduction

Advanced instrumentation and control tools are keys in enhancing energy production and resource recovery during methanogenic anaerobic digestion (AD) processes. Process monitoring can help to identify instabilities during AD, react on time before a severe crash happens, re-stabilize crashed plants, give an overall view of the biogas process, accomplish a successful start-up of a plant, increase gas production, and control odor-related problems to name but a few. There are quite a few possible reasons for process instabilities ranging from the changes in the feedstock and the temperature to the trace metal limitation. Parameters characterizing the processes and the early indicators representing them are the possible subdivided categories of the monitoring parameters in industrial processes.

Control devices to monitor different aspects of the processes are ubiquitous in industrial plants. Many different parameters are involved in industrial processes, as the lack of their monitoring will lead to process failure. For example, gas composition measurements are one of the required steps in monitoring the processes. Carbon dioxide and methane concentration, two major components of the biogas, are measured and monitored by the use of various sensors, including gas analyzers and infrared absorption. In the case of gas composition monitoring, controlling the presence of hydrogen sulfide and the explosive character of the biogas are the other important issues that need to be considered. The pH, volatile fatty acids, alkalinity, total and volatile solids, solid and hydraulic retention time, organic loading rate, feeding schedule, mixing and the range of inhibitors, as well as foam and scum are

H. Ghanavati (✉)
Agricultural Biotechnology Research Institute of Iran (ABRII),
Agricultural Research, Education and Extension Organization (AREEO), Karaj, Iran
e-mail: ghanavatih@abrii.ac.ir

M. Tabatabaei and H. Ghanavati (eds.), *Biogas*, Biofuel and Biorefinery
Technologies 6, https://doi.org/10.1007/978-3-319-77335-3_8

the other important issues, having the key roles in controlling the processes. It is also worth mentioning that the above parameters may be fluctuating in a way that their troubleshooting must be taken into account. It is crucial to know that many industrial problems are originated from not only industrial discharges and process failures, but also equipment malfunction, inadequate maintenance, or design deficiencies. Hence, having an adequate acquaintance with the structure of major equipment in an industrial plant will help technicians and operators to keep the plant in a stable condition. In this chapter, major operation and process control equipment alongside with operational parameters, troubleshooting issues, mostly from a practical point of view, are thoroughly reviewed and discussed.

8.2 Operation and Controlling Equipment in Anaerobic Digestion Process

8.2.1 Mechanical Pretreatment

Mechanical pretreatment, as a critical unit in industrial biogas plants, may consist of shredders and pulpers. These equipment are generally used in order to enhance the (specific) surface area of hard-solid substrates (such as municipal solid wastes—MSW, waste wood, mixed industrial wastes, bulky refuse, waste tyres, waste papers and cardboard, etc.) through breaking down and/or crushing, leading to their more efficient digestion and enhanced AD process. In fact, the larger the particle size, the lower the amount of methane produced during an AD process, that is, the particle size is inversely proportional to the maximum substrate utilization rate (Jain et al. 2015). In addition to size reduction, removal of impurities along with lowering the cost of operation and maintenance, e.g., pumps plugging and blades corrosion, as well as impurities settlement in digesters are among the other benefits of mechanical treatment.

Shredders are often equipped with a magnetic separator to remove metal contaminations from the inlet stream. Since plastics cannot be removed through shredding, separating them prior to the process is essential in order to achieve a reasonably favorable performance of the unit (Hansen et al. 2007). Pulpers (equipped with hydrocyclones) are basically used for the dissolution and defiberation of fermentable organic matters contained in mixed MSW for subsequent biogas production. Pulpers are also important for the efficient separation of non-fermentable "heavy" (e.g., bones, stones, glass, batteries, metal objects, etc.) and "light" impurities (e.g., textile, wood, plastics, foil, etc.) through a rake located at the top of the pulper (for light fraction) and a sluice located at the bottom of the pulper followed by a classifying screw for dewatering and collection of the heavy fraction (Fig. 8.1) (BTA international 2017). A perforated plate is also located at the outlet in order for the organic suspension to be withdrawn. Overall, an easily manageable and pumpable organic suspension including (literally) only pure fermentable components of the waste stream is the main output of this unit operation.

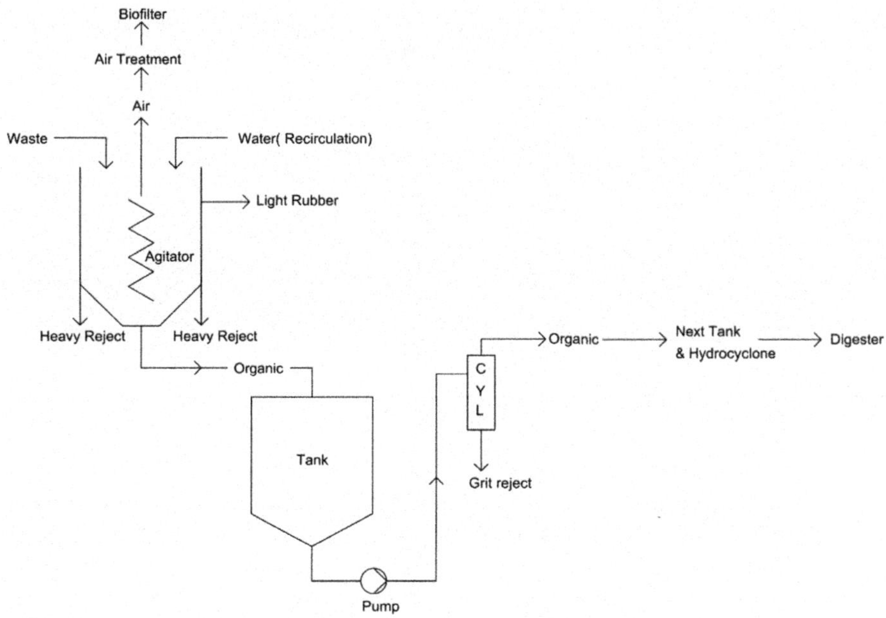

Fig. 8.1 Schematic view of a pulper and its hydrocyclone part together with the waste streams

Figure 8.1 shows a schematic presentation of a pulper and its hydrocyclone part together with the streams of waste.

8.2.2 *Temperature Control*

Temperature is one of the most important and must-be controlled parameters, which plays a crucial role in the anaerobic process stability. There are three general ranges of temperature each favoring a specific type of microorganisms including (1) psychrophilic: about 10–20 °C, (2) mesophilic: about 35–40 °C (3), and thermophilic: about 50–55 °C (Jain et al. 2015). There are a couple of factors, which contribute to heat generation or transfer in a digester including process reaction, mixing (impellers), as well as heat exchangers (hot water or steam). When scaling up digesters from laboratory scale to industrial scale, controlling the temperature at a desired level becomes more difficult, since the ratio of the surface area of the digester to its volume will become smaller. It should be noted that this ratio is an important parameter in determining the speed of cooling by heat exchangers in industrial digesters as well. As a result, a very accurate and effective design of heat exchangers is needed to hold the temperature of digesters at a desired level. An internal loop (warm water) of heat exchanger used for heating digesters is shown in Fig. 8.2.

Fig. 8.2 Internal loop (warm water) for heating the digester

The mixing process itself can also generate heat. Usually, 10–30% of the heat required for the AD process could be contributed by mixing. In cases where systems are overheated, e.g., in hot regions of the world, alternative mixing such as bubble column or air-lift loop would be a better option in order to avoid extra heat generation by conventional mixing. Naturally-occurring evaporation can also exert a cooling effect. Nevertheless, if there is still surplus heat in the system, different types of heat exchangers namely cooling coil, vessel wall, cooling baffles, external loop, etc. could be used.

Based on field experiences, the optimum temperature for mesophilic digesters (42 °C) can be achieved if a fluid temperature range of 65–70 °C is maintained in heat exchangers. An applicable parameter, which plays a critical role in determining how well a heat exchanger has been designed, is the Stanton Number. It is described as the ratio of the heat transfer capacity through coils to the convection capacity in cooling water. Figure 8.3 illustrates the different ranges of this number along with their interpretations. Equation 8.1 also shows the relation through which one may derive the number. Although implementing a heat exchanger in an industrial digester will have such advantages as greater design freedom and faster heat transfer, there will also be such challenges as cold shocks, shear stress in the pump, and oxygen and substrate depletion (in an external loop exchanger).

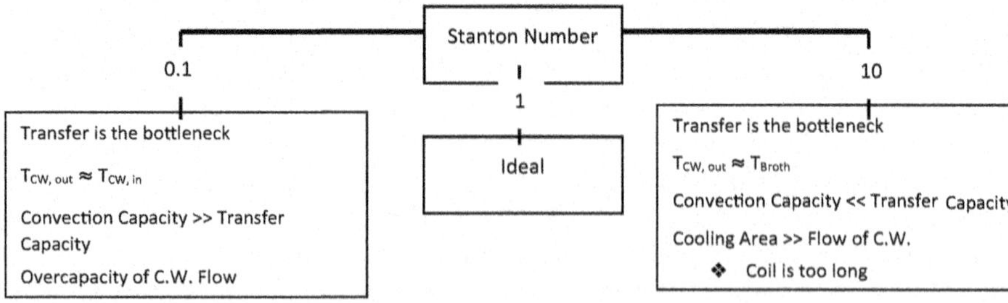

Fig. 8.3 Different ranges of Stanton number along with their interpretations

$$St_{Heat} = \frac{U \cdot A_t}{\rho_{cw} \cdot C_p \cdot \varnothing_{cw}} \tag{8.1}$$

There are a couple of devices used for monitoring the process temperature including liquid thermometers, bimetal thermometers, pressure on liquid or gas expansion bulbs, thermistors, resistance temperature detectors (RTDs), infrared detectors, and crystal window tape. In order for having a stable temperature in biogas digesters, it is crucial to continuously control the temperature with the help of accurate temperature control devices (Clemens 2013). Each control device is made for a unique application. For instance, the RTD is typically used on lower, ambient-range temperatures, while gas- and liquid-filled temperature sensors and thermistors are frequently used for equipment-protection and cooling systems. In digesters, mostly with the help of proportional integral derivative (PID) controllers, the temperature is monitored by controlling a shunt valve that regulates the flow of heating water ("Operational guidelines for Plönninge biogas plant," n.d.). In biogas production processes, controlling the temperature is much more important than the other operational parameters considering the presence of microorganisms, each with their own working temperature range (Esteves et al. 2013). Hence, operators have to periodically calibrate the instruments using a standard and accurate temperature measuring device. In conclusion, controlling the temperature in an industrial scale process is one of the most important and difficult tasks, while it should be noted that adding more heat is easier than removing it from the unit (Atiemo-Obeng and Calabrese 2004). Thus, controlling the temperature, as an operational parameter, is a crucial step in making a stable industrial process.

From the troubleshooting perspective, if the temperature of the slurry under digestion drops, the inlet and outlet pressure of the exchangers and warming system pumps, the health of the thermometer devices, as well as temperature profile in the digester should primarily be checked out. Regarding the origin of problem, opening and cleaning the heat exchangers, measures to mitigate line plugging, or implementation of an increase in the mixing procedure of the digesters may be taken into account. If the slurry temperature increases, it may be ascribed to the improper operation of temperature controller, in which the wise prescription is repairing or replacing the controller. Additionally, it would be wise to check the CHP heat exchanger setting, the health of the warming pipes in the digesters by checking the make-up water, and the accumulation of solid materials at around the sensing device. Furthermore, checking the manual to regulate the setting, closing the valve of the damaged pipe, and opening and cleaning the device, respectively, should be considered. As another worth-mentioning point, in order to prevent any occurrence of deterioration in warming system pipes, aliphatic polyamine should be added 20–30 L per megawatt (MW) of plant capacity to circulating water annually.

8.2.3 Pressure Control

Many instruments have been introduced, for both static fluids and a moving streams, to control and monitor the pressure (Potter and Voss 1975). In an industrial biogas plant, it is crucial to continuously control the pressure for produced biogas. The digesters gas pressure can be measured using both digital and analog pressure gauges. The pressure/vacuum relief valve (safety valve), which is a must-control device and mounted on top of the digester, reacts to the exceedance level of the biogas pressure as well as possible vacuum conditions, leading to a guaranteed safe operation (Fig. 8.4). As another application of pressure measuring, slurry level can be determined using specific liquid gauge pressure devices, which is capable of determining tiny fluctuations.

In terms of pressure control and monitoring, the following components are commonly used, while should precisely be tuned:

Fig. 8.4 A typical mechanical-based pressure control safety valve

- Check valve, to prevent higher pressure gas to come back to the digester;
- Pressure regulator, to maintain a constant pressure at the point of use (such as engine);
- Pressure/vacuum relief valve, to prevent digester structural damages due to a pressure build-up or vacuum condition;
- Manometer, to monitor and control the pressure at collection and storage systems, along the transfer pipes, as well as the inlet and outlet points in units (e.g., scrubber and chiller) (Marx et al. 2010).

8.2.4 Pumps Management

Transferring feedstock into the digesters is one of the primary roles of pumps in an industrial biogas plant. As an important central unit, almost all biomass tanks are connected to pump room, mainly consisting of valves, pumps, and different transmitters. In order for the operators to have an easy access to the control system close to the working area, there is also a small control room in the pump room (Water and Engines 2014). Some of the most important components of the pump room can be seen in Fig. 8.5.

Designing a pump to transmit a specified amount of fluid over a given distance directly relates to the design of the pipelines, as one may use a large-diameter pipe with a small pressure drop, which involves a higher capital cost with lower running costs, or a smaller-diameter pipe with a greater pressure drop, which involves a

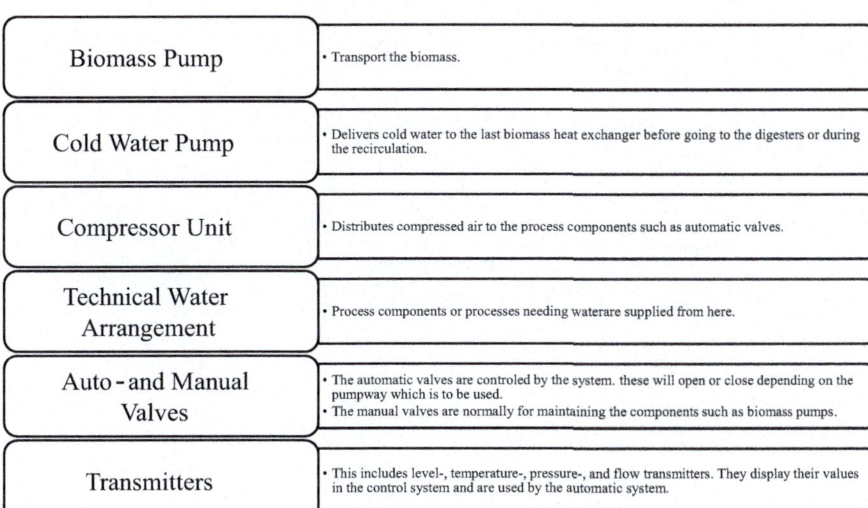

Fig. 8.5 Some of the most important components of the pump room. Adopted from Water and Engines (2014)

lower capital cost with higher running costs due to the need for more pumps (Abdel-Aal 2016). Figure 8.6 shows the most influential parameters in selection of a pump while Fig. 8.7 presents pump classifications.

Practically, horsepower of a pump can be calculated using the following relationship (Eqs. 8.2 and 8.3) (Abdel-Aal 2016).

$$Hydraulic\ Horsepower\ (HP) = \frac{H.l.Q}{3960} = \frac{P.Q}{1714} \tag{8.2}$$

$$Brake\ HP\ (actual) = \frac{HP\ (hydraulic)}{\alpha} \tag{8.3}$$

Where

H is the head in ft;
l is the specific gravity;
Q is the flow rate (gpm);
P is the gauge pressure in lb/in^2;
α is the pump efficiency; 60% is used for centrifugal pumps.

In spite of the fact that there are various types of pumps in industrial processes, it is common to use piston and centrifugal pumps in industrial biogas plants.

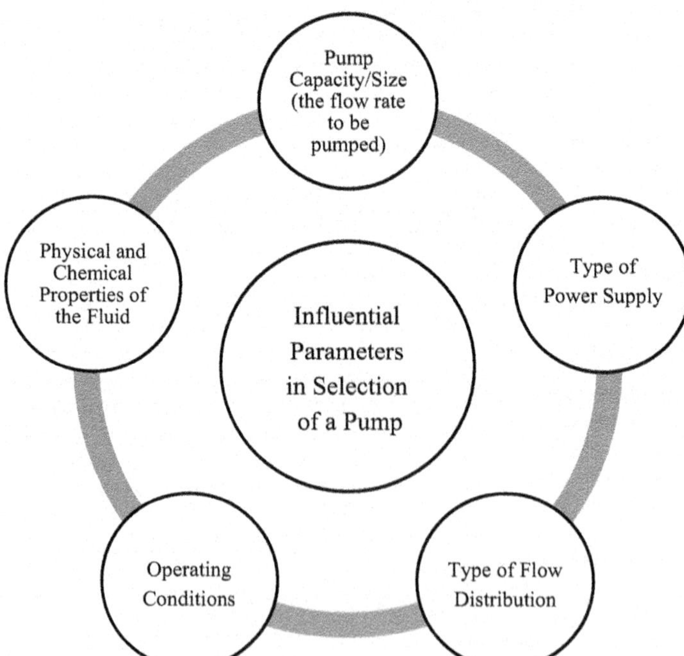

Fig. 8.6 Influential parameters in selection of a pump. Adopted from Abdel-Aal (2016)

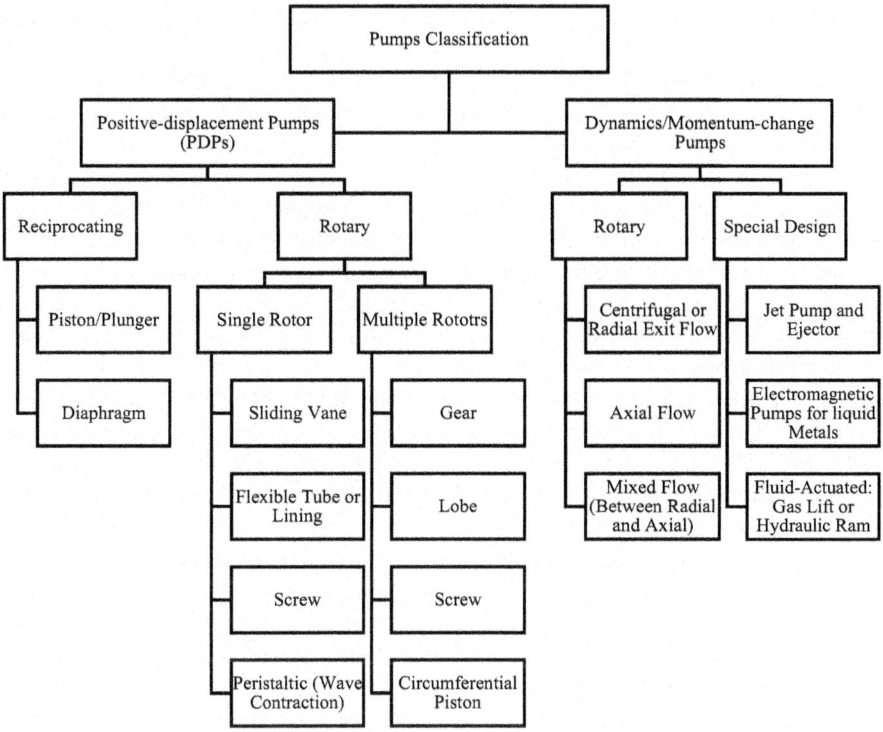

Fig. 8.7 Classification of pumps. Adopted from Potter and Voss (1975)

Piston pumps are usually used for pumping viscous sludge into digesters while centrifugal ones are used for the recirculation of the sludge. Alongside these two common types, progressive cavity and diaphragm pumps are also used to some extent in such industrial processes (Marx et al. 2010).

8.2.5 Mixers and Mixing Management

Uniform concentration, a particular scale of segregation (e.g. particle size, drop size, or striation thickness), and providing a required rate of mixing are the three possible objectives of a good mixing (Kresta et al. 2015). An adequate mass transfer and sufficient contact time between substrate and bacteria need to be guaranteed by a precise design of the reactor. While doing so, a good mixing design can also reduce the digester content, avoid the settlement of such more dense materials as inert ones at the bottom of the digester, and also the formation of floating layer at the top of it. It is necessary to adjust the rate of mixing accurately in different operational circumstances, as this rate will directly affect the quality of the process, especially when there are some sensitive agents like microorganisms, play a key role in it (Esteves et al. 2013).

Different types of mixers that may be used in an industrial biogas production plant are flow accelerators, turbulence generators, and hybrid mixers (Liebetrau and Jacobi 2016). Small propellers running at a relatively high speed (300 rpm and above), a low ratio of thrust and power, lower energy efficiency in comparison to the other ones, and being prone to recirculation behavior are the major characteristics of the turbulent generators, which make them less applicable for industrial biogas plants. Large and low-speed (30–50 rpm) propellers, high volume flow, very high thrust, very low power consumption, energy-efficient, and being able to make an adequate turbulence are the major characteristics of flow accelerators, which are widely used in industrial biogas plants. Good energy efficiency, impressive depth of penetration, and a low tendency in recirculation accompanying with the beneficial features of turbulent generators and flow accelerators make hybrid mixers a good option for the industrial purposes (Liebetrau and Jacobi 2016). The common structure of mixers in industrial biogas plants is a top mixer with two impellers at different levels with a controllable speed. Specially designed rotor blades at the top of the mixer will prevent the formation of floating hard layers in the digester (Esteves et al. 2013).

Based on field experiences, in some of the industrial biogas digesters, the digester itself consists of two upper-mounted mixers alongside with another lower-mounted one, which should be used periodically (e.g., weekly). The mixers should be implemented vertically or, at most, with an angle of 45° with respect to the wall. The longer the propeller, the higher the power of the mixer, for instance, 90 and 140 cm propeller in diameter will lead to a 3000 and 4000 N, respectively. Mixing must be scheduled based on process conditions and features to reach the optimum and desirable outcomes. For instance, in a digester equipped with three mixers, mixing is applied by two of the mixers in a circulation choice of manner. In this case, the schedule is in a way that mixing is performed for a period of 20 min/h.

Overall, appropriate mixing—as a vital tool—will result in a better quality of the following features:

(1) Eliminating any concentration and temperature gradients, (2) improving the AD process by the creation of a sufficient contact between microorganisms and substrates throughout the digester, (3) preventing stratification, (4) reducing the build-up of scum in the digester, (5) promoting the release of biogas, (6) enhancing the heat transfer, and (7) creating uniform and homogenized fluid.

In terms of inhibition, strong mixing causes a disruption in syntrophic and methanogenic bacteria, which, could ultimately lead to the inhibition of syntrophic oxidation of volatile fatty acids (VFAs). It has been observed that mixing intensity has an effect on process inhibition and recovery from organic overload. Intensive mixing during high overload will result in acidification and process failure, i.e., successful digestion will be achieved by low mixing intensity. It should be noted that continuous mixing does not take place in such processes, as it will lead to a reduction in biogas production.

8.2.6 *Gas Composition Monitoring*

Two main components of biogas are CH_4 and CO_2. The ratio of these two major components is stable in a digester except for the occurrence of any imbalance in the biogas production process. The optimal composition for biogas is 100% methane, which seems to be far away from practical approaches (Esteves et al. 2013). In another word, higher concentrations of methane in biogas will lead to a better combustion, cleaner emissions, and a higher output power. In order for controlling the biogas composition, it is better to monitor it by using a continuous gas analyzer, while it is possible to use a portable biogas analyzer as well. This system could be accompanied by a controllable air pump, which finally leads to the implementation of bio-desulfurization by the present sulfur oxidizing bacteria (SOB). More specifically, desulfurization is carried out by a control loop, in which if the gas analyzer detects a deviation from the safe range of oxygen concentration in the gas phase, it will order the air pump to blow into the digester making the digester's environment favorable (i.e., by controlling the oxygen concentration) to the particular bacteria. The importance of frequent gas composition analyzing is that an exceedance of 6% v/v in oxygen concentration in the digester will lead to a higher risk of explosion, as well as an inhibition in the methanogenic and other anaerobic bacteria. In addition, an oxygen concentration below 1% v/v causes an increase in the H_2S concentration.

Biogas analyzers measure CH_4, CO_2, H_2S and O_2, by sampling from the digester discharge pipe every 30 min, and then the concentration for each gas is logged, displayed, and transferred. Operators use the data to make sure about the process preventing any possible shutdown or financial losses. Figure 8.8 shows the schematic presentation and the actual view of an automated industrial gas analyzer together with a desulfurization system (AwiFLEX, n.d.).

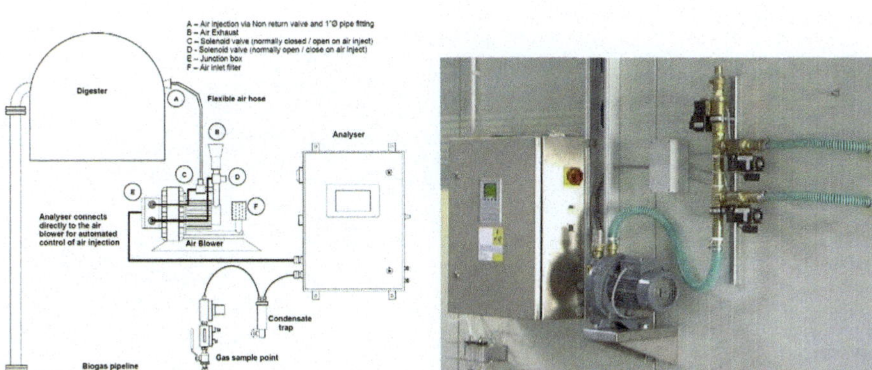

Fig. 8.8 Schematic presentation and the actual view of an automated industrial gas analyzer together with a desulfurization system. Courtesy of AwiFLEX

8.2.7 Digester Covers

Digester cover should be able to keep the digester gas in and prevent air penetration into the digester, preventing the possibility of an explosive condition and gases and odor vitalization. There are two common types of digester covers, i.e., fixed and flexible. Limiting the downward movement of the digester cover as well as protecting the internal equipment from damage, corbels play a key role as another essential part of the flexible covers in digesters (Marx et al. 2010). The flexible covers usually consist of two membrane layers, between which air is continuously pumped. The outer membrane will protect the inner one from different risks of damage, e.g., negative sun radiation effects and rupture. In addition to this, pumping air aims to hold the digester at an inflated shape. In terms of pressure, 4–8 cm of water pressure is the usual pressure range of the gas available on the weight of the gasholder per unit area. Approximately, 50% of the total daily gas production is considered as the volume of the gas cover (Vindis et al. 2014). Fixed holding covers, consists of a concrete dome on top of the digester itself. Long life of utilization (over 20 years) and being devoid of costlier mild steel gas holder which is susceptible to corrosion, are the prominent characteristics of fixed gas holders (Jain et al. 2015). Figure 8.9 shows a flexible gas holder in an industrial biogas plant.

8.2.8 Visual Control

Visual controls are the easiest and most cost-effective way to ensure the safety of workers and industrial processes, improve the overall efficiency of the processes, as well as save time and money. Visual controls can be either a single tag on a pipe or a mounted glass along a pipeline or on a vessel (Carmichael, n.d.). As mentioned above,

Fig. 8.9 Flexible cover digester

Fig. 8.10 Digester-mounted glass. Courtesy of PROGECO S.r.l

pressure, level, and temperature gauges are also among the most prominent visual controls mounted on must-control devices in the field. It is the common of many industries to use different types of visual controls. Figure 8.10 shows a biogas sight glass equipped with wiper, and water spray system. This kind of sight glass is specifically designed for installation in digester tanks used in biogas production with an innovative expanding gasket that conforms to irregularities in the cement walls and ports used in digester tank construction, creating a gas-tight seal. With the help of sight glasses, operators set mixing devices at optimum positions and investigate the overall condition in the digester, e.g., the possibility of scum or foam creation. The equipped wiper will also help to wipe out any vapor on the glass for a clear vision. The intensity of the installed illumination system near glass may also be adjusted by the operators to the desirable extent. Additionally, the camera makes recording particular circumstances in the digester possible for the future investigation.

8.3 Bioprocess Operational Parameters and Troubleshooting

AD is a complex biochemical process in which methane rich biogas produced from organic material. The microbial community in the AD process, such as methanogens, are quite sensitive to environmental conditions and different elements within the digester. This part focuses on general information about the desired range of important parameters and their troubleshooting in AD process in order to secure a stable and efficient process.

Moreover, inhibitors are also reviewed and discussed. Inhibitors, as the major obstacles in running an AD process, may be divided into two categories namely the ones which already present in the substrate (organic or inorganic compounds in the feedstock), and the end-products which are resulted from microbial reactions during a part of AD process. The former consists of many compounds such as high salt loads, antibiotics, long chain fatty acids (LCFA), heavy metals, or other toxic organic substances, while the latter, which are the most common inhibitors, may be VFA, LCFA, ammonia, sulfide, etc. (Boe 2006).

8.3.1 pH

Among many important bioprocess parameters in a biogas production plant, pH is one of the must-control ones. The optimal pH ranges of the hydrolysis and acidogenesis stages are 5.0–6.0 and 5.5–6.5, respectively, whereas the ideal pH range for methanogenic bacteria is 6.8–7.2. In terms of inhibitory effects, the lower the pH, the lower the activity of Methanogens, which will lead to the accumulation of VFAs and subsequently a sour medium. A sour digester can be fixed by the addition of some alkaline chemical materials to the digester (Table 8.1). The key point in adding the materials is that they should be added slowly to avoid inhibition of bacterial activity. Moreover, in the utilization of pH adjusting chemicals, all the safety considerations should be implemented by operators.

8.3.2 Electrical Conductivity

Electrical conductivity (EC), as an estimation of salt content, is determined by measuring the content of dissolved solutes in aqueous extracts of organic solid products by an EC meter. The importance of EC is rooted in the accumulative inhibitory effect of metal and non-metal ionic contents in the biogas processes, due to digesters wastewater recirculation, entrance of highly contented salt feeds, or the

Table 8.1 Common chemicals used to adjust pH in biogas digesters (Marx et al. 2010)

Chemical name	Formula	Common name
Calcium oxide	CaO	Unslaked or quicklime
Calcium hydroxide	$Ca(OH)_2$	Slaked or hydrated lime
Anhydrous ammonia	NH_3	Agricultural fertilizer
Ammonium hydroxide	NH_4OH	Liquid ammonia
Sodium carbonate	$NaCO_3$	Soda ash
Sodium bicarbonate	$NaHCO_3$	Bicarbonate of soda (baking soda)
Sodium hydroxide	$NaOH$	Lye or caustic soda

addition of excessive amounts of alkaline materials. The maximum possible amount of EC could be 25–30 dS/m. In order to eliminate the effect of salinity and its subsequent inhibition, dilution with water with a low EC or even fresh water is recommended as the best practical way. This procedure may also lead to some negative impacts such as the reduction of organic dry matter (ODM) and buffering capacity, but, in the long and controlled run, the biogas process will be stabilized.

8.3.3 Volatile Fatty Acids, Alkalinity, and the Respective Ratio

The levels of VFAs and total alkalinity (TAK) or buffering capacity are two of the typical fast indicators for monitoring the digestion process. As intermediates and potential inhibitory compounds, VFAs, i.e., acetic, propionic, butyric, valeric acids, etc., are produced during the hydrolysis and acidogenesis stages of AD process. VFAs are then utilized as substrate for the acetogenesis and methanogenesis stages to produce methane. The substrates containing bicarbonate buffering capacity and high ammonia content, keep the pH stable around weak alkaline condition, and the digester can tolerate high value of VFA and prevent pH drops (Boe 2006; Falk 2011; Lahav and Morgan 2004). In spite of the presence of such buffering systems, high organic loads of easy degradable carbohydrates and/or introduction of toxic substances could disturb pH stability, and cause accumulation of VFA (Falk 2011). The VFA/TAK ratio reveals the quantity of volatile organic acids to the buffer capacity of carbonate (total alkaline carbonate) in a digester (Deublein and Steinhauser, n.d.).

Based on field experiences in AD plants running on organic fraction of municipal solid waste (OFMSW), the VFA and alkalinity should be in ranges between 7200–12,000 and 12,000–20,000 mg/L, respectively. Accordingly, the VFA/TAK ratio is normally in the range of 0.2–0.6. During start-up, the VFA, alkalinity, and, subsequently, their ratio, are a little different, commonly ranged <4000, 10,000–15,000 mg/L, and 0.3–0.5, respectively.

With a change in the environmental conditions such as substrate types, fast biodegradability, inhibitory effects due to substrate overload, digesters sludge and bacterial biomass removal, and temperature instability, the concentrations of VFAs may increase.

If the ratio of VFA/TAK increases, the following measures could be take: (1) lowering or stopping feeding rate, (2) addition of a secondary sludge or external microbial seed, (3) lowering sludge removal rate, (4) enhancing mixing time, (5) checking digester temperature (6) addition of alkaline materials.

In order to prevent the inhibition of methane production by VFAs accumulation during the AD process, co-digestion or two/three-stage digestion systems have been proven to be effective (Jain et al. 2015).

8.3.4 Ammonia

Ammonia, which is a significant factor affecting the process stability, mainly comes from the degradation of protein wastes. Its toxicity goes up at high pH and high temperature values because of the higher concentration of free ammonia generated under such conditions. Considering the effects of ammonia, the higher the concentration, the lower the methanogenic activity. Methanogens have a higher sensitivity towards ammonia in comparison with the other types of anaerobic microorganisms existing in anaerobic digesters. Regarding different pH values and the need for achieving temperature acclimation, a wide range of inhibitory concentrations of ammonia exists. This range of ammonia must be from 1 to 4 g/L for mesophilic AD processes (also till 6 g/L in appropriate pH values) and from 1 to 2.5 g/L for thermophilic AD processes.

There are three possible actions to reduce NH_3 inhibition, i.e., (1) reducing the input of N-rich substrates (e.g., slaughter house substrates, rape, clover, poultry manure), (2) adding substrates with high C/N ratio, (3) Add Fe^{3+} ($Fe(OH)_3$), and (4) adding clay minerals (Clemens 2013). Other procedures may be as follows: pH reduction; co-digestion with other compounds; addition of Ca^{2+} and Na^{+} rich bentonite (Boe 2006); air stripping and chemical precipitation (Kabdasli et al. 2000); biomass retention enhancement; dilution (Chen et al. 2008); microorganisms immobilization with different types of inert material (clay, activated carbon, or zeolite) (Hansen et al. 1998); addition of ionic exchangers or adsorbents; addition of activated carbon (because of its indirect impact on reducing the inhibition of ammonia); and the addition of antagonistic cations such as Mg^{2+} or Ca^{2+} as process stabilizers (Boe 2006). Nevertheless, the general prescriptions for ammonia toxicity are liquid dilution, solid recycling, and a possible reduction in the amount of entering ammonia-rich feed.

8.3.5 Sulfur Compounds

Sulfate and sulfur compounds, also present in protein wastes, affect both acetogenic and methanogenic organisms because of sulfate reducing bacteria (SRB) being metabolically versatile. With lower concentrations of sulfate, there would be a competition between sulfate-reducing bacteria and methanogenic archaea for hydrogen and acetate. Likewise, with higher concentrations of sulfate, there would also be a competition between SRB and acetogenic bacteria for propionate and butyrate. Sulfide has inhibitory effects in the AD process at even low concentration as 0.003–0.006 M total sulfur or 0.002–0.003 M H_2S (O'Flaherty et al. 1998). In general, sulfate reduction inhibition can be divided into two stages: the competition for common organic and inorganic substrates by SRB suppresses methane production and the toxicity of sulfide to anaerobic bacteria (Harada et al. 1994).

Dissolved sulfate can be removed through various processes including: diluting the wastewater stream (undesirable approach due to higher total volume of wastewater required to be treated), reducing the input of S-rich substrates (e.g., slaughter house substrates as well as paper and leather industry wastewater), precipitation using iron salts, and an anaerobic treatment system coupled with a sulfide removal step over the whole process. The techniques used for sulfide removal may be categorized into physico-chemical techniques (stripping), chemical reactions (coagulation, oxidation, precipitation), or biological conversions (partial oxidation to elemental sulfur (Chen et al. 2008; Li 2015). Nevertheless, the general prescriptions for sulfide inhibitory effects are liquid dilution, precipitation using iron salts, and controlled rate of feeding.

8.3.6 Heavy Metals

As one of the major causes of digester upset or failure, heavy metals such as cobalt, copper, iron, nickel, and zinc, are potential inhibitory compounds in AD processes (Sanchez et al. 1996). Heavy metals can pose serious threat, i.e., toxicity, at their high concentrations, while some of them, e.g., nickle and copper, at low concentrations (below 10–4 M), are vital for enzymatic activity of anaerobic bacteria. The toxicity is applied to the process by replacing naturally occurring metals in enzyme prosthetic groups, inactivating enzymatic system, or, in another word, through the disruption of enzyme function and structure (Chen et al. 2008). An important point that is worth mentioning is that heavy metals, unlike many other toxic substances, are not biodegradable and have the possibility to reach toxic concentrations swiftly (Sterritt and Lester 1980). It is worth mentioning that heavy metals originate from organics accompanied by impurities such as batteries, electronic devices, and hazardous or industrial wastes. The minimum inhibitory concentrations of some heavy metals are shown in Table 8.2.

The methods for the reduction of heavy metal inhibitory effects are modification in the separation of impurities, liquid dilution, precipitation using sulfur and iron compounds (but pH should be controlled above 7).

Table 8.2 Heavy metals and their inhibitory concentration to the AD process

Heavy metal	Minimum inhibitory concentration (mg/L)
Cr	~300
Fe	1750 (carbonate)
Ni	~300
Cu	~300
Zn	~400
Cd	~600
Pb	~340

8.3.7 Long-Chain Fatty Acids

LCFAs are produced through the degradation of lipids and are absorbed on the bacterial cell wall limiting the transportation of essential nutrients. For instance, 18-C LCFA such as oleic and linoleic acid, have inhibitory effects even at concentrations as low as 1.5 g/L. Likewise, hydrogenotrophic methanogens are progressively negatively affected by these compounds in the following order: linoleic acid (18:2) > oleic acid (18:1) > stearic acid (18:0). Dilution of LCFAs, i.e., adding new substrate free of LCFAs, can be considered as a useful method in order to give the microorganism the opportunity to recover (Boe 2006; Templer et al. 2006).

8.3.8 Organics

In order to have an efficient biogas production process, it is essential to control many parameters such as organic loading rate (OLR). OLR is defined as the amount of organic feed, introduced daily into the digester. In another word, OLR is the quantity of volatile solids fed per working volume of digester per day and is expressed as kg VS/m^3 digester/d. An optimum OLR value can be different depending on the type of substrate (Esteves et al. 2013). An overload of organic materials in the digester will lead to a possible acidification and, as a result, a reduction in methane production. During start-up, the OLR should be slowly increased, commonly starting at an OLR of 0.5 kg VS/m^3/d, till reaching the desire range to ensure the adaptation of microorganisms and a perfect biogas production process. Monitoring sludge pumping volume and the amount of volatile solids in the feed sludge have been suggested as the first steps in troubleshooting OLR-related problem. If the VFA/TAK ration increases to 0.3, the addition of a secondary sludge (if available) is recommended. In a mesophilic continuously stirred tank reactor (CSTR), the appropriate range of ORL is 3–5 kg VS/m^3/d with regard to the type of substrate. It has been observed that the inhibitory effect of ORL for microbial populations is when it exceeds 6.4 kg VS/m^3/d (Moriarty 2013). Additionally, to minimize the process errors, many biogas plants operate under low organic material loadings, therefore ignoring potentials for higher biogas production rates (Falk 2011). If any problem occurs during operation by excessive loading rates, the main remedy is to reduce or stop loading. In addition to this, providing longer mixing time and controlling temperature might be applicable.

8.3.9 Foam and Scum

Excessive chemicals addition during start-up or a sour condition in a digester can cause foaming. Additionally, biological foam can also be generated in the digesters.

Brown colored foam is typically associated with having more biomass in the tank than necessary for the influent waste load. In order to fix the problem, increasing chemical feeding together with adequate mixing, while monitoring the other parameters, i.e., volatile acids, alkalinity, pH, and gas production, are suggested. Moreover, performing a cleaning procedure including all gas lines, gas meters, manometer lines, check valves, pressure/vacuum relief valves, and any other gas handling equipment that was affected by the foaming event, is recommended.

In digesters, scum baffle in used to prevent the production of a scum layer on the top of the liquid. Scum may be produced in response to a reduction in the digester's temperature, an insufficient mixing, the presence of light and fibrous feeding substrates, and a low TS value (below 4%) in the digester (the ease of the presence of light undigested materials at the surface of the digester in comparison with high TS). Some of disadvantages associated with scum formation include losing the digester capacity, blocking the liquid and even gas pipes, the reduction of biogas released from liquid, as well as scum penetration into the inner cover layer (Fig. 8.11). Mixing enhancement, putting the impeller at the surface, digester liquid recirculation, and oil addition (70 L/d/MW) directly into the digester could be used as methods for scum troubleshooting.

Fig. 8.11 Scum formation and its penetration into the inner cover layer of the digester

8.4 Conclusions

As two of the most important parts of the AD systems playing key roles in enhancing the energy production and resource recovery from raw materials, advanced instrumentation and control tools should always be considered. Operational and controlling equipment including mechanical pretreatment (shredders and pulpers), temperature and pressure controllers, pumps, mixers, gas composition analyzers, as well as common types of visual controls were briefly reviewed in the this chapter. Likewise, some of the bioprocess operational parameters and their respective troubleshooting strategies were also investigated. These include pH, EC, VFAs, alkalinity, ammonia, sulfate and sulfur compounds, heavy metals, LCFAs, organics, as well as foam and scum. In terms of troubleshooting, many of the procedures or solutions presented are based on the author's field experiences, making the chapter of practical benefit to both the scientific and industrial communities.

References

Abdel-Aal HK (2016) Chemical engineering primer with computer applications. CRC Press, London

Atiemo-Obeng VA, Calabrese RV (2004) Rotor-stator mixing devices, handbook of industrial mixing science and practice. Wiley, Hoboken. https://doi.org/10.1002/0471451452

AwiFLEX (n.d.) Biogas analysers control desulphurisation [WWW Document]. URL http://www.awite.com/products/gasanalyse.html

Boe K (2006) Online monitoring and control of the biogas process. Environment. Technical University of Denmark. http://www.nei-dk-4757.pdf

BTA International (2017). Core component of the hydromechanical processing [WWW Document]

Carmichael D (n.d.) Visual controls and process safety management in the petrochemical industry: safety, knowledge transfer, and OSHA's NEP

Chen Y, Cheng JJ, Creamer KS (2008) Inhibition of anaerobic digestion process: a review. Bioresour Technol 99:4044–4064. https://doi.org/10.1016/j.biortech.2007.01.057

Clemens J (2013) How to optimize the biogas process according to process control monitoring data in biogas plants. Bonn, Germany

Deublein D, Steinhauser A (n.d.) Biogas from waste and renewable resources: an introduction, 2nd edn. Wiley-VCH Verlag GmbH & Co. KGaA

Esteves S, Miltner M, Puchas K (2013) Monitoring review and guide: for the optimization of anaerobic digestion and biomethane plants. Full report. Intelligent Energy Europe Programme. https://www.aile.asso.fr/wp-content/uploads/2013/03/rapport_wp5-2_optimisation-des-unites.pdf

Falk HM (2011) Monitoring the anaerobic digestion process. Jacobs University, Germany

Hansen K, Angelidaki I, Ahring B (1998) Anaerobic digestion of swine manure: inhibition by ammonia. Water Res 32:5–12

Hansen TL, Jansen J, Davidsson Å, Christensen TH (2007) Effects of pretreatment technologies on quantity and quality of source-sorted municipal organic waste for biogas recovery. Waste Manag 27:398–405. https://doi.org/10.1016/j.wasman.2006.02.014

Harada H, Uemura S, Momonoi K (1994) Interaction between sulfate-reducing bacteria and methane-producing bacteria in UASB reactors fed with low strength wastes containing different levels of sulfate. Water Res 28:355–367

Jain SS, Jain SS, Tim I, Lee J, Wah Y, Wolf IT, Lee J, Tong YW (2015) A comprehensive review on operating parameters and different pretreatment methodologies for anaerobic digestion of municipal solid waste. Renew Sustain Energy Rev 52:142–154. https://doi.org/10.1016/j.rser.2015.07.091

Kabdasli I, Tunay O, Ork I, Yilmaz S, Arikan O (2000) Ammonia removal from young landfill leachate by magnesium ammonium phosphate precipitation and air stripping. Water Sci Technol 41:237–240

Kresta SM, Etchells AW, Dickey DS, Atiemo-Obeng VA (2015) Advances in industrial mixing: a companion to the handbook of industrial mixing. Wiley. https://doi.org/10.1017/CBO9781107415324.004

Lahav O, Morgan BE (2004) Titration methodologies for monitoring of anaerobic digestion in developing countries—a review. J Chem Technol Biotechnol 79:1331–1341. https://doi.org/10.1002/jctb.1143

Li C (2015) Wet and dry anaerobic digestion of biowaste and of co-substrates. Karlsruher Institute of Technology (KIT)

Liebetrau J, Jacobi F (2016) Monitoring and process control of anaerobic digestion plants. Chem Eng Technol 39:598

Marx C, Schmidt M, Flanagan J, Hanson G (2010) Introduction to activated sludge study guide

Moriarty K (2013) Feasibility study of anaerobic digestion of food waste in St. Bernard, Louisiana

O'Flaherty V, Mahony T, O'Kennedy R, Colleran E (1998) Effect of pH on growth kinetics and sulphide toxicity thresholds of a range of methanogenic, syntrophic and sulphate-reducing bacteria. Process Biochem 33:555–569

Operational guidelines for Plönninge biogas plant (n.d.)

Potter MC, Voss JF (1975) Fluid mechanics. Wiley, New York. John Wiley & Sons Canada, Limited, Canada

Sanchez JM, Valle L, Rodriguez F, Morinigo MA, Borrego JJ (1996) Inhibition of methanogenesis by several heavy metals using pure cultures. Lett Appl Microbiol 23:439–444

Sterritt RM, Lester JN (1980) Interaction of heavy metals with bacteria. Sci Total Environ 14:5–17

Templer J, Lalman JA, Jing N, Ndegwa PM (2006) Influence of C18 long chain fatty acids on hydrogen metabolism. Biotechnol Prog 22:199–297

Vindis P, Stajnko D, Lakota M (2014) Options for reduction of maize silage in biogas plant Drazenci, In: Katalinic B (ed) DAAAM international scientific book. DAAAM International, pp 121–130. https://doi.org/10.2507/daaam.scibook.2014.09

Water D, Engines CO (2014) CSTR operation and maintenance manual. Ver. 2. Bioprocess control Sweden AB. http://www.bioprocesscontrol.com/media/1323/bioprocess-control-cstr.pdf

Chapter 9
Analytical Methods in Biogas Production

Peyman Mahmoodi, Sara Farmanbordar and Keikhosro Karimi

9.1 Analysis of Total Solid (TS) and Volatile Solid (VS)

Biomass samples subjected to anaerobic digestion for biogas production contain different amounts of moisture, whereas the biogas yield is directly proportional to the dry matters available in the samples. The biogas yield and the results of chemical analyses of samples are generally reported on the basis of dry weight. Total solid (TS) is the amount of solid remaining after heating the biomass sample. TS analysis is rather simple and requires ordinary equipment, a balance and an oven, available in most laboratories. The amount of sample typically required is only 0.25–1 g for solid samples, 1–2.5 g for slurry samples, and 5 mL for liquor samples (should be filtered using a 0.2 μm pore size filter prior to TS analysis) (Sluiter et al. 2008). To be accurate enough, each sample should be run at least in duplicates. A certain amount of the biomass sample should be placed inside a dry aluminum dish and oven dried at 105 °C for at least 4 h. After cooling in a desiccator, the final weight will be measured. TS is calculated as follows (Eq. 9.1) (Sluiter et al. 2008):

$$\mathit{Total\,solid}(\%) = \frac{\mathit{weight\,of\,dried\,sample\,with\,dish} - \mathit{weight\,of\,dish}}{\mathit{weight\,of\,initial\,sample}\,(\mathit{without\,dish})} \times 100 \quad (9.1)$$

Volatile solids (VS) is the amount of organic solids in a biomass sample. For determining the amount of VS, an aluminum dish containing a biomass sample

P. Mahmoodi · S. Farmanbordar · K. Karimi (✉)
Department of Chemical Engineering, Isfahan University of Technology, Isfahan, Iran
e-mail: karimi@cc.iut.ac.ir

K. Karimi
Industrial Biotechnology Group, Research Institute for Biotechnology and Bioengineering, Isfahan University of Technology, Isfahan, Iran

M. Tabatabaei and H. Ghanavati (eds.), *Biogas*, Biofuel and Biorefinery Technologies 6, https://doi.org/10.1007/978-3-319-77335-3_9

(which has been pre-dried in an oven at 105 °C for 4 h), is placed in a (electric) furnace at 550 °C for 30 min. Then, the sample is weighed after cooling in a desiccator and the VS is calculated according to the following equation (Eq. 9.2) (Sluiter et al. 2008):

$$Volatile\ solids(\%) = \frac{weight\ of\ dried\ sample\ with\ dish\ -\ weight\ of\ ash\ with\ dish}{weight\ of\ initial\ sample\ (without\ dish)} \times 100 \quad (9.2)$$

Before cooling in the desiccator, the desiccator absorbent should be freshened by heating, so that the adsorption of water would be avoided, since the hot biomass is highly susceptible to absorbing water from air.

9.2 Determination of Structural Carbohydrates and Lignin in Lignocellulosic Biomass

Carbohydrates are either structural or non-structural. Prior to the analysis of lignin and carbohydrates contained in lignocellulosic materials, non-structural carbohydrates must be removed using water and ethanol extraction (Karimi and Taherzadeh 2016a, b). Organic materials, non-structural carbohydrates, and nitrogen components can be dissolved in water. Components such as waxes and chlorophylls can be dissolved in ethanol. A certain quantity of biomass is placed in a filtering crucible and then, the filtering crucible is placed in a Soxhlet extractor system. Water extraction is carried out for 24 h, by circulation for 4–5 times per hour. Subsequently, the remaining solids dried in an oven at 105 °C for 4 h will be weighed. Moreover, the remaining solids will be subjected to ethanol extraction.

Determination of structural carbohydrates and lignin in lignocellulosic biomass is carried out according to the NREL method (Sluiter et al. 2008). According to the method, the hydrocarbon polymers are converted into monomeric sugars using acid hydrolysis. The percentage of the polymeric components in the biomass will then be determined using the analysis of the monomeric sugars. Briefly, a glass bottle with blue screw cap containing 0.3 g biomass (based on TS) and 3 mL 72% (w/w) sulfuric acid is placed in a water bath at 30 °C for 60 min. The content of the glass bottle are mixed by a Teflon stir rod every 5–10 min. After 60 min of hydrolysis, 84 mL deionized water is added to the sample so that the acid concentration reaches 4% (w/w). The glass bottle (with the blue screw cap closed) will then be autoclaved at 121 °C for 1 h. After cooling, the content of the glass bottle is filtered using an ash-less filter paper. The solid phase, separated from the liquid phase, must be washed with water to achieve a neutral pH. The remaining solids are placed in an oven at 105 °C for 4 h for determining the ash and acid insoluble lignin (AIL). The weight of the remaining solids after cooling in a desiccator will also be recorded. Then, the remaining solids are placed in an electric furnace at 575 °C for 24 h

(Sluiter et al. 2008). The ash and AIL are calculated according to the following equations (Eqs. 9.3 and 9.4, respectively):

$$Ash(\%) = \frac{weight\ of\ sample + crucible\ (after\ furnace) - weight\ of\ crucible}{0.3\text{g TS}} \times 100 \tag{9.3}$$

$$AIL(\%) = \frac{weight\ of\ sample\ (after\ oven) - weight\ of\ sample\ (after\ furnace) - weight\ of\ protein\ of\ sample}{0.3\text{g TS}} \tag{9.4}$$

The liquid phase passed through an ash-less filter paper is used for determining the amount of acid soluble lignin (ASL) using UV-Vis spectroscopy at a specific wavelength such as 320 nm for corn stover and 240 nm for bagasse (Sluiter 2008). The ASL is calculated according to the following equation (Eqs. 9.5 and 9.6):

$$ASL(\%) = \frac{UV_{abs} \times 87(\text{mL}) \times Dilution}{0.3\text{ g TS} \times \varepsilon} \times 100 \tag{9.5}$$

$$Dilution = \frac{volume\ of\ sample + volume\ of\ diluting\ solvent}{volume\ of\ sample} \tag{9.6}$$

where ε is the absorptivity of biomass at specific wavelength, which is specific for each substrate such as 30 L/g cm for corn stover and 15 L/g cm for bagasse.

Total amount of lignin is calculated according to the following equation (Eq. 9.7):

$$lignin(\%) = ASL(\%) + AIL(\%) \tag{9.7}$$

For calculating sugar percentage, the pH of the liquor is increased to about 6 with calcium carbonate. After centrifugation, liquor is analyzed using a high performance liquid chromatography (HPLC) equipped with a refractive index (RI) detector for determining sugars concentration following acid hydrolysis. Glucose, xylose, mannose, arabinose, and galactose are detected on an Aminex HPX-87P column at 80 °C with an eluent of deionized water (0.6 mL/min) Sugar concentration could be calculated according to Eq. 9.8:

$$concentration\ of\ sugar\left(\frac{\text{mg}}{\text{mL}}\right) = \frac{concentration\ of\ sugar\ (from\ HPLC)(\text{mg/mL}) \times dilution\ factor}{R(average\ recovery\ of\ specific\ sugar)\ (\%)/100} \tag{9.8}$$

The calculation of sugar percentage can be done according to the following equation (Eq. 9.9):

$$Sugar(\%) = \frac{concentration\,of\,sugar \times 87(\mathrm{mL}) \times \frac{1\,\mathrm{g}}{1000\,\mathrm{mg}}}{0.3\ \mathrm{g\ TS}} \times 100 \qquad (9.9)$$

9.3 Elemental Analysis (CHNSO Analysis)

Theoretical biogas production can be calculated based on carbon, hydrogen, nitrogen, sulfur, and oxygen (CHNSO) content analysis by an elemental analyzer. Elemental analyzer, coupled with a flame photometric detector (FPD), is operated by combusting a sample at high temperatures in the presence of oxygen. High temperature is used for the thermal decomposition of sample and the separation of CHNS elements. At these conditions, CHNS elements in the presence of oxygen are converted into CO_2, H_2O, NO_2, and SO_3 gases, respectively. The resultant gases, carried with a helium flow to a layer filled with copper, are measured by a gas chromatography (GC) equipped with a thermal conductivity detector (TCD). Whereas, the resultant gases for trace sulfur determination through water trap are measured using the FPD. For oxygen determination, the elemental analyzer is operated at the pyrolysis mode. The reactor, which contains nickel coated carbon, is operated at 1060 °C. The oxygen in the sample that is combined with the carbon is converted into carbon monoxide. The carbon monoxide is then detected by the GC with a TCD detector.

9.4 Analysis of Chemical Oxygen Demand (COD)

The chemical oxygen demand (COD) test procedure is based on the chemical decomposition of organic and inorganic pollutants. The COD test is analyzed based on two methods: titrimetric method and colorimetric method. In the titrimetric method, a reflux apparatus equipped with a 500 or 250 mL Erlenmeyer flask and a condenser, is used. 20 mL sample, 30 mL concentrated sulfuric acid, and 10 mL 0.25 N potassium dichromate are poured inside the Erlenmeyer flask. Then, 0.2–0.5 g mercuric sulfate is added to remove the nitrite and chloride interference. Silver sulfate is used as a catalyst. The reflux apparatus is heated at 150 °C for 2 h. After cooling to room temperature, the sample volume is increased to 100 mL with distilled water. The excess potassium dichromate is titrated with ferrous ammonium sulfate (FAS) by 2–3 drops ferroin indicator solution (Federation and Association 2005). All these steps are also carried out for 20 mL distilled water as a control sample. The COD based on mg/L is calculated according to the following equation (Eq. 9.10) (Federation and Association 2005):

$$COD\left(\frac{\mathrm{mg}}{\mathrm{L}}\right) = \frac{(mL\,FAS\,used\,for\,control\,sample\ -\ mL\,FAS\,used\ for\ sample) \times molarity\,of\,FAS \times 8000}{mL\,sample} \tag{9.10}$$

9.5 Volatile Fatty Acid Determination by High Performance Liquid Chromatography

When digesters are supplemented by extra amounts of substrate, called overloading, the yield of produced methane is decreased or the process may be stopped. The accumulation of volatile fatty acid (VFA) interferes with the performance balance of the microorganisms involved in the decomposition of the organic materials and methane product (methanogens). The high concentration of VFA decreases pH, resulting in the inhibition of the methanogens. Therefore, the analysis of VFA inside the digestion is important (Lossie and Pütz 2008). Although total VFA can be measured by titration, accurate analysis of VFA is carried out by HPLC (Zeppa et al. 2001). HPLC pumps a liquid sample in a solvent, known as the mobile-phase, at high pressure through a pack column. The column has the ability to separate components that are in the sample mixture. These components are detected using a RI detector. The concentration of the components is calculated based on a calibration curve, obtained according to the concentrations and peak areas of pure components. Analysis of the organic acids, the ethanolic and butanolic fermentation metabolites is carried out by HPLC with RI or UV detectors. An example of suitable columns is Aminex HPX-87H that operates at 60 °C with 0.005 M sulfuric acid as mobile-phase (0.6 mL/min). Besides VFA, all monomeric carbohydrates can also be analyzed by HPLC. An example of suitable column for measurement of sugar mixtures, including glucose, xylose, mannose, arabinose and galactose, is Aminex HPX-87P column, operated at 80 °C with an eluent of deionized water (0.6 mL/min).

9.6 Methods for Determining Methane Potential

Biochemical methane potential (BMP) is a measure of the substrate's potential to generate methane gas. BMP test provides a criterion representing the biodegradable fraction contained in a given substrate which can be anaerobically converted into methane. The results of the BMP test can be used as a means of comparison in terms of potential of biogas production among various substrate. BMP can be measured using both theoretical and experimental techniques. Experimental data, obtained using the BMP test, can be used to optimize the design and operation of an anaerobic digester (Moody et al. 2009).

9.6.1 *Theoretical Biochemical Methane Potential (TBMP)*

When some characteristics of a given substrate including COD, elemental composition (C, H, O, and N), and organic fraction composition (OFC) are known, the methane productivity of the substrate can be determined by the methods described below. These methods are applied by assuming that all VS presented in the substrate are biodegradable. Therefore, it should be considered that TBMP is higher than the real amount of BMP and it is necessary to adjust this value by using the biodegradability index obtained from the experimental BMP tests (Nielfa et al. 2015).

9.6.1.1 TBMP Based on Elemental Compositional Analysis

Various organic fractions including VFAs, proteins, lipids and carbohydrates have individual TBMP which can normally be calculated from Buswell equations (Eqs. 9.11 and 9.12) regardless of the degradability of the materials (Symons and Buswell 1933; Møller et al. 2004).

$$C_nH_aO_bN_c + \left(n - \frac{a}{4} - \frac{b}{2}\right)H_2O \rightarrow \left(\frac{n}{2} - \frac{a}{8} + \frac{b}{4}\right)CO_2 + \left(\frac{n}{2} + \frac{a}{8} - \frac{b}{4}\right)CH_4 \tag{9.11}$$

$$TBMP = \frac{(a/2 + a/8 - b/4)22.4}{(12n + a + 16b)} \tag{9.12}$$

When proteins are present in a substrate, NH_3 is released and must be considered for calculating the TBMP according to Boyle equations (Eqs. 9.13 and 9.14) as follows (Nielfa et al. 2015; Raposo et al. 2011).

$$\begin{aligned} C_nH_aO_bN_c &+ \left(n - \frac{a}{4} - \frac{b}{2} + \frac{3c}{4}\right)H_2O \rightarrow \left(\frac{n}{2} - \frac{a}{8} + \frac{b}{4} + \frac{3c}{8}\right)CO_2 \\ &+ \left(\frac{n}{2} + \frac{a}{8} - \frac{b}{4} - \frac{3c}{8}\right)CH_4 + cNH_3 \end{aligned} \tag{9.13}$$

$$\text{TBMP} = \frac{22.4(n/2 + a/8 - b/4 - 3c/8)}{(12n + a + 16b + 14c)} \tag{9.14}$$

9.6.1.2 TBMP Based on Organic Fraction Composition (OFC)

TBMP can also be calculated by determining the amounts of several organic components in a given substrate. If the OFC of a given substrate is known, TBMP can be estimated as a weighted average of individual TBMP for organic components using the following equation (Eq. 9.15):

$$\text{TBMP} = 415(\%\text{Carbohydrates}) + 496(\%\text{Proteins}) + 1014(\%\text{Lipids}) + 370(\%\text{VFA}) \tag{9.15}$$

Whereas individual TBMPs are derived from the elemental composition method (Eqs. 9.12 and 9.14) by considering the average formulas for lipids ($C_{57}H_{104}O_6$), proteins ($C_5H_7O_2N$), VFA ($C_2H_4O_2$), and carbohydrates ($C_6H_{10}O_5$) (Nielfa et al. 2015; Møller et al. 2004; Raposo et al. 2011).

9.6.1.3 TBMP Based on Chemical Oxygen Demand (COD)

Maximum amount of methane can be calculated according to Eqs. 9.16 ad 9.17 when the COD concentration and VS of substrate are specified.

$$TBMP = \frac{n_{CH_4}RT}{PVS_{added}} \tag{9.16}$$

$$n_{CH_4} = \frac{COD}{64(\text{g/mol})} \tag{9.17}$$

where R is the gas constant, T is the temperature of process, p is atmospheric pressure (1 atm), and VS (g) is the organic total solid supplemented to the anaerobic digestion process (Nielfa et al. 2015). TBMP is reported as mL CH_4 at standard temperature and pressure conditions per amount of added VS.

9.6.2 Experimental BMP Test

All previously explained methods calculate the BMP regardless of the degradability of the materials. As a result, real BMP should be determined using laboratory-based incubation tests. There are several batch protocols for estimating methane potential of various substrates. In all these methods, a small amount of substrate is incubated with an anaerobic inoculum and methane generation is measured by simultaneous measurement of gas volume and gas composition. Adani et al., Harries et al., Owens et al., and Heerenklage introduced protocols for estimating BMP. However, the basic approach of these methods is the same, while the technical approach is significantly different (Adani et al. 2001; Harries et al. 2001; Heerenklage and Stegmann 2001; Owens and Chynoweth 1993). The most widely used methods for the determination of BMP experimentally are described below.

9.6.2.1 German Standard Procedure (VDI4630)

Batch fermentation can be applied to all solid and liquid organic substrates. This procedure provides some information about the digestibility and possible biogas yield of a given substrate. Furthermore, it can offer qualitative information about the speed of anaerobic degradation and inhibitory effect of materials under investigation conditions. However, batch fermentation test cannot provide any information about the process stability in reactors, biogas yield under practical conditions, the mono-fermentability of the substrate, and the limits of organic loading rate per unit volume.

- ***Technical considerations***

Glass is the preferred material for the fermentation test apparatus which is in contact with biogas and sludge. All equipment should be tested to ensure that there is no leak. Figure 9.1 shows the schematic of the apparatuses used in the batch fermentation test.

Substrate and sludge are incubated under mesophilic (37 ± 2 °C) or thermophilic (55 ± 1 °C) conditions. A climatic chamber or water bath can be used to obtain a constant temperature control during the batch process duration. When water bath is used, it should be noted that the level of water in the bath should always be higher than the content of the fermentation vessel. Although continuous mixing is not necessary for batch fermentation, a single mixing per day is sufficient to prevent the formation of dry and inactive floating layers. A 0.5, 1, and 2 L bottle can be used as fermentation flask. Larger fermentation flask (10–20 L) may be better to be used when the substrate is nonhomogeneous.

Seeding sludge, used as inoculum in the fermentation process, should be untreated digested sludge from a municipal sewage treatment. The seeding sludge should have a VS content greater than 50% of TS content. The seeding sludge should be adapted at the fermentation temperature for a week to minimize its own gas production by means of a hunger phase. Large contaminants should be

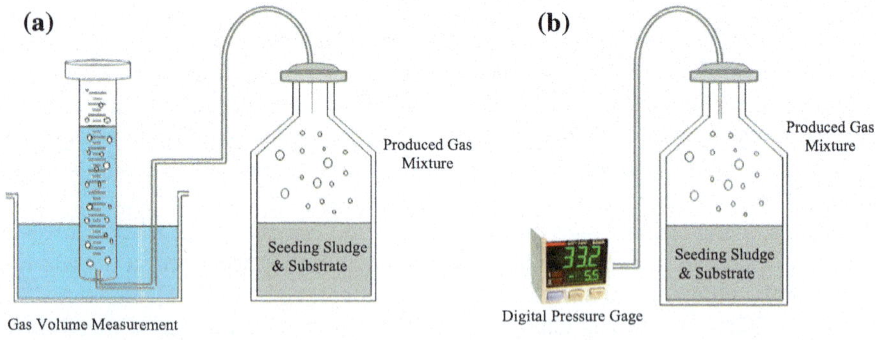

Fig. 9.1 The schematic representation of some apparatuses used in the batch fermentation process based on the VDI 4630 protocol, both systems **a** and **b** should be placed in a climate chamber at a desired temperature

separated from the seeding sludge before it use by filtering. The fermentation broth should contain 1.5–2% w/w of organic mass from the seeding sludge to standardize the fermentation protocol. For example, 500 mL fermentation batch should contain 7.5–10 g VS from the seeding sludge. There are some limitations to the weight of substrate and seeding sludge in the fermentation batch. The mass of total organic solid substrate should not be over 50% organic total solid sludge ($\frac{oTS_{substrate}}{oTS_{sludge}} \leq 0 \cdot 5$). Total solid content of fermentation batch should be lower than 10% to ensure adequate mass transfer during fermentation. The gas yield during batch fermentation from substrate should be more than 80% of the theoretical BMP. A reference material, which has known biogas potential, should be used as substrate in the batch fermentation as control test to ensure that the seeding sludge used has an adequate biological activity. One option is microcrystalline cellulose (with 100% conversion) leading to 740–750 mL methane/g OTS, according to the elemental composition analysis method. When 80% of this value has been reached in the control test, it could be assumed that the biological activity is adequate. All batch fermentation test including main samples, reference, and zero sample (control sludge) should be carried out in duplicates or triplicates (Standard 2006).

- ***Test procedure***

The substrate is weighed, added to the fermentation bottle, and mixed with water if necessary. Then, the bottle is carefully filled with a sufficient amount of already-adapted seeding sludge. In the following, the crater of the bottles is closed and sealed. The gas phase in the bottle is purged with nitrogen to prevent aerobic degradation processes which have negative effects on biogas yield. Regular mixing (shaking bottles each day during the batch fermentation) should be conducted to ensure full suspension of the sediments. The quantity and quality of the produced gas should be measured periodically to make the gas formation perfectly recognizable. At the beginning of the process, it is necessary to perform daily measurements of gas production. When the daily gas production declines, the frequency of gas measurements can be reduced to once per two or three days. The test is continued until the daily gas production is lower than 1% the total volume of produced gas up to that time. The majority of the gas is usually produced within the first week. Mostly the biological degradations are finished after 20–40 d. Finally, the pH of the content of the fermentation batch must be determined and recorded (Standard 2006).

9.6.2.2 Hansen et al. Method

An easy to operate method was modified by Hansen et al. (2004) to estimate the potential of methane production from solid waste samples generating high amounts of methane. The procedure is adapted and modified from the existing methods especially from the one described by Angelidaki and Ahring (1997). The Hansen method was originally used for 100 samples during a 2-year period (Hansen et al. 2004).

- ***Technical considerations***

An active inoculum should be transferred from a thermophilic biogas plant in 25-L containers. The temperature of the inoculum drops to the ambient temperature during the delivery process. Therefore, it must be re-adapted to the desired temperature. In this way, the required amount of inoculum is filtered and transferred to a glass bottle. The headspace of the bottle is purged by N_2-gas. Then, the inoculum should be stored in a 55 °C incubator for 3 d to ensure that the remaining easily degradable materials still present in the inoculum are removed (Fig. 9.2).

In the biogas production procedure, only 10 g dry matter is used in order to manage the production in the batch equipment. Therefore, the homogeneity of substrate is an important factor to ensure a representative sampling. When heterogeneous substrates, e.g., municipal solid waste is used, it should be carefully homogenized by multiple sampling and carefully blending of the substrate. Dilute solution with 10% DM content is used as substrate. Main tests are carried out in a 2 L glass bottle with a thick rubber septum (Fig. 9.3). The exact volume of the bottle is measured by weighing the water contained in each bottle. Headspace volume is calculated by subtracting the volume of added inoculum and substrate (it is assumed that the density of both inoculum and substrate is 1 g/mL) from the total volume.

All tests are conducted as triplicate batch fermentations to minimize the unfavorable effects of varying quality of the inoculum as well as possible non-homogeneity of the substrate. A mixture of cellulose and avicel with equal weight proportions should be used as the control test to ensure that the inoculum has an adequate biological activity (Hansen et al. 2004).

- ***Test procedure***

An amount of 400 mL of the re-adapted inoculum is added to each batch reactor while stirring. Then, each reactor is supplied with 100 mL of the substrate containing 10% DM with 80–90% VS. Thereafter, bottles should be sealed and purged with 80% N_2 and 20% CO_2 gas to ensure anaerobic conditions in the headspace.

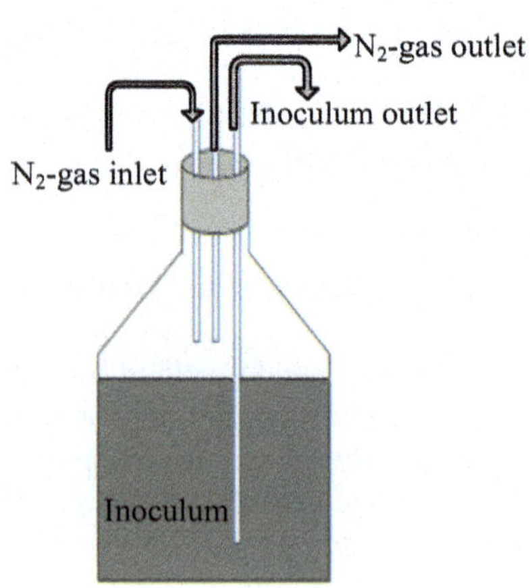

Fig. 9.2 Schematic presentation of the bottle used for the re-adaptation of inoculum

Fig. 9.3 Illustration of the glass bottle used as batch reactor and the pressure lock syringe

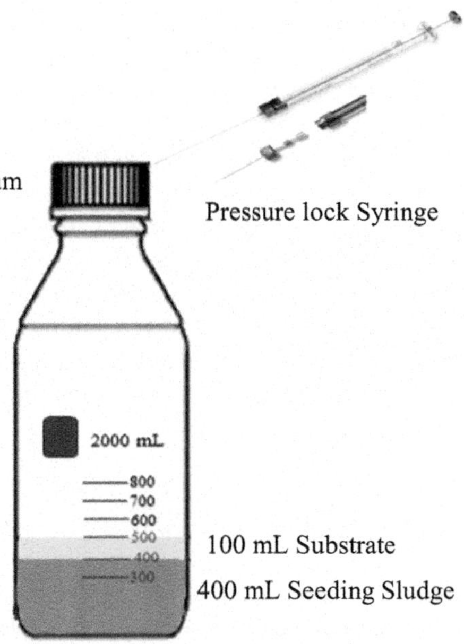

Under these conditions, the batch biogas reactor with an anaerobic headspace contains 2 g VS/100 mL solution. Finally, sealed reactors should be incubated at 55 °C (± 1 °C) for 50 d. The process is carried out over 50 d to ensure that the biodegradable materials have been completely degraded. During the fermentation period, all reactors must be regularly shaken and moved around the incubator to avoid any probable temperature difference in the incubator. The produced biogas is measured 25–20 times during the procedure. Daily monitoring is necessary in the first week. Then, it is sufficient to measure once a week. In each monitoring, 200 μL gas sample is withdrawn from the headspace of the reactors by a pressure lock syringe through the septum. The pressure lock must be closed when the syringe is still penetrated into the septum. The usage of the pressure lock syringe makes it possible to sample with a fixed volume of gas at the actual pressure in the reactor. The sample should be injected into a GC for measuring the mass of methane. It should be noted that the volume of sample is lower than (the volume of all samples taken should be lower than 0.7% of the headspace volume) the headspace in each reactor, so that sampling does not cause any significant effects on actual headspace pressure.

Six L biogas can be approximately produced from the amount of substrate added in each reactor which is higher than the free capacity of the bottles. Therefore, the produced gas should be regularly released during the experiment to avoid high pressures and consequent leakages (it is preferred that the pressure is always kept lower than 2 bar). The pressure can be released by inserting a needle into the septum of the reactors. The volume of the released gas can be accurately calculated by measuring the methane content in the headspace of the reactor after and before the release. Although the highest amounts of methane (80–90%) may be produced during the first 8–10 days, measurements should be continued for 50 d to ensure that all organic wastes, which may be slowly degradable, are converted into biogas. Nevertheless, when theoretical methane potential has been obtained within a short time, the process can be terminated.

For each run, triplicate blank samples with only water and inoculum should be included to estimate the amounts of methane produced by the inoculum. At the end of the experiment, i.e., on day 50, a sample should be taken from each reactor to measure the VFA and nitrogen contents. These tests will be carried out if the methane potential is low and inhibition by ammonia or VFA accumulation may occur. Finally, the accumulated produced methane is shown as a function of the fermentation time to estimate some parameters including inhibition effect and lag phase (Hansen et al. 2004).

9.6.2.3 Moller Method

Moller method is another most widely used method described according to the international standard ISO 11734. The BMP is measured in batch experiments performed in 1100 mL injectable bottles. Inoculum is obtained from farm-scale biogas plants and is re-adapted a 3t5 ± 0.5 °C for two weeks before use. The adaptation is used to ensure that the amount of methane originated from the inoculum is reduced as much as possible. The inoculum and substrate are added to the biogas reactors at a certain ratio (ranging from 0.7 to 7). The bottles should be sealed by butyl rubber stoppers and aluminum crimps. Then, the batch reactors are flushed with N_2 gas and incubated at mesophilic conditions (35 ± 0.5). Each sample is digested in triplicates. Three reactors containing only the inoculum and water should be considered as blank test similar to the other protocols. Produced gas in the blank test must be subtracted from the amount of gas produced in main samples. Produced gas can be measured either by connecting gas collection bag, or by monitoring the pressure of the headspace. It should be noted that both methods are applicable because the difference between their reported results are negligible. The concentration of CO_2 and methane should be determined by GC analysis (Møller et al. 2004).

9.7 Methods to Measure the Volume of the Produced Biogas

9.7.1 Liquid Replacement System (LRS)

Biogas volume can be estimated by liquid replacement system (LRS) technics at intervals. This apparatus is connected to a bioreactor by a needle during each gas volume measurement interval. There are three types of liquid replacement gasometer including height and weight type, which are shown in Fig. 9.4. In the height liquid replacement gasometer, one opened cylinder is reversibly submerged in an open container of special liquid (Fig. 9.4a).

Headspace of the bioreactor is connected to the top of the cylinder; the produced gas then flows to the cylinder and replaces the filled liquid. The volume of produced gas can be determined by measuring the height in the cylinder and the container. In the weight liquid replacement gasometer type; the produced biogas is injected into the column of the liquid located on the container. Then, an amount of liquid drained to the container is weighed and the volume of the produced gas can be estimated from both the weight and the differences of height in the column (Fig. 9.4b) (Walker et al. 2009).

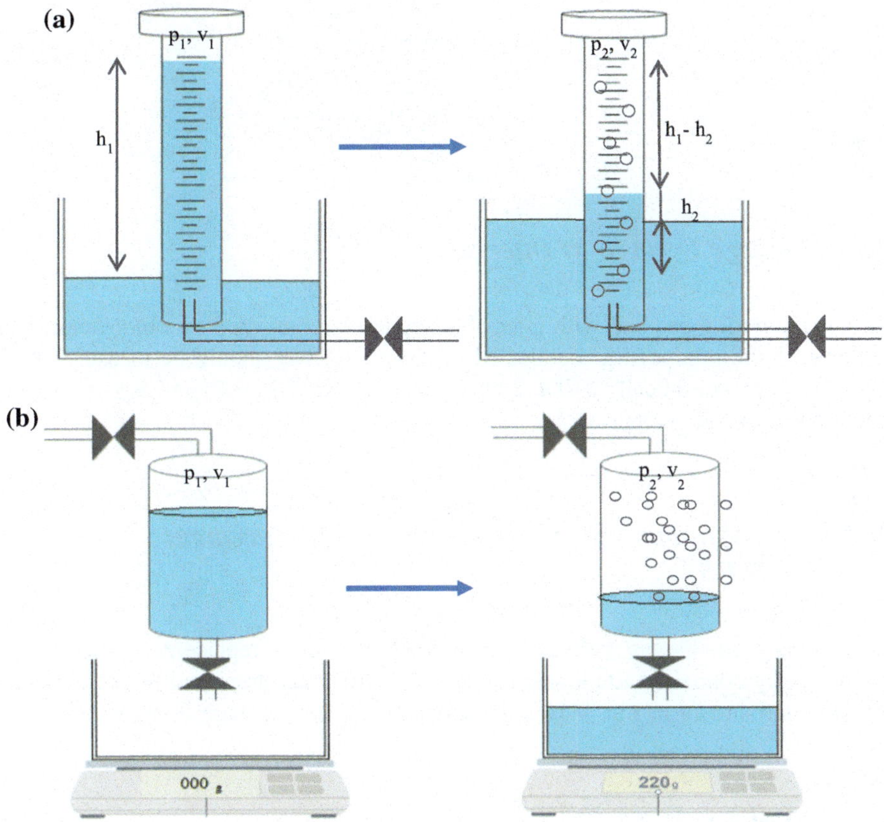

Fig. 9.4 The Illustration of the height **a** and weight **b** type of the liquid replacement gasometers

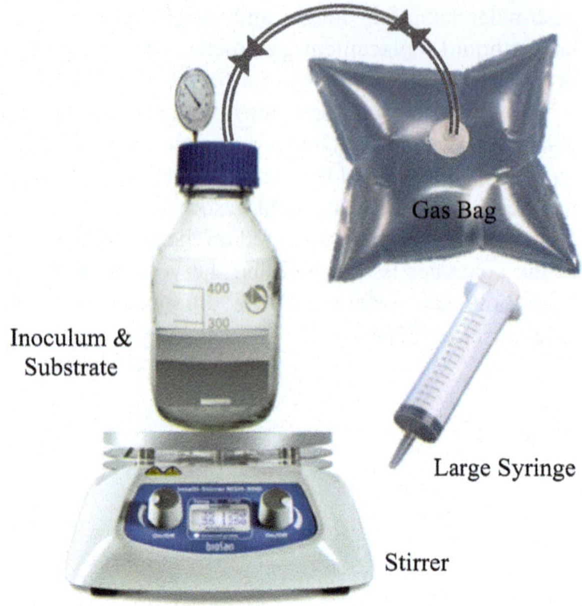

Fig. 9.5 The illustration of the large syringe at intervals method. The apparatus should be placed in a climate chamber at a desired temperature

9.7.2 Large Syringe at Intervals

In this method, a gas bag with injectable septum is connected to a bioreactor. The volume of the produced gas collected in the plastic bag should be measured by a large syringe periodically (VDI 2006; Triolo et al. 2011). Figure 9.5 shows the illustration of this method.

9.7.3 Continuous Liquid Replacement Measurement (CLRS)

In this method, biogas volume is measured by liquid replacement apparatuses permanently connected to bioreactors. Some of them are described in the VDI4630 method and shown in Fig. 9.1a, b (Standard et al. 2006; Pham et al. 2013).

9.8 Methods to Measure Methane Concentration

9.8.1 Adsorption of CO_2 in Alkaline Liquid

Alkaline solutions, including NaOH and KOH solutions, can adsorb CO_2 and H_2S gas from the biogas mixture (Lasocki et al. 2015). CO_2 adsorption in alkaline liquid is mostly used to estimate methane concentration in several biomethane potential tests (Raposo et al. 2011; Triolo et al. 2011; Guwy 2004; Rozzi and Remigi 2004). Figure 9.6 schematically shows this measurement method. According to the figure, a scaled cylinder is filled by an acid solution (0.5 M HCl) and is reversibly submerged in a container containing the same liquid (Fig. 9.6a). A tube connected to a syringe in inserted into the headspace of the bioreactor while the other end of the tube in place in the cylinder. At the connection to the bioreactor, there exists a pin which will be opened when the tube is located at right spots at both end. Then, the produced biogas will flow to the cylinder and replace an amount of liquid (Fig. 9.6b, c). The volume of the produced gas (V_1 mL) can be measured accordingly. To measure methane concentration, KOH should be added to the container to increase the pH (above 9). At this basic condition, CO_2 and H_2S gas is adsorbed and the volume of the gas is decreased (V_2 mL). Therefore, the measured volume represents CH_4 content in the biogas mixture. It should be noted that, the different between the first and the second volume (V_1–V_2) shows the CO_2 content of the biogas because H_2S concentration is completely negligible in the adsorbed gas (Pham et al. 2013).

9.8.2 Gas Chromatography (GC)

Biogas is a gas mixture of methane, carbon dioxide, hydrogen, and hydrogen sulfide, the main components of biogas are methane and carbon dioxide though. The analysis of the produced biogas components (methane and carbon dioxide) can

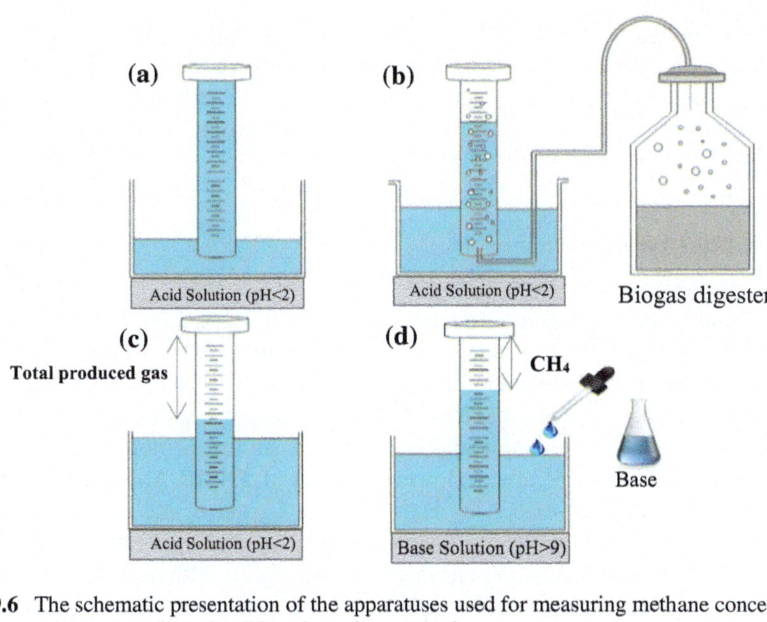

Fig. 9.6 The schematic presentation of the apparatuses used for measuring methane concentration in biogas mixture based on the CO_2 adsorption method

be carried out by GC, which is an ideal analytical instrument (Andersen et al. 2010; Kolb 2006). TCD is less sensitive than FID, but TCD is mostly used for the detection of light compounds (Poole 2003). For the analysis of biogas, GC is equipped with a packed column and a TCD detector. The carrier gas is nitrogen at a flow rate of 50 mL/min. The column, injector, and detector temperatures are 40, 100, and 150 °C, respectively. Biogas sample is injected into the GC by a pressure-lock syringe. Certain volumes of pure methane gas and carbon dioxide gas are injected into GC to establish calibration curves. For each gas, a calibration curve is obtained based on the gas volume versus peak area. According to the calibration curve, the gas volume and the percentage of gas composition are obtained. Biogas samples are first taken at the beginning of each interval at the environmental pressure mode (the pressure in the digester is released by inserting a needle in the septum while the other end of the tube connected to the needle is placed in a water container to avoid air introduction into the digester). At the end of each interval, samples are also taken at high pressure mode. The samples are analyzed by GC and the amount of the produced biogas is calculated according to the following equations (Eqs. 9.18 and 9.19):

$$\mathit{Gas\ volume\ in\ the\ digester}\ (\mathrm{mL}) = \frac{\mathit{Syringe\ volume}(\mu\mathrm{L})}{\mathit{Sampling\ volume}(\mu\mathrm{L})} \times \mathit{Free\ volume\ of\ the\ digester}\ (\mathrm{mL}) \quad (9.18)$$

$$\mathit{Produced\ gas\ volume\ during\ one\ interval} = \mathit{gas\ volume\ in\ the\ digester\ at\ high\ pressure}_{\mathit{end\ of\ interval}} - \mathit{gas\ volume\ in\ the\ digester\ at\ environmental\ pressure}_{\mathit{beginning\ of\ interval}} \quad (9.19)$$

For the analysis of the produced biogas from biomass, the produced biogas volume from the control sample (inoculum and water) should be deduced from the amount of the produced biogas from the sample. A pure cellulose or Avicel sample could be used as a control sample for ensuring the activity of the inoculum used.

9.9 Conclusions

Different substrate analyses are required to design and operate anaerobic digestion systems for efficient biogas production. Some of these analytical methods including CHNSO elemental analysis and COD analysis are used to estimate the theoretical biochemical methane potential (TBMP). Using these procedures, the amount of the produced biogas can be theoretically estimated. It should be noted that the experimental amount of BMP is always lower than that of TBMP. The determination of BMP via the method described by Hansen et al. is one of the most widely used methods. In this approach, the methane concentration and the volume of produced biogas are estimated using gas chromatography (GC). Although GC analysis is more expensive than the other volumetric methods, it has a significantly higher accuracy comparatively. It should be considered that the occurrence of experimental errors including gas leakages and personal errors in addition to systematic errors are the reason behind the lower accuracy of the other methods compared with GC.

Refrences

Adani F, Calcaterra E. Malagutti L (2001) Preparation of a test for estimating biogas production from pretreated urban waste. In: Proceeding Sardinia 2001 eight international waste management and landfill symposium, Cagliari, Italy

Andersen A, Seeley J, Aurandt J (2010) The use of gas chromatography for biogas analysis. In: APS Ohio Sections Spring Meeting Abstracts

Federation WE, Association APH (2005) Standard methods for the examination of water and wastewater. American Public Health Association (APHA), Washington, DC, USA

Guwy A (2004) Equipment used for testing anaerobic biodegradability and activity. Rev Environ Sci Biotechnol 3(2):131–139

Hansen TL et al (2004) Method for determination of methane potentials of solid organic waste. Waste Manag 24(4):393–400

Harries C, Cross C, Smith R (2001) Development of a BMP test and application to testing of MSW samples. In: Proceeding Sardinia 2001 eight international waste management and landfill symposium, Cagliari, Italy

Heerenklage J, Stegmann R (2001) Comparison of test systems for the determination of the gas potential from waste. In: Proceeding Sardinia 2001 eight international waste management and landfill symposium, Cagliari, Italy

Karimi K, Taherzadeh MJ (2016a) A critical review on analysis in pretreatment of lignocelluloses: degree of polymerization, adsorption/desorption, and accessibility. Biores Technol 203:348–356

Karimi K, Taherzadeh MJ (2016b) A critical review of analytical methods in pretreatment of lignocelluloses: composition, imaging, and crystallinity. Biores Technol 200:1008–1018

Kolb B, Ettre LS (2006) Static headspace-gas chromatography: theory and practice. Wiley

Lasocki J, Kołodziejczyk K, Matuszewska A. (2015) Laboratory-scale investigation of biogas treatment by removal of hydrogen sulfide and carbon dioxide. Polish J Environ Stud 24(3)

Lossie U, Pütz P (2008) Targeted control of biogas plants with the help of FOS. TAC, Practicle report: laboratory analysis, titration, FOS/TAC, Brochure of the company Hach-Lange

Møller HB, Sommer SG, Ahring BK (2004) Methane productivity of manure, straw and solid fractions of manure. Biomass Bioenerg 26(5):485–495

Moody L et al (2009) Use of biochemical methane potential (BMP) assays for predicting and enhancing anaerobic digester performance. In: Proceedings of the 44th croatian and the 4th international symposium on Agriculture, Opatija, Croatia, 16–20 February 2009

Nielfa A, Cano R, Fdz-Polanco M (2015) Theoretical methane production generated by the co-digestion of organic fraction municipal solid waste and biological sludge. Biotechnol Rep 5:14–21

Owens J, Chynoweth D (1993) Biochemical methane potential of municipal solid waste (MSW) components. Water Sci Technol 27(2):1–14

Pham C et al (2013) Validation and recommendation of methods to measure biogas production potential of animal manure. Asian-Australasian J Animal Sci 26(6):864

Poole CF (2003) The essence of chromatography. Elsevier

Raposo F et al (2011) Biochemical methane potential (BMP) of solid organic substrates: evaluation of anaerobic biodegradability using data from an international interlaboratory study. J Chem Technol Biotechnol 86(8):1088–1098

Rozzi A, Remigi E (2004) Methods of assessing microbial activity and inhibition under anaerobic conditions: a literature review. Re/Views Environ Sci Bio/Technol 3(2):93–115

Sluiter A et al (2008) Determination of total solids in biomass and total dissolved solids in liquid process samples. National Renewable Energy Laboratory, Golden, CO, NREL Technical Report No. NREL/TP-510-42621, p 1–6

Symons G, Buswell A (1933) The methane fermentation of carbohydrates1, 2. J Am Chem Soc 55 (5):2028–2036

Triolo JM et al (2011) A new algorithm to characterize biodegradability of biomass during anaerobic digestion: influence of lignin concentration on methane production potential. Biores Technol 102(20):9395–9402

Verein Deutcher Ingenieure VDI (2006) VDI 4630 Fermentation of organic materials. Characterization of the substrate, sampling, collection of material data, fermentation tests, p 92

Walker M et al (2009) Potential errors in the quantitative evaluation of biogas production in anaerobic digestion processes. Biores Technol 100(24):6339–6346

Zeppa G, Conterno L, Gerbi V (2001) Determination of organic acids, sugars, diacetyl, and acetoin in cheese by high-performance liquid chromatography. J Agric Food Chem 49(6):2722–2726

Chapter 10
Biogas Purification and Upgrading Technologies

María Rosario Rodero, Roxana Ángeles, David Marín, Israel Díaz, Alexandre Colzi, Esther Posadas, Raquel Lebrero and Raúl Muñoz

10.1 The Need for Biogas Upgrading

Apart from a nutrient-rich digestate, the anaerobic digestion of organic matter (either from energy crops or residues such as the organic fraction of municipal solid waste or domestic, agroindustrial and livestock wastewaters, etc.) generates a CH_4-rich biogas that can be used as a substitute of natural gas (Muñoz et al. 2015). Both the flowrate and composition of this renewable energy vector depends on the oxidation-reduction state and biodegradability of the organic carbon present in the organic matter and the type of anaerobic digestion process, i.e., closed digester or open landfill (Jönsson et al. 2003). Thus, biogas from closed digesters exhibits higher CH_4 contents and significantly lower O_2 and N_2 levels than landfill biogas. The anaerobic digestion of livestock manure, sewage sludge or municipal organic waste produces a biogas containing CH_4 at 50–70%, CO_2 at 30–50%, N_2 at <3%, O_2 at <1%, H_2S at <10,000 ppm_v, NH_3 at <100 ppm_v, hydrocarbons at <200 mg m^{-3}, H_2O at 5–10%, and siloxanes at <40 mg m^{-3}. Likewise, landfill biogas (extracted using forced ventilation) contains CH_4 at 35–65%, CO_2 at 15–50%, N_2 at 5–40%, H_2O at <5%, O_2 at <5%, H_2 at <3%, CO at <3%, H_2S at <100 ppm_v, NH_3 at <10 ppm_v, halogenated hydrocarbons at <200 ppm_v Cl^-/F^-, volatile organic contaminants at <4500 mg m^{-3}, and siloxanes at <50 mg Si m^{-3} (Bailón and Hinge 2012) (Table 10.1).

Most biogas applications require a minimum level of CH_4, which makes CO_2 and N_2 the main biogas contaminants due to their high concentrations. In addition, their presence also lowers the Wobbe index of biogas, a parameter targeted in most international biomethane regulations (Ryckebosch et al. 2011). O_2 is also considered

M. R. Rodero · R. Ángeles · D. Marín · I. Díaz · A. Colzi · E. Posadas
R. Lebrero · R. Muñoz (✉)
Department of Chemical Engineering and Environmental Technology,
University of Valladolid, Doctor Mergelina S/N, 47011 Valladolid, Spain
e-mail: mutora@iq.uva.es

M. Tabatabaei and H. Ghanavati (eds.), *Biogas*, Biofuel and Biorefinery Technologies 6, https://doi.org/10.1007/978-3-319-77335-3_10

Table 10.1 Typical composition of different biogas and natural gas types, and technical specifications of biomethane in France, Germany, and The Netherlands (Persson et al. 2006; Ryckebosch et al. 2011; Bailon and Higen 2012; Yang et al. 2014; Sun et al. 2015; Awe et al. 2017)

Parameter	Unit	Landfills gas	Biogas from anaerobic digestion	North sea natural gas	Dutch natural gas	Upgraded Biogas quality for injection in the natural gas grid					Possible impact
						France		Germany		The Netherlands	
						Low quality gas	High quality gas	Low quality gas grid	High quality gas grid		
Lower heating value	MJ/Nm^3	16	23	40	31.6	34.4–37.8	38.52–46.08	30.2–47.2		31.6–38.7	–
	kWh/Nm^3	4.4	6.5	11	8.8	–	–	–	–	–	–
	MJ/kg	12.3	20.2	47	38	–	–	–	–	–	–
Density	Kg/Nm^3	1.3	1.2	0.84	0.8	–	–	–	–	–	–
Higher Wobbe index	MJ/Nm^3	18	27	55	43.7	42.48–46.8	48.24–56.52	37.8–46-8	46.1–56.5	43.46–44.41	–
CH_4 number		>130	>135	70	–	–	–	–	–		–
CH_4	vol.%	45	63	87	81	–	–	–	–	>80	–
$CH_{4,}$ variation	vol.%	36–65	53–70	–	–	–	–	–	–	–	–
Higher hydrocarbons	vol.%	0	0	12	3.5	–	–	–	–	–	–
H_2	vol.%	0–3	0	0	–	<6		≤5		<12	–

(continued)

Table 10.1 (continued)

Parameter	Unit	Landfills gas	Biogas from anaerobic digestion	North sea natural gas	Dutch natural gas	Upgraded Biogas quality for injection in the natural gas grid					Possible impact
						France		Germany		The Netherlands	
						Low quality gas	High quality gas	Low quality gas grid	High quality gas grid		
CO	vol.%	0	0	0	0	–	–		–	–	–
CO_2	vol.%	40	47	1.2	1	<2.5		<6		<6 (<10–10.3 for regional grid)	Decreased calorific value, anti-knock properties of engines, and corrosion
CO_2, variation	vol.%	15–50	30–47	–	–	–	–		–	–	–
N_2	vol.%	15	0.2	0.3	14	–	–		–	–	Decreased calorific value, anti-knock properties of engines, corrosion
N_2, variation	vol.%	5–40	–	–	–	–	–		–	–	–
O_2	vol.%	1	0	0	0	<0.01		<3		<0.5	Corrosion, fooling in cavern storage, creation of explosive mixtures
	ppm_V	–	–	–	–	<100		–	–	–	–
	mol%							–	–	<0.5	–
O_2, variation	vol.%	0–5	–	–	–	–	–	–	–	–	–

(continued)

Table 10.1 (continued)

Parameter	Unit	Landfills gas	Biogas from anaerobic digestion	North sea natural gas	Dutch natural gas	Upgraded Biogas quality for injection in the natural gas grid					Possible impact
						France		Germany		The Netherlands	
						Low quality gas	High quality gas	Low quality gas grid	High quality gas grid		
H_2S	ppm_v	<100	<1000	1.5	–	–	–	–	–	≤5	Corrosion in compressors, gas storage tanks, and engines. Toxic concentration of H_2S (>5 cm^3 m^{-3}) remain in the biogas SO_2 and SO_3 are formed during to combustion, which are more toxic than H_2S and cause corrosion in the presence of water
H_2S, variation	ppm_v	0–100	0–1000	1–2	–	–	–	–	–	–	–
Sulphur	mg/Nm^3	–	–	–	–	<100 <75		<30		<45	Corrosion in the presence of water
NH_3	ppm_v	5	<100	0	–	–	–	–	–	≤3	Corrosion in the presence of water
Total chlorine (as Cl^-)	mg/Nm^3	20–200	0–5	0	–	<1		–	–	<50	Corrosion in combustion engines

a biogas impurity above a critical threshold concentration based on its comburent activity. H_2S in combination with O_2 and water generates H_2SO_4, which can corrode engines, gas storage tanks, pipelines and compressors. The presence of halocarbons and NH_3 can also cause corrosion in pipelines and engines during biogas combustion (Petersson and Wellinger 2009). Finally, the presence of methylsiloxanes is attracting increasing attention due to the undesired formation of silicone oxide deposits during combustion, which can cause abrasion, overheating, and malfunctioning of engines and valves (Abatzoglou and Boivin 2009).

Biogas valorization has traditionally involved the generation of heat and power via biogas combustion in turbines, micro-turbines or internal combustion engines. Recent investigations have also highlighted the potential of reformed biogas as substrate in fuel cells for the combined production of heat and electricity. The use of biogas as autogas or as a green substitute of natural gas via injection into natural gas grids requires a more stringent purification compared to its direct combustion in the above mentioned energy-valorization units (IEA-Bioenergy 2014). Biogas production in Europe at the end of 2014 was based on >17,000 digesters and accounted for a total installed capacity of 8339 MW_{el} (equivalent to the electricity generated by eight 1000 MW nuclear reactors) (EBA 2015). According to the latest estimations of the European Biogas Association, biogas production will increase up to 20 billion Nm^3 by 2020 and will represent a significant share of Europe's natural gas consumption.

The global market of conventional and emerging technologies for biogas upgrading is nowadays changing as a result of the stricter biogas composition specifications (Bauer et al. 2013b). Physical/chemical biogas upgrading methods present a high energy and/or chemical demand, which has triggered the development of biotechnologies as alternative biogas upgrading platforms due to their low operating costs and environmental impacts. In this context, both aerobic and anoxic biotrickling filtration or microaerobic anaerobic digestion can support biogas desulfurization as efficiently as activated carbon filtration or chemical precipitation. Similarly, algal-bacterial photobioreactors can carry out a simultaneous removal of CO_2, H_2S, NH_3, and VOCs from biogas in a single step-process at low operating costs and environmental impacts. This book chapter will critically review the state of the art of the main physical/chemical and biological technologies available nowadays in the market with a special focus on the removal of CO_2, H_2S, siloxanes, halocarbons, O_2, and N_2.

10.2 Technical Specifications for Upgrading

The specifications of the final application of biogas determine its required composition and therefore, the type of upgrading to be applied (Table 10.1).

The selection of the most appropriate biogas upgrading technology involves factors such as investment and operating costs, recovery and loss of methane, and removal efficiency for the biogas impurities as described above (Persson 2003; Sun et al. 2015). Based on the promising potential of the upgraded biogas as a renewable energy for substituting natural gas, manufacturers and countries have set standards for the utilization of the upgraded biogas as a fuel for stoves and boilers, engines and gas turbines, gas grid injection, substrate for fuel cells, and use as a vehicle fuel (Bailon and Hinge 2012).

The use of biogas in boilers for heat generation only requires gas pressurization at 8–25 mbar, as well as the removal of H_2S <1000 ppm_v and water prior to combustion. When biogas is to be used in domestic stoves, the concentration of H_2S should be <10 ppm_v (IEA bioenergy 2000; Bailon and Hinge 2012). H_2S concentration in biogas used for electricity generation in internal combustion engines should be reduced to 200–1000 ppm_v along with water in order to avoid the condensation of acid aqueous solutions in gas lines, which could cause corrosion (Bailon and Hinge 2012). Internal combustion engines also require levels of NH_3 below 32–50 mg m^{-3}, siloxanes below 5–28 mg m^{-3}, and halocarbons below 65–100 mg m^{-3} prior to combustion. The size of the turbine determines the maximum allowed concentrations of H_2S (micro-turbines can tolerance <70,000 ppm_v and turbines <10.000 ppm_v), siloxanes (0.03–0.1 ppm_v), and halocarbons (200–1500 ppm_v Cl^-/F^-) when biogas is used for the combined production of heat and electricity on site (Sun et al. 2015; EPRI 2006).

On the other hand, when biogas is to be used as a substitute of gas natural (often named as biomethane) for vehicle fueling or via injection into natural gas grids for further use in domestic gas appliances, cogeneration plants or industry, technical specification are higher than those set by boiler, engines, or turbine manufacturers. The standards and specifications of biomethane for grid injection are country specific, with a European draft under development nowadays. Table 10.1 displays the required biomethane composition in countries like France, Germany, and the Netherlands. Interestingly, biomethane is divided into high calorific (H) and low calorific (L) gas based on its Wobbe index, and must contain CH_4 concentrations higher than 80–96%, CO_2 contents <2–3%, O_2 levels <0.2–1%, H_2S <5–15 mg m^{-3}, NH_3 <3–20 mg m^{-3}, and methylsiloxanes <5–10 mg Si m^{-3} (Persson et al. 2006; Bailon and Hinge 2012). Biogas as a vehicle fuel uses the same engine and vehicle configuration as natural gas. However, a higher concentration of H_2 is allowed in biomethane for vehicle fueling (0.1 H_2%, mol) (Sun et al. 2015).

Finally, fuel cells offer a high flexibility in term of biogas composition due to their high operating temperatures (up to 1000 °C). However, H_2S levels in biogas used as a substrate in fuel cells must be <5 ppm_v and siloxane removal complete to prevent long-term damage in heat exchangers, catalysts and sensors (Lampe 2006; Haga et al. 2008).

10.3 Physical/Chemical Biogas Upgrading Technologies

10.3.1 CO_2 Removal Technologies

CO_2 removal from biogas, which accounts for 25–50% on a volume basis, is mandatory in order to increase the biogas energetic content and calorific value, to reduce the transportation costs, and eventually, to partially mitigate greenhouse gas emissions from biogas production plants.

Physical/chemical technologies for the removal of CO_2 from raw biogas are based on the transfer of this target compound to another gas, liquid or solid phase, where it may further undergo a chemical reaction. Their current application in real biogas upgrading plants is way ahead of that of their biological counterparts, mainly due to their high efficiency and wide field experience (partly coming from the chemical industry). In this sense, water scrubbing accounts for ~41% of the global biogas upgrading market, followed by chemical scrubbing and pressure swing adsorption with 22 and 21%, respectively. Other mature technologies such as organic solvent scrubbing or membrane separation represent 6 and 10% of the market share, respectively. Finally, cryogenic CO_2 separation, still not reliably commercialized at full scale, accounts for only 0.4% of the upgrading market share at a global level (Thrän et al. 2014). The main features of these physical/chemical technologies are discussed below:

Operating Principles CO_2 separation by absorption is nowadays the most widely implemented technology. It is based on the transfer of this compound from the biogas to a liquid scrubbing solution, which can be water, an organic solvent or a chemical solution. While the two first rely only on CO_2 mass transfer and physical absorption of the molecule to the scrubbing liquid, a chemical reaction takes place between the solvent and the absorbed CO_2 in the latter. The absorption process takes place in a packed column with random packing materials such as Pall or Rasching rings to promote gas-liquid contact and reduce the risk of biomass growth. Operation under a counter-current configuration is preferred regardless of the scrubbing configuration.

In the particular case of **water scrubbing**, the higher aqueous solubility of CO_2 compared to that of CH_4 allows for the selective removal of CO_2 using water as absorbent. Depending on the application, water scrubbing can be carried out in single-pass scrubbers with low-quality water or in sequential units of pressurized CO_2 absorption followed by a two-stage stripping process for water (tap quality) regeneration. The absorption process is usually carried out at high pressures (6–10 bar) while the amount of water required depends on the operating pressure and temperature (typically ranging between $0.1 - 0.2\text{m}^3_{\text{water}}\text{Nm}^{-3}_{\text{biogas}}$) (Bauer et al. 2013a, b; Persson 2003).

In **organic solvent scrubbing**, water is substituted by solvents such as methanol or dimethyl ethers of polyethylene glycol, with higher affinity for CO_2 than water. While the use of organic solvents allows for lower liquid recycling rates and reduced plant sizes, a preconditioning step to remove the moisture of the biogas is required (due to the hygroscopic nature of the solvents). Moreover, cooling and heating stages are usually implemented to promote CO_2 absorption (at low temperature of 20 °C) and its subsequent desorption for organic solvent regeneration (at ~40 °C) (Muñoz et al. 2015).

A further enhanced scrubbing performance is obtained in **chemical scrubbing** when a CO_2-reactive absorbent is employed, since the reaction of the absorbed CO_2

with the chemical reagent results in higher absorption capacities and process operation at maximum CO_2 concentration gradients (Ryckebosch et al. 2011). This allows for more compact units, lower liquid recycling rates and process operation at low absorption and stripping pressures (between 1 and 2 bar and 1.5–3 bar, respectively). Moreover, recovery of the absorbent is accomplished in a desorption unit equipped with a reboiler, which simplifies the scrubbing unit configuration (Patterson et al. 2011). Chemicals such as alkanol amines or alkali aqueous solutions (KOH, K_2CO_3, NaOH, $Fe(OH)_3$ or $FeCl_3$) are frequently used in chemical scrubbing (Eqs. 10.1–10.3) (Awe et al. 2017; Salihu and Alam 2015).

$$CO_2 + 2OH^- \rightarrow CO_3^{2-} + H_2O \tag{10.1}$$

$$CO_2 + CO_3^{2-} + H_2O \rightarrow 2HCO_3^- \tag{10.2}$$

$$CO_2 + R - NH_2 + H_2O \rightarrow R - NH_3^+ + 2HCO_3^- \tag{10.3}$$

CO_2 can be also selectively transferred to a solid phase via adsorption, in a process so called **pressure swing adsorption (PSA)**. The selected porous adsorbent must present a high specific surface area, exhibit a linear adsorption isotherm, be non-hazardous, readily available, and stable under long-term operation. PSA adsorbents are selective to CO_2 molecules due to their lower size compare to that of CH_4 and a higher adsorption affinity. Typical adsorbents meeting these requirements are activated carbon, silica gel, activated alumina, zeolite, or polymeric sorbents (Bauer et al. 2013b; Patterson et al. 2011; Ryckebosch et al. 2011). The adsorbents are packed in vertical columns, with commonly implemented PSA systems consisting of 4 stages that take place in 4 interconnected columns operating in parallel. These stages alternate pressurization and CO_2 adsorption at 4–10 bars to increase CO_2 retention inside the pores, and subsequent depressurization and regeneration of the saturated column by venting to ambient pressure and purging with upgraded biogas for desorption (Muñoz et al. 2015).

Membrane separation has gained importance in the last decades due to the advances in nanotechnology, which have resulted in the development of membrane materials with enhanced selectivity factors, lower costs, easier manufacturing, higher stability at high operating pressures, and easier scalability. In particular, polymeric materials such as cellulose acetate are preferred for biogas upgrading applications (Basu et al. 2010; Muñoz et al. 2015). The selective permeation of CO_2 from the raw biogas to a gas or a liquid phase is based on the permeability of the membrane material, with CO_2/CH_4 selectivity factors of up to 1000/L. Other biogas components such as O_2, H_2O, or H_2S might also preferentially permeate over CH_4. Gas-gas membrane systems operate under high pressures (20–40 bars) in single-pass or multiple stage membrane units, with further recirculation of the permeates or the retentates to increase the purity of the upgraded biomethane. Conversely, the operation of the gas-liquid membranes is performed at atmospheric pressure using CO_2-liquid absorbents such as alkanol amines or alkali aqueous solutions to improve the overall CO_2 selectivity (Muñoz et al. 2015).

In addition to these physical/chemical technologies based on mass transfer, the selective recovery of the biogas components can rely on their different condensation temperatures as implemented in **cryogenic separation**. In this sense, the raw biogas temperature is sequentially decreased at constant pressure (~10 bar) to remove water, H_2S, siloxanes and halogens (−25 °C), liquid CO_2 (−55 °C), and the remaining CO_2 in solid phase (−85 °C), obtaining a CH_4-enriched stream. Further cooling to temperatures below −160 °C generates a N_2 and O_2-free liquefied biomethane (Awe et al. 2017; Ryckebosch et al. 2011; Bauer et al. 2013b). However, despite the satisfactory separation efficiencies achieved and the possibility to obtain liquid biomethane as a final product, cryogenic separation is still an energy-intensive emerging technology with few plants under operation at a global scale.

Upgrading Capacities and Pre-treatment Requirements Overall, high CO_2 removal efficiencies and low CH_4 losses are achieved in these mature physical/chemical technologies, with final CO_2 recoveries always >95%. An efficient CO_2 separation allows to obtain a biomethane complying with the most stringent quality requirements for its final use, which correspond to biomethane injection into natural gas grids (CH_4 concentrations of 80–96% and CO_2 <2–3%) (Table 10.2). The highest CH_4 recoveries are reached with chemical scrubbing, although values of up to 99–99.5% have been recorded for membrane separation when complex designs with recirculation of both the permeate and retentate are used (Benjaminsson 2006).

Table 10.2 Upgrading capacities, investment costs, and energy use of physical/chemical technologies for CO_2 removal from biogas (adapted from Muñoz et al. (2015) and Awe et al. (2017))

Technology	CH_4 loss (%)	Final CH_4 concentration (%)	Capital costs (€ $(Nm^3h^{-1})^{-1}$) [Plant capacity $(Nm^3\ h^{-1})$]	Energy consumption $(kWh\ Nm^{-3})^a$
Scrubbing				
Water-Scrubbing	<2	>96	5500–2500–2000 (100–500 → 1000)	0.2–0.3
Solvent-Scrubbing	<2	96–98.5	4500–2000–1500 (250–1000 → 1500)	0.2–0.51
Chemical-Scrubbing	0.1-1.2	>99	3200–1500 (600–1800)	0.12–0.15*
PSA	–	96–98	2700-1500 [600-2000]	0.25–0.6
Membrane separation	–	96–98	6000–2500-2000 (100–400 → 1000)	0.2–0.38
Cryogenic separation	<2	>97	No data available	0.42–1

[a]Only gas compression and liquid pumping requirements are considered

Other compounds can also be removed via physical/chemical technologies from raw biogas together with CO_2, however, biogas pre-treatment for the removal of some of them prior to CO_2 separation is highly recommended in order to avoid operating problems. For instance, water scrubbing may cope with H_2S concentrations of 300–2500 ppm_v in the gas phase, although its presence might result in elemental sulfur accumulation, corrosion or odour nuisance (Muñoz et al. 2015). The presence of high concentrations of H_2S also entails the regeneration of the organic solvent via steam or inert gas stripping to avoid its degeneration, and might also trigger amine poisoning in chemical scrubbing. In the particular case of PSA, H_2S and siloxanes are irreversibly adsorbed onto the molecular sieves, reducing the lifetime of the adsorbent. Therefore, a removal of these compounds together with moisture is commonly accomplished by activated carbon filters and condensation prior to PSA upgrading (Bauer et al. 2013b). Finally, whereas conventional membrane-based upgrading facilitates the permeation of several biogas components such as O_2, H_2O, and H_2S along with CO_2, membrane clogging and deterioration is triggered by the presence of particles, H_2S, H_2O, VOCs, NH_3, and siloxanes in the raw biogas, and therefore, their previous removal by condensation and carbon filtration is highly recommended (Patterson et al. 2011; Bauer et al. 2013b).

Investment and Operating Costs The investment costs of physical/chemical CO_2-upgrading technologies are characterized by the economy of scale, thus the larger the upgrading plant capacity, the lower the cost per unit of biogas flow treated (Table 10.2). This is more significant in membrane separation, where the capital costs rapidly increase when scaling down the process. In general, investment costs of 1500–2000 € $(Nm^3\ h^{-1})^{-1}$ are reported for design flowrates over 1000 $Nm^3\ h^{-1}$ regardless of the technology.

Process operating costs for physical absorption mainly derive from the electricity used for biogas compression and liquid pumping, while in chemical scrubbing, these costs are moderate (due to the lower operating pressure required) and are mainly governed by the energy required for solvent regeneration at high temperatures (120–150 °C). In this context, the cost of consumables associated with CO_2 absorption are marginal, which involves ~20–200 L h^{-1} of water in water scrubbing; minor organic solvent make-up in organic solvent scrubbing (due to the low vapor pressures of the absorbents employed); or the supplementation of $\sim 3\ mg\ Nm^{-3}_{biogas}$ of amines, antifoam, and water in chemical scrubbing (Muñoz et al. 2015; Patterson et al. 2011). Main PSA operating costs are also associated with the electricity requirements for biogas compression and pre-conditioning, with no costs from water make-up addition or heat for adsorbent regeneration. On the other hand, membrane replacement (with 5–10 years of lifetime) and biogas pre-treatment are together with gas compression, the main contributors to the operating costs in membrane separation. Moreover, while maintenance costs of absorption and adsorption technologies usually range between 2 and 3% of the investment costs, this value is slightly higher (3–4%) for membrane-based biogas upgrading (Muñoz et al. 2015).

10.3.2 H_2S Removal Technologies

The concentrations of H_2S in biogas are commonly reduced nowadays by physical/chemical technologies. The most popular physical/chemical technologies are adsorption onto activated carbon, adsorption using iron oxide or hydroxide, membrane separation, absorption and in situ precipitation in the digester via iron salt addition (Pettersson and Wellinger 2009; Muñoz et al. 2015).

Adsorption on Activated Carbon
H_2S removal by activated carbon filtration can be carried out either by simple physical adsorption onto the carbon surface or by catalytic conversion, where H_2S is converted into sulphur and water. The latter mechanism requires a high temperature (50–70 °C) and pressure (7–8 bar), and the addition of 4–6% of air in the biogas to support the partial oxidation of H_2S (Eq. 10.4) (Ryckebosch et al. 2011). The lifetime of the activated carbon ranges from 4000 to 8000 h depending on the H_2S loading rate applied to the filter (Wellinger et al. 2005), a regeneration or replacement of the carbon being necessary after carbon saturation.

$$2H_2S + O_2 \rightarrow 2S + 2H_2O \tag{10.4}$$

The partial oxidation of H_2S to elemental sulphur is supported by the impregnation of the carbon with potassium iodide (KI), permanganate ($KMnO_4$), or zinc oxide (ZnO) as catalysers. KI and $KMnO_4$ are the most commonly used catalysts when the biomethane is to be injected into the natural gas grid, while the use of ZnO is often limited as a result of its high cost (Petersson and Wellinger 2009).

Adsorption Using Iron Oxide or Hydroxide
Process operation is based on the selective adsorption of H_2S in adsorbent modules operated in parallel using an adsorption-regeneration configuration. These modules contain an organic packing material impregnated with iron oxide (Fe_2O_3), hydroxide oxide ($Fe(OH)_3$), or zinc oxide (ZnO) that retain H_2S during biogas circulation (Muñoz et al. 2015; Iovane et al. 2014; Abatzoglou et al. 2009).

The chemical reactions involved in H_2S oxidation and adsorbent regeneration are described by Eqs. 10.5–10.7:

$$Fe_2O_3 + 3H_2S \rightarrow Fe_2S_3 + 3H_2O \tag{10.5}$$

$$2Fe(OH)_3 + 3H_2S \rightarrow Fe_2S_3 + 6H_2O \tag{10.6}$$

$$2Fe_2S_3 + 3O_2 \rightarrow 2Fe_2O_3 + 6S \tag{10.7}$$

The two first reactions are endothermic and need a moderate temperature between 25 and 50 °C to occur, while the third reaction is exothermic and temperature is not controlled. Adsorption using iron oxide or hydroxide operates at a biogas residence time ranging from 1 to 15 min and can cope with H_2S concentrations of up to 100 ppm_v (Muñoz et al. 2015; Ryckebosch et al. 2011). This physical/chemical technology is widely implemented because it is efficient (reduction rates of 99% are typically reported) and exhibit moderate operating costs (Iovane et al. 2014; Rutledge 2005).

Membrane Separation

H_2S can be removed from biogas by permeation (along with CO_2) through a semipermeable membrane that retains CH_4. Similar to the membrane units for CO_2 separation, H_2S removal can be carried out at high pressure in gas: gas modules or at low pressure in gas: liquid modules with a CO_2 absorbent on the other side of the membrane (Iovane et al. 2014). For biogas containing H_2S concentrations of 2%, removal efficiencies of 98% and 58–94% have been reported (Ryckebosch et al. 2011; Iovane et al. 2014). This technology is not suitable for medium-high strength biogas streams, and the presence of O_2 can cause operational problems during liquid regeneration if water is used as absorbent.

Absorption

Absorption is the most common technology for the removal of H_2S in the chemical industry. This method is based on the mass transfer of H_2S from the biogas into a liquid solvent such as water or organic solvents. One of the most popular processes is based on the chemical absorption of H_2S into a catalytic solution of Fe^{3+}-EDTA, where the hydrogen sulphide is removed (via oxidation to S) and separated by sedimentation from the Fe^{3+}-EDTA solution. Then, this solution with the reduced catalyst is regenerated by oxidation with O_2 (Eqs. 10.8–10.9) (Horikawa et al. 2004; Demmink and Beenackers 1998):

$$2Fe^{3+} + S^{2-} \rightarrow 2Fe^{2+} + S \tag{10.8}$$

$$2Fe^{2+} + 0.5O_2 + H_2O \rightarrow 2Fe^{3+} + 2OH^- \tag{10.9}$$

This technology is designed to operate at ambient temperature and pressure, under low biogas residence time and low liquid/gas ratios. Removals of H_2S between 90 and 100% are typically achieved in this technology (Ryckebosch et al. 2011).

In Situ Precipitation

Addition of Fe^{2+} or Fe^{3+} ions in the form of $FeSO_4$, $FeCl_3$, and $FeCl_2$ salts into the organic feed or directly into the digester represents an efficient mechanism for the control of H_2S concentration in biogas. These ions reacts with the dissolved H_2S, generating the insoluble FeS and/or elemental S following the reactions shown in Eqs. 10.10–10.11 (Ryckebosch et al. 2011; Muñoz et al. 2015; Pettersson and Wellinger 2009):

$$Fe^{2+} + S^{2-} \rightarrow FeS \tag{10.10}$$

$$2Fe^{3+} + 3S^{2-} \rightarrow 2FeS + S \tag{10.11}$$

Table 10.3 Advantages and disadvantages of physical/chemical H_2S removal technologies

Technology	Advantages	Disadvantages
Adsorption on activated carbon	• High removal efficiencies of H_2S	• High temperature is needed for carbon regeneration • Short lifetime of the activated carbon
Adsorption using iron oxide or hydroxide	• Removal efficiencies >99% • Low operating cost	• Adsorbent regeneration is expensive • Concentration of H_2S to be treated must be <100 ppm_v • Temperature must be controlled
Membrane separation	• Can remove CO_2 simultaneously • High H_2S removal efficiencies	• Concentration of H_2S to be treated must be <2%
Absorption	• Water can be used as solvent • Removal efficiencies 90–100% • Operation at ambient temperature and pressure	• Low liquid/biogas ratios needed
In-situ precipitation	• Low investment cost • Efficient at high H_2S concentrations	• Not efficient at low concentrations of H_2S • H_2S in treated biogas >100–150 ppm_v • High operating costs

This method is efficient to reduce H_2S concentration in moderate to high strength biogas, but cannot decrease H_2S levels below 100–150 ppm_v. While its operation is simple and the investment costs are low, the widespread use of this technology is limited by the high operating cost derived from the purchase of the iron salts (Persson et al. 2006).

Table 10.3 summarizes the main advantages and disadvantages of the physical/chemical H_2S removal technologies discussed in Sect. 10.3.2.

10.3.3 Siloxanes, VOC, and Halocarbons Removal Technologies

Siloxanes are compounds containing a silicon-oxygen bond (Si–O) that are used in multiple cleaning products and cosmetics. These products are the source of siloxanes in biogas from landfills and wastewater treatment plant, which are responsible for damages to engines, valves, cylinder heads, etc. (Ryckebosch et al. 2011; Soreanu et al. 2011). The most important siloxane removal technology is **adsorption on activated carbon**, which depends on the water content in biogas and is often combined with a pre-treatment for water removal. The removal efficiency of this technology can reach 95% (Ryckebosch et al. 2011) and 74–83% (Schweigkofler and Niessner 2001). This technology is limited by the need of process operation at high pressure and low moisture contents, and by the technical difficulties associated with the regeneration of the activated carbon. Siloxane **adsorption** can also be conducted **on silica gel**, which is a granular form of silicon dioxide (SiO_2) made from sodium silicate (Na_2O_3Si) and commonly used as a desiccant. Silica gel presents a polar nature that allows the adsorption of siloxanes molecules up to removal efficiencies of 95% (Ryckebosch et al. 2011). The main disadvantage of this technology is the need for process operation at high pressure, which increases both investment and operating costs. **Cryogenic separation** or cryogenic condensation of siloxanes can support removal efficiencies of 99.3% when the temperature of biomethane is decreased to −70 °C and of 25.9% when the temperature of biomethane reaches −25 °C (Hagmann et al. 2001). The widespread implementation of this technology is limited by its high investment and operating costs.

Halogenated compounds and VOCs are typically removed by activated carbon adsorption in two packed bed columns operated in parallel in a sequential adsorption-regeneration mode (Ryckebosch et al. 2011; Muñoz et al. 2015). Regeneration of the activated carbon is performed at 200 °C (Wellinger et al. 2005).

10.3.4 O_2 and N_2 Removal Technologies

The O_2 and N_2 present in biogas are not biologically generated during anaerobic digestion. These gases are typically present at high concentrations in landfill gas when biogas is collected by vacuum generation as a result of air infiltration. Despite O_2 levels <0.5% in biomethane are admissible, biogas in air is explosive at concentrations ranging from 6–12% at biogas methane contents of 60%, depending on the temperature (Petersson and Wellinger 2009; Bailón and Hinge 2012). On the other hand, N_2 is an inert gas difficult to remove during upgrading, with a limited impact on the applications of biogas except for a decreased calorific value and CH_4 content (Wellinger and Lindberg 2005).

Table 10.4 shows the technologies for both O_2 and N_2 removal from biogas along with their main advantages and disadvantages. **Pressure swing adsorption (PSA)** is based on the differences in gas adsorption rates to capture these biogas contaminants at a high pressure in vertical columns packed with absorbents under a sequence of adsorption, depressurization, desorption, and pressurization. **Membrane separation** is based on the selective permeability of O_2 and N_2 across membranes under a gas-gas configuration at high pressure. Finally, **cryogenic separation** uses temperature difference to separate O_2 and N_2 from the rest of the biogas components. Overall, the technologies for the removal of O_2 and N_2 require high investment and operating costs, high energy demands, and a complex process control (Persson et al. 2007; Muñoz et al. 2015; Awe et al. 2017).

Table 10.4 Overview of the advantages and disadvantages of physical/chemical O_2 and N_2 removal technologies (Wellinger and Lingberg 2005; Ryckebosch 2011; Yang et al. 2014)

Technologies	Advantages	Disadvantages
PSA	• Removes CO_2, N_2 and O_2 • Low energy demand • Low level of emissions	• H_2S and water removal is needed before PSA • Periodical regeneration of the adsorbent needed
Membrane separation	• Compact and light in weight • Easy operation and maintenance • Low energy requirements	• Relatively low CH_4 purity and high CH_4 losses • High cost derive from membrane purchase • Complex maintenance
Cryogenic separation	• Produces CO_2 as by product • Removal of multiple impurities	• High energy demand • High capital cost

10.4 Biological Biogas Upgrading Technologies

10.4.1 CO_2 Removal Technologies

10.4.1.1 Hydrogenotrophic CO_2 Removal (Biological Methanation of CO_2)

Biogas upgrading through biological methanation of CO_2 consists of the utilization of H_2 (electron donor) by hydrogenotrophic methanogens to transform CO_2 (carbon source and electron acceptor) into CH_4 according to Eq. 10.12 (Rittmann 2015):

$$4H_2 + CO_2 \rightarrow CH_4 + 2H_2 \quad \Delta G^0 = -130.7\,\mathrm{kJ\,mol^{-1}} \tag{10.12}$$

The development of this technology is linked to the shift of electricity production from fossil fuels towards renewable sources. Electricity generation from renewable energies is not coupled to demand, hence biological methanation is an option to chemically store renewable energy as CH_4, by H_2 production (via water electrolysis) and the subsequent upgrading of biogas (Rittmann et al. 2013). This technological platform, with 17 projects of biogas upgrading currently under development in Europe (Bailera et al. 2017), is commonly named power-to-gas. Power-to-gas technologies are not only applied for biogas upgrading but also to convert different sources of CO_2 (or even CO) to CH_4. However, biogas upgrading using a power-to-gas approach exhibits low operating costs and the opportunity to use the heat from methanation and the O_2 produced during electrolysis for microaerobic removal of H_2S. In any cases, the costs of H_2 production during electrolysis govern the overall process costs (Götz et al. 2016).

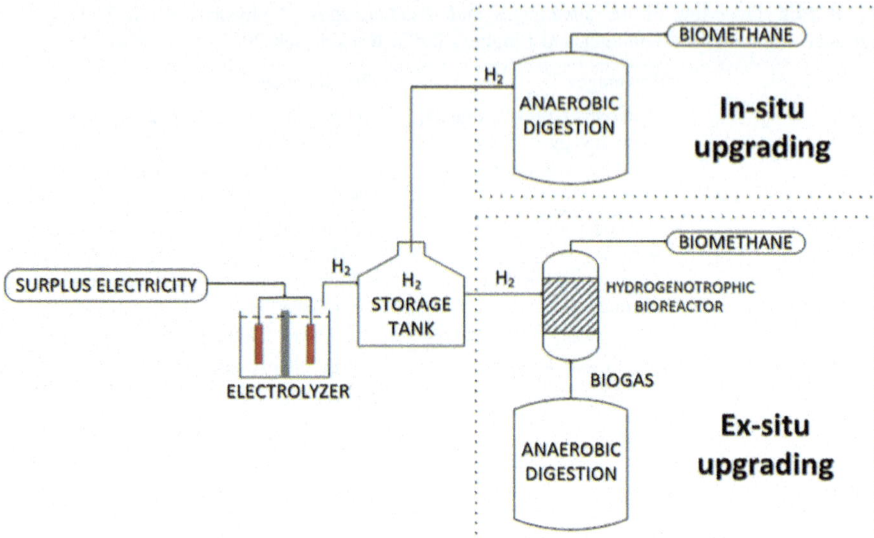

Fig. 10.1 Main set-ups for the biological conversion of CO_2 into CH_4. Adapted from Rittmann (2015)

Biomethanation can be performed in two different configurations (Fig. 10.1). On one side, in situ upgrading implies the supply of H_2 to the anaerobic digester so that biomethane is directly produced within the digester. On the other side, ex situ upgrading entails the supply of H_2 and biogas to a specific bioreactor, containing an enriched hydrogenotrophic *archaea* community, where CO_2 conversion to CH_4 takes place. In both upgrading configurations, *archaea* from the species *Methanobacterium*, *Methanococcus*, *Methanoculleus*, *Methanosaeta*, *Methanosarcina*, *Methanospirilum*, and *Methanothermobacter* have been repeatedly found (Bassani et al. 2015; Kougias et al. 2016; Luo and Angelidaki 2013a; Wang et al. 2013).

The poor solubility of H_2 in water (dimensionless Henry's constant of 52 at 35 °C) always limits the bioconversion of CO_2 to CH_4. In this context, a limiting H_2 mass transfer as the rate-determining step results in an enhanced biomass formation with a concomitant reduction in the production of CH_4 (Strevett et al. 1995). Indeed, several bioreactor configurations have been tested to enhance H_2 mass transfer and to maximize CO_2 conversion to CH_4 using stoichiometric mixtures of H_2 and CO_2 (80:20 by volume) such as CSTR (Kim et al. 2013), fixed-bed bioreactors (Lee et al. 2012), bubble columns (Díaz et al. 2015), and biotrickling filters (Burkhardt et al. 2015).

Ex situ bioreactors devoted to biogas upgrading have been shown to efficiently convert H_2 (1.9–97 $L_{H_2}L_R^{-1}d^{-1}$) and CO_2 to CH_4 with a final CH_4 concentration in the treated biogas up to 98% (Kougias et al. 2016; Luo and Angelidaki 2012; Martin et al. 2013; Rachbauer et al. 2016). However, most studies have been carried out only at lab-scale (0.6–58 L) and those reaching a final CH_4 concentration suitable for biomethane injection in the grid (> 95%) are limited to a load of 7.2 $L_{H_2}L_R^{-1}d^{-1}$. In this context, pressurized bioreactors have been proposed to compensate for the decrease in conversion efficiency at high H_2 and biogas load rates (Seifert et al. 2014). An increase in the pressure of the bioreactor would result in higher H_2 transfer rates mediated by a higher H_2 solubility according to Henry's Law. Ex situ upgrading by directly contacting H_2 and biogas in a bioreactor independent from the digester presents the benefit of implementing individual control strategies and advanced bioreactors configurations to increase the H_2 mass-transfer coefficient ($k_{LH_2}a$) in contrast to in situ bioreactors (Rittmann 2015).

In situ upgrading is limited by the fact that anaerobic digesters are not designed to maximize gas-liquid mass transfer, but to provide optimal organic matter removal instead. Hence, the H_2 gas-liquid mass transfer coefficients ($k_{LH_2}a$) in anaerobic digesters are low. Nevertheless, the low CO_2 productivity per digester volume results in both lower H_2 load ($L_{H_2}L_R^{-1}d^{-1}$) and $k_{LH_2}a$ requirements. In this sense, the adaptation of the ADM1 to in situ H_2 injection showed that $k_{LH_2}a$ of ~30 h^{-1} should be applied for an efficient CO_2 bioconversion (Bensmann et al. 2014). The few studies evaluating the performance of a direct supply of H_2 to the anaerobic digester (in situ upgrading) have been performed in CSTR (Bassani et al. 2015; Luo et al. 2012; Luo and Angelidaki 2013a, b; Wang et al. 2013) and UASB digesters (Bassani et al. 2015) treating manure, sewage sludge or potato-starch wastewater. H_2 loading rates between 0.19 and 1.8 $L_{H_2}L_R^{-1}d^{-1}$ have been reported at lab-scale along with biomethane concentrations of up to 99%. Additionally, the consumption of CO_2 within the digester can induce inhibitory pH increases if the alkalinity of the organic fed is not properly controlled (Luo et al. 2012).

To summarize, biological methanation of CO_2 with H_2 produced with excess of electricity from renewable sources (ex situ upgrading) or through direct production of biomethane within anaerobic digesters (in situ upgrading) at a low cost, while storing surplus electricity in the form of CH_4, is a promising technology to upgrade biogas. However, only lab-scale bioreactors tests have been reported so far and further developments are required to increase the H_2 gas-liquid transfer. In this context, incipient research on pressurized bioreactors and novel configurations that result in high gas-liquid transfer rates will play a key role in the progress and commercial application of the biological methanation of CO_2.

10.4.1.2 Photosynthetic CO_2 Removal

Photosynthetic CO_2 removal is based on the bioconversion of CO_2 into microalgae biomass through oxygenic photosynthesis, which is carried out by eukaryotic microalgae and prokaryotic cyanobacteria (from now on referred as microalgae) (López et al. 2013; Muñoz et al. 2015). Light-mediated water photolysis is needed for the redox reduction of CO_2, where the electrons are transferred from the water to CO_2. This transportation takes place against an electrochemical gradient and occurs via a series of reactions that require a source of energy (light) to proceed. Overall, CO_2, water, nutrients, and mineral salts are converted into energy-rich compounds contained in the microalgal biomass and oxygen (Tredici 2009; Muñoz et al. 2015). These processes can be simplified as follows (Eq. 10.13):

$$\begin{aligned} & CO_2 + H_2O + \text{sunlight (photons)} + \text{nutrients} \\ & \quad \rightarrow O_2 + CH_{1.63}N_{0.14}O_{0.43}P_{0.006}S_{0.005} + \text{waste heat} \end{aligned} \tag{10.13}$$

CO_2 absorption from raw biogas into an aqueous cultivation broth is required prior removal by microalgal photosynthesis (Posadas et al. 2015). In this context, the moderate aqueous solubility of CO_2 (Henry's Law constant ≈0.83 at 20 °C) supports an efficient mass transfer to the cultivation broth (Sander 1999). The optimum growth of microalgae requires 1.8 g CO_2 $g_{microalgae}^{-1}$, and therefore, an adequate CO_2 mass transfer to the cultivation medium is mandatory during photobioreactor design and operation. Similarly, nutrients supplementation is also necessary (mainly nitrogen (N), phosphorus (P) and sulphur (S) along with other trace elements) in order to guarantee a successful CO_2 biofixation (Wang et al. 2008; Trobajo et al. 2014). In this context, anaerobic effluents, which are characterized by a high nutrient content (mainly N and P), have emerged as a sustainable alternative to synthetic culture medium. The use of these effluents as nutrient media allows the recovery of nutrients from digestates in the form of a valuable algal-bacterial biomass (Bahr et al. 2014; Posadas et al. 2015) (Table 10.5).

Despite the fact that several microalgae species can support photosynthetic biogas upgrading, the most commonly reported ones are *Chlorella*, *Arthospira*, *Spirulina*, and *Scenedesmus*, which are characterized by their tolerance to high CO_2 concentrations and pH values (Muñoz et al. 2015; Wang et al. 2008) (Table 10.5). In this regard, few authors have isolated microalgae species capable of withstanding CO_2 concentrations of up to ≈40–60% (Wang et al. 2008). H_2S concentrations of around 100 ppm_v have been found to inhibit microalgae-based processes (Kao et al. 2012). However, the fast H_2S oxidation to SO_4^{2-} by the sulphur oxidizing bacteria (SOB) naturally present in biogas upgrading photobioreactors and/or the high dissolved oxygen (DO) concentrations in the cultivation broth (≈2–25 mg O_2 L^{-1}) avoid any potential inhibitory effects by H_2S (Toledo-Cervantes et al. 2016). Finally, CH_4 does not entail any inhibition of microalgae activity at concentrations ranging from 20 to 80% mainly due to its low aqueous solubility. This low CH_4 solubility constitutes an important advantage in terms of biogas upgrading by scrubbing because it allows the absorption of CO_2 and H_2S while minimizing CH_4 losses (Kao et al. 2012; Posadas et al. 2015).

Parameters such as light availability, temperature, pH, and DO determine the rates of biological CO_2 biofixation by microalgae, assuming that no other limiting or inhibitory parameters impact on the process once the transfer of CO_2 and mixing are optimized (Posadas et al. 2016). Photosynthesis in most microalgae species gets saturated at ≈250 μmol m^{-2} s^{-1}, which corresponds to ≈10% and 17% of the summer and winter peak outdoors light irradiances (2500 and 1200 μmol m^{-2} s^{-1}, respectively). Under these conditions, and based on the fact that ≈10–20% of the

Table 10.5 Experimental studies on photosynthetic biogas upgrading under different configurations of photobioreactors

Photobioreactor design and volume (L)	Input biogas composition (%)	CO_2-RE (%)	Output biogas composition (%)	Culture conditions	Microalgae population	References
Set of bubbled columns: 0.4-0.6 L	Real biogas = CH_4: 38–81 CO_2: 19–62 H_2S: 0.0–0.2	–	$O_2 \leq 3.5$	Indoors at continuous light irradiance of 1000 µmol m^{-2} s^{-1}/ Cultivation in synthetic medium	*Chlorella vulgaris*	Duosková et al. (2010)
Set of bubbled columns of 50 L	Real biogas = CH_4: 69 ± 1 CO_2: 20 ± 1 H_2S: 0.000–0.005	73–86	CH_4: 86–91	Outdoors (Taiwan 24°09′N 120° 40′E)/Cultivation in artificial seawater	*Mutant Chlorella sp. strain*	Kao et al. (2012)
Bubble column of 0.8 L	Real biogas = CH_4: 42 CO_2: 49 CO: 2 Others: 7	–	–	Indoors at continuous light irradiance of 18 W (10-15 cm of the culture)/Cultivation in synthetic medium	*Spirulina platensis*	Sumardiono et al. (2014)
Enclosed photobioreactor of 1 L	Real biogas = CH_4: 67–82 CO_2: 8–23	100	O_2: 10–24	Indoors at continuous light irradiance of 36 µmol m^{-2} s^{-1}/ Cultivation in synthetic medium	*Spirulina platensis*	Converti et al. (2009)
Enclosed tubular photobioreactor of 5.3 L	Real biogas = CH_4: 60 CO_2: 40	10–90	CO_2: 0–0	Indoors at light:dark cycles of 16 h:8 h at irradiance of 80 µmol m^{-2} s^{-1}/Cultivation in synthetic medium	*Scenedesmus obliquus*	Thiansathit et al. (2015)

(continued)

Table 10.5 (continued)

Photobioreactor design and volume (L)	Input biogas composition (%)	CO_2-RE (%)	Output biogas composition (%)	Culture conditions	Microalgae population	References
Enclosed tubular photobioreactor of 0.45 L	CH_4: 58 CO_2: 42 H_2S: 0.05	≈94–98	CH_4: 50.1–53.3 CO_2: 1.2–2.5 O_2: 18.2–23.4 H_2S: 0	Indoors at continuous light irradiance of 35, 60 and 100 µmol m^{-2} s^{-1}/Cultivation in synthetic medium	*Chlorella vulgaris*	Mann et al. (2009)
Algal pond of 15 L with a biolift absorption unit inside the pond	Real biogas = CH_4: 55–71 CO_2: 44–48 H_2S: 1	74–95	CH_4: 88–97 CO_2: 2.5–11.5 H_2S <0.5	Indoor cultivation/Cultivation in synthetic medium and settled diluted piggery wastewater	*Chlorella vulgaris*	Conde et al. (1993)
180 L raceway interconnected to a 0.8 L bubble column (external absorption unit)	Synthetic biogas: N_2: 69.5 CO_2: 30 H_2S: 0.5	40–98	H_2S: 0 O_2 <1	Indoors at continuous light irradiance of 80 µmol m^{-2} s^{-1}/ Cultivation in synthetic medium and in real diluted centrate	*Spirulina platensis, Phormidium, Oocystis, Microspora*	Bahr et al. (2014)
180 L raceway interconnected to a 2.5 L bubble column (external absorption unit)	Synthetic biogas: CH_4: 70 CO_2: 29.5 H_2S: 0.5	72–79	CH_4: 81 ± 2 CO_2: 6.8–8.8 H_2S: 0 O_2: 0.7 ± 0.2 N_2: 5.9–7.2	Indoors at light:dark cycles of 16 h:8 h at irradiance of 104 ± 25 µmol m^{-2} s^{-1}/ Cultivation in diluted anaerobically digested vinasse and diluted raw vinasse	*Geitlerinema* sp., *Limnothrix planktonica, Pseudoanabaena minima, Stigeoclonium tenue, Leptolyngbya benthonica, Planktolyngvya brevicellularis, Staurosira* sp.	Posadas et al. (2015)
180 L raceway interconnected to		40–100	O_2: 2–20	Indoors at continuous light irradiance of 75 ± 5 µmol m^{-2}		Posadas et al. (2016)

(continued)

Table 10.5 (continued)

Photobioreactor design and volume (L)	Input biogas composition (%)	CO_2-RE (%)	Output biogas composition (%)	Culture conditions	Microalgae population	References
a 2.5 L bubble column (external absorption unit)	Synthetic biogas: CO_2: 30 N_2: 70			s^{-1}/Cultivation in real diluted centrate	*Microspora sp., Scenedesmus, Synechocysitis aquatilis, Woronichia sp.*	
180 L raceway interconnected to a 2.5 L bubble column (external absorption unit)	Synthetic biogas: CH_4: 70 CO_2: 29.5 H_2S: 0.5	50–95	CH_4: 70–94 CO_2: 9–25 H_2S: 0 O_2: 0.1–2.0 N_2: 0.6–5.0	*Outdoors (Valladolid (Spain)-summer; 41°39′N 4°44′ W)*/ Cultivation in real centrate	*Chlorella sp., Pseudanabaena sp.,Chloroidium saccharophilum*	Posadas et al. (2017)
180 L raceway interconnected to a 2.5 L bubble column (external absorption unit)	Synthetic biogas: CH_4: 70 CO_2: 29.5 H_2S: 0.5	80	H_2S: 0 O_2: 0.3–3 N_2: 6–10	Indoors at light:dark cycles of 16 h:8 h at irradiance of 104 ± 25 μmol m^{-2} s^{-1}/ Cultivation in diluted anaerobically digested vinasse	*Chlorella sp., Chloromonas sp., Geitlerinema sp., Microspora sp., Pseudanabaena sp., Stigeoclonium* sp., *Planktolyngbya* sp. and others	Serejo et al. (2015)
180 L raceway interconnected to a 2.5 L bubble column (external absorption unit)	Synthetic biogas: CO_2: 30 N_2: 70	55 ± 6	N_2 (CH_4 hypothetical scenario): 87	Indoors at continuous light irradiance of 75 μmol m^{-2} s^{-1}/ Cultivation in diluted centrate	*Phormidium sp., Oocystis, Microspora sp.*	Alcántara et al. (2015)
180 L raceway interconnected to a 2.5 L bubble column (external absorption unit)	Synthetic biogas: CH_4: 70 CO_2: 29.5 H_2S: 0.5	≈100	CH_4: 97.2 ± 0.2 CO_2: 0.4 ± 0.1 H_2S: 0 O_2: 0.03 ± 0.04 N_2: 2.4 ± 0.2	Indoors at light:dark cycles of 16 h:8 h at irradiance of 420 ± 105 μmol m^{-2} s^{-1}/ Cultivation in real centrate	*Geitlerinema sp., Limmothrix planktonica, Acutudesmus obliquus, Chlorella vulgaris, Mychonastes homosphaera, Navicula, Phormidium sp., Stigeoclonium tenue*	Toledo-Cervantes et al. (2016)

(continued)

Table 10.5 (continued)

Photobioreactor design and volume (L)	Input biogas composition (%)	CO_2-RE (%)	Output biogas composition (%)	Culture conditions	Microalgae population	References
180 L raceway interconnected to a 2.5 L bubble column (external absorption unit)	Synthetic biogas: CH_4: 70 CO_2: 29.5 H_2S: 0.5	≥95	CH_4: 96.2 ± 0.7 CO_2: 0.1–2 H_2S: 0 O_2: 0.1–1 N_2: 1–4	Indoors at light:dark cycles of 14 h:10 h at irradiance of 1500 ± 600 μmol m^{-2} s^{-1}/ Cultivation in real centrate	*Chlorella minutissima*	Toledo-Cervantes et al. (2017)
75 L raceway interconnected to a 0.7 L bubble column (external absorption unit)	Real biogas: CH_4: 72 ± 2 CO_2: 28 ± 2	93	O_2: 1.2	Indoors at light:dark cycles of 14 h:10 h at irradiance of 1500 ± 600 μmol m^{-2} s^{-1}/ Cultivation in synthetic medium	*Nannochloropsis gaditana*	Meier et al. (2015)
50 L raceway interconnected to a 0.3 L bubble column (external absorption unit)	Real biogas: CH_4: 65 ± 1.5 CO_2: 32.0 ± 1.9	89–93	CO_2: 2–4.5 O_2: <1	Indoors light:dark cycles of 12 h:12 h at irradiance of 25, 50, 75 and 100 μmol m^{-2} s^{-1}/ Cultivation in synthetic medium	*Chlorella sorokiniana*	Meier et al. (2017)

total solar irradiation is lost by reflection, the maximum light irradiance that can be fixed by microalgae ranges from 1 to 7% (depending on the photobioreactor configuration), which limits microalgae productivity to 10–35 g m^{-2} d^{-1} (Park et al. 2011). Optimum temperature for microalgae activity ranges from 15 to 30 °C, despite the fact that some authors have reported successful algae growth at 35 °C (Muñoz et al. 2015). Although the optimum pH for microalgae ranges from 7 to 8, values of ≈9–10 are required to maintain a high CO_2 gradient between the gas and liquid phase (Posadas et al. 2016). Several microalgae species are able to growth at these high pHs (Toledo-Cervantes et al. 2016). Finally, DO concentrations > 25 mg L^{-1} can inhibit microalgae activity. In this regard, oxygen removal constitutes an important issue in the design of photobioreactors (Mendoza et al. 2013).

Similar physical/chemical and biological mechanisms to those supporting CO_2 removal from flue gases in photobioreactors take place during photosynthetic CO_2 removal from biogas. In this regard, photobioreactors are designed to maximize CO_2 absorption, mixing, nutrients supply, light distribution, pH control, and oxygen removal (Muñoz et al. 2015). The main difference with photobioreactors designed to treat flue gases is that the CO_2-laden gas is bubbled into a sump in open photobioreactors and discharged to the atmosphere during flue gas treatment, while enclosed photobioreactors (bubble columns or tubular systems) and raceways with an additional biogas scrubbing unit are used when biomethane is recovered (López et al. 2013) (Muñoz et al. 2015) (Table 10.5).

Enclosed photobioreactors are characterized by capital cost investments of 500–3000 € m^{-2}, energy consumptions of ≈50–100 W m^{-3}, and illuminated area to volume ratios of ≈30–70 m^{-1} (Muñoz et al. 2015). Their design allows a high light utilization efficiency (≈4–6%), which results in biomass concentrations of up to ≈6 g total suspended solids (TSS) L^{-1} and biomass productivities of ≈25–45 g m^{-2} d^{-1} (Acién et al. 2012). CO_2 removal efficiencies in a set of bubble columns accounted for 86%, while a complete CO_2 removal has been reported in enclosed tubular photobioreactors (Table 10.5). Despite the high CO_2 removal efficiency achieved in this photobioreactor configuration, further improvements should be focused on the reduction of O_2 concentration (below 1%) and the increase in CH_4 concentration (> 95%) in the upgraded biogas (Table 10.5; Muñoz et al. 2015). Furthermore, the use of enclosed systems to treat biogas has been mainly carried out at laboratory scale indoors (Table 10.5). Therefore, a systematic evaluation of their performance at pilot scale under outdoors conditions must be conducted.

Raceways are open photobioreactors characterized by their simplicity in construction and operation, and their lower capital cost investments (2–20 € m^{-2}), energy requirements (2–10 W m^3), and illuminated area to volume ratios (3–10 m^{-1}) compared to their enclosed counterparts (Tredici 2009; López et al. 2013). A high microalgae diversity and risk of external contamination with local microalgae species or predators is expected in high rate algal ponds (HRAPs) due to their open design (Posadas et al. 2014). The main disadvantage of these photobioreactors is their low light utilization efficiency (≈2%), which results in low biomass concentrations in the cultivation broth (≈0.3–1 gTSS L^{-1}) and low biomass productivities (5–20 g m^{-2} d^{-1}) (Tredici 2009; Posadas 2016). Another limitation is their high water footprint (up to 17.5 L m^{-2} d^{-1} in arid areas (Guieysse et al. 2013)], which can compromise their environmental sustainability. HRAPs treating biogas have been used via integration with an external absorption unit (biolift or column) in order to recover and reuse the treated biomethane (Table 10.5). The interconnection between the HRAP and the absorption column is carried out by an external liquid recirculation, which minimizes CH_4 losses to the atmosphere ($\leq 2\%$ mass basics) (Toledo-Cervantes et al. 2017). A successful treatment of anaerobic effluents coupled to photosynthetic biogas upgrading has been reported at pilot scale under laboratory and outdoors conditions in these photobioreactors (Table 10.5). Thus, Toledo-Cervantes et al. (2016) found O_2 concentrations of $0.03 \pm 0.04\%$ in the upgraded biogas, along with CH_4 purities of $97.2 \pm 0.2\%$, while minimizing the effluent flowrate and maximizing nutrients recovery in the harvested biomass. However, the reduction in N_2 concentration in the upgraded biogas, which is continuously stripped out from the recycling algal broth as a result of the open nature of the HRAPs and its equilibrium with the atmosphere ($\approx$14 mg N_2 L^{-1}), is mandatory to achieve high CH_4 concentrations. Recently, Posadas et al. (2017) evaluated the performance of this photobioreactor configuration for the simultaneous treatment of centrate and biogas at pilot scale under outdoors conditions (Table 10.5). The results showed that at a low alkalinity in the cultivation broth (inorganic carbon concentration $\approx$500 mg L^{-1}), CO_2 removal efficiency was highly influenced by the temperature. The maximum CO_2 removal efficiencies and CH_4 purities recorded in their study were 95 and 94%, respectively. Nevertheless, further research should be focused on the optimization of this pilot scale system under outdoors conditions to maintain a year-round CH_4 content >95%.

10.4.2 H_2S Removal Technologies

10.4.2.1 Microaerobic H_2S Removal

A limited amount of oxygen/air can be introduced as a strategy to reduce sulfide levels in anaerobic reactors (Sousa et al. 2016).

A detailed knowledge of the nature of both biological and chemical oxidation of sulfide is necessary to understand the effect of oxygen dosage. The most important bioconversions involved in aerobic sulfide removal are presented in the following equations (Eqs. 10.14–10.16) (Janssen et al. 1995):

$$2HS^- + O_2 \rightarrow 2S^0 + 2OH^- \quad \Delta G = -169.35\,\mathrm{kJ \cdot mol^{-1}} \tag{10.14}$$

$$2HS^- + 4O_2 \rightarrow 2SO_4^{2-} + 2H^+ \quad \Delta G = -732.58\,kJ \cdot mol^{-1} \tag{10.15}$$

$$2HS^- + 2O_2 \rightarrow S_2O_3^{2-} + H_2O \quad \Delta G = -387.35\,kJ \cdot mol^{-1} \tag{10.16}$$

Microaerobic H_2S removal in anaerobic digesters relies on the action of SOB, which grow lithoautotrophically on H_2S while producing elemental sulphur (S^0) instead of sulfate under O_2-limited conditions according to Eq. 10.14 (Madigan et al. 2009). SOB show different morphologies, physiological and ecological characteristics, and use primarily O_2 as the terminal electron acceptor (Tang et al. 2009).

While in situ microaerobic H_2S removal has been traditionally used in anaerobic digesters treating agricultural wastes based on the economic benefits of on-site biogas exploitation (Muñoz et al. 2015), recent research has extended its application to anaerobic reactors treating industrial wastewaters (Rodríguez et al. 2012), wastewater treatment plants (WWTPs) sludge, cow manure (Jenicek et al. 2008; Kobayashi et al. 2012) or sewage (Sousa et al. 2016). In this technology, the headspace of anaerobic digesters acts as a H_2S abatement unit where multiple microaerophilic SOBs such as *Acidithiobacillus* sp., *Arcobacter* sp., *Sulfuricuvum* sp., *Sulfurimonas* sp., *Thiobacillus* sp., *Thiofaba* sp., and *Thiomonas* sp. develop upon a restricted quantity of O_2 is provided (Díaz et al. 2011; Kobayashi et al. 2012; Rodríguez et al. 2012). SOB grow over the headspace walls and ceiling of anaerobic digesters as a result of the lack of any biomass support, superimposed layers of S^0 representing the support material for SOB growth (with a high specific surface area that facilitates O_2 transfer) (Díaz et al. 2011; Kobayashi et al. 2012). The main advantage of in situ H_2S removal is the absence of end-of-pipe units for desulfurization. Nevertheless, an excessive S^0 deposition on the digester's headspace would possibly impair the removal performance over time by reducing the residence time of biogas and, consequently, the O_2 transfer rate to the microorganisms. Therefore, a periodical cleaning of the reactor headspace is critical to maintain the H_2S removal efficiency (Krayzelova et al. 2015). Several authors have observed that the low O_2 supply rates required for H_2S abatement do not significantly compromise the performance of organic matter removal or CH_4 productivity (Díaz et al. 2010; Rodríguez et al. 2012). On the contrary, enhanced organic matter hydrolysis and methanogenic activity as a result of the suppression of sulfide toxicity have been reported (Jenicek et al. 2010, 2011).

Several authors have comparatively evaluated the efficiency of air dosing into the headspace or into the liquid phase of anaerobic digesters (Krayzelova et al. 2015). When applied to the headspace, oxygen will directly react with gas sulfide, which entails lower quantities of air dosing (Díaz et al. 2011; Ramos et al. 2012). In this context, dosing lower quantity of air causes lower contamination of biogas by nitrogen (Krayzelova et al. 2015). On the other hand, when air is overdosed to assure complete H_2S removal, the surplus oxygen can contaminate biogas (Díaz et al. 2010, 2011) increasing the risk of explosion.

The biogas residence time is a key parameter controlling H_2S removal efficiencies (Muñoz et al. 2015). H_2S removal efficiencies over 97% are typically met when operating at biogas residence times over 5 h (Muñoz et al. 2015). In addition, higher O_2 to H_2S molar ratios are required to maintain H_2S removal efficiencies over 99%, when decreasing the biogas residence time in the headspace (Muñoz et al. 2015).

In this context, the O_2 (or equivalent air) supply rate can be adjusted to 0.3–3% of the biogas production rate depending on the H_2S concentration and the aforementioned biogas residence time (Muñoz et al. 2015). A variable O_2/air dosing is often required in most digesters in order to minimize the residual O_2 in the upgraded biogas as a result of the variable biogas production rates. Hence, a residual O_2 concentration of 1–1.8% in the biogas can be achieved by controlling the oxygen reduction potential (ORP) in the anaerobic mixed liquor, while 0.3–0.5% residual O_2 concentrations were recorded when employing biogas production as the control variable, both operational strategies supporting H_2S removal efficiencies larger than 99% (Ramos and Fdz-Polanco 2014).

Dosing air into the anaerobic broth also causes a decrease of the sulfide concentration in the liquid phase (Díaz et al. 2011; Krayzelova et al. 2014; van der Zee et al. 2007; Zhou et al. 2007). However, this decrease is often not larger than 20–30% (Krayzelova et al. 2014) and cannot explain the large decrease in H_2S concentration in the biogas (Krayzelova et al. 2015). This implies that most H_2S oxidation takes place in the digester headspace even if air is dosed into the liquid phase (Krayzelova et al. 2015). Along with the reduction in the H_2S levels in biogas, the decrease in the sulfide concentration in the liquid has the additional positive effect of decreasing sulfide toxicity towards methanogens. The mass transfer of oxygen into the liquid phase is intensified in digesters by mixing using biogas recirculation. However, air dosing into the mixed liquor will increase the consumption of O_2 due to the oxidation of biodegradable organic compounds (Díaz et al. 2011; Fdz-Polanco et al. 2009).

Finally, a recent economic evaluation of the in situ H_2S treatment of 550 m^3/h of biogas in full-scale WWTP sludge digesters showed that the total cost of H_2S removal using a PSA O_2 generator (92–98% O_2) was lower than process operation with air or pure O_2. The utilization of an oxygen generator entailed the lowest operational costs (0.82 € kg S^{-1} or 0.0018 € m^{-3} of biogas treated) compared to air and pure O_2 supply (1.18 € kg S^{-1} and 1.72 € kg S^{-1}, respectively) (Díaz et al. 2015).

10.4.2.2 Biotrickling Filtration

Biotrickling filters (BTF) consists of a packed bed column (where biomass growth occurs as a biofilm) sprayed by a recirculating aqueous phase that contains the essential nutrients for microbial growth. An efficient H_2S and oxygen transport between the gas and liquid phases, pH and temperature control, nutrient supply, and a controlled washout of accumulated metabolites are the main advantages of this biotechnology (Dumont 2015; Muñoz et al. 2015). Biotrickling filtration for H_2S treatment is based on the action of SOB (Gabriel et al. 2013). In aerobic BTF, lithoautotrophic bacteria can use H_2S as the energy source while O_2 is used as the electron acceptor according to Eqs. 10.17 and 10.18:

$$H_2S + 0.5O_2 \rightarrow S + H_2O \tag{10.17}$$

$$H_2S + 2O_2 \rightarrow SO_4^{2-} + 2H^+ \tag{10.18}$$

The control of the oxygen dosage into the BTF is critical due to both safety concerns (explosion risks) and to the need to avoid biogas dilution (Lebrero et al. 2016). NO_3^- or NO_2^- can also be used in anoxic BTFs as electron acceptor for the biological oxidation of H_2S, which would contribute to a concomitant nitrogen removal from digestates via denitrification (Li et al. 2016). The stoichiometry of H_2S removal via nitrate or nitrite reduction is described by Eqs. 10.19 and 10.20 (Lebrero et al. 2016; Dumont 2015).

$$5H_2S + 2NO_3^- \rightarrow 5S + N_2 + 4H_2O + 2OH^- \tag{10.19}$$

$$5H_2S + 8NO_3^- \rightarrow 5SO_4^{2-} + 4N_2 + 4H_2O + 2H^+ \tag{10.20}$$

Elemental sulfur might be preferred over sulfate formation in order to avoid trickling liquid acidification (Fortuny et al. 2008). However, the accumulation of elemental sulfur under oxygen or nitrate limiting conditions increases the risk of BTF clogging (Montebello et al. 2012). Indeed, elemental sulfur accumulation is nowadays considered as the bottleneck limiting the applicability of this biotechnology. Several bacterial genera such as *Thiothrix*, *Thiobacillus*, *Thiomonas*, *Acidithiobacillus*, and *Sulfurimonas* are capable of oxidizing H_2S under neutral/basic pH conditions using the CO_2 present in biogas as a carbon source (De Arespacochaga et al. 2014; Maestre et al. 2010). Process operation under acidic pH conditions do not entail a reduction in the H_2S removal capacity as a result of the development of acidophilic bacterial biofilms of *Acidithiobacillus thioxidants, Acidiphilium* sp., and *Thiobacillus ferrooxidans* able to grow at a pH of 2–4 (Montebello et al. 2013; Syed et al. 2006). Temperature ranges between 28 and 35 °C are desirable to maximize bacterial activity (Muñoz et al. 2015), 30 °C being the most typical optimum temperature for anoxic H_2S removal (Fernández et al. 2014).

Research studies on BTFs have been performed using HD-Q-PAC, polyurethane foam, pall rings, and polypropylene carriers as packing materials under anoxic and aerobic conditions with inlet concentrations in the range 500–10,000 ppm_v (Table 10.6). High removal efficiencies of 80–100% were achieved at biogas residence times ranging from 2 to 79 min. In this regard, biogas residence times below 2 min often result in a significant deterioration in H_2S removal efficiency during the treatment of inlet H_2S concentrations of 2000 ppm_v due to mass transfer limitations (Fortuny et al. 2011). Trickling liquid velocity (TLV) regulation may improve the oxygen gas-liquid mass transfer and influence the gas-liquid fluid dynamics in the

Table 10.6 Experimental studies on H_2S biotrickling filtration units under anoxic and aerobic conditions

Type process	Packing material	pH	$[H_2S]_{inlet}$ (ppmv)	Gas residence time (min)	Elimination capacity (g H_2S m^{-3} h^{-1})	H_2S-RE (%)	References
Microaerobic	Cylindrical polypropylene carrier	2.6	7800	5.9	121.7	100	Zhou et al. (2015)
Aerobic	Polypropylene Pall rings	6.5–7.0	2000–10,000	1.9	56.3–262.7	92.7–100	López et al. (2016)
Aerobic	Polypropylene Pall rings	1.7	2107 ± 151	3.8–5.9	54.0	99	Rodríguez et al. (2014)
Aerobic	Stainless steel Pall rings	2.5	2000–10,000	2.2	160.0–223.0	80–100	Montebello et al. (2014)
Aerobic	HD Q-PAC	1.5–2.0	2200–4350	1.3–1.4	169.0	84	De Arespacochaga et al. (2014)
Aerobic	Stainless steel Pall rings	6.0–6.5	2000	3	52.5	99	Montebello et al. (2012)
Aerobic	HD-Q-PAC	6.0–6.5	2000	2-3	55.0–82.0	98	Fortuny et al. (2011)
Aerobic	HD-Q-PAC	6.5–7.0	2000	3	55.6	99	Maestre et al. (2010)
Anoxic	Polypropylene	7.5–8.0	1000	2.0–5.0	54.5	100	Li et al. (2016)
Anoxic	Polyurethane foam cubes inserted into plastic curls	7.0	–	30-79	4.3–26.2	99.1	Lebrero et al. (2016)
Anoxic	Polyurethane foam cubes	7.3–7.5	–	2.4–3.4	99.8–130.0	99	Fernández et al. (2014)
Anoxic	Polyurethane foam cubes	7.4–7.5	2000	2.7	60.0	99	Montebello et al. (2012)
Anoxic	Polyester fibre and lava rocks	6.5	500–1500	5–16	177–182	93–96	Soreanu et al. (2009)

BTF (López et al. 2016). Overall, an increase in the TLV enhances both biomass and S flushing and the wettability of the biofilm, although in some BTF packed with lava rocks, a TLV increase might result in a decrease in the mass transfer coefficients due to media flooding (Kim and Deshusses 2008). O_2/H_2S ratios of 2–41 and NO_3^-/H_2S ratios of 0.25–1.6 are recommended for an efficient H_2S oxidation in aerobic and anoxic BTFs, respectively (Muñoz et al. 2015; Li et al. 2016; Soreanu et al. 2008).

H_2S biofiltration exhibits a better environmental performance and lower operating cost than physical/chemical technologies. Thus, aerobic and anoxic BTFs can provide a cost-competitive H_2S removal at 0.013 and 0.016 € m^{-3}, operating costs significantly lower than those of chemical precipitation ($FeCl_3$) and chemical scrubbing (0.024 and 0.30 € m^{-3}, respectively) (Fernández et al. 2014; Tomàs et al. 2009; Miltner et al. 2012). Packing material replacement represents the main cost during the operation of this biotechnology (up to 44% of the total operating cost) (Estrada et al. 2012).

10.4.2.3 Photosynthetic H_2S Removal

Biogas upgrading in algal–bacterial photobioreactors constitutes a promising alternative for the simultaneous removal of H_2S and CO_2 in a single-step process (Bahr et al. 2014). This biotechnology is based on the oxidation of H_2S to sulfate by SOB using the oxygen photosynthetically produced during CO_2 biofixation by microalgae. This process can be described by Eq. 10.21 (Syed et al. 2006):

$$H_2S + CO_2 + \text{nutrients} + O_2 \rightarrow \text{biomass} + SO_4^{2-}/S + H_2O \tag{10.21}$$

The excess of oxygen in the cultivation broth mediated by photosynthesis promotes the complete oxidation of H_2S to sulfate. A recent study identifying bacteria from the genus *Thioalbus* in the algal-bacterial broth supported the biological nature of H_2S oxidation in photobioreactors devoted to biogas upgrading (Toledo-Cervantes et al. 2016).

The transport of H_2S from the biogas to the algal-bacterial broth is necessary prior to H_2S oxidation. This is carried out in an external biogas absorption column interconnected to the photobioreactor or directly via biogas sparging in the photobioreactor. The limiting step during photosynthetic biogas upgrading is CO_2 removal due to the fact that the H_2S aqueous solubility is three times higher than the aqueous solubility of CO_2 according to their Henry's Law constants ($H_{H2S} \approx 2.44$ at ≈ 20 °C) (Sander 1999) and to the rapid biological H_2S oxidation (Muñoz et al. 2015). The fact that H_2S and CO_2 are acidic gases supports process operation at high pH values (9–10) in the cultivation broth along with alkaliphilic SOB and high pH-tolerant microalgae in order to enhance CO_2 and H_2S mass transfer (Bahr et al. 2014).

The absence of packing material in the biogas scrubbing unit together with the high O_2 concentration prevailing in algal-bacterial photobioreactors during biogas upgrading (Posadas et al. 2016) will prevent the clogging problems typically encountered in BTFs due to elemental sulfur accumulation (Lebrero et al. 2016). H_2S removal efficiencies of 100% concomitant with CO_2 removals of 80–95% are typically reported during photosynthetic biogas upgrading (Bahr et al. 2014; Serejo et al. 2015; Posadas et al. 2015; Toledo-Cervantes et al. 2017; Posadas et al. 2017).

10.4.3 Siloxanes, VOC, and Halocarbons Removal Technologies

Contrary to the well accepted non-biodegradability of siloxanes, microorganisms such as *Pseudomonas*, *Arthrobacter* and *Fusarium Oxysporium* have been recently shown to degrade octamethylcyclotetraxilosane (D4), hexamethylcyclotrisiloxane (D3) and dimethylsilanediol (Wasserbauer and Zadák 1990; Accettola et al. 2008).

However, the biological removal of siloxanes has been poorly explored to date. Accetola et al. (2008) investigated the removal of D3 in a BTF packed with Pall rings treating an air emission contaminated with D3 at 45 mg m^{-3} under a gas residence time of 2.1 min. Removal efficiencies of up to 20% were recorded in this study. Likewise, Poppat and Dessuses (2008) studied the removal of D4 under aerobic (air emission) and anaerobic (N_2 emission) conditions in a BTF packed with Cattle bone Porcelite at inlet D4 concentrations of 45 mg m^{-3}. Removal efficiencies of 50–60% at a gas residence time of 30–40 min were reported under aerobic conditions, while the maximum D4 removal under anaerobic conditions accounted for 15% at a residence time of 4 min. Similarly, Li et al. (2014) recorded D4 removal efficiencies of 74% at a residence time of 13.2 min in an aerobic BTF packed with lava rocks. The high gas residence times required and low siloxane removal efficiencies typically reported can be explained by the low solubility of siloxanes in water. In this context, biosurfactant excretion to the recycling liquid solution by some microorganisms could increase the solubility of siloxanes (Accetola et al. 2008; Soreanu et al. 2011; Li et al. 2014). Overall, the mass transfer capacity of conventional bioprocesses has to be significantly improved in order to make biotechnologies a cost-effective alternative to conventional physical-chemical technologies (Muñoz et al. 2015).

Finally, halogenated compounds such as methylene chloride, 1,1,1-trichloroethane, dichlorodifluoromethane, carbon tetrachloride and tetrachloroethylene, typically found in landfill biogas (Rasi et al. 2011), can be biodegraded at trace level concentrations under aerobic and anaerobic conditions (Lollar et al. 2010; Grostern and Edwards 2006). Similarly, the biodegradability of most VOCs identified in biogas (benzene, toluene, volatile fatty acids, etc.) has been consistently reported in literature (Muñoz et al. 2012). Unfortunately, to the best of our knowledge, there is no experimental study assessing the fate of VOCs and halogenated present in biogas in biological systems.

10.5 Conclusions

Today, physical/chemical methods for biogas purification such as adsorption, chemical precipitation, water/chemical/organic solvent scrubbing, membrane separation, or cryogenic separation are mature technologies able to produce a biomethane complying with most international regulations for injection into natural gas networks or use as a vehicle fuel. Unfortunately, the room for technical and economic optimization of these technologies is nowadays very limited (except for membrane or cryogenic separation). The high energy and chemical requirements restrict the cost-competitive use of biogas as an energy vector in spite of its environmental benefits. In this regard, biological methods such as photosynthetic upgrading can support a cost-efficient and simultaneous elimination of CO_2 and H_2S, with a concomitant conversion of CO_2 into a microalgal biomass for the production of biofertilizers or high added value products. The storage of the grid excess renewable electricity as H_2 can support the biological reduction of CO_2 into CH_4 in chemolitotroph-based bioreactors. On-going R&D projects are up-scaling these technologies in Spain and Denmark to validate their technical and economic viability at semi-industrial scale. These projects are addressing key process limitations such as the gas-liquid mass transfer of CO_2 and H_2 in algal-bacterial photobioreactors and chemolitotrophs-based bioreactors, respectively. Similarly, biotrickling filtration and anaerobic digestion under microaerobic conditions are capable of supporting H_2S removals > 99% at much lower operational costs than activated carbon filtration or in situ chemical precipitation. However, despite the fact that biological biogas desulfurization technologies have been optimized in the past decades and are commercially available today, they still suffer from severe operational problems caused by the accumulation of elemental sulfur in the packing material or in the digester's headspace. Finally, lab scale assays have confirmed the feasibility of the microbial biodegradation of both methyl siloxanes and halocarbons from biogas under aerobic and anaerobic conditions; the mass transfer of these biogas contaminants from the gas phase to the microorganisms has been identified as a potential process limitation during process implementation in continuous bioreactors.

Acknowledgements This work was supported by MINECO via the FEDER program (CTM2015-70442-R and Red Novedar), the European Commission (INCOVER project, this project has received funding from the European Union's Horizon 2020 research and innovation programme under grant agreement No. 689242), the Regional Government of Castilla y León (Project VA024U14 and UIC 71) and INIA (RTA2013-00056-C03-02). CONACyT-SENER is also gratefully acknowledged for the PhD grant of Roxana Ángeles Torres.

References

Abatzoglou N, Boivin S (2009) A review of biogas purification processes. Biofuels Bioprod Bioref 3:42–71. https://doi.org/10.1002/bbb.117

Accettola F, Guebitz G, Schoeftner R (2008) Siloxane removal from biogas by biofiltration: biodegradation studies. Clean Tech Environ Policy 10:211–218. https://doi.org/10.1007/s10098-007-0141-4

Acién FG, Fernández JM, Magán JJ, Molina E (2012) Production cost of a real microalgae production plant and strategies to reduce it. Biotechnol Adv 30:1344–1353. https://doi.org/10.1016/j.biotechadv.2012.02.005

Alcántara C, García-Encina P, Muñoz R (2015) Evaluation of simultaneous biogas upgrading and treatment of centrates in a HRAP through C, N and P mass balances. Water Sci Technol 72:150–157. https://doi.org/10.2166/wst.2015.198

Awe OW, Zhao Y, Nzihou A, Minh DP, Lyczko N (2017) A review of biogas utilisation, Purification and upgrading technologies. Rev Waste Biomass Valor 8(2):267–283. https://doi.org/10.1007/s12649-016-9826-4

Bahr M, Díaz I, Dominguez A, González Sánchez A, Muñoz R (2014) Microalgal-biotechnology as a platform for an integral biogas upgrading and nutrient removal from anaerobic effluents. Environ Sci Technol 48:573–581. https://doi.org/10.1021/es403596m

Bailera M, Lisbona P, Romeo LM, Espatolero S (2017) Power to Gas projects review: lab, pilot and demo plants for storing renewable energy and CO_2. Renew Sustain Energy Rev 69:292–312. https://doi.org/10.1016/j.rser.2016.11.130

Bailón L, Hinge J (2012) Report: biogas and bio-syngas upgrading. Danish Technological Institute. http://www.teknologisk.dk/_root/media/52679_Report-Biogasandsyngasupgrading.pdf. Accessed Dec 8 2014

Bassani I, Kougias PG, Treu L, Angelidaki I (2015) Biogas Upgrading via Hydrogenotrophic Methanogenesis in two-stage continuous stirred tank reactors at mesophilic and thermophilic conditions. Environ Sci Technol 151001074540007. https://doi.org/10.1021/acs.est.5b03451

Basu S, Khan A, Cano-Odena A, Liu C, Vankelecom I (2010) Membrane-based technologies for biogas separations. Chem Soc Rev 39:750–768. https://doi.org/10.1039/b817050a

Bauer F, Persson T, Hulteberg C, Tamm D (2013a) Biogas upgrading—technology overview, comparison and perspectives for the future. Biofuels Bioprod Biorefin 7:499–511

Bauer F, Hulteberg C, Persson T, Tamm D (2013b) Biogas upgrading—Review of commercial technologies. SGC Rapport 2013:270. SGC. http://vav.griffel.net/filer/C_SGC2013–270.pdf. Accessed Oct 10 2014

Benjaminsson J (2006) NYA Renings—Och Uppgraderingstekniker för biogas: Rapport SGC 163. Svenskt Gastekniskt Center

Bensmann A, Hanke-rauschenbach R, Heyer R, Kohrs F, Benndorf D, Reichl U, Sundmacher K (2014) Biological methanation of hydrogen within biogas plants: a model-based feasibility study. Appl Energy 134:413–425. https://doi.org/10.1016/j.apenergy.2014.08.047

Burkhardt M, Koschack T, Busch G (2015) Biocatalytic methanation of hydrogen and carbon dioxide in an anaerobic three-phase system. Bioresour Technol 178:330–333. https://doi.org/10.1016/j.biortech.2014.08.023

Conde JL, Moro LE, Travieso L, Sánchez EP, Leiva A, Dupeirón R, Escobedo R (1993) Biogas purification process using intensive microalgae cultures. Biotechnol Lett 15:317–320

Converti A, Oliveira RPS, Torres BR, Lodi A, Zilli M (2009) Biogas production and valorization by means of a two-step biological process. Bioresour Technol 100:5771–5776. https://doi.org/10.1016/j.biortech.2009.05.072

De Arespacochaga N, Valderrama C, Mesa C, Bouchy L, Cortina JL (2014) Biogas biological desulphurisation under extremely acidic conditions for energetic valorisation in solid oxide fuel cells. Chem Eng J 255:677–85. https://doi.org/10.1016/j.cej.2014.06.073

Demmink F, Beenackers A (1998) Gas desulfurization with ferric chelates of EDTA and HEDTA: new model for the oxidative absorption of hydrogen sulfide. Ind Eng Chem Res 37:1444–1453

Diaz I, Lopes AC, Perez SI, Fdz-Polanco M (2010) Performance evaluation of oxygen, air and nitrate for the microaerobic removal of hydrogen sulphide in biogas from sludge digestion. Biores Technol 101:7724–7730. https://doi.org/10.1016/j.biortech.2010.04.062

Díaz I, Lopes AC, Perez SI, Fdz-Polanco M (2011) Determination of the optimal rate for the microaerobic treatment of several H_2S concentrations in biogas from sludge digesters. Water Sci Technol 64(1):233–238. https://doi.org/10.2166/wst.2011.648

Díaz I, Pérez C, Alfaro N, Fdz-Polanco F (2015) A feasibility study on the bioconversion of CO_2 and H_2 to biomethane by gas sparging through polymeric membranes. Bioresour Technol 185:246–253. https://doi.org/10.1016/j.biortech.2015.02.114

Dousková I, Kastánek F, Maléterová Y, Kastánek P, Doucha J, Zachleder V (2010) Utilization of distillery stillage for energy generation and concurrent production of valuable microalgal biomass in the sequence: biogas-cogeneration-microalgae-products. Energ Convers Manage 51:606–611. https://doi.org/10.1016/j.enconman.2009.11.008

Dumont E (2015) H_2S removal from biogas using bioreactors: a review. Int J Energy Environ 6 (5):479–498

EPRI (2006) Assessment of fuel gas clean up systems for waste gas fueled power generation. California, USA. http://www.epri.com/abstracts/Pages/ProductAbstract.aspx?ProductId=000000000001012763. Accessed Mar 13 2017

Estrada JM, Kraakman NJR, Lebrero R, Muñoz R (2012) A sensitivity analysis of process design parameters, commodity prices and robustness on the economics of odour abatement technologies. Biotechnol Adv 30:1354–1363. https://doi.org/10.1016/j.biotechadv.2012.02.010

European Biogas Association. http://european-biogas.eu/2015/12/16/biogasreport2015/

Fdz-Polanco M, Diaz I, Perez SI, Lopes AC, Fdz-Polanco F (2009) Hydrogen sulphide removal in the anaerobic digestion of sludge by micro-aerobic processes: pilot plant experience. Water Sci Technol JInt Assoc Water Poll Res 60:3045–3050. https://doi.org/10.2166/wst.2009.738

Fernández M, Ramírez M, Gómez JM, Cantero D (2014) Biogas biodesulfurization in an anoxic biotrickling filter packed with open-pore polyurethane foam. J Hazard Mater 264:529–535. https://doi.org/10.1016/j.jhazmat.2013.10.046

Fortuny M, Baeza JA, Gamisans X, Casas C, Lafuente J, Deshusses MA, Gabriel D (2008) Biological sweetening of energy gases mimics in biotrickling filters. Chemosphere 71:10–17. https://doi.org/10.1016/j.chemosphere.2007.10.072

Fortuny M, Gamisans X, Deshusses MA, Lafuente J, Casas C, Gabriel D (2011) Operational aspects of the desulfurization process of energy gases mimics in biotrickling filters. Water Res 45:5665–5674. https://doi.org/10.1016/j.watres.2011.08.029

Gabriel D, Deshusses MA, Gamisans X (2013) Desulfurization of biogas in biotrickling filter. In: John Wiley & Sons (ed) Air pollution prevention and control: bioreactors and bioenergy, 1st edn. Wiley, p 513–523

Götz M, Lefebvre J, Mörs F, McDaniel Koch A, Graf F, Bajohr S, Reimert R, Kolb T (2016) Renewable power-to-gas: a technological and economic review. Renew Energy 85:1371–1390. https://doi.org/10.1016/j.renene.2015.07.066

Grostern A, Edwards EA (2006) A 1,1,1-Trichloroethane-degrading anaerobic mixed microbial culture enhances biotransformation of mixtures of chlorinated Ethenes and Ethanes. Appl Environ Microbiol 72(12):7849–7856. https://doi.org/10.1128/AEM.01269-06

Guieysse B, Béchet Q, Shilton A (2013) Variability and uncertainty in water demand and water footprint assessments of fresh algae cultivation based on case studies from five climatic regions. Bioresour Technol 128:317–323. https://doi.org/10.1016/j.biortech.2012.10.096

Haga K, Adachi S, Shiratori Y, Itoh K, Sasaki K (2008) Poisoning of SOFC anodes by various fuel impurities. Solid State Ion 179(27):1427–1431

Hagmann M, Hesse E, Hentschel P, Bauer T (2001) Purification of biogas removal of volatile silicones. In: 8th international waste management and landfill symposium, Sardinia, pp 641–666

Horikawa M, Rossi F, Gimenes M, Costa C, da Silva M (2004) Chemical absorption on H_2S for biogas purification. Braz J Chem Eng 21(3):415–422

IEA Bioenergy (2000) Biogas upgrading and utilisation. In: Wheeler, P. Jaatinen, T. Lindberg, A. Lundeberg, S. Holm-Nielsen, J. B. Wellinger, A. et al (eds) Task 24: energy from biological conversion of organic waste. Paris, France. http://www.biores.eu/docs/BIOGASFUNDAMENTALS/Biogas_upgrading__utilisation.pdf

IEA Bioenergy (2014) Task 37. http://www.iea-biogas.net/_download/publi-task37/upgrading_rz_low_final.pdf. Accessed June 5 2014

Iovane P, Nanna F, Ding Y, Bikson B, Molino A (2014) Experimental test with polymeric membrane for the biogas purification from CO_2 and H_2S. Fuel 135:352–358. https://doi.org/10.1016/j.fuel.2014.06.060

Janssen AJH, Sleyster R, Van der Kaa C, Jochemsen A, Bontsema J, Lettinga G (1995) Biological sulphide oxidation in a fed-batch reactor. Biotechnol Bioeng 47(3):327–333. https://doi.org/10.1002/bit.260470307

Jenicek P, Keclik F, Maca J, Bindzar J (2008) Use of microaerobic conditions for the improvement of anaerobic digestion of solid wastes. Water Sci Technol 58:1491–1496. https://doi.org/10.2166/wst.2008.493

Jenicek P, Koubova J, Bindzar J, Zabranska J (2010) Advantages of anaerobic digestion of sludge in microaerobic conditions. Water Sci Technol 62(2):427–434. https://doi.org/10.2166/wst.2010.305

Jenicek P, Celis CA, Koubova J, Pokorna D (2011) Comparison of microbial activity in anaerobic and microaerobic digesters. Water Sci Technol 63(10):2244–2249. https://doi.org/10.2166/wst.2011.579

Jönsson O, Polman E, Jensen J, Eklund R, Schyl H, Ivarsson S (2003) Sustainable gas enters the European gas distribution system. In: world gas conference, Tokio

Kao CY, Chiu SY, Huang TT, Dai L, Hsu LK, Lin CS (2012) Ability of a mutant strain of the microalgae *Chlorella* sp. to capture carbon dioxide for biogas upgrading. Appl Energ 93:176–183. https://doi.org/10.1016/j.apenergy.2011.12.082

Kim S, Choi K, Chung J (2013) Reduction in carbon dioxide and production of methane by biological reaction in the electronics industry. Int J Hydrogen Energy 38:3488–3496. https://doi.org/10.1016/j.ijhydene.2012.12.007

Kim S, Deshusses MA (2008) Determination of mass transfer coefficients for packing materials used in biofilters and biotrickling filters for air pollution control. 1. Experimental results. Chem Eng Sci 63:841–855. https://doi.org/10.1016/j.ces.2007.10.011

Kobayashi T, Li YY, Kubota K, Harada H, Maeda T, Yu HQ (2012) Characterization of sulfide-oxidizing microbial mats developed inside a full-scale anaerobic digester employing biological desulfurization. Appl Microbiol Biotechnol 93:847–857. https://doi.org/10.1007/s00253-011-3445-6

Kougias PG, Treu L, Benavente DP, Boe K, Campanaro S, Angelidaki I (2016) Ex-situ biogas upgrading and enhancement in different reactor systems. Bioresour Technol 225:429–437. https://doi.org/10.1016/j.biortech.2016.11.124

Krayzelova L, Bartacek J, Kolesarova N, Jenicek P (2014) Microaeration for hydrogen sulphide removal in UASB reactor. Biores Technol 172:297–302. https://doi.org/10.1016/j.biortech.2014.09.056

Krayzelova L, Bartacek J, Díaz I, Jeison D, Volcke EIP, Jenicek P (2015) Microaeration for hydrogen sulfide removal during anaerobic treatment: a review. Rev Environ Sci Biotechnol 14:703–725. https://doi.org/10.1007/s11157-015-9386-2

Lampe S (2006) Assessment of fuel gas clean up systems for waste gas fueled power generation. Palo Alto, CA, USA

Lebrero R, Toledo-Cervantes A, Muñoz R, del Nery V, Foresti E (2016) Biogas upgrading from vinasse digesters: a comparison between an anoxic biotrickling filter and an algal-bacterial photobioreactor. J Chem Technol Biotechnol 91(9):2488–2495. https://doi.org/10.1002/jctb.4843

Lee JC, Kim JH, Chang WS, Pak D (2012) Biological conversion of CO_2 to CH_4 using hydrogenotrophic methanogen in a fixed bed reactor. J Chem Technol Biotechnol 87:844–847. https://doi.org/10.1002/jctb.3787

Li X, Jiang X, Zhou Q, Jiang W (2016) Effect of S/N ratio on the removal of hydrogen sulfide from biogas in anoxic bioreactors. Appl Biochem Biotechnol 180:930–944. https://doi.org/10.1007/s12010-016-2143-3

Li Y, Zhang W, Xu J (2014) Siloxanes removal from biogas by a lab-scale biotrickling filter inoculated with Pseudomonas aeruginosa S240. J Hazard Mater 275:175–184. https://doi.org/10.1016/j.jhazmat.2014.05.008

Lollar B, Hirschorn S, Mundle S, Grostern A, Edwards E, Lacrampe-Couloume G (2010) Insights into enzyme kinetics of chloroethane biodegradation using compound specific stable isotopes. Environ Sci Technol 44:7498–7503. https://doi.org/10.1021/es101330r

López JC, Quijano G, Souza TSO, Estrada JM, Lebrero R, Muñoz R (2013) Biotechnologies for greenhouse gases (CH_4, N_2O, CO_2) abatement: state-of-the-art and challenges. Appl Microbiol Biot 97:2277–2303. https://doi.org/10.1007/s00253-013-4734z

López LR, Bezerra T, Mora M, Lafuente J, Gabriel D (2016) Influence of trickling liquid velocity and flow pattern in the improvement of oxygen transport in aerobic biotrickling filters for biogas desulfurization. J Chem Technol Biotechnol 91(4):1031–1039. https://doi.org/10.1002/jctb.4676

Luo G, Angelidaki I (2013a) Co-digestion of manure and whey for in situ biogas upgrading by the addition of H_2: process performance and microbial insights. Appl Microbiol Biotechnol 97:1373–1381. https://doi.org/10.1007/s00253-012-4547-5

Luo G, Angelidaki I (2013b) Hollow fiber membrane based H_2 diffusion for efficient in situ biogas upgrading in an anaerobic reactor. Appl Microbiol Biotechnol 97:3739–3744. https://doi.org/10.1007/s00253-013-4811-3

Luo G, Angelidaki I (2012) Integrated biogas upgrading and hydrogen utilization in an anaerobic reactor containing enriched hydrogenotrophic methanogenic culture. Biotechnol Bioeng 109:2729–2736. https://doi.org/10.1002/bit.24557

Luo G, Johansson S, Boe K, Xie L, Zhou Q, Angelidaki I (2012) Simultaneous hydrogen utilization and in situ biogas upgrading in an anaerobic reactor. Biotechnol Bioeng 109:1088–1094. https://doi.org/10.1002/bit.24360

Madigan MT, Martinko JM, Dunlap PV, Clark DP (2009) Brock biology of microorganisms, 12th edn. Pearson Benjamin-Cummings, San Francisco

Maestre JP, Rovira R, Álvarez-Hornos FJ, Fortuny M, Lafuente J, Gamisans X, Gabriel D (2010) Bacterial community analysis of a gas-phase biotrickling filter for biogas mimics desulfurization through the rRNA approach. Chemosphere 80:872–880. https://doi.org/10.1016/j.chemosphere.2010.05.019

Mann G, Schlegel M, Schumann R, Sakalauskas A (2009) Biogas conditioning with microalgae. Agron Res 7:33–38

Martin MR, Fornero JJ, Stark R, Mets L, Angenent LT (2013) A single-culture bioprocess of Methanothermobacter thermautotrophicus to upgrade digester biogas by CO_2 -to-CH_4 conversion with H_2. Archaea 2013:157529. https://doi.org/10.1155/2013/157529

Meier L, Pérez R, Azócar L, Rivas M, Jeison D (2015) Photosynthetic CO_2 uptake by microalgae: an attractive tool for biogas upgrading. Biomass Bioenerg 73:102–109. https://doi.org/10.1016/j.biombioe.2014.10.032

Meier L, Barros P, Torres A, Vílchez C, Jeison D (2017) Photosynthetic biogas upgrading using microalgae: effect of light/dark photoperiod. Renew Energy. https://doi.org/10.1016/j.renene.2017.01.009

Mendoza JL, Granados MR, De Godos I, Acién FG, Molina E, Heaven S, Banks CJ (2013) Oxygen transfer and evolution in microalgal culture in open raceways. Bioresour Technol 137:188–195. https://doi.org/10.1016/j.biortech.2013.03.127

Miltner M, Makaruk A, Krischan J, Harasek M (2012) Chemical-oxidative scrubbing for the removal of hydrogen sulphide from raw biogas: potentials and economics. Water Sci Technol 66:1354–1360. https://doi.org/10.2166/wst.2012.329

Montebello AM, Fernández M, Almenglo F, Ramírez M, Cantero D, Baeza M, Gabriel D (2012) Simultaneous methylmercaptan and hydrogen sulfide removal in the desulfurization of biogas in aerobic and anoxic biotrickling filters. Chem Eng J 200–202:237–246. https://doi.org/10.1016/j.cej.2012.06.043

Montebello AM et al (2013) Operational aspects, pH transition and microbial shifts of a H2S desulfurizing biotrickling filter with random packing material. Chemosphere 93:2675–2682. https://doi.org/10.1016/j.chemosphere.2013.08.052

Montebello A, Mora M, López L, Bezerra T, Gamisans X, Lafuente J, Baeza M, Gabriela D (2014) Aerobic desulfurization of biogas by acidic biotrickling filtration in a randomly packed reactor. J Hazard Mater 280:200–208. https://doi.org/10.1016/j.jhazmat.2014.07.075

Muñoz R, Daugulis AJ, Hernández M, Quijano G (2012) Recent advances in two-phase partitioning bioreactors for the treatment of volatile organic compounds. Biotechnol Adv 30 (6):1707–1720. http://dx.doi.org/10.1016/j.biotechadv.2012.08.009

Muñoz R, Meier L, Diaz I, Jeison D (2015) A review on the state-of-the-art of physical/chemical and biological technologies for an integral biogas upgrading. Rev Environ Sci Bio/Technol 14:727–759. https://doi.org/10.1007/s11157-015-9379-1

Park JBK, Craggs RJ, Shilton AN (2011) Wastewater treatment high rate algal ponds for biofuel production. Bioresour Technol 102:35–42. https://doi.org/10.1016/j.biortech.2010.06.158

Patterson T, Esteves S, Dinsdale R, Guwy A (2011) An evaluation of the policy and techno-economic factors affecting the potential for biogas upgrading for transport fuel use in the UK. Energ Policy 39:1806–1816. https://doi.org/10.1016/j.enpol. 2011.01.017

Persson M (2003) Evaluation of upgrading techniques for biogas. Rapport SGC 142. Swedish Gas Center

Persson M, Jönsson O, Wellinger A (2006) Biogas upgrading to vehicle fuel standards and grid injection. IEA Bioenergy. Task 37. http://www.energyineducation.ie/Renewables/Bioenergy/Biogas_upgrading_to_vehicle_fuel_standards_and_grid_connection_IEA.pdf. Accessed Feb 24 2017

Persson M, Jonsson O, Wellinger A (2007) Biogas upgrading to vehicle fuel standards and grid. IEA Bioenerg. 1–32

Petersson A, Wellinger A (2009) Biogas upgrading technologies—developments and innovations. IEA Bioenergy. Task 37. https://www.iea-biogas.net/files/daten-redaktion/download/publi-task37/upgrading_rz_low_final.pdf. Accessed Mar 12 2017

Popat S, Deshusses M (2008) Biological removal of siloxanes from landfill and digester gases: opportunities and challenges. Environ Sci Technol 42:8510–8515. https://doi.org/10.1021/es801320w

Posadas E, Serejo ML, Blanco S, Pérez R, García Encina PA, Muñoz R (2015) Minimization of biomethane oxygen concentration during biogas upgrading in algal-bacterial photobioreactors. Algal Res 12:221–229. https://doi.org/10.1016/j.algal.2015.09.002

Posadas E, Szpak D, Lombó F, Domínguez A, Díaz I, Blanco S, García-Encina PA, Muñoz R (2016) Feasibility study of biogas upgrading coupled with nutrient removal from anaerobic effluents using microalgae-based processes. J Appl Phycol. https://doi.org/10.1007/s10811-015-0758-3

Posadas E (2016) Innovative algal-bacterial processes for wastewater treatment: a further step towards full scale implementation: http://uvadoc.uva.es/handle/10324/18777

Posadas E, Marín D, Blanco S, Lebrero R, Muñoz R (2017) Simultaneous biogas upgrading and centrate treatment in an outdoors pilot scale high rate algal pond. Technol, Bioresour. https://doi.org/10.1016/j.biortech.2017.01.071

Rachbauer L, Voitl G, Bochmann G, Fuchs W (2016) Biological biogas upgrading capacity of a hydrogenotrophic community in a trickle-bed reactor. Appl Energy 180. https://doi.org/10.1016/j.apenergy.2016.07.109

Ramos I, Diaz I, Fdz-Polanco M (2012) The role of the headspace in hydrogen sulfide removal during microaerobic digestion of sludge. Water Sci Technol 66(10):2258–2264. https://doi.org/10.2166/wst.2012.457

Ramos I, Fdz-Polanco M (2014) Microaerobic control of biogas sulphide content during sewage sludge digestion by using biogas production and hydrogen sulphide concentration. Chem Eng J 250:303–311. https://doi.org/10.1016/j.cej.2014.04.027

Rasi S, Läntelä J, Rintala J (2011) Trace compounds affecting biogas energy utilisation—A review. Energy Convers Manag 52:3369–3375. https://doi.org/10.1016/j.enconman.2011.07.005

Rittmann S, Seifert A, Herwig C (2013) Essential prerequisites for successful bioprocess development of biological CH_4 production from CO_2 and H_2. Crit Rev Biotechnol 8551. https://doi.org/10.3109/07388551.2013.820685

Rittmann SKMR (2015) A critical assessment of microbiological biogas to biomethane upgrading systems, In: Biogas Science and Technology. Springer International Publishing, pp 117–135. https://doi.org/10.1007/978-3-319-21993-6_5

Rodriguez E, Lopes A, Fdz-Polanco M, Stams AJ, Garcia-Encina PA (2012) Molecular analysis of the biomass of a fluidized bed reactor treating synthetic vinasse at anaerobic and micro-aerobic conditions. Appl Microbiol Biotechnol 93:2181–2191. https://doi.org/10.1007/s00253-011-3529-3

Rodríguez G, Dorado AD, Fortuny M, Gabriel D, Gamisans X (2014) Biotrickling filters for biogas sweetening: oxygen transfer improvement for a reliable operation. Process Saf Environ 92:261–268. https://doi.org/10.1016/j.psep.2013.02.002

Rutledge B (2005) California Biogas industry assessment white paper. WestStart-CALSTART. http://www.calstart.org/Libraries/Publications/California_Biogas_Industry_Assessment_White_Paper.sflb.ashx. Accessed Feb 22 2017

Ryckebosch E, Drouillon M, Vervaeren H (2011) Techniques for transformation of biogas to biomethane. Biomass Bioenerg 35:1633–1645. https://doi.org/10.1016/j.biombioe.2011.02.033

Salihu A, Alam MdZ (2015) Upgrading strategies for effective utilization of biogas. Environ Prog Sustain Energy 34:1512–1520

Sander R (1999) Compilation of Henry's Law constants for inorganic and organic species of potential importance in environmental chemistry http://www.mpchmainz.mpg.de/sander/res/henry.html Accessed Jan 27 2017

Schweigkofler M, Niessner R (2001) Removal of siloxanes in biogases. J Hazard Mater 83 (3):183–196

Seifert AH, Rittmann S, Herwig C (2014) Analysis of process related factors to increase volumetric productivity and quality of biomethane with Methanothermobacter marburgensis. Appl. Energy 132. https://doi.org/10.1016/j.apenergy.2014.07.002

Serejo M, Posadas E, Boncz M, Blanco S, Garcia-Encina PA, Muñoz R (2015) Influence of biogas flow rate on biomass composition during the optimization of biogas upgrading in microalgal-bacterial processes. Environ Sci Technol 49(5):3228–3236. https://doi.org/10.1021/es5056116

Soreanu G, Béland M, Falletta P, Edmonson K, Seto P (2008) Laboratory pilot scale study for H2S removal from biogas in an anoxic biotrickling filter. Water Sci Technol 57:201–207. https://doi.org/10.2166/wst.2008.023

Soreanu G, Béland M, Falletta P, Ventresca B, Seto P (2009) Evaluation of different packing media for anoxic H_2S control in biogas. Environ Technol 30:1249–1259. https://doi.org/10.1080/09593330902998314

Soreanu G, Béland M, Falletta P, Edmonson K, Svoboda L, Al-Jamal M, Seto P (2011) Approaches concerning siloxane removal from biogas—A review. Can Biosyst Eng 53:8.1–8.18

Sousa MS, Oliveira CJS, Lopes AC, Rodríguez ER, Holanda GBM, Landim PGC, Firmino PIM, dos Santos AB (2016) Technical, economical, and microbiological aspects of the microaerobic process on H_2S removal for low sulfate concentration wastewaters. Appl Biochem Biotechnol 180:1386–1400. https://doi.org/10.1007/s12010-016-2174-9

Strevett KA, Vieth RF, Grasso D (1995) Chemo-autotrophic biogas purification for methane enrichment: mechanism and kinetics. Chem Eng J Biochem Eng J 58:71–79. https://doi.org/10.1016/0923-0467(95)06095-2

Sumardiono S, Budiyono Syaichurrozi I, Sasongko SB (2014) Utilization of biogas as carbon dioxide provider for *Spirulina platensis* culture. Curr Res J Biological Sci 6(1):53–59

Sun Q, Li H, Yan J, Liu L, Yu Z, Yu X (2015) Selection of appropriate biogas upgrading technology-a review of biogas cleaning, upgrading and utilisation. Renew Sustain Energy Rev 51:521–532. https://doi.org/10.1016/j.rser.2015.06.029

Syed M, Soreanu G, Falletta P, Béland M (2006) Removal of hydrogen sulfide from gas streams using biological processes—A review. Can Biosystems Eng 48:2.1–2.14

Tang K, Baskaran V, Nemati M (2009) Bacteria of the sulphur cycle: an overview of microbiology, biokinetics and their role in petroleum and mining industries. Biochem Eng J 44:73–94. https://doi.org/10.1016/j.bej.2008.12.011

Thiansathit W, Keener TC, Khang S-J, Ratpukdi T, Hovichitr P (2015) The kinetics of Scenedesmus obliquus microalgae growth utilizing carbon dioxide gas from biogas. Biomass Bioenerg 76:79–85. https://doi.org/10.1016/j.biombioe.2015.03.012

Thrän D et al (2014) Biomethane—status and factors affecting market development and trade. IEA Task 40 and Task 37

Toledo-Cervantes A, Serejo M, Blanco S, Pérez R, Lebrero R, Muñoz R (2016) Photosynthetic biogas upgrading to bio-methane: Boosting nutrient recovery via biomass productivity control. Algal Res 17:46–52. https://doi.org/10.1016/j.algal.2016.04.017

Toledo-Cervantes A, Madrid-Chirinos C, Cantera S, Lebrero R, Muñoz R (2017) Influence of the gas-liquid flow configuration in the absorption column on photosynthetic biogas upgrading in algal-bacterial photobioreactors. Bioresour Technol 225:336–342. https://doi.org/10.1016/j.biortech.2016.11.087

Tomàs M, Fortuny M, Lao C, Gabriel D, Lafuente J, Gamisans X (2009) Technical and economical study of a full-scale biotrickling filter for H_2S removal from biogas. Water Pract Technol 4. https://doi.org/10.2166/wpt.2009.026

Tredici MR (2009) Photobiology of microalgae mass cultures: understanding the tools for the next green revolution. Biofuels 1:143–162. https://doi.org/10.4155/bfs.09.10

Trobajo R, Ibáñez C, Clavero E, Salvadó J, Jorgensen SE (2014) Modelling the response of microalgae to CO_2 addition. Ecol Model 294:42–50. https://doi.org/10.1016/j.ecolmodel.2014.08.020

van der Zee FP, Villaverde S, Garcia PA, Fdz-Polanco F (2007) Sulfide removal by moderate oxygenation of anaerobic sludge environments. Biores Technol 98:518–524. https://doi.org/10.1016/j.biortech.2006.02.011

Wang B, Li Y, Wu N, Lan CQ (2008) CO_2 bio-mitigation using microalgae. Appl Microbiol Biotechnol 79:707–718. https://doi.org/10.1007/s00253-008-1518-y

Wang W, Xie L, Luo G, Zhou Q, Angelidaki I (2013) Performance and microbial community analysis of the anaerobic reactor with coke oven gas biomethanation and in situ biogas upgrading. Bioresour Technol 146:234–239. https://doi.org/10.1016/j.biortech.2013.07.049

Wasserbauer R, Zadák Z (1990) Growth of Pseudomonas putida and *P. fluorescens* on silicone oils. Folia Microbiol 35:384–393

Wellinger A, Lindberg A (2005) Biogas upgrading and utilisation. IEA Bioenergy Task 24: energy from biological conversion of organic waste. http://www.biogasmax.eu/media/biogas_upgrading_and_utilisation__018031200_1011_24042007.pdf

Yang L, Ge X, Wan C, Yu F, Li Y (2014) Progress and perspectives in converting biogas to transportation fuels. Renew Sustain Energy Rev 40:1133–1152

Zhou Q, Liang H, Yang S, Jiang X (2015) The removal of hydrogen sulfide from biogas in a Microaerobic Biotrickling filter using polypropylene carrier as packing material. Appl Biochem Biotechnol 175(8):3763–3777. https://doi.org/10.1007/s12010-015-1545-y

Zhou W, Imai T, Ukita M, Li F, Yuasa A (2007) Effect of limited aeration on the anaerobic treatment of evaporator condensate from a sulfite pulp mill. Chemosphere 66(5):924–929. https://doi.org/10.1016/j.chemosphere.2006.06.004

Chapter 11
Biorefineries: Focusing on a Closed Cycle Approach with Biogas as the Final Step

Benyamin Khoshnevisan and Irini Angelidaki

11.1 Introduction

The increasing energy demands as a consequence of fast-growing global population and higher living standards over the last few decades have triggered huge interest in finding new energy resources. Currently, most chemicals, materials, and energy carriers originate from fossil resources (Brar et al. 2016). Concerns over the rising global temperature, the depletion of fossil fuels, and increased environmental pollution, as well as the fluctuations of oil prices have encouraged researchers and energy policy makers to explore practical solutions for generating bioenergy and bioproducts with less environmental impacts (Parajuli et al. 2015). Today, about 80% of the global energy demand is supplied through fossil resources and the global energy demand in 2035 is still projected to rise by 40% with fossil fuels contributing 75% (Parajuli et al. 2015). Therefore, it is anticipated that sooner or later there will be no more fossil fuel to extract in an economical fashion and the world has to adapt to this new paradigm (Sharara et al. 2012).

While it is less complicated to provide future renewable electricity and heat due to the availability of a variety of renewable alternatives (i.e., wind, solar, hydro, biomass, and others), major challenges still exist regarding supplying of biochemicals and biofuels. In this context, biomass has a huge potential to play a pivotal role due to the fact that both biochemicals and biofuels can be extracted from biomass resources. Biorefineries which are analogous to today's petroleum

B. Khoshnevisan
Department of Mechanical Engineering of Agricultural Machinery,
Faculty of Agricultural Engineering and Technology,
College of Agriculture and Natural Resources, University of Tehran, Karaj, Iran

B. Khoshnevisan · I. Angelidaki (✉)
Department of Environmental Engineering, Technical University of Denmark,
DK-2800 Kgs Lyngby, Denmark
e-mail: iria@env.dtu.dk

M. Tabatabaei and H. Ghanavati (eds.), *Biogas*, Biofuel and Biorefinery Technologies 6, https://doi.org/10.1007/978-3-319-77335-3_11

refineries are identified as processing facilities capable of using biomass as feedstock to produce fuels, power, and chemicals (Yang and Yu 2013). This chapter aims at introducing the concept of biorefineries, the necessity of moving towards biorefineries, their opportunities and challenges, and potential feedstocks which can be used within the biorefinery concept. Moreover, the benefits of coupling biogas production with biorefineries are discussed and the problems and challenges are evaluated.

11.2 Biorefinery: Definitions and Perspectives

The increased awareness on the need to use biomass resources as well as the growing interest in upgrading more low-quality lignocel-lulosic biomass to valuable products along with the increased attention to the production of starch for energy applications led to the establishment of the term "biorefinery" in the 1990s (Berntsson et al. 2012; Kamm et al. 2006). Among the first definitions presented for the term "biorefinery", the term "Green biorefinery" was presented in 1997 in which biorefinery was referred to as technologies (Soyez et al. 1997). The definition offered was as follows, "Green biorefineries represent complex (to fully integrated) systems of sustainable, environmentally, and resource-friendly technologies for the comprehensive (holistic) material and energetic utilization as well as exploitation of biological raw materials in the form of green and residue biomass from a targeted sustainable regional land utilization" (Soyez et al. 1997).

The US Department of Energy considered biorefineries as an overall concept of a processing plant where a spectrum of valuable products are produced out of biomass feedstocks (Energy 1997). The American National Renewable Energy Laboratory (NREL) referred to biorefinery as a "facility" that integrates biomass conversion processes and equipment with the aim of providing fuels, power, and chemicals from biomass (NREL 2005). In this definition, the biorefineries are regarded as facilities developed to fulfil today's petroleum refineries' functions.

Among the distinctive definitions frequently observed in the literature for the term "biorefinery" (Berntsson et al. 2012; Demirbas and Demirbas 2010; Mansoornejad et al. 2010), the most comprehensive one was offered by the IEA Bioenergy Task 42: "Biorefining is the sustainable processing of biomass into a spectrum of marketable products and energy" (Cherubini 2010; Cherubini et al. 2007). This can be considered as the most exhaustive definition because it simultaneously aggregates the sustainability issues, the types of feedstocks, broad spectrum of obtained products, and economic considerations.

The economic aspects of biorefineries are important because it is often difficult to get positive economy balance, as the production cost of biomass-based fuels is often high. Therefore, integrating biomaterial and biochemical production (i.e., higher-value products) with generation of biofuels (i.e., higher-volume products) can potentially result in increased overall profitability. Although, in petroleum refineries, a wide range of processes can be employed, e.g., fluid catalytic cracking, thermal cracking, hydrocracking, etc., to produce a large number of valuable products out of crude oil, only few petroleum refineries employ all available conversion platforms. Due to the fact that biorefineries are aimed at competing with today's petroleum refineries, it is important to reduce the production costs, and therefore, they should only use the most cost effective conversion technologies to increase the overall profitability.

Biorefineries are imposing significant environmental impacts since they can simultaneously reduce our dependence on fossil resources and alleviate the environmental pollutions caused by the high consumption of fossil-based fuels. Generally speaking, biorefineries can be considered as multiple production of biofuel and biomaterials from various biomass feedstocks with the objectives of decreased non-renewable energy resources utilization, minimized related environmental impacts, and maximized efficient use of biomass. These objectives can be met if the following ecological perspectives are taken into consideration (Cherubini 2010; Gravitis and Suzuki 1999):

- Carbon, water and nitrogen cycles of agricultural and forestry plants.
- Technical and economic evaluations of existing and pilot biorefineries.
- Environmental impact evaluations through the whole life cycle of bio-products.

11.3 Types of Biorefineries and Their Classifications

The first serious attempts made at large-scale utilization of biomass-based resources were made in the 19th and at the beginning of the 20th centuries when the production of bio-based products like pulp, paper, guncotton and viscose silk, soluble cellulose, and furfural was reported (Kamm et al. 2006). The technological and scientific progress achieved during recent decades resulted in a wide range of biomass-based fuel and materials in the context of biorefineries. Due to a variety of distinctive technologies and platforms used in biorefining of biomass feedstocks, along with a broad spectrum of products and different feedstocks employed, several schemes have been proposed in the literature (Chambost and Stuart 2007; Huber 2008; van Ree and Annevelink 2007) to classify biorefineries and to make a systematic arrangement among them (Fig. 11.1).

In one of these systematic arrangements, the biorefineries are classified as generation-I, generation-II, and generation-III on the basis of the technologies employed. "Dry milling ethanol plant", as an example of the first generation, can be mentioned whose outputs are ethanol, feed co-products, and carbon dioxide. The second generation category has strived to overcome the intronsic inflexibility of the first generation using wet milling technology to produce a variety of end products including starch, high-fructose corn syrup, ethanol, corn oil, plus corn gluten feed, and meal. The final product of the generation-II biorefineries depends upon demands, market prices, contract obligations, and management considerations. The first and the second generations typically use grains as feedstock. The third

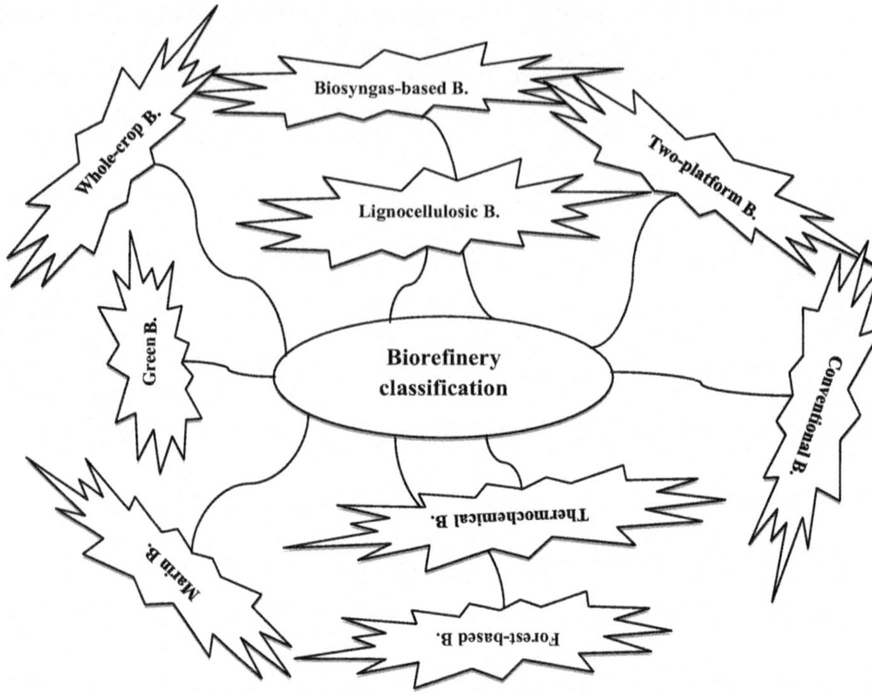

Fig. 11.1 Attempts made to classify biorefineries as observed in the published literature and the different terms introduced

generation biorefineries are the most advanced aimed at using agricultural or forest lignocellulosic biomass to produce multiple product streams, for example ethanol, chemicals, and plastics (Kamm et al. 2006).

However, many more classifications of biorefineries have been defined in the literature, such as the "lignocellulosic feedstock biorefinery", "whole crop biorefinery", "green biorefineries", and "biorefinery two platforms concept" (Kamm and Kamm 2004a, b; Werpy and Petersen 2004). Moreover, Demirbas and Demirbas (2010) added some new terms to this type of classification such as "oilseed biorefinery" and "forest biorefinery". "Lignocellulosic biorefineries" employ nature-dry biomass such as cellulose-containing biomass and wastes (Table 11.1) while in "green biorefineries", nature-wet raw materials including green grass, alfalfa, clover, or immature cereal are utilized.

The green biorefineries include two main pathways following a wet fractionation step. The outputs of these two steps are fiber-rich press cake and nutrient-rich green juice. The former contains cellulose, starch, dyes and pigments, crude drugs, and other chemicals; and can be used to produce biogas or syngas. The nutrient-rich green juice undergoes a fermentation process leading to the production of biogas, amino and organic acids, proteins, enzymes, etc. In the "whole crop biorefinery", the feedstock including wheat, rye, triticale, etc., undergo biorefining process and

Table 11.1 Potential products of a lignocellulosic feedstock biorefinery (adapted from Kamm et al. (2006))

Lignocellulose												
Cellulose					Hemicellulose					Lignin		
Glucose					Xylose							
5-Hydroxymethyl furfural					Furfural							
Softener	Lubricants	Chemicals and polymers	Fermentation products	Cellulose applications	Furan resins	Chemical products	Nylon	Xylite	Plant gum	Sulphur-free Solid fuel	Sub-bituminous coal	Natural Binder

both seeds and straw is employed to produce a wide range of products. Straw can be treated under a decomposition stage and converted into principle components, i.e., lignin, hemicellulose, and cellulose. Instead of the decomposition process, gasification can be employed to produce syngas. In contrast, seeds can be either used in the grinding phase whose outputs can be binder, adhesive, and cement, or processed in the starch producing step. The extracted starch under chemical or biotechnological conversion as well as extrusion processes can generate valuable final products such as methanol, acetate starch, bioplastic, co- and mix-polymerisate.

"Two platforms concept" consists of the sugar platform and the syngas platform. However, NREL has suggested four different platforms i.e., sugar, thermochemical, biogas carbon-rich chains, and plant products platforms.

The conversion route is another criterion by which the biorefineries can be classified into five groups as follows (Demirbas and Demirbas 2010):

- Biosyngas-based
- Pyrolysis-based
- Hydrothermal-upgrading-based
- Fermentation-based
- Oil-plant-based

Efforts have been made to adapt a systematic approach for biorefinery classification, since the aforementioned classifications are broad, arbitrary and generic, and in some cases, heterogeneous. Moreover, currently used classifications can be combined by linking different technologies. Cherubini et al. (2009) chose five criteria, i.e., platforms, products, feedstocks, and processes to form five groups, each one consisting of some sub-categories (Fig. 11.2). Accordingly, they

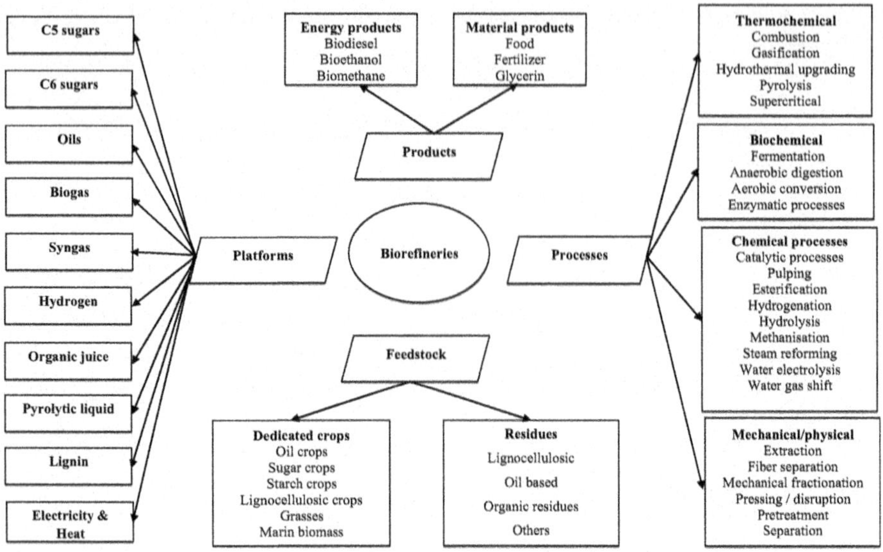

Fig. 11.2 Features and subgroups involved in proposed classification approach based on (Cherubini et al. 2009)

suggested that the biorefineries be classified by listing the main features of the biorefinery system, drawing a scheme of the features identified, and labeling the system by quoting the involved number of platforms, products, feedstocks, along with the processes.

11.4 Barriers and Obstacles to Biorefineries

As discussed earlier, biorefineries are aiming at producing bulk chemicals, biomaterials, and bio-energies from biomass feedstock for overall improving the economy of biomass use. In order to implement a commercial biorefinery, all technical and non-technical barriers should be overcome. Since the products of biorefineries are derived from biomass, the production cost can be mentioned as the most important barriers followed by the transportation cost of biomass-based feedstock. It is worth quoting that, biomass as the feedstock for biorefineries experiences seasonal changes. These seasonal diversities can lead to the need for storage facilities causing storage cost to be added to the total production cost. One important aspect of upscaling biorefineries is the infrastructure required for collection and storage of a large amount of biomass. Such an integrated feedstock supply system need to be constructed at a sustainable fashion and at a reasonable cost (Demirbas and Demirbas 2010). It is worth mentioning that by combining different technologies, based on the pathways shown in Fig. 11.3, for simultaneous

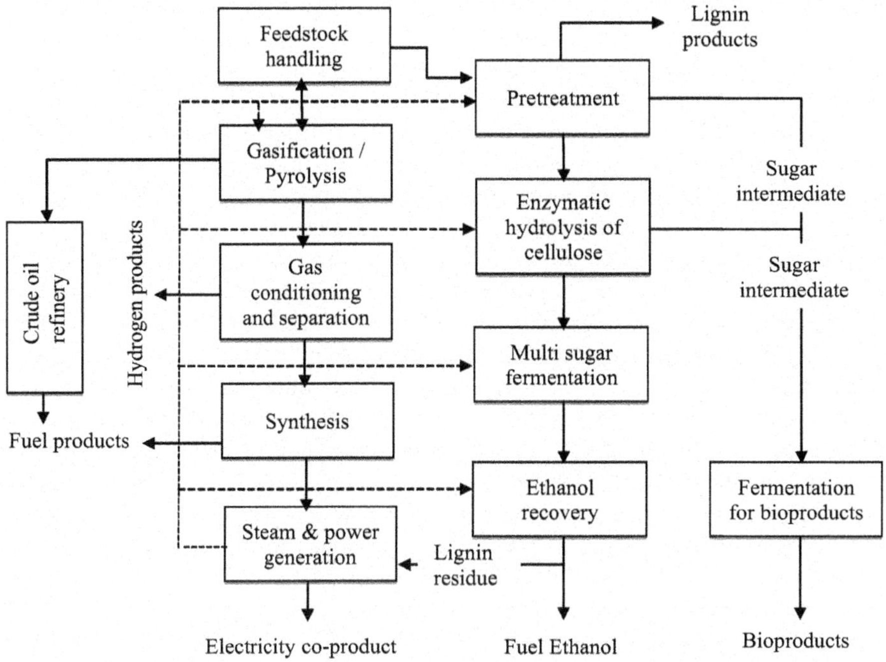

Fig. 11.3 Block diagram of an integrated biorefinery to use different platforms and produce different products (Fernando et al. 2006; Wang et al. 2004)

production of bioenergies and biomaterials, the overall production cost can be reduced and more flexibility in product generation can be offered.

The composition of biomass, undergoing a biorefining process, varies enormously. This can be regarded as a benefit for biorefineries because it enables them to produce a more diverse spectrum of products, even more than those generated by petroleum refineries. However, this compositional variation in biomass feedstock can also result in some disadvantages which need to be overcome. The economic and sustainable processing of raw materials in such a biorefinery requires advanced and sophisticated technologies most of which are still at a pre-commercial stage (Dale and Kim 2006).

Another major non-technical barrier which should be discussed herein is the use of land for production of biorefineries' feedstock. The competition between food production sector *vs.* raw materials supply for biorefineries over land and even other limited resources such as water can be taken into account as a serious limitation towards developing future biorefineries. This point of view has led to a serious discussion in scientific communities. While some believe that use of biomass as a feedstock for biorefineries can create jobs and boost economic growth (Negash and Swinnen 2013), others insist that it might reduce food availability and increase its price, thereby posing a real threat to food security, especially in the developing countries (Janssen and Rutz 2011). On the other hand, both direct and indirect land use change (LUC) effects cannot be ignored when imposing restrictions on the use of land. LUC effects refer to change in soil carbon pools caused by human activities which have huge impacts on the global carbon cycle and can potentially bring about climate change effects. Moreover, the indirect land use change (ILUC) cannot be disregarded in this context because it is responsible for global warming effects. When a piece of land, used for agricultural purposes such as growing food or feed, is now dedicated to biorefinery purposes, another non-cropland—such as grasslands and forests—somewhere else should be devoted to agricultural purposes. This transformation is known as ILUC effects and can neutralize the greenhouse gas savings resulted from replacing fossil-based fuels with the biofuels generated in biorefineries.

Deforestation, defined as "conversion of forest land to non-forest land" (DeFries et al. 2007) has been identified as a serious problem originating from emerging future biorefineries. This is in parallel with LUC effects because deforestation decreases the carbon sequestration. For example, it has been well-documented that the production of soybean-based biodiesel in Brazil and Argentina has contributed to deforestation (Janssen and Rutz 2011). This is due to the fact that the increasing demands for soybean has brought about the conversion of forest land to soybean farms (Nepstad et al. 2006). In spite of these on-going debates and concerns, some reports have shown that simultaneous production of biomass-based products and forest protection are possible depending on policies adopted (Demirbas 2009; Ravindranath et al. 2011).

Generally, it can be concluded that although biorefineries can take advantage of several benefits including energy security, climate change benefits, sustainable management of wastes, coproduction of valuable biochemicals, and rural economic development (Brar et al. 2016), there are still drawbacks and challenges which need to be effectively dealt with. Some of these challenges are summarized in Fig. 11.4.

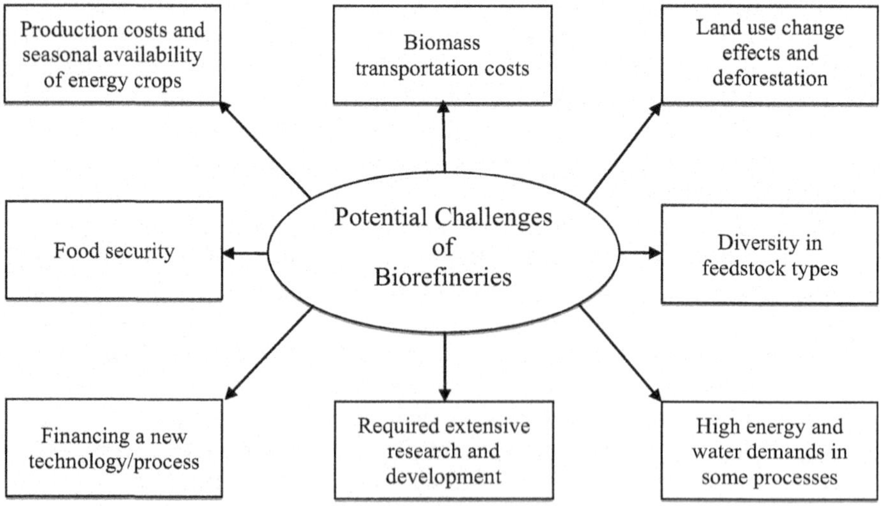

Fig. 11.4 Potential Challenges of the Biorefinery Concept (Brar et al. 2016)

11.5 Feedstock

Petrochemical industry has been playing a pivotal role in the livelihood of mankind by fulfilling needs for energy, material, and chemicals. In order to replace fossil-based refineries by biomass-based refineries, our today's requirements for energy and non-energy products should be completely met by future biorefineries. There is a huge potential to supply both bioenergy and biochemicals from biomass-based feedstock. Taking a look at outputs of today's refineries. i.e., textile goods, housing products, transportation products, etc. reveals that most of the products from the petrochemical industry are derived from 8 to 9 foundation chemicals (Werpy et al. 2004). Accordingly, the US Department of Energy (DOE) endeavored to identify twelve building block chemicals that can be produced from sugars via biological or chemical conversions (Fig. 11.5). Building block chemicals are molecules with multiple functional groups which can be transformed into new families of useful molecules (Werpy et al. 2004). This term is generally used to describe a virtual molecular fragment or a real chemical compound whose molecules possess reactive functional groups (Szmant 1989). They are employed to show how molecules can be assembled in a bottom-up modular order, i.e., nano-particle, metal-organic frameworks, organic molecular constructs, and supra-molecular complexes, ensuring the final compound or a (supra) molecular construct will be generated (Tu and Tirrell 2004). Thirty potential candidates out of 300 initially evaluated were introduced and by an iterative process based on the petrochemical model using building blocks, chemical data, known market data, properties, performance of the potential candidates, and the prior industry

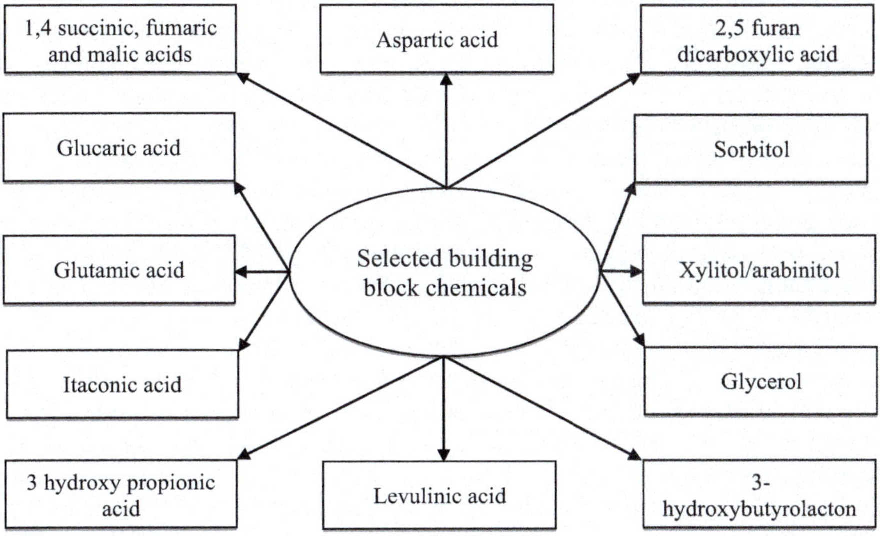

Fig. 11.5 Twelve sugar based building block chemicals (Werpy et al. 2004)

experiences, the top twelve final candidates were chosen. According to the DOE, our requirements for different chemical products can be met by these building block chemicals. Moreover, the final candidates are more appropriate than the other competitors in terms of feedstock costs, estimated processing cost, current market volumes, and prices. These features make the selected building block chemicals capable of competing directly against fossil-based chemicals, as well benefiting from chemical functionality, and improved properties (Werpy et al. 2004). Conversion pathways, derivatives, and potential applications of some most important building block chemicals identified in the literature are tabulated in Table 11.2.

While these building block chemicals can be extracted from various feedstock, attempts have been made to identify easily fermentable substrates to decrease the process costs and increase the total profitability. In parallel with decreasing the process cost, increasing the production level has also been a focus in research. Some researchers have come to the conclusion that productivity and increases in yield can be achieved by using engineered microorganisms, minimizing the production of undesired by-products, and the use of neutralizing agents (Engel et al. 2008; Jiang et al. 2009; Pachapur et al. 2016; Ye et al. 2013). Moreover, improving the product recovery step and increasing product purity can also help to achieve higher product quality and thereby, better prices along with reduction of process costs (Dan et al. 2010; Misra et al. 2011; Pachapur et al. 2016).

As elaborated earlier, there is a huge potential to fulfil our future requirements for chemicals and energy through biorefineries. A great deal of feedstock has been already examined and their advantageous and drawbacks have been discussed. Forest-based feedstock for biorefinery purposes seem to be appropriate feedstock as they cover about 32% of the land area but account for 89.3% of the total standing biomass (Brar et al. 2016). Forest-based biomass is composed of cellulose, hemicellulose, and lignin of which cellulose and hemicellulose can be converted to sucrose, xylose, glucose, galactose, and arabinose. These intermediate products can be converted to a range of platform chemicals through fermentation pathway including propanediol, ethanol, lactic acid, ethylene, succinic acid, glycerol, propane, etc. It is worth mentioning that the commercial production of some of these platform chemicals seems nonviable at the current state (Danner and Braun 1999). For example, 1 ton of fermented hexose (glucose or fructose) using the well-studied organism, *Saccharomyces cerevisiae,* according to the stoichiometric product yield, can result in the production of 511 kg ethanol. While, 8 ton of sugarcane biomass is needed to achieve 1 ton of hexose (Brar et al. 2016).

Animal fat and vegetable oils can also be added to the list of biorefinery feedstock because a number of platform chemicals including glycerol, succinic acid, propionic acid, butanol, and ethanol can be obtained from such feedstock. While animal fat and vegetable oils have long been evaluated for biodiesel production purposes, the use of the resulted by-products i.e., platform chemicals with adhesives, paints, lubricants, food additives, and biopolymers applications, can lead to a biorefinery approach. It is worth quoting that the simultaneous production of biodiesel and platform chemicals from animal fat and vegetable oils should be

Table 11.2 Conversion pathways, derivatives, and potential applications of some most important building block chemicals

Building blocks	Pathways	Derivatives/derivatives family	Potential application of derivatives
Four Carbon 1,4-Diacids[a]	Fermentation Chemical process	THF, BDO, GBL family Pyrrolidinone Family Straight chain polymers Branched polymers	Solvents, fibers such as lycra Green solvents, water soluble polymers (water treatment) Fibers (lycra, others) TBD
2,5-Furan dicarboxylic acid[a]	Enzymatic conversions	Diols and Aminations Levulinic and Succinic Acids Polyethylene terephthalate analogs Furanoic Polyamines	Polyesters and nylons All uses of succinic and levulinic Furanoic polyesters for bottles, containers, films Polyamide market for use in new nylons
3-Hydroxypropionic acid[a]	Fermentation	1, 3 propane diol Acrylate family	Sorona fiber Contact lenses, diapers
Aspartic acid[a]	Chemical process Fermentation Enzymatic conversions	Amine butanediol, amine tetrahydrofuran, amine Aspartic anhydride Polyaspartic	Amino analogs of C4 1,4 dicarboxylic acids
Glucaric acid[a]	Chemical process Catalytic oxidation	Lactones Polyglucaric esters and amides	Solvents Nylons or different properties
Glutamic acid[a]	Chemical process Fermentation	Diols (1,5-propanediol) Diacids (1,5-propanediacid) Aminodiol (5-amino, 1-butanol)	Monomers for polyesters and polyamides
Itaconic acid[a]	Chemical process Fermentation Aerobic fungal fermentation	Methyl butanediol, butyrolactone, tetrahydrofuran family Pyrrolidinones Polyitaconic	May confer new useful properties for the BDO, GBL, and THF family of polymers New polymer opportunity
Levulinic acid[a]	Chemical process	Methyl tetrahydrofuran γ- butyrolactone Acetyl acrylates Acetic-acrylic succinic acids Diphenolic acid	Fuel oxygenates, solvents Copolymerization with other monomers for property enhancement Replacement for bisphenol A used in polycarbonate synthesis

(continued)

Table 11.2 (continued)

Building blocks	Pathways	Derivatives/derivatives family	Potential application of derivatives
3-Hydroxybutyrolactone[a]	Chemical process	Furans. Analogs of pyrrolidones Amino analogs to tetrahydrofuran	Solvents Amino analogs to lycra fibers
Glycerol[a]	Chemical process Enzymatic conversions	PLA Analogs Glyceric Acid Propylene glycol 1,3-propanediol Branched polyesters and polyols	PLA with better polymeric properties Polyester fibers with new properties Antifreeze, humectant, etc. Sorona fiber Unsaturated Polyurethane Resins for use in insulation
Sorbitol[a]	Chemical process	Isosorbide, anhydrosugars Propylene glycol, lactic acid Branched polysaccharides	PET Antifreeze, PLA Water soluble polymers
Xylitol/arabinitol[a]	Chemical process Fermentation Enzymatic transformation	Xylaric and Xylonic acids Arabonic acid and Arabinoic acid Xylaric and Xylonic acids Arabonic acid and Arabinoic acid	Antifreeze Unsaturated polyester resins New polymer opportunities
Butanol[b]	Fermentation	2-Methyl-2-butanol, 2-butanol	As alternative fuel
Butyric acid[c]	Fermentation	R)-3-(Boc-amino)-4-(4-bromophenyl)butyric acid	Cosmetics Pharmaceuticals, As a “natural preservative” in the food industry
Lactic acid[d]	Fermentation	Lactate ester Polylactic acid (PLA) Acrylic acid 1,2-Propanediol Pyruvic acid	Hygroscopic and emulsifying properties, solvents Biodegradable plastic Acrylate polymers, biochemical intermediate Commodity chemical

[a]Werpy et al. (2004) [b]Cooksley et al. (2012) [c]Zhang et al. (2009) [d]Gao et al. (2012)

carefully suggested due to the fact that biotechnological progress has led to the direct production of these platform chemicals with decreased investment costs and increased total yield.

Microalgae are also considered as another potential feedstock for future biorefineries. From one hand, a vast number of researchers have shown that microalgae species have technical potential to produce lipid or carbohydrate biofuel precursors taking into account greenhouse gas and land use sustainability metrics, rapid biomass production rates, and high solar conversion efficiencies (Lardon et al. 2009; Melis 2009; Reijnders 2008; Stephenson et al. 2010). On the other hand, the economic analysis of algal biofuel production has proven that there is still a long way before achieving economic algal biofuel production, capable of competing with petroleum-based fuels (Brar et al. 2016; Sheehan et al. 1998; Williams and Laurens 2010). Nevertheless, the biorefinery approach has been suggested as a practical solution to achieve commercially relevant rates of return because it can result in simultaneous production of algal biofuels and value-added products. Pigments, vitamins, phytosterols, polysaccharides, organic acids, lipids, and miscellaneous algal compounds are high-value platform chemical which can be extracted from algae.

Chlorophylls, carotenoids, and phycobiliproteins are among the large number of pigments which can be extracted from algae. They are also rich in vitamin. It has been well-documented that different combinations and concentrations of vitamin B12 (cobalamin), vitamin B1 (thiamine), and vitamin B7 (biotin) can be found in algae (Brar et al. 2016; Croft et al. 2006; Provasoli and Carlucci, 1974). Moreover, there are several metabolic pathways in distinctive algae species resulting in synthesizing other vitamins, including vitamins A, C, and E (Hirschberg 1999). Phytosterols known as steroid alcohols are valuable platform chemicals owing to their medical applications, i.e., potential for lowering total and LDL cholesterol. They are also employed as therapeutic agents to treat hypercholesterolemia (Francavilla et al. 2010; Ostlund et al. 2003; St-Onge et al. 2003). Polysaccharides have been reported as a possible platform for the production of biofuels while they also have high values in the marketplace in terms of their applications in the food industry (Brar et al. 2016; Wargacki et al. 2012). Production of succinic and malic acids, two organic acids listed among the top 15 building block chemicals, from algae is anticipated to increase progressively in the near future in response to an additional market size of 25×10^6 ton per year for succinic acid-derived polymers (Bozell and Petersen 2010). Algal lipids have high values in the marketplace and they can be employed for biofuel production, nutritional supplements, and pharmaceutical applications. Microalgae are capable of bio-synthesizing lipids by diverting their central metabolic pathways when they are under certain stress conditions (Brar et al. 2016).

The utilization of agro-industrial waste for energy and biochemical production has gained lots of interests and the conducted studies have shown a great potential to revolutionize the chemical industry (Chandra et al. 2012; Octave and Thomas 2009). Agro-industrial wastes are important feedstock within the biorefinery concept since they are produced in huge amounts and a wide spectrum of valuable platform chemicals can be produced from them. The use of waste as biorefinery feedstock can decrease the total production cost and increase the total profitability. However, the challenges, i.e., non-uniformity, social perspectives, technology issues, collection, storage, and segregation, regarding the use of agro-industrial waste in biorefineries cannot be ignored. Currently a considerable deal of efforts has been concentrated on the production of bioethanol, as well as cogeneration of biofuels and adsorbents.

11.6 Biogas Production and Biorefinery Approach

Energy recovery and more specifically biogas production under anaerobic conditions plays a key role in developing future biorefineries because they contribute to a more sustainable performance of the whole system under consideration. Energy recovery in the form of biogas is a way to close the cycle and use the residual organic matters which have not been recovered. Anaerobic digestion (AD) is a versatile process, by which different types of organic matters are converted into biogas. On the contrary, many other bioconversion processes have a much narrower substrate preference, leaving unutilized a large portion of the organic matters. Therefore, biogas can be seen not only as an effluent purification process, but also energy producing path. Most of the biorefinery concepts have AD as a part of the proposed processes. In better words, integrating AD into some current technologies has been proposed as a practical solution which can simultaneously increase the total profitability and overcome some challenges involved. For example, biogas production from pre-hydrolysate under a biorefinery approach has been proposed to maximize the profitability resulted from the conversion of available sugars in woods (Safari et al. 2017). Softwood pine for example, due to its lignocellulosic structure, requires a pretreatment step prior to enzymatic or biological conversion. After completion of the pretreatment, the solid fraction is filtrated from the pre-hydrolysate, i.e., the liquid fraction, and undergoes enzymatic hydrolysis for ethanol fermentation (FazeliNejad et al. 2016; Khoshnevisan et al. 2016; Shafiei et al. 2015). To make this process economically viable, separate hydrolysis and co-fermentation (SHCF) or simultaneous saccharification and co-fermentation (SSCF) have long been used to convert pre-hydrolysate to ethanol (Dien et al. 2003; McMillan et al. 1999). The proposed methods bring about some new challenges including low ethanol yield, differences in the optimal fermentation conditions of the involved strains, etc. Accordingly, the integration of ethanol and biogas production from softwood has been evaluated and reported as a practical solution to overcome the aforementioned problems (Safari et al. 2017).

On the other hand, the economic profitability when using biomass in a biorefinery approach can be improved compared with using it for biogas production alone. As an example, Santamaría-Fernández et al. (2017) reported that the combination of protein refining and biogas production could be more economically favorable compared with sole biogas production from green biomass crops.

It should be noted that biogas production has been introduced as an economic solution for many industries because it can easily contribute to decreasing the costs associated with energy consumption and wastewater treatment (McKendry 2002; Schmidt et al. 2013; Wilkie et al. 2000). Nevertheless, recent studies have argued that the most interesting and impactful contribution of biogas solutions are their potential for product valorization and material upcycling (Batista et al. 2017; Begum et al. 2016; Hagman et al. 2017).

11.6.1 Microalgal as Biogas Feedstock

Microalgal feedstock has been widely considered for biofuel and biochemical production, there are several challenges to overcome though. The high accumulation of lipids in microalgae makes them attractive feedstock for biodiesel production. Moreover, different kinds of metabolites including pigments, fatty acids, proteins, and nutritional supplements for human consumption can be obtained from microalgae (Ramos-Suárez et al. 2014; Spolaore et al. 2006). Coupling of AD to the extraction of such metabolites from microalgae has also been examined as a potential way to improve the economics of the process. It has been shown that metabolites extraction could function as a pretreatment method for increasing the biodegradability of microalgal cells (González-Fernández et al. 2011; Mussgnug et al. 2010). Moreover, it can simultaneously decrease the C/N ratio and thereby, alleviate potential inhibition of methanogenesis due to increased ammonia levels (Zhong et al. 2012). The biogas potential from microalgae has been reported in several publications pioneering by Golueke and Oswald (1959). Table 11.3 tabulates a summary of biogas potential from different microalgae species.

Several challenges have been discussed by researchers as major factors affecting biogas production from microalgae including high capital cost, low algae productivity, slow conversion rate, and high sensitivity of AD process (Roy and Das 2015). Low concentration of biomass has been identified as one of the limiting factors because solid biomass content of most uncontrolled outdoor microalgae cultures is less than 1 g L^{-1} (Golueke and Oswald 1959; Stephens et al. 2010). Concentrating and dewatering of microalgae cultures have been suggested as practical solutions to the aforementioned problem, they are expensive and time-consuming procedures though (Harun et al. 2010; Pragya et al. 2013; Stephens et al. 2013; Ward et al. 2014). Integrating AD process into microalgae production can potentially offset the energy requirements with respect to the resultant methane production (Sialve et al. 2009).

The rigid cell wall structure is another problematic issue because it hinders accessibility of the AD microorganisms to the algal biomass. The increased process cost makes pretreatment methods as inappropriate approach to break down the rigid cell wall structure. Ramos-Suárez et al. (2014) integrated AD with amino acid extraction and reported improved economics of the process. Another dilemma in AD of microalgae is ammonia inhibition. The significant protein and lipid

Table 11.3 Methane biogas production through anaerobic digestion of different species of microalgae biomass

Microalgae species	C/N ratio	Methane yield	Loading rate
Tetraselmis	N/A	252 L kg^{-1} VS	5400 mg VS^{-1} L^{-1}
Scenedesmus	7.3	291.5-409.3 L kg^{-1} VS	3.85 g VS^{-1} L^{-1}
Chlorella vulgaris	N/A	403 L kg^{-1} VS	2 g VS^{-1} L^{-1}
Microspora	N/A	413 L kg^{-1} VS_{algae}	N/A
Chlamydomonas	N/A	310 L kg^{-1} VS_{algae}	N/A
Acutodesmus	N/A	223 L kg^{-1} VS_{algae}	N/A
Nannochloropsis oculata	N/A	204 L kg^{-1} VS	N/A
Lake Chaohu natural population consortium	N/A	295 L kg^{-1} VS	N/A
Nannochloropsis salina (lipid extracted biomass)	4.4	130 L kg^{-1} VS	2000 mg VS^{-1} L^{-1}
Arthrospira maxima	4.3–5.33	173 L kg^{-1} VS	500 mg TS^{-1} L^{-1}
Phaeodactylum tricornutum	N/A	350 L kg^{-1} COD	1.3 ± 0.4–5.8 ± 0.9
Scenedesmus obliquus	N/A	240 L kg^{-1} VS	2000 mg VS^{-1} L^{-1}
Scenedesmus sp.	N/A	170 L kg^{-1} COD	1000 mg COD^{-1} L^{-1}
Scenedesmus sp. (single stage)	N/A	290 L kg^{-1} VS	18,000 mg VS^{-1} L^{-1}
Scenedesmus sp. (two stage)	N/A	354 L kg^{-1} VS	18,000 mg VS^{-1} L^{-1}
Scenedesmus obliquus	N/A	287 L kg^{-1} VS	2000 mg TS^{-1} L^{-1}
Microcystis sp.	N/A	0.070–0.153 L	1500–6000 mg VS^{-1}
Nanochloropsis oculata	N/A	390 L kg^{-1} VS	N/A

concentrations of microalgae lead to the formation of ammonia when these compounds are broken down during the hydrolysis stage. The extraction of protein and lipid for further use in biochemical industries, or lipid extraction for biofuel production purposes can decrease the possibility of ammonia inhibition in the subsequent AD process (Spolaore et al. 2006). Protein and lipid extraction followed by AD can also help to achieve increased C/N ratios when considering microalgae for biogas production. Last but not least among the challenges discussed herein is the high nutrient requirement of microalgae for their mass production (Collet et al. 2011). This requirement, particularly for nitrogen and phosphorous, has been met by employing huge amounts of chemical fertilizers causing a serious competition with the agricultural sector (Fenton 2012; Stephenson et al. 2010; Ward et al. 2014).

11.6.2 Lignocellulosic Biomass for Biogas Production

Lignocellulosic biomass also holds a huge potential for being used as feedstock for biogas production due to their abundance, availability, and their high carbohydrate content. Although lignocellulosic materials generally cover two groups of feedstock, i.e., energy crops and lignocellulosic residues, this section only deals with the second generation biomass, i.e., wastes and agricultural residues such as straw and woody biomass. As presented in Table 11.4, energy crops also have a significant potential for biogas production but due to their competition with conventional crops production over water resources and land use, their application as feedstock for AD will not be discussed herein.

Lignocellulosic materials can be divided into four different groups, i.e., agricultural residues (straw), fruit and vegetable waste, woody residues, and paper waste. Although being appropriate for AD, the major disadvantage of lignocellulosic residuals is their high amount of lignin content, which can be regarded as a serious obstacle for AD process. In general, those lignocellulosic residues, containing a higher amount of volatile solids and a lower amount of refractory volatile solids, are more preferable for AD process (Monnet 2003).

Table 11.4 Methane yield of various energy crops (Deublein and Steinhauser 2011; Kabir et al. 2015)

Crop	Methane yield
Maize (whole crop)	205–405 (m^3 CH_4 kg^{-1} VS)
Potatoes	276–400 (m^3 CH_4 kg^{-1} VS)
Wheat (grain)	384–426 (m^3 CH_4 kg^{-1} VS)
Oats (grain)	283–492 (m^3 CH_4 kg^{-1} VS)
Triticale	337–555 (m^3 CH_4 kg^{-1} VS)
Sorghum	295–372 (m^3 CH_4 kg^{-1} VS)
Barley	353–658 (m^3 CH_4 kg^{-1} VS)
Red clover	300–350 (m^3 CH_4 kg^{-1} VS)
Alfalfa	340–500 (m^3 CH_4 kg^{-1} VS)
Sunflower	154–400 (m^3 CH_4 kg^{-1} VS)
Oilseed rape	240–340 (m^3 CH_4 kg^{-1} VS)
Peas	390 (m^3 CH_4 kg^{-1} VS)
Ryegrass	390–410 (m^3 CH_4 kg^{-1} VS)
Fodder beet	420–500 (m^3 CH_4 kg^{-1} VS)
Nettle	120–420 (m^3 CH_4 kg^{-1} VS)
Hemp	355–409 (m^3 CH_4 kg^{-1} VS)
Grass ensilage	0.6–0.7 (m^3 CH_4 kg^{-1} DM)
Leaves of sugar beet	0.4–0.8 (m^3 CH_4 kg^{-1} DM)
Sugar beet	0.4–1.0 (m^3 CH_4 kg^{-1} DM)
Clover	0.6–0.8 (m^3 CH_4 kg^{-1} DM)
Diverse kinds of cereals	0.4–0.9 (m^3 CH_4 kg^{-1} DM)
Barley ensilage	0.75–0.99 (m^3 CH_4 kg^{-1} DM)
Rye ensilage	0.57–0.79 (m^3 CH_4 kg^{-1} DM)

11.6.3 Wood Residues

It has been well established that biogas production from woody residues is not economically feasible due to factors affecting the efficiency of the AD process including low moisture content, high lignin content, cellulose crystallinity, and degree of association between lignin and carbohydrates (Kabir et al. 2015). Recent research studies have shown that coupling biomaterial with biomethane production form woody residues would result in better economic and environmental benefits (Khoshnevisan et al. 2016; Safari et al. 2017; Shafiei et al. 2011). Biogas production from woody residues necessitates a pretreatment step. A large number of pretreatment steps such as alkaline pretreatment, N-methylmorpholine-N-oxide, untreated, steam explosion, and fungal treatment have been identified and evaluated. Based on the substrate employed and the pretreatment method conducted, different methane production rates have been reported (Mirahmadi et al. 2010; Take et al. 2006; Teghammar et al. 2010).

11.6.4 Agricultural Residues

Agricultural residues are among lignocellulosic materials with significant potential for biogas and biomaterial production. The straw-based lignocellulosic residues of agricultural origin can undergo AD and produce huge amounts of biogas. As elaborated earlier, gas production rates reported in the literature varies depending on the kinds of cereals used in AD system. The main obstacle using straw-based lignocellulosic residues for biogas production is the pretreatment step (Rahimi et al. 2018; Khoshnevisan et al. 2017). However, extracting building block chemicals from straw-based lignocellulosic materials can compensate for the pretreatment step. Although a large number of studies have been conducted to evaluate different pretreatment methods, it is difficult to conclude which pretreatment method works best and produces the highest level of gas. This is due to the fact that most studies failed to address economic and environmental perspectives. Table 11.5 tabulates methane potential of various kinds of straw under different pretreatment methods.

11.6.5 Paper Wastes

Paper waste, a lignocellulosic material, has also been a focus for AD. Biological methane potential of paper waste hugely depends on the type of the paper, i.e., pulp and paper sludge, paper tube residues, etc. Moreover, the pretreatment method applied and the inoculum used could influence the specific methane yield. It has been well-established that the specific methane yield of untreated paper ranges between 100 and 200 L kg^{-1} VS (Wellinger et al. 2013). Pretreatment can significantly improve AD of paper waste leading to higher specific methane

Table 11.5 Methane potential of different kinds of straw (Odhner et al. 2012; Wellinger et al. 2013)

Type of straw	Pretreatment	Digestion type	AD temperature (°C)	Organic loading	Specific methane yield	AD time
Wheat	Untreated	Mesophilic			189 L kg^{-1} VS	
	Milled	Mesophilic	37.5	SI[1] ratio 1:3	275 L kg^{-1} VS	
	Steam explosion	Mesophilic	37.5	SI[1] ratio 1:3	331 L kg^{-1} VS	
	Physical pretreatment 30 × 50 mm	Mesophilic	37	89 g VS + 2 L water + 2 L slurry	162 L kg^{-1} VS	60
	Physical pretreatment 0.088 mm	Mesophilic	37	89 g VS + 2 L water + 2 L slurry	249 L kg^{-1} VS	60
Rice	Untreated	Mesophilic	35	400 ml swage + 1 g straw	54 L kg^{-1} straw	30
	Untreated	Psychrophilic	22	12.6 g VS L^{-1}	240 L kg^{-1} VS	120
	Untreated	Thermophilic	55	40 ml Inoc. + 0.2 substrate	30 NL kg^{-1} VS	45
	Untreated	Mesophilic	35	50 g solid L^{-1}	190 L kg^{-1} VS	24
	Acetic + propionic acids (1:1); solid acid ratio (1:20)	Mesophilic	35	400 ml swage + 1 g straw	213.5 L kg^{-1} straw	30
	Phosphate supplementation 155 mg-P L^{-1}	Psychrophilic	22	12.6 g VS L^{-1}	250 L kg^{-1} VS	120
	Grounded 25 mm	Mesophilic	35	50 g solid L^{-1}	200 L kg^{-1} VS	24
	Grounded 25 mm 110 °C + NH_3 20 mg g^{-1}	Mesophilic	35	50 g solid L^{-1}	245 L kg^{-1} VS	24
	NMMO	Thermophilic	55	40 ml Inoc. + 0.2 substrate	212 NL kg^{-1} VS	45
	Physical pretreatment 30 × 50 mm	Mesophilic	37	79.4 g VS + 2 L water + 2 L slurry	241 L kg^{-1} VS	60
	Physical pretreatment 0.088 mm	Mesophilic	37	79.4 g VS + 2 L water + 2 L slurry	365 L kg^{-1} VS	60

(continued)

Table 11.5 (continued)

Type of straw	Pretreatment	Digestion type	AD temperature (°C)	Organic loading	Specific methane yield	AD time
Corn	Untreated	Mesophilic	35	40.25 g VS L^{-1}	153.7 L kg^{-1} VS	30
	NaOH 8% Wt	Mesophilic	35	40.25 g VS L^{-1}	472 L kg^{-1} VS	30
	Ammonia 5% Wt	Mesophilic	35	40.25 g VS L^{-1}	243.5 L kg^{-1} VS	30
	Urea 4% Wt	Mesophilic	35	40.25 g VS L^{-1}	178 L kg^{-1} VS	30
	Pleurotus florida	Mesophilic	35	40.25 g VS L^{-1}	380 L kg^{-1} VS	30
	Pleurotus florida 300 g ground straw +225 g water 121 °C for 2 h	Mesophilic	35	40.25 g VS L^{-1}	404.8 L kg^{-1} VS	30

[a]Substrate Inoculum dry matter ratio

production. The untreated pulp and paper sludge under mesophilic condition reportedly produced 190 L CH_4 kg^{-1} VS, while in contrast, a pretreatment with 0.6% NaOH at 37 °C water bath for 6 h increased the specific methane production by 68.5% (Lin et al. 2009). Simultaneous pretreatment with steam explosion and sodium hydroxide has shown better results than sole sodium hydroxide when treating paper tube residues under thermophilic conditions. The specific methane yield resulted from pretreatment of paper tube residues with steam explosion and 2% NaOH at 220 °C was estimate at 403 L kg^{-1} VS. Adding 2% H_2O_2 to the mentioned pretreatment method increased the specific methane yield by 22%. Untreated paper tube residues and the one treated with 2% NaOH at 190 °C produced 222 and 269 L CH_4 kg^{-1} VS, respectively (Teghammar et al. 2010).

11.6.6 Industrial Waste

The high potential of industrial waste for biogas production cannot be ignored. Biofuel plants and biorefineries are among the distinctive industries where very large amounts of organic by-products are accumulated. These organic by-products are appropriate feedstock for the AD process. For example, the silage fractions remain after bio-ethanol production in grain-processing bio-ethanol plants can undergo the AD process (Cassidy et al. 2008; Drosg et al. 2008; Rosentrater et al. 2006). Moreover, it has been well established that, cane juice silage is anaerobically degradable, and so, it is a suitable substrate for AD (Cail and Barford 1985; Callander and Barford 1983; Russo et al. 1985). In biodiesel plants, the glycerol and the wastewaters generated along with the oil extraction residual cake can also undergo the AD process (Wellinger et al. 2013). Nevertheless, the limitations regarding AD of industrial organic wastes should be neglected. More specifically, these feedstock can potentially contain a huge amount of undesirable compounds such as biological, physical or even chemical pollutants. Physical impurities, pathogens, heavy metals and/or persistent organic compounds found in industrial organic wastes can neutralize the environmental benefits of AD and pose health risks to humans and animals. This problem is more critical when the produced digestate is used as fertilizer (Wellinger et al. 2013).

11.7 Summary and Concluding Remarks

Providing energy and materials through biorefineries has attracted an increasing deal of interest and this popularity is mainly attributed to the positive sustainability impacts of biorefineries. In better words, biorefineries are meant to treat biomass feedstock and deliver a spectrum of products with positive effects while displacing their fossil-fuel originated counterparts. This approach makes biorefineries capable of competing with today's petroleum refineries. While the development of biorefineries for supplying bioenergy and biomaterials for coming decades seems promising and the current examples of biorefineries can be found all around the world, there is still a long way to go before biorefineries can be considered as comprehensive alternative to petroleum refineries.

To satisfy the future demands for bioenergy and biochemical, a substantial amount of biomass from agriculture, forestry, and waste need to be dedicated to biorefineries. From the sustainability point of views, the allocation of the available biomass resources to different types of biorefineries should be judiciously managed. Otherwise, it can possess negative ecological impacts, socio-economic consequences, and other environmental burdens. Although a wide range of biomass feedstock can undergo biorefining process, the selection of feedstock, processing pathways and final products should be done wisely by following a systematic approach. For instance, if the biorefineries are meant to supply future block building chemicals, the top twelve final candidates already identified through an iterative process based on the petrochemical model using building blocks, chemical data, known market data, properties, performance of the potential candidates, and the prior industry experiences, should be considered.

Multi-criteria assessment can also be employed to determine the overall sustainability of biorefineries due to the fact that it can simultaneously combine the physical, ecological, environmental, and socio-economic considerations. For instance, when facing a dilemma between two alternatives, e.g., lignocellulosic versus macroalgae biorefineries for producing specific types of biomaterials and bioenergies, the question to be answered would be which feedstock could better satisfy mass and energy balance, economic balance, employment opportunities, environmental issues, and technical possibilities. The economic aspects of biorefineries are also important because it is often difficult to get positive economy balance, as the production cost of biomass-based fuels is often high. The competition between food production sector versus raw materials supply for biorefineries over land and even other limited resources such as water must also be taken into account as a serious limitation in developing future biorefineries. Direct and indirect LUC effects should also incorporated into any final decisions.

Finally, the review of already published studies well shows that the integration of AD units with various biorefinery platforms or even currently-existing biofuel plants holds a huge potential to produce a positive economic, as well as energy and mass balance, with lower environmental intensity.

References

Batista AP, López EP, Dias C, da Silva TL, Marques IP (2017) Wastes valorization from Rhodosporidium toruloides NCYC 921 production and biorefinery by anaerobic digestion. Biores Technol 226:108–117

Begum S, Golluri K, Anupoju GR, Ahuja S, Gandu B, Kuruti K, Maddala RK, Venkata SY (2016) Cooked and uncooked food waste: a viable feedstock for generation of value added products through biorefinery approach. Chem Eng Res Des 107:43–51

Berntsson T, Sandén BA, Olsson L, Åsblad A (2012) What is a biorefinery?
Bozell JJ, Petersen GR (2010) Technology development for the production of biobased products from biorefinery carbohydrates—the US Department of Energy's "Top 10" revisited. Green Chem 12:539–554
Brar SK, Sarma SJ, Pakshirajan K (2016) Platform chemical biorefinery: future green chemistry. Elsevier
Cail R, Barford J (1985) A comparison of an upflow floc (tower) digester and UASB system treating cane juice stilage. Agric Wastes 14:291–299
Callander I, Barford J (1983) Anaerobic digestion of high sulphate cane juice stillage in a tower fermenter. Biotech Lett 5:755–760
Cassidy D, Hirl P, Belia E (2008) Methane production from ethanol co-products in anaerobic SBRs. Water Sci Technol 58:789–793
Chambost V, Stuart PR (2007) Selecting the most appropriate products for the forest biorefinery. Industrial Biotechnol 3:112–119
Chandra R, Takeuchi H, Hasegawa T (2012) Methane production from lignocellulosic agricultural crop wastes: A review in context to second generation of biofuel production. Renew Sustain Energy Rev 16:1462–1476
Cherubini F (2010) The biorefinery concept: using biomass instead of oil for producing energy and chemicals. Energy Convers Manag 51:1412–1421
Cherubini F, Jungmeier G, Mandl M, Philips C, Wellisch M, Jrgensen H, Skiadas I, Boniface L, Dohy M, Pouet J-C, (2007) IEA Bioenergy Task 42 on Biorefineries: co-production of fuels, chemicals, power and materials from biomass. IEA
Cherubini F, Jungmeier G, Wellisch M, Willke T, Skiadas I, Van Ree R, de Jong E (2009) Toward a common classification approach for biorefinery systems. Biofuels, Bioprod Biorefin 3:534–546
Collet P, Hélias A, Lardon L, Ras M, Goy R-A, Steyer J-P (2011) Life-cycle assessment of microalgae culture coupled to biogas production. Biores Technol 102:207–214
Cooksley CM, Zhang Y, Wang H, Redl S, Winzer K, Minton NP (2012) Targeted mutagenesis of the *Clostridium acetobutylicum* acetone–butanol–ethanol fermentation pathway. Metab Eng 14:630–641
Croft MT, Warren MJ, Smith AG (2006) Algae need their vitamins. Eukaryotic Cell 5:1175–1183
Dale BE, Kim S (2006) Biomass refining global impact–the biobased economy of the 21st century. In: Biorefineries-industrial processes and products: status quo and future directions, 41–66
Dan W, Hao C, Jiang L, Jin C, Zhinan X, Peilin C (2010) Efficient separation of butyric acid by an aqueous two-phase system with calcium chloride. Chin J Chem Eng 18:533–537
Danner H, Braun R (1999) Biotechnology for the production of commodity chemicals from biomass. Chem Soc Rev 28:395–405
DeFries R, Achard F, Brown S, Herold M, Murdiyarso D, Schlamadinger B, de Souza C (2007) Earth observations for estimating greenhouse gas emissions from deforestation in developing countries. Environ Sci Policy 10:385–394
Demirbas A (2009) Political, economic and environmental impacts of biofuels: a review. Appl Energy 86:S108–S117
Demirbas A, Demirbas MF (2010) Biorefineries, Algae Energy. Springer, pp 159–181
Deublein D, Steinhauser A (2011) Biogas from waste and renewable resources: an introduction. Wiley
Dien B, Cotta M, Jeffries T (2003) Bacteria engineered for fuel ethanol production: current status. Appl Microbiol Biotechnol 63:258–266
Drosg B, Wirthensohn T, Konrad G, Hornbachner D, Resch C, Wäger F, Loderer C, Waltenberger R, Kirchmayr R, Braun R (2008) Comparing centralised and decentralised anaerobic digestion of stillage from a large-scale bioethanol plant to animal feed production. Water Sci Technol 58:1483–1489
Energy, U.S.D.O (1997) Energy, environmental, and economics (E3) handbook—a resource tool to aid the office of industrial technologies

Engel CAR, Straathof AJ, Zijlmans TW, van Gulik WM, van der Wielen LA (2008) Fumaric acid production by fermentation. Appl Microbiol Biotechnol 78:379–389

FazeliNejad S, Ferreira JA, Brandberg T, Lennartsson PR, Taherzadeh MJ (2016) Fungal protein and ethanol from lignocelluloses using Rhizopus pellets under simultaneous saccharification, filtration and fermentation (SSFF). Biofuel Res J 3:372–378

Fenton O (2012) Agricultural nutrient surpluses as potential input sources to grow third generation biomass (microalgae): a review. Algal Res 1:49–56

Fernando S, Adhikari S, Chandrapal C, Murali N (2006) Biorefineries: current status, challenges, and future direction. Energy Fuels 20:1727–1737

Francavilla M, Trotta P, Luque R (2010) Phytosterols from *Dunaliella* tertiolecta and *Dunaliella* salina: a potentially novel industrial application. Biores Technol 101:4144–4150

Gao T, Wong Y, Ng C, Ho K (2012) L-lactic acid production by *Bacillus subtilis* MUR1. Biores Technol 121:105–110

Golueke CG, Oswald WJ (1959) Biological conversion of light energy to the chemical energy of methane. Appl Microbiol 7:219–227

González-Fernández C, Molinuevo-Salces B, García-González MC (2011) Evaluation of anaerobic codigestion of microalgal biomass and swine manure via response surface methodology. Appl Energy 88:3448–3453

Gravitis J, Suzuki M (1999) Biomass refinery–a way to produce value added products and base for agricultural zero emissions system. In: Proceedings of the international conference on agricultural engineering for 21st century, pp. 14–17

Hagman L, Blumenthal A, Eklund M, Svensson N (2017) The role of biogas solutions in sustainable biorefineries. J Cleaner Production

Harun R, Singh M, Forde GM, Danquah MK (2010) Bioprocess engineering of microalgae to produce a variety of consumer products. Renew Sustain Energy Rev 14:1037–1047

Hirschberg J (1999) Production of high-value compounds: carotenoids and vitamin E. Curr Opin Biotechnol 10:186–191

Huber GW (2008) Breaking the chemical engineering barriers to Lignocellulosic Biofuels Citeseer

Janssen R, Rutz DD (2011) Sustainability of biofuels in Latin America: risks and opportunities. Energy Policy 39:5717–5725

Jiang Y, Xu C, Dong F, Yang Y, Jiang W, Yang S (2009) Disruption of the acetoacetate decarboxylase gene in solvent-producing *Clostridium acetobutylicum* increases the butanol ratio. Metab Eng 11:284–291

Kabir MM, Forgács G, Horváth IS (2015) Biogas from lignocellulosic materials, Lignocellulose-Based Bioproducts. Springer, pp. 207–251

Kamm B, Gruber PR, Kamm M (2006) Biorefineries-industrial processes and products. Biorefineries industrial processes and products. Wiley-VCH, Weinheim

Kamm B, Kamm M (2004a) Biorefinery-systems. Chem Biochem Eng Q 18:1–7

Kamm B, Kamm M (2004b) Principles of biorefineries. Appl Microbiol Biotechnol 64:137–145

Khoshnevisan B, Shafiei M, Rajaeifar MA, Tabatabaei M (2016) Biogas and bioethanol production from pinewood pre-treated with steam explosion and N-methylmorpholine-N-oxide (NMMO): A comparative life cycle assessment approach. Energy 114:935–950

Khoshnevisan B, Rafiee S, Tabatabaei M, Ghanavati H, Mohtasebi SS, Rahimi V, … & Karimi K (2017) Life cycle assessment of castor-based biorefinery: a well to wheel LCA. Int J Life Cycle Assess 1–18

Lardon L, Helias A, Sialve B, Steyer J-P, Bernard O (2009) Life-cycle assessment of biodiesel production from microalgae. ACS Publications

Lin Y, Wang D, Wu S, Wang C (2009) Alkali pretreatment enhances biogas production in the anaerobic digestion of pulp and paper sludge. J Hazard Mater 170:366–373

Mansoornejad B, Chambost V, Stuart P (2010) Integrating product portfolio design and supply chain design for the forest biorefinery. Comput Chem Eng 34:1497–1506

McKendry P (2002) Energy production from biomass (part 2): conversion technologies. Biores Technol 83:47–54

McMillan JD, Newman MM, Templeton DW, Mohagheghi A (1999) Simultaneous saccharification and cofermentation of dilute-acid pretreated yellow poplar hardwood to ethanol using xylose-fermenting *Zymomonas mobilis*. Appl Biochem Biotechnol 79:649–665

Melis A (2009) Solar energy conversion efficiencies in photosynthesis: minimizing the chlorophyll antennae to maximize efficiency. Plant Sci 177:272–280

Mirahmadi K, Kabir MM, Jeihanipour A, Karimi K, Taherzadeh M (2010) Alkaline pretreatment of spruce and birch to improve bioethanol and biogas production. BioResources 5:928–938

Misra S, Gupta P, Raghuwanshi S, Dutt K, Saxena R (2011) Comparative study on different strategies involved for xylitol purification from culture media fermented by *Candida tropicalis*. Sep Purif Technol 78:266–273

Monnet F (2003) An introduction to anaerobic digestion of organic wastes. Remade Scotland, 1–48

Mussgnug JH, Klassen V, Schlüter A, Kruse O (2010) Microalgae as substrates for fermentative biogas production in a combined biorefinery concept. J Biotechnol 150:51–56

Negash M, Swinnen JF (2013) Biofuels and food security: micro-evidence from Ethiopia. Energy Policy 61:963–976

Nepstad DC, Stickler CM, Almeida OT (2006) Globalization of the Amazon soy and beef industries: opportunities for conservation. Conserv Biol 20:1595–1603

NREL (2005) Conceptual biorefinery. Available from http://www.nrel.gov/biomass/biorefinery.html. (Last seen August 1, 2005)

Octave S, Thomas D (2009) Biorefinery: toward an industrial metabolism. Biochimie 91:659–664

Odhner PB, Horváth I, Kabir MM, Schabbaeur A (2012) Biogas from lignocellulosic biomass. Svenskt Gastekniskt Center

Ostlund RE, Racette SB, Stenson WF (2003) Inhibition of cholesterol absorption by phytosterol-replete wheat germ compared with phytosterol-depleted wheat germ. The American journal of clinical nutrition 77:1385–1389

Pachapur V, Sarma S, Brar S, Chaabouni E (2016) Platform chemicals: significance and need

Parajuli R, Dalgaard T, Jørgensen U, Adamsen APS, Knudsen MT, Birkved M, Gylling M, Schjørring JK (2015) Biorefining in the prevailing energy and materials crisis: a review of sustainable pathways for biorefinery value chains and sustainability assessment methodologies. Renew Sustain Energy Rev 43:244–263

Pragya N, Pandey KK, Sahoo P (2013) A review on harvesting, oil extraction and biofuels production technologies from microalgae. Renew Sustain Energy Rev 24:159–171

Provasoli L, Carlucci A (1974) Vitamins and growth regulators. Botanical monographs

Rahimi V, Karimi K, Shafiei M, Naghavi R, Khoshnevisan B, Ghanavati H, … & Tabatabaei M (2018) Well-to-wheel life cycle assessment of Eruca Sativa-based biorefinery. Renew Energy 117:135–149

Ramos-Suárez JL, Cuadra FG, Acién FG, Carreras N (2014) Benefits of combining anaerobic digestion and amino acid extraction from microalgae. Chem Eng J 258:1–9

Ravindranath N, Lakshmi CS, Manuvie R, Balachandra P (2011) Biofuel production and implications for land use, food production and environment in India. Energy Policy 39:5737–5745

Reijnders L (2008) Do biofuels from microalgae beat biofuels from terrestrial plants? Trends Biotechnol 26:349–350

Rosentrater KA, Hall HR, Hansen CL (2006) Anaerobic digestion potential for ethanol processing residues. In: 2006 ASAE annual meeting. american society of agricultural and biological engineers, p 1

Roy S, Das D (2015) Gaseous fuels production from Algal biomass, Algal Biorefinery: an integrated approach. Springer, Berlin, pp 297–319

Russo C, Sant'Anna G, de Carvalho Pereira SE (1985) An anaerobic filter applied to the treatment of distillery wastewaters. Agricu Wastes 14:301–313

Safari A, Karimi K, Shafiei M (2017) Dilute alkali pretreatment of softwood pine: a biorefinery approach. Biores Technol 234:67–76

Santamaría-Fernández M, Molinuevo-Salces B, Lübeck M, Uellendahl H (2017) Biogas potential of green biomass after protein extraction in an organic biorefinery concept for feed, fuel and fertilizer production. Renew Energy
Schmidt T, Pröter J, Scholwin F, Nelles M (2013) Anaerobic digestion of grain stillage at high organic loading rates in three different reactor systems. Biomass Bioenerg 55:285–290
Shafiei M, Karimi K, Taherzadeh MJ (2011) Techno-economical study of ethanol and biogas from spruce wood by NMMO-pretreatment and rapid fermentation and digestion. Biores Technol 102:7879–7886
Shafiei M, Kumar R, Karimi K (2015) Pretreatment of lignocellulosic biomass, Lignocellulose-based bioproducts. Springer, pp 85–154
Sharara MA, Clausen EC, Carrier DJ (2012) An overview of biorefinery technology. Biorefinery co-products: phytochemicals, primary metabolites and value-added biomass processing, 1–18
Sheehan J, Dunahay T, Benemann J, Roessler P (1998) A look back at the US Department of Energy's aquatic species program: biodiesel from algae. Nat Renew Energy, Laboratory, p 328
Sialve B, Bernet N, Bernard O (2009) Anaerobic digestion of microalgae as a necessary step to make microalgal biodiesel sustainable. Biotechnol Adv 27:409–416
Soyez K, Kamm B, Kamm M (1997) The green biorefinery. In: Proceedings of 1st international green biorefinery conference, Neuruppin, Germany
Spolaore P, Joannis-Cassan C, Duran E, Isambert A (2006) Commercial applications of microalgae. J Biosci Bioeng 101:87–96
St-Onge M-P, Lamarche BT, Mauger J-F, Jones PJ (2003) Consumption of a functional oil rich in phytosterols and medium-chain triglyceride oil improves plasma lipid profiles in men. J Nutr 133:1815–1820
Stephens E, Ross IL, Hankamer B (2013) Expanding the microalgal industry–continuing controversy or compelling case? Curr Opin Chem Biol 17:444–452
Stephens E, Ross IL, Mussgnug JH, Wagner LD, Borowitzka MA, Posten C, Kruse O, Hankamer B (2010) Future prospects of microalgal biofuel production systems. Trends Plant Sci 15:554–564
Stephenson AL, Kazamia E, Dennis JS, Howe CJ, Scott SA, Smith AG (2010) Life-cycle assessment of potential algal biodiesel production in the United Kingdom: a comparison of raceways and air-lift tubular bioreactors. Energy Fuels 24:4062–4077
Szmant HH (1989) Organic building blocks of the chemical industry. Wiley
Take H, Andou Y, Nakamura Y, Kobayashi F, Kurimoto Y, Kuwahara M (2006) Production of methane gas from Japanese cedar chips pretreated by various delignification methods. Biochem Eng J 28:30–35
Teghammar A, Yngvesson J, Lundin M, Taherzadeh MJ, Horváth IS (2010) Pretreatment of paper tube residuals for improved biogas production. Biores Technol 101:1206–1212
Tu RS, Tirrell M (2004) Bottom-up design of biomimetic assemblies. Adv Drug Deliv Rev 56:1537–1563
van Ree R, Annevelink E (2007) Status report biorefinery 2007. Agrotechnology & Food Sciences Group
Wang M, Hess R, Wright C, Ibsen K, Ruth M, Jechura J, Spath P, Graham R, Sokhansanj S, Perlack R (2004) Office of the Biomass Program: Multi-Year Analysis Plan, FY04-FY08. National Renewable Energy Lab, Golden, CO (US)
Ward A, Lewis D, Green F (2014) Anaerobic digestion of algae biomass: a review. Algal Res 5:204–214
Wargacki AJ, Leonard E, Win MN, Regitsky DD, Santos CNS, Kim PB, Cooper SR, Raisner RM, Herman A, Sivitz AB (2012) An engineered microbial platform for direct biofuel production from brown macroalgae. Science 335:308–313
Wellinger A, Murphy JD, Baxter D (2013) The biogas handbook: science, production and applications. Elsevier
Werpy T, Petersen G (2004) Top value added chemicals from biomass. US Department of Energy, Office of Scientific and Technical Information, No.: DOE/GO-102004–1992. https://www.osti.gov/

Werpy T, Petersen G, Aden A, Bozell J, Holladay J, White J, Manheim, A, Eliot D, Lasure L, Jones S (2004) Top value added chemicals from biomass. Volume 1-Results of screening for potential candidates from sugars and synthesis gas. DTIC Document

Wilkie AC, Riedesel KJ, Owens JM (2000) Stillage characterization and anaerobic treatment of ethanol stillage from conventional and cellulosic feedstocks. Biomass Bioenerg 19:63–102

Williams PJLB Laurens LM (2010) Microalgae as biodiesel & biomass feedstocks: review & analysis of the biochemistry, energetics & economics. Energy Environ Sci 3:554–590

Yang S-T, Yu M (2013) Integrated biorefinery for sustainable production of fuels, chemicals, and polymers. Bioprocessing Technologies in Biorefinery for Sustainable Production of Fuels, Chemicals, and Polymers 1

Ye L, Zhou X, Hudari MSB, Li Z, Wu JC (2013) Highly efficient production of L-lactic acid from xylose by newly isolated Bacillus coagulans C106. Biores Technol 132:38–44

Zhang C-H, Ma Y-J, Yang F-X, Liu W, Zhang Y-D (2009) Optimization of medium composition for butyric acid production by Clostridium thermobutyricum using response surface methodology. Biores Technol 100:4284–4288

Zhong W, Zhang Z, Luo Y, Qiao W, Xiao M, Zhang M (2012) Biogas productivity by co-digesting Taihu blue algae with corn straw as an external carbon source. Biores Technol 114:281–286

Chapter 12
Waste Management Strategies: Life Cycle Assessment (LCA) Approach

Benyamin Khoshnevisan, Shahin Rafiee and Meisam Tabatabaei

12.1 Introduction

Increasing world's population along with vast urbanization have brought about challenges for waste management systems in terms of waste generation, waste collection, waste treatment/disposal, waste recycling, as well as recovering energy from waste (Pharino 2017). It has been reported that the rate of municipal solid waste (MSW) generation is even faster than the rate of urbanization; 0.64 kg of MSW generation per person per day in about a few decades ago reached 1.2 kg per person per day in the year 2012 (Hoornweg and Bhada-Tata 2012). It has been anticipated that by the year 2025, 4.3 billion urban residents in the world will generate about 1.42 kg MSW per person per day equaling 2.2 billion tonnes per year (Laurent et al. 2014a). Due the fact that waste generation and management significantly contribute to human health and environmental preservation, and also have enormous impacts on the economy, creating regional integrated waste management strategies are inevitable.

Among distinctive definition presented for "waste", lack of use/value, or "useless remains" are the simplest ones. Cheremisinoff (2003) referred to waste as

B. Khoshnevisan · S. Rafiee (✉)
Department of Mechanical Engineering of Agricultural Machinery, Faculty of Agricultural Engineering and Technology, College of Agriculture and Natural Resources, University of Tehran, Karaj, Iran
e-mail: shahinrafiee@ut.ac.ir

M. Tabatabaei (✉)
Biofuel Research Team (BRTeam) / Agricultural Biotechnology Research Institute of Iran (ABRII), Agricultural Research, Education and Extension Organization (AREEO), Karaj, Iran
e-mail: meisam_tabatabaei@abrii.ac.ir; meisam_tab@yahoo.com

M. Tabatabaei and H. Ghanavati (eds.), *Biogas*, Biofuel and Biorefinery Technologies 6, https://doi.org/10.1007/978-3-319-77335-3_12

materials resulting from inefficient production processes on the industrial side, and low durability of goods and unsustainable consumption patterns on the consumer side. McDougall et al. (2008) considered waste as a by-product of human activity. Pharino (2017) defined the term "waste" as any substances or objects which the holder discards, or intends or is required to discard. Based on the definitions presented in the literature, it can be inferred that becoming 'waste' depends on many factors including time, location, state, income level, and personal preferences (Christensen 2011). In order to be able to thoroughly develop waste management strategies, it is necessary to appropriately distinguish different sources and types of waste and have a good knowledge about different types of waste and their composition. Waste classification facilitates creating waste management strategies and helps better achieve the most sustainable waste scenario under regional conditions. Tables 12.1 and 12.2 summarize sources and types of solid waste as well as waste composition linked to different sources, respectively.

A waste management system generally consists of collection, transportation, biological (i.e., composting or anaerobic digestion) or thermal (i.e., incineration) treatment, and disposal (i.e., landfilling). Concerns over the Earth's finite material and energy resources along with the generation of pollution and wastes that exceed the ability of the planet's natural sinks to absorb and convert them into harmless compounds have resulted in shifting away from disposal methods to waste minimizing and reuse strategies. The Waste Management Hierarchy, an internationally recognized strategy, stresses on maximizing the upstream waste management hierarchy towards reduce, reuse, and recycle (Fig. 12.1). However, categorizing waste management methods as the best to the worst or from the most environmentally-preferred to least one does not practically help with determining the option of the lowest environmental burdens and the highest economical-sustainability under different circumstances. In fact, different waste management options should be considered to effectively deal with various waste materials at a given condition (Rajaeifar et al. 2017). This necessitates developing integrated waste management strategies capable of adapting with regional conditions.

The design and implementation of integrated sustainable waste management is highly challenging because it must fully covers many dimensions including financial, technical, legal, environmental, and sociocultural perspectives (Pharino 2017). Van de Klundert et al. (2001) developed a framework consisting of four principles; (1) Equity, (2) Effectiveness, (3) Efficiency, and (4) Sustainability (Fig. 12.2). While the "equity", referring to the accessibility of the system for all citizens, can be simply evaluated in this framework, the assessment of effectiveness, efficiency, and sustainability of the targeted integrated system could be time consuming and more complicated. Due to the fact that solid waste management is known to be an important contributor to many different environmental problems, creating an integrated management system which would be comprehensively capable of addressing environmental problems and contributing to moving towards a more environmentally sustainable society should be given priority before implementing any strategies (Laurent et al. 2014b; Rajaeifar et al. 2015b).

Table 12.1 Sources and types of waste (Hoornweg and Bhada-Tata 2012)

Source	Typical waste generators	Types of solid waste
Residential	Single and multifamily dwellings	Food waste, paper, cardboard, plastics, textiles, leather, yard waste, wood, glass, metals, ash, special waste (e.g., bulky items, consumer electronics, white goods, batteries, oil, tires), household hazardous waste (e.g., paint, aerosols, gas tanks, waste containing mercury, motor oil, cleaning agents), e-waste (e.g., computers, phones, TVs)
Industrial	Light and heavy manufacturing, fabrication, construction sites, power and chemical plants (excluding specific process waste when the municipality does not oversee their collection)	Housekeeping waste, packaging, food waste, construction and demolition materials, hazardous waste, ash, special waste
Commercial	Stores, hotels, restaurants, markets, office buildings	Paper, cardboard, plastics, wood, food waste, glass, metals, special waste, hazardous waste, e-waste
Institutional	Schools, hospitals (non-medical waste), prisons, government buildings, airports	Same as commercial
Construction and demolition	New construction sites, road repair, renovation sites, demolition of buildings	Wood, steel, concrete, dirt, bricks, tiles
Municipal services	Street cleaning, landscaping, parks, beaches, other recreational areas, water and wastewater treatment plants	Street sweeping, landscape and tree trimmings, general waste from parks, beaches, and other recreational areas, sludge
Process	Heavy and light manufacturing, refineries, chemical plants, power plants, mineral extraction and processing	Industrial process waste, scrap materials, off-specification products, slag, tailings
Medical waste	Hospitals, nursing homes, clinics	Infectious waste (bandages, gloves, cultures, swabs, blood and body fluids), hazardous waste (sharps, instruments, chemicals), radioactive waste from cancer therapies, pharmaceutical waste
Agricultural	Crops, orchards, vineyards, dairies, feedlots, farms	Spoiled food waste, agricultural waste (e.g., rice husks, cotton stalks, coconut shells, coffee waste), hazardous waste (e.g., pesticides)

Life cycle assessment (LCA) methodology, despite some underlying problems, has been shown to be an appropriate approach to quantify environmental impacts caused by distinctive solid waste management scenarios, by identifying the hot

Table 12.2 Type of waste composition linked to different sources (Hoornweg and Bhada-Tata 2012)

Composition	Sources
Organic	Food scraps, yard (leaves, grass, brush) waste, wood, process residues
Paper	Paper scraps, cardboard, newspapers, magazines, bags, boxes, wrapping paper, telephone books, shredded paper, paper beverage cups
Plastic	Bottles, packaging, containers, bags, lids, cups
Glass	Bottles, broken glassware, light bulbs, colored glass
Metal	Cans, foil, tins, non-hazardous aerosol cans, appliances (white goods), railings, bicycles
Other	Textiles, leather, rubber, multilaminates, e-waste, appliances, ash, other inert materials

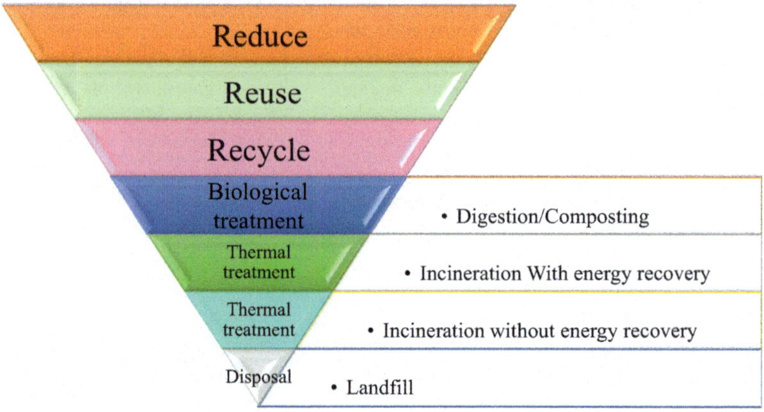

Fig. 12.1 A hierarchy of waste management

spots through different waste management practices and introducing the most eco-friendly scenarios (Bisinella et al. 2017; Mercante et al. 2012; Rajaeifar et al. 2015a; Saner et al. 2012; Yıldız-Geyhan et al. 2017). Despite the fact that LCA was initially introduced to take into account the environmental impacts related to a product or service, it has been also employed in evaluating the waste management systems to provide new insights into their environmental aspects. Although in the product-based LCA a cradle to grave approach is commonly employed to focus on the production and use stages, in the waste-based LCA, the end-of-life of products is in the spotlight (Christensen 2011). This principal difference brings about deviations in the definition of system boundary for a waste LCA compared with a product LCA (Fig. 12.3). Accordingly, the objective of this chapter was to provide insights about the general principles and methods of LCA followed by providing guidelines on the application of LCA in waste management systems.

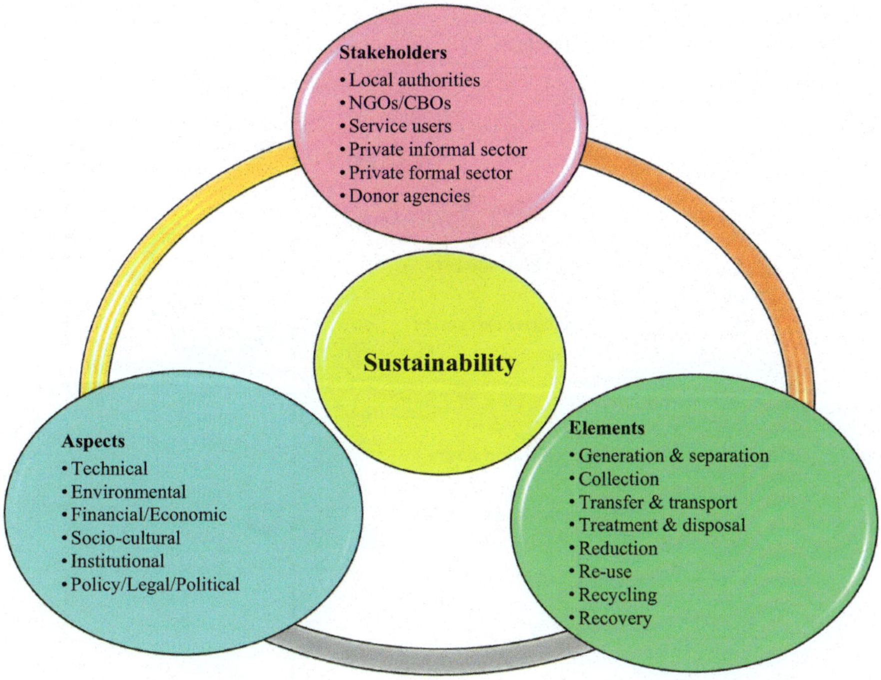

Fig. 12.2 The integrated sustainable waste management model (Van de Klundert et al. 2001)

12.2 General Principles of Life Cycle Assessment

12.2.1 Origins and Background of LCA

The first mention of the life cycle concept, by this name, was presented by Novick (1959) as a report to take into consideration the cost LCA of RAND Corporation. In the late 1960s and 1970s, Hunt (1974) led a research study to investigate the resource and emission profiles of nine beverage container alternatives ignoring the quantitative assessment of the associated impacts on the environment or resources. By increased public environmental concerns in the 1990s, it was proved that a more strategic and systematic approach to environmental challenges would be necessary, therefore, the International Organization for Standardization's ISO 14001 in 1996 indicated that environmental management would be no longer an option (Horne 2009). The urgent need to increase the interpretability of LCA results showed that an assessment method would be required to make the environmental impacts of the inventory results apparent (Hauschild and Huijbregts 2015). Accordingly, in the early 1990s, the first methods for the assessment of environmental impacts in the concept of LCA were presented by Ahbe et al. (1990) and Heijungs et al. (1992).

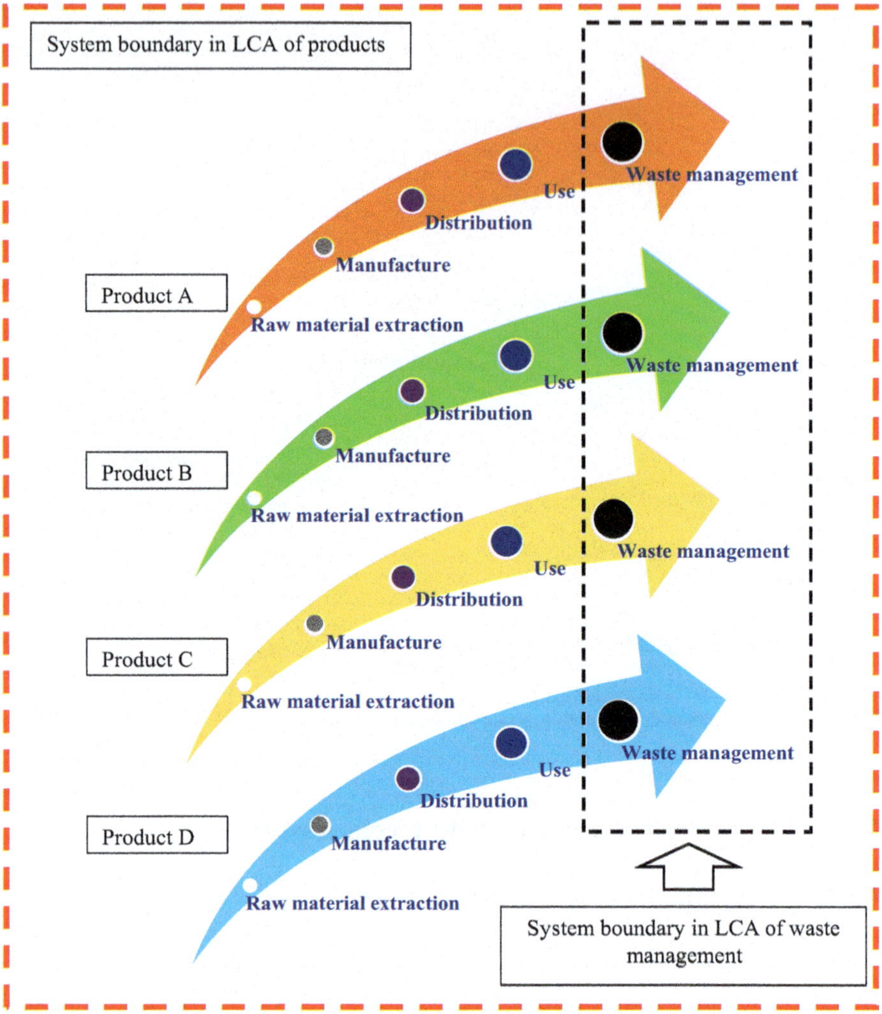

Fig. 12.3 The difference between waste-based LCA and product-based LCA. Adapted from (Christensen 2011)

In parallel with these attempts, the International Organization for Standardization also developed a series of framework and fundamental principle to systematically describe LCA and its main elements (Fig. 12.4) to provide the minimum requirements for performing an LCA study. Along with the development of the ISO 14040 series, a large number of impact assessment methods including EDIP (Wentzel et al. 1997), CML 2002 (Guinée et al. 2002), Ecoindicator 99 (Goedkoop and Spriensma 2000), and ReCiPe (Goedkoop et al. 2009) were also established in the form of national projects. Despite providing an indispensable framework for LCA, in many cases ISO 14040 series suggest a range of choices to practitioners with significant

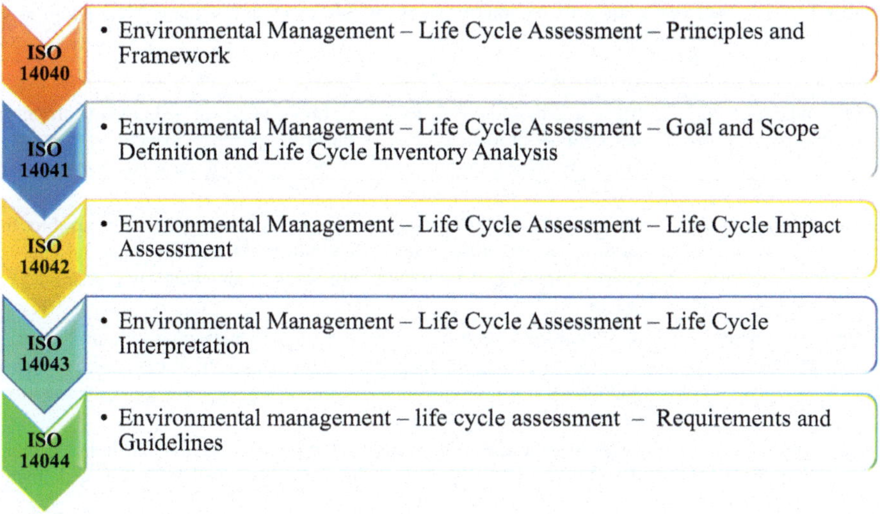

Fig. 12.4 ISO 14040 series on LCA

impacts on the final results. The International Reference Life Cycle Data System (ILCD) was therefore developed by the European Commission to provide guidance on planning, developing, and reporting both life cycle emission and resource consumption inventory (LCI) data sets and LCA studies (Wolf et al. 2012).

12.2.2 Definitions for LCA, Benefits, and Limitations

LCA is an environmental management tool aimed at supporting decisions and policies rather than being a decision making tool. While different definitions for LCA are reportedly seen in the literature, the most comprehensive one can be found in the introductory part of ISO 14040 as follows:

> LCA studies the environmental aspects and potential impacts throughout a product's life (i.e. cradle-to-grave) from raw material acquisition through production, use and disposal. The general categories of environmental impacts needing consideration include resource use, human health, and ecological consequences.

A similar definition was also adopted by the Society of Environmental Toxicology and Chemistry (SETAC) (Consoli 1993) and Scandinavian Ministers of the Environment DIN-NAGUS (Nordic Guidelines) (Christiansen et al. 1995). Klöpffer (2014) referred to LCA as a science-based method aimed at evaluating the environmental impacts of product systems. Disregarding the different terms used to define LCA, all definitions unanimously consider LCA as a technique established to assess the environmental impacts associated with all the stages of a product's life from raw material extraction through materials processing, manufacture,

Table 12.3 A brief explanation of mandatory stages of life cycle assessment (McDougall et al. 2008)

Terms	Definitions
Goal and scope definition	Stage at which the functional unit for comparison is defined (normally per equivalent use), as well as the study purpose, system boundaries, life cycle stages, unit processes, types of data, geographical scope, and scope of the assessment
Life cycle inventory	Process of accounting for all the inputs and outputs of the product system over the life cycle. Will result in a list of raw material and energy inputs, and individual emissions to air, water and as solid waste
Life cycle impact assessment	Associates the inputs and outputs with particular environmental issues, e.g., ozone depletion, and converts the inventory of materials, energy, and emissions into representative indicators, e.g., an aggregate loading of ozone-depleting chemicals
Life cycle interpretation	Evaluation of the significance of the inputs, outputs, and indicators of the system life cycle. This stage is the least well accepted or defined

distribution, use, repair and maintenance, and final disposal or recycling. ISO 14040 divides the entire LCA procedure into four distinct stages: goal and scope definition, life cycle inventory (LCI) analysis, life cycle impact assessment (LCIA), and interpretation. A brief explanation of these terms and their important differences is demonstrated in Table 12.3.

In the second phase of each LCA study, i.e., LCI, a mass balance is performed in which all necessary inputs to the system and outputs leaving the system boundary as well as emissions taking place in any stages and operations of the life cycle are accounted. One of the main benefits of LCA is incorporating both direct and indirect inputs as well as emissions for production, distribution, use, and disposal. Direct inputs and emissions refer to data from the foreground system where inputs are employed to generate a product or deliver a service. Contrary to direct data, indirect inventory data are collected from the background system, such as from the initial production of the energy, where raw materials are manufactured and energy carriers are extracted/processed/produced. As can be noticed in each LCA study, data belonging to different time spans are aggregated and included over space regardless of when they occur and where they are located.

Another important feature of LCA is its ability in aggregating data from separate unit processes and operations. In better words, LCA connects different unit processes to a system (Heijungs et al. 2014). More specifically, for each unit process, material and energy flows as well as emissions, products, and wastes of the total system are mapped. Such a detailed system allows the identification of the hotspots where the greatest environmental burdens are taking place. Moreover, in comparative studies, such a systematic approach facilitates investigating the environmental impacts of different alternatives and help find areas where environmental improvement can be made.

Besides what have already been mentioned as benefits of LCA approach, its ability in expressing the environmental impacts in terms of either individual indices called "midpoints" or aggregated index called "endpoints" cannot be ignored. This aspect of LCA enables practitioners to either deeply investigate about a particular impact category, e.g., global warming potential, or instead, simultaneously focus on an area of protection (damage category), e.g., ecosystem quality.

In spite of the benefits discussed herein, there are also some limitations which should not be ignored. One limitation of LCA studies arises when the environmental burdens through a given life cycle should be attributed to the co-products. In many cases, the system under consideration is multi-functional meaning that the processes produce more than one valuable output (product or service). The environmental burdens associated with a particular multi-functional process then need to be partitioned over the various functional flows of that process posing challenges related to multi-functionality and allocation (Heijungs and Guinée 2007). While several alternatives have been suggested to deal with multi-functionality issues including system expansion, economic, mass, exergy and energy allocation, selection of each method can significantly affect the final results. This limitation will be discussed in details in the subsequent section.

The idea that LCA is capable of delivering a comprehensive and overall assessment is not correct because it employs an overall system balance and functional unit to aggregate data over time and space. Consequently, it is not able to determine the actual environmental effects of a system (McDougall et al. 2008). In better words, the precise effects of such an environmental impact are time and cite-dependent as well being affected by the origin of emissions. For instance, the environmental impacts caused by a specific amount of emission released in a particular event, e.g., in a single factory, would be completely different from the same amount of emissions released continuously over years from several sources.

The other important limitation that needs to be taken into account is the type of LCA employed, i.e., attributional or consequential. As a brief description, the attributional approach considers the flows in the environment within a chosen temporal window while in contrast, the consequential method deals with changes in flows in response to decisions (Ekvall et al. 2016). According to the ILCA handbook, the attributional model describes the actual or forecasted specific or average supply chain plus the use and end-of-life value chain, all embedded into a static technosphere (Wolf et al. 2012); meaning that a linear relationship is laid under the concept of attributional approach. However, in practice, the interrelationship within a production system is, in most cases, non-linear. Plevin et al. (2014) well argued that attributional LCA is not predictive of real-world impacts and the results cannot be employed for policy decision making. They discussed that more production of product A with partially lower greenhouse gas (GHG) emissions than product B does not necessarily mean a similar reduction in carbon emissions.

12.2.3 Different Phases of LCA

According to ISO 14040, LCA studies are comprised of four compulsory steps, i.e., goal and scope definition, life cycle inventory (LCI) analysis, life cycle impact assessment (LCIA), and interpretation, along with some optional steps, i.e., normalization, classification, and weighting. Different stages of the LCA framework and their interconnections are illustrated in Fig. 12.5.

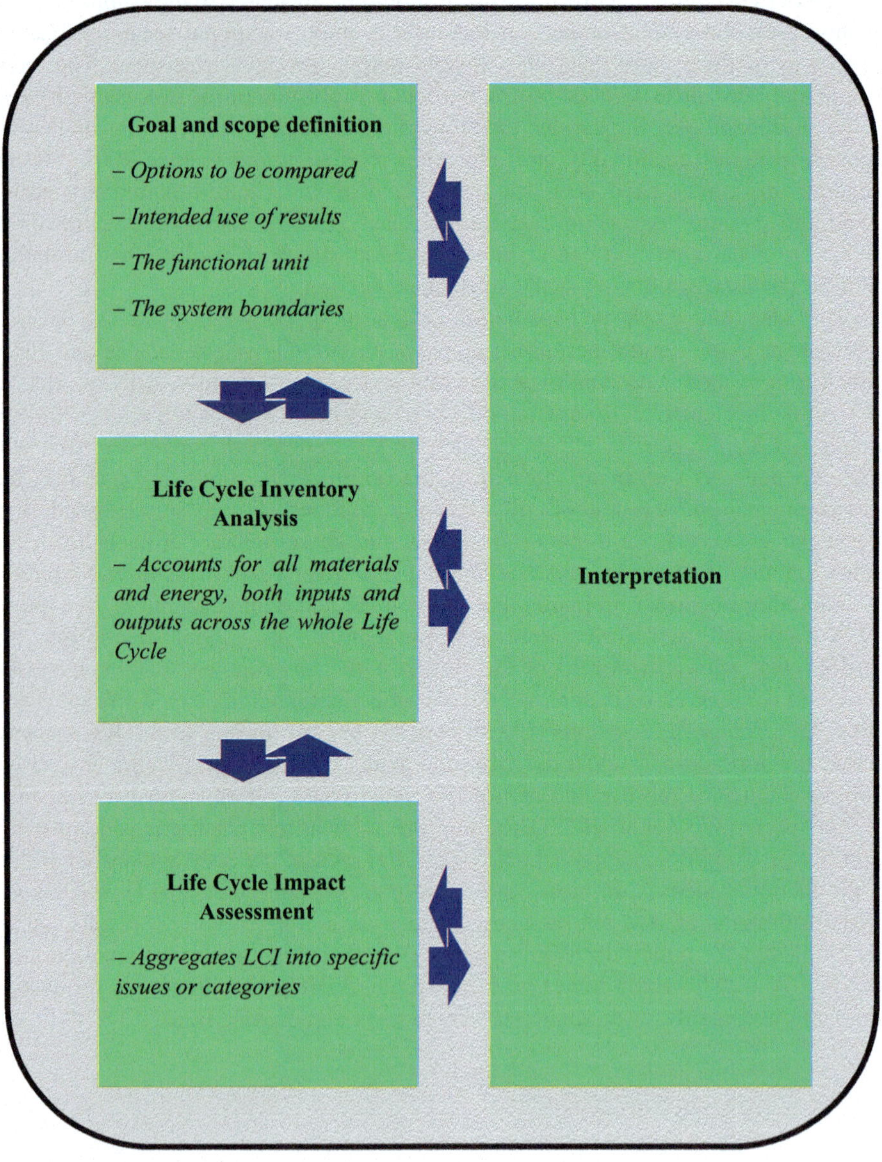

Fig. 12.5 LCA phases according to ISO 14040

12.2.3.1 Goal and Scope Definition

At the first step, the "Goal and Scope" of the study shall be determined. The goal definition includes specifying the fundamental concepts of the study, the reason for which the study is performed, the options/alternatives that should be compared, the intended use of the results, and the audience for whom the LCA study is conducted. The ISO 14044 states that:

> The goal and scope of an LCA shall be clearly defined and shall be consistent with the intended application. Due to the iterative nature of LCA, the scope may have to be refined during the study.

This statement emphasizes that the LCA has an iterative nature as shown by double arrows in Fig. 12.5, therefore, any modifications of goal and scope must be documented and reported. Moreover, it can be inferred that, despite the fact that the goal and scope of the study are determined prior to data collection, any modifications during the course of the study can take place as data are collected and new information is revealed (Curran 2017).

Scope definition outlines the most important parameters by which the study is performed. As instructed by ISO 14040, this step shall explicitly describe the following aspects:

- Product system and its functions
- Technical, geographical, and temporal system boundaries
- Functional unit
- Allocation methods
- Data availability and depth of study
- Assumptions, limitations, and restrictions
- Data quality requirements
- Type of critical review (if any) and the type and format of the report required for the study.

The product system represents the functions of the system by illustrating the unit processes and the collection of operations connected by flows of intermediate products. The unit processes are the smallest parts of a product system and are used to manifest the activities of a single operation or a group of operations. The product system is used as a basis for the definition of functional unit. A flow chart can simply and precisely describes the product system under consideration and reflects the interrelations of unit processes within the product system. The system boundary then separates those unit processes that will be included in the system from the rest of technosphere and shows what is omitted from the assessment (Fig. 12.6). It should be born in mind that if a comparative LCA study is intended to be performed, adapted boundaries must be balanced between different systems in terms of inclusion and omission.

Besides technical aspects of system boundary, geographical and temporal system boundaries are also important issues. Geographical system boundary provides information about the place where the product(s) is manufactured or service(s) is

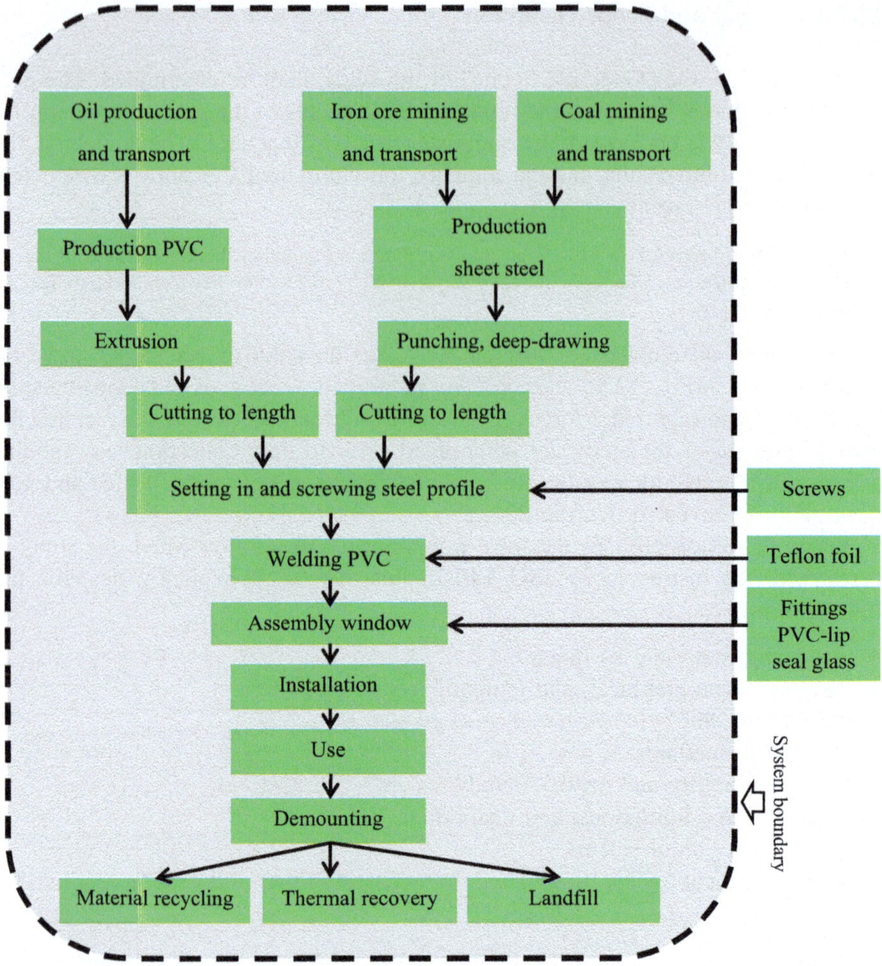

Fig. 12.6 Product system and system boundary (Klöpffer 2014)

delivered. It is worth quoting that, the geographical system boundary always has extensions beyond the selected range because all input materials and energy carriers are not supplied from the concerned country. Therefore, in the life cycle impact assessment (LCIA) stage, for some impact categories (e.g., climate change/greenhouse effect, stratospheric ozone depletion), global effects are considered, while, the regional or local effects are taken into account for the other impact categories e.g., eutrophication potential (Klöpffer 2014).

Once the goal of the study is assessed, the functional unit (FU) shall be determined in order to describe the product or process under study. This is performed by reflecting the function that the system delivers at the product unit level. The prominent role of such a measure relates to the LCI stage since FU will be used as a reference point to which inputs to the system boundary and outputs of the system are related. In better words, FU demonstrates the function(s) to which the environmental assessment is implemented, so that, the amount of input materials and energy carriers used within the system boundary are calculated for the preferred function(s).

Early in the scope definition of each LCA study, it must be determined which modeling approach will be employed: attributional or consequential modeling. This decision will affect the final results and has implications for many of the later choices including the choice of allocation method or inventory data collection. The type of LCI modeling used depends on several factors, among them, the decision-context reproducibility and robustness, practical feasibility, and stakeholder acceptance can be referred to.

As discussed earlier, in the attributional approach, inputs and outputs of a product system are attributed to the functional unit by linking and/or partitioning the unit processes of the system. In the consequential modeling, the objective is to identify the environmental consequences of a decision or a proposed change in a system under study (oriented to the future), which means that market and economic implications of a decision must to be taken into account. One of the most important distinctions between the attributional and consequential modeling is the type of data used in LCI analysis. In the attributional approach, "average data" is used while in the consequential modeling, "marginal data" are employed. Due to the fact that in the attributional approach, efforts are made to relate the environmental impacts caused by some specific activities that have contributed to the production, consumption, and disposal of a targeted product, it will be relevant to collect data on specific or market average suppliers. In contrast, the consequential LCA is used to find out the environmental impacts related to those activities that are expected to change when producing, consuming, and disposing of a product. Therefore, data from marginal suppliers will better fit the objectives.

Allocation method, as another distinction between the attributional and consequential modeling, shall be chosen in the goal and scope definition phase. As discussed earlier, many product systems have multi-functional performance, so that, the environmental burdens need to be partitioned over the various functional flows. Because allocation is a challenging issue in each LCA study, ISO 14041 emphasizes that allocation should be avoided by either dividing the unit process into some sub-processes and collecting the input and output data related to these sub-processes, or expanding the product system in order to include the additional functions related to the co-products (ISO 14041 1998). To handle the allocation problem, the ISO hierarchy is as follows:

1. Subdivision of multi-functional processes
2. System expansion (with/without substitution)
3. Allocation.

In the subdivision methodology, the objective is to break down the targeted unit process with several outputs or functionality (Fig. 12.7a) into some single output

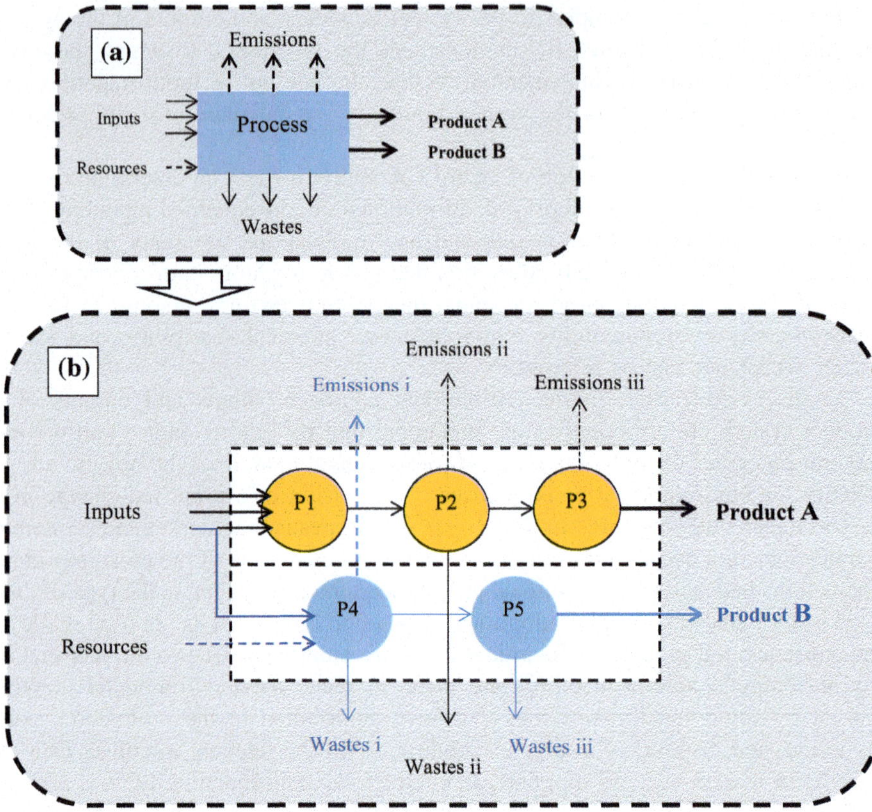

Fig. 12.7 Solving the multi-functionality problem by subdivision (Wolf et al. 2012)

unit processes (Fig. 12.7b). It is worth mentioning that subdivision is often but not always applicable. This approach can be considered if the separated unit processes will not be multi-functional.

If subdivision cannot be performed, system expansion is the best option to avoid the need for allocation. System expansion can be performed in two different ways. In the first approach, so called "system expansion with subtraction", the additional function(s) of the system is subtracted. Imagine system A has two co-products, e.g., wheat and straw, meaning that environmental burdens should be divided between these co-products. Straw is intended to be combusted and used to replace fossil-based heating. To perform "system expansion with subtraction", a substituted product must exist. In this case, the same amount of heat can be generated out of burning fossil based fuel. When the required heat is generated out of straw combustion, the associated emissions related to fossil-based heating system are avoided. Accordingly, the associated resource use and emissions of fossil-based heating system are also subtracted from system A as shown in Fig. 12.8. The second approach of system expansion is performed when among the two systems under

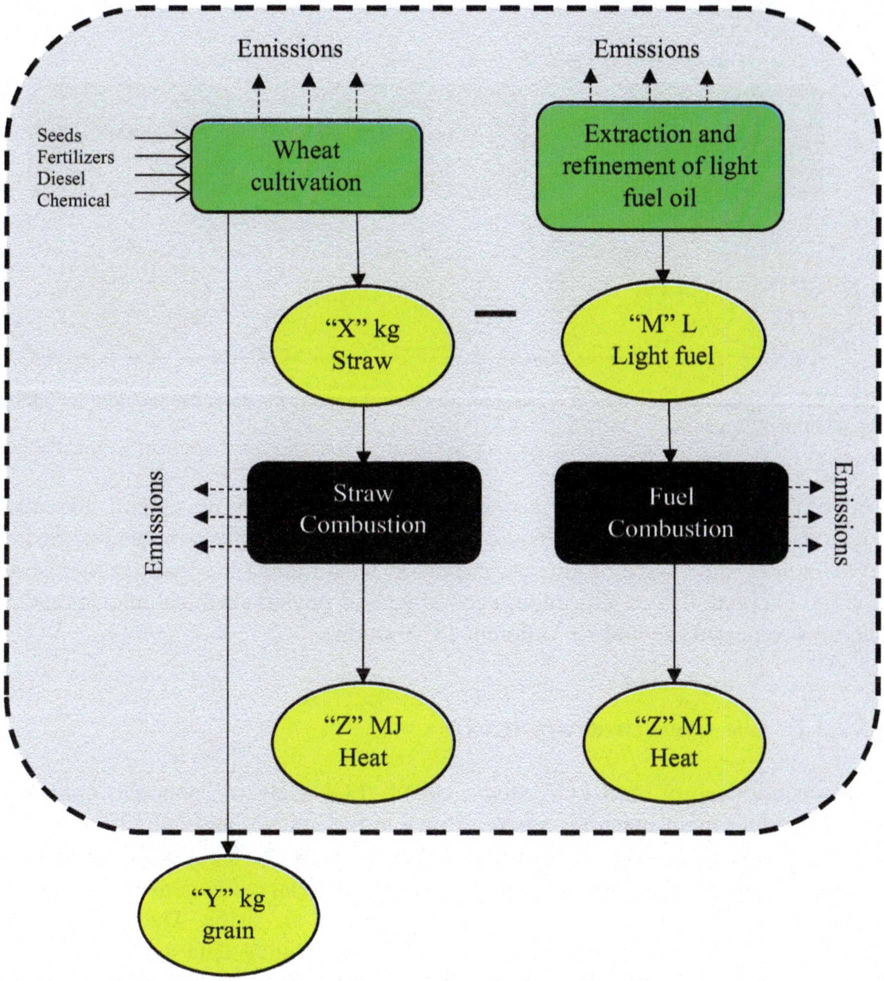

Fig. 12.8 Application of "system expansion with subtraction" to solve the multi-functionality problem by substituting the unnecessary co-functions

comparison, one has some additional functions. This would be done by expanding the system boundaries and adding missing functions and the inventories of the respective mono-functional products to the given case (Fig. 12.9).

Allocation is suggested to be performed as the last option for multi-functional systems. ISO 14041 states that "where allocation cannot be avoided, the inputs and outputs of the system should be partitioned between its different products or functions in a way which reflects the underlying physical relationships between them". Moreover, it continues that "where physical relationship alone cannot be established or used as the basis for allocation, the inputs should be allocated between the products and functions in a way which reflects other relationships

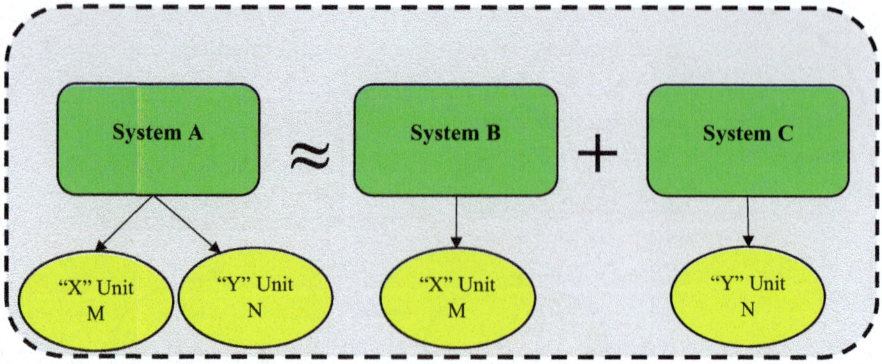

Fig. 12.9 Application of "system expansion" to solve the multi-functionality problem by adding the additional functions

between them, e.g., based on economic value". Accordingly, several allocation methods have been introduced and evaluated and their results were also compared to show how the choice of allocation method would affect the final results. Mass, energy, exergetic-based allocation, economic, and physio-chemical allocations are the most reported methods in different LCA studies.

12.2.3.2 Life Cycle Inventory (LCI)

The second stage of each LCA study, called "LCI analysis", concerns collecting data regarding input materials, energy flows to the system, and outputs as well as pollutants leaving the system boundary. In better words, the focus of LCI analysis is to quantify the inventory of the various flows of material extractions and substance emissions crossing the system boundary (Jolliet et al. 2015). The LCI analysis iteratively includes data collection and validation, relating data to unit processes and to the adopted functional unit, data aggregation, and refining the system boundaries. Due to the fact that data collection may span several reporting locations and published references, ISO 14041 suggests some several steps to ensure uniform and consistent understanding of the product systems will be modeled. These steps are as follows (ISO 14041 1998):

- Drawing of specific process flow diagrams that outline all unit processes to be modeled, including interrelationships;
- Description of each unit process in detail and listing data categories associated with each unit process;
- Development of a list that specifies the units of measurement;
- Description of data collection techniques and calculation techniques for each data category, to assist personnel at the reporting locations with understanding the type of information needed for the LCA study; and

- Provision of instructions to reporting locations to document clearly any special cases, irregularities or other items associated with the data provided.

Two approaches have been identified for LCI analysis, i.e., process-based and input-output approach. In the first approach, inventory is calculated by multiplying the reference flows and corresponding intermediary flows assessed per FU by the direct emission or extraction factors of each unit process. In the input-output approach, the inventory is calculated by using economic data in order to relate the direct demand of a good or service to the total demand within the entire economy. The inventory of emissions and extractions is then calculated by multiplying the total demand per FU by their corresponding emissions per dollar spent in each sector (Jolliet et al. 2015).

When collecting data, double counting should be avoided. Also, the time when data have been collected and the place where data have been obtained shall be documented. In this context, data aggregation over time and space is proposed to be performed to solve the problem arising from data collected from several locations at different time spans. As an example, imagine a situation in which the production process of product “X” emits 100 kg carbon dioxide into the air. This total emission may consist of 30 kg CO_2 emission in Denmark in 2012, 20 kg CO_2 emission in Netherlands in 2014, 40 kg in Germany in 2015, and 10 kg emission in Poland in 2017. Therefore, accurately accounting for the specific time and place of every emission necessitates providing an immense amount of data and calculations. To deal with this problem and to avoid huge deal of work, data aggregation over time can be considered. It means that we assume the effect caused by such an emission is time independent. Moreover, all emissions are aggregated regardless of the location at which the emission took place.

If data from the literature are used, the employed literature shall be referenced. The consistency through an LCI analysis must be kept. If the higher heating value, for instance, is used in a part of an LCI calculation, the same approach should be used for the other processes as well. In each iteration of inventory data collection, the validity of the collected data should be justified by establishing, for example, mass balances, energy balances and/or comparative analyses of emission factors (ISO 14041 1998).

12.2.3.3 Life Cycle Impact Assessment (LCIA)

Based on the ISO 14040, the LCIA is the third phase of an LCA study in which the inventory data are converted into some selected environmental issues, called impact categories, to better understand their environmental significance ISO 14042 (2000). This conversion is done by multiplying emissions by their corresponding predefined characterization factors. Characterization factors refer to some figures derived from characterization models which are applied to convert the assigned LCI results to the common unit of a given impact category. As depicted in Fig. 12.10, the LCIA phase is composed of some mandatory and optional elements.

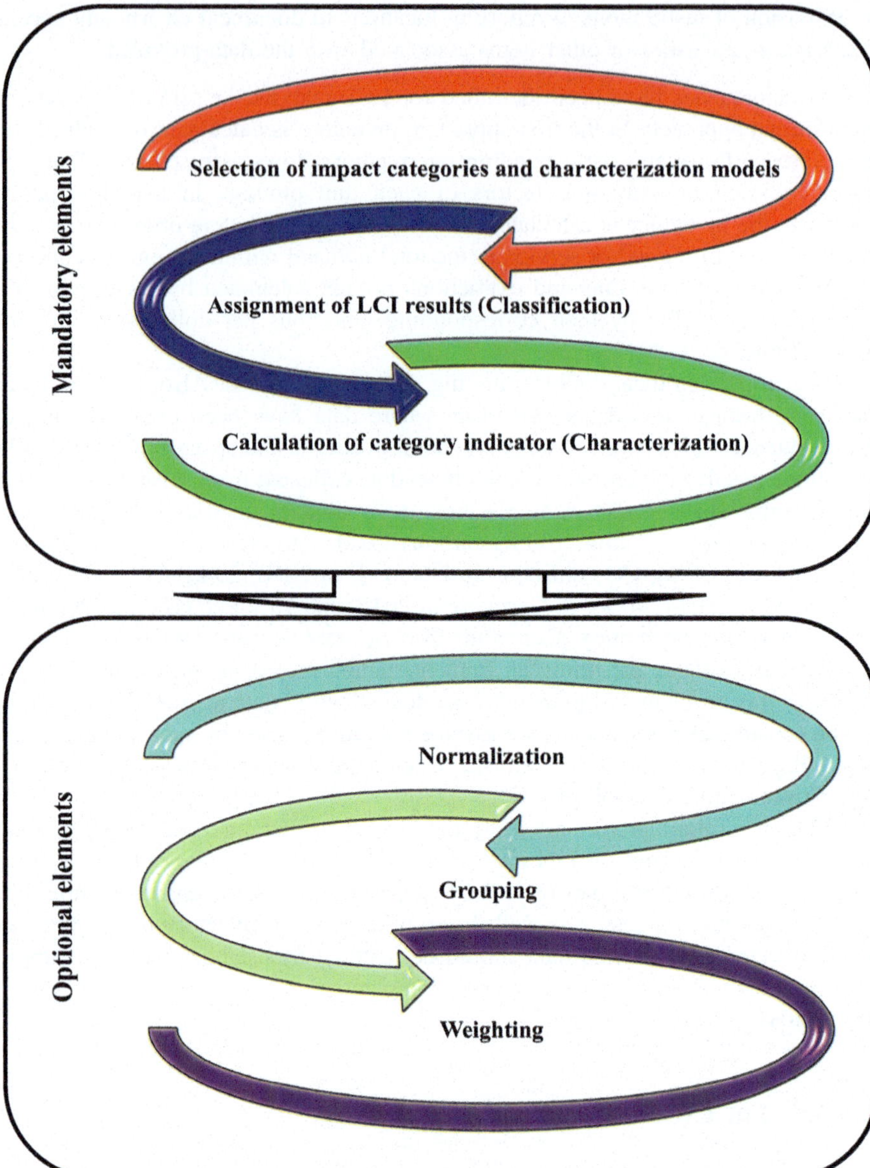

Fig. 12.10 Elements of the LCIA according to ISO 14042

To link LCI results to the impact categories, the inventory data having similar effects are aggregated into an intermediary level called midpoint categories or impact categories. The inventory data with similar effects are multiplied by their corresponding characterization factors to demonstrate their contributions to that specific impact category. A well-known impact category is global warming by which the impact of GHGs is attributed to warming the Earth. Each impact category or midpoint category affects one or more areas of protection. The areas of protection are also called damage categories such as ecosystem quality, human health, and climate change.

Each impact category has its own specific unit; therefore, the magnitude of each calculated indicator cannot be understood when it is compared with other indicators. Normalization can be used to present the magnitude of the category indicator results relative to some reference information. Moreover, identifying "important" impact categories, understanding the meaning of results by comparing with more familiar references or solving tradeoffs between the results are among the most important reasons for employing normalization (Pizzol et al. 2017). In normalization, the numerical values resulted from the characterization step are generally divided by some reference values. Several approaches have been introduced for this optional step. External normalization relates the results of an LCI analysis to an external database or normalization reference. By contrast, internal normalization utilizes values within the study and shows the relative significance of an impact with regards to the other competing alternatives (Curran 2017). For example, in the external approach, the total GHGs emissions of a specific product is related to the total regional GHGs emissions but in the internal method the significance of GHGs of product "A" are expressed relative to those of product "B" and "C". If the use of normalization is inevitable, the following recommendation shall be taken into account (Curran 2017):

- To document and justify the choice of any normalization references
- To communicate the results clearly by, e.g., reporting units and explaining their meaning
- To integrate uncertainty assessment
- To perform scenario analysis whenever possible e.g., by applying more than one method.

The other optional element of LCIA is grouping by which results are prioritized by sorting or ranking. Grouping can be carried out with two possible procedures as mentioned by ISO:

- Sorting by which the impact categories are sorted on a nominal basis
- Ranking by which the impact categories are ranked on ordinal basis or a given hierarchy.

Weighting is the process of converting indicator results of different impact categories to a single score. By this approach, the multiple scores of each scenario is converted to a single score by using numerical factors based on value-choices. The weighting methods are generally classified based on distance to target weighting approach, panel weighting approach, Monetary weighting approach, and binary weighting approach.

12.2.3.4 Sensitivity and Uncertainty Analysis

The intrinsic nature of LCA brings about a magnitude of uncertainties. According to ISO 14043, LCIs should also undergo sensitivity and uncertainty analysis to better understand their quality and limitations. Sensitivity analysis is a systematic approach to show how the choices made regarding methods and data affect the outcomes of a study. In better words, sensitivity analysis identifies the assumptions with significant influence on the final outcome of an LCI. The procedure includes comparison of the results obtained using certain given assumptions, methods or data with the new results obtained by variations in these parameters. The sensitivity can be performed over following variations (ISO 14043 2000);

- Allocation method;
- Cut-off criteria;
- Boundary setting and system definition;
- Judgements and assumptions concerning data;
- Selection of impact category;
- Assignment of inventory results (classification);
- Calculation of category indicator results (characterization);
- Normalized data;
- Weighted data;
- Weighting method;
- Data quality.

Ideally, all parameters contributing to an LCI analysis should be considered when sensitivity analysis is performed, but in practice, it rarely happens and the procedure is often limited to a number of selected parameters. One of the systematic methods for performing sensitivity analysis, developed by Heijungs et al. (1992), uses confidence limits for all input parameters and identifies those parameters for which the margins of uncertainty have large influences on the final results. Subsequently, the margins of uncertainty for these parameters can be improved.

Uncertainty analysis is a systematic procedure to quantify the uncertainty introduced in the results of a life cycle inventory analysis due to the cumulative effects of model imprecision, input uncertainty, and data variability. Uncertainty can be categorized as stochastic uncertainty, choice uncertainty, and knowledge-based uncertainty.

12.2.4 Special Types of LCA

Beside the conventional LCA discussed herein, there are new branches of this well-established and widely used environmental management tool. The new trends in LCA can be categorized as; Life cycle sustainability assessment, Life cycle costing, Carbon footprint, Water footprint, Social LCA, Eco-efficiency assessment, Input-output (IO) and hybrid LCA, Material flow analysis, and Resource efficiency assessment. Dealing with these types of LCA are beyond the scope of the current chapter and more information about these types of LCA can be found in (Finkbeiner 2016).

12.2.5 LCA in Waste Management

Waste management is a large and complex system and its complexity will also be extended when it is linked to other sectors such as energy supply chain, agriculture, and industry. In fact, waste management is a challenging issue because achieving a sustainable solution necessitates considering a multi-dimensional relationship, i.e., technological, economic, environmental, and social dimensions. In this context, LCA can contribute to addressing the concerns about sustainability of waste management strategies. In another word, LCA can be used as a decision support tool in waste management revealing the appropriateness of implemented waste management strategies.

12.2.5.1 System Boundary in Waste Management Strategies

The first step of a successful waste-LCA performed is to accurately define the goal and scope of the study including a proper system boundary, an appropriate functional unit, allocation method/system expansion, etc. This step begins with selecting the system boundary of the study. When the system boundary of waste management scenarios is defined, with respect to the goal of the study, the following activities along with the outputs of the system could be included within the system; collection, transportation, recovery and separation of materials, treating and disposal of waste, as well as energy and nutrient recovery. Regarding the capital and equipment required such as vehicles employed for waste collection and equipment used for composting or incinerating, their resource and energy consumption as well as their associated emissions are also included in LCI analysis but manufacturing, fabrication, construction, and maintenance are not taken into account. Moreover, the energy used in the office buildings where the operations are supervised can also be included in the scope of the study (Curran 2017).

While in product-LCA studies, inputs and outputs are followed from “cradle to grave”, in waste-LCA, not all inputs can be followed from the “cradle”. The waste materials shall be considered from the point when they are discarded as waste. This concept, called zero-burden, shall always be followed unless the scenario under consideration affects the amount of waste generation such as what happens in waste minimization strategies. Similarly, the end point of all recovered materials, energies, and nutrients cannot be set at “grave”, thereby the downstream processes of these materials are not included in the system boundary. This always applies for recycled materials which are used to replace their equivalent virgin materials. Therefore, the subsequent use of these materials is excluded from the system boundary.

12.2.5.2 Time Horizon in Waste LCA

Time horizon is also important in the LCA of waste management. The focus of most LCA studies is on annual waste generation and management. When Landfill is part of the scenario under evaluation, it must be noticed that emissions from landfill take place in a rather long time horizon. Under conditions laid forth, a time boundary of around 100 years can be assumed to estimate the landfill related emissions. Then, the estimated emissions can be attributed to the year of the study.

12.2.5.3 Functional Unit in Waste-LCA

The selection of FU in waste-LCA studies must be done in accordance with the goal of the study. In the LCA of waste management strategies, the objective is to find out which alternative is more environmentally beneficial. Therefore, the FU can be defined as the quantity of a given amount of waste which will be managed under a specific waste management strategy. If energy recovery from waste management options is targeted, the FU would be different. In this context, the FU would be defined, for example, as "X" MWh produced.

12.2.5.4 Open-Loop Versus Closed-Loop Recycling

When dealing with recycling strategies, distinctions should be made between two types of recycling. In closed-loop recycling, the recycled materials are sent back to the same production process. In this approach, the recovered materials go back into the production process of the same type of product, so that there are repeated recovery and reuse cycles. In the second approach, i.e., open-loop recycling, the recovered materials from waste stream are sent to a different type of product system. Making distinction between open-loop and closed-loop is sometimes misleading. As an example, the post-consumer plastic waste is incinerated to generate electricity. This recovered electricity may be used in the analyzed system but this electricity as the secondary product is completely different from the original material, i.e., polymer. While such a recycling system is sometimes mistakably interpreted as closed-loop, it must be categorized as open-loop recycling. Accordingly, two sub categories of open-loop recycling have been identified. The first sub-category is called "open-loop with same primary route" in which the recovered materials with no changes in their inherent properties are used in a different type of product system. The second sub-category is called "open-loop with different primary route" in which the inherent properties of the recycled materials are changed. Therefore, it would be better to use the function of the primary and secondary goods as a basis to find out if an open-loop or closed-loop modeling should be chosen.

12.2.5.5 Multi-functionality and Allocation in Waste-LCA

Regardless either open-loop recycling or closed-loop approach is considered for LCI analysis, the recycled materials will offset some similar virgin materials. It means that a new function has been provided by waste management system. Therefore, to have a fair assessment, the environmental burdens and the impacts caused by the waste management strategies should be partitioned between the two new functions. As discussed earlier, the system expansion should be given priority to allocation methods as instructed by ISO 14040 and ILCD Handbook.

System expansion is performed to take into account or subtract all associated burdens which are avoided due to replacing virgin materials by recovered ones within a production system. Despite being the first alternative and so straightforward, system expansion will not be feasible in some cases and for some specific product systems. For instance, performing system expansion in some cases requires excessive data collection from downstream processes outside the scope of the study. On the other hand, it is sometimes impossible to assess what impacts or burdens are actually avoided when products are recycled. This could be due to the fact that some recovered materials do not have the same quality compared with their virgin materials, and so, they will be used in another production process which requires a lower quality material. Under these circumstances, allocation can be performed.

To correctly perform the allocation in recycling systems, the following question should be answered: Does the material under consideration have any market values at the end of its life? This question shall be answered because market value above zero and below zero need to be differentiated when dealing with allocation problem.

If the material has a market value above zero at the end of its life, it would be called co-product rather than waste. This co-product can further be recycled, as expected, turning into a secondary good. This secondary good at the end of its life can also be recycled again and again; hence, many co-products will be available and all should be considered when the inventory for the second goods is identified. A three step procedure has been recommended by the ILCD handbook and will be briefly explained herein. At the first step, the total amount of uses is determined. The total amount of use refers to the sum of the amount of primary use plus the amount obtained after each recycling round. For instance, if 1 kg of material "X" is recycled with 90% recycling rate at the end of its life, it means that 0.9 kg secondary material is recovered. This amount of secondary material can also be incorporated into a new product and be recycled at the same recycling rate resulting in 0.81 kg recovered end-of-life product. Therefore, the total amount of uses from the primary materials is calculated as 1 kg + 0.9 kg + 0.81 kg, etc. At the second step, the total LCI associated with the total amount of uses must be estimated. This includes the LCI of primary production plus the LCI of effort for reuse/recycling/recovery, plus the LCI of final waste management but the processes related to the manufacture and use of the products made from these materials are not included. At the final stage, the average inventory per unit is calculated by dividing the total life cycle inventory of the total amount of use by the total amount of uses.

When the end-of-life product has a negative market value, it would be called waste rather than co-product. However, there are two types of cases to be differentiated:

(1) No valuable product is produced during waste treatment, thereby the inventory is fully attributed to the first system that has generated the waste.
(2) A valuable product is produced during waste treatment. This valuable product, called "secondary good" would be a co-product for the first system so that allocation is to be applied. To perform allocation under such a circumstance, all treatment processes that are necessary until the treated end-of-life product achieves a market value of zero are assigned to the first system.

12.2.5.6 Using LCA Results for Making Decisions

When LCA results are obtained, they can be combined with other aspects of sustainability issues including technical, economic, and social information to reach the best decision. Key issues which must be taken into account include cost, accessibility to new facilities, technical feasibility, environmental performance, market behavior about recovered energy and materials, and public acceptance (Weitz 2012). These results are further implicitly or explicitly combined to fairly compare distinctive integrated waste management strategies and wisely choose the best alternative.

A large number of studies performed to compare different integrated waste management strategies, unanimously argued that landfill disposal is not the preferred option. While recycling could be the most favorable waste management strategy, it is not possible to recycle some waste fractions due to their low quality or high level of contaminations. Accordingly, integrated waste management strategies consisting of recycling and waste to energy options would be the most environmentally friendly and cost effective alternative. Briefly, recycling is given priority as far as recycled materials can replace an equivalent amount of virgin materials. The heating value of non-recycled materials, the environmental performance, and the efficiency of energy recovery shall be considered when deciding over the use of waste to energy option. Materials which can be biologically treating are highly recommended to undergo anaerobic digestion process with the aim of energy and nutrient recovery.

12.3 Conclusions

With the increased awareness on waste-related environmental issues, LCA has been increasingly employed as a decision support tool to help achieve the best integrated waste management strategies. Despite its popularity in waste management studies, faulty assumptions and incorrect methodologies can still be found in performing waste LCA studies leading to scientific errors and erroneous conclusions. A well performed LCA study begins with a clear goal definition covering the intended applications, the limitations and restrictions, the drivers and motives, the target audience, the potential disclosure to the public, and the commissioner of the study. Following the goal definition, the context of the study should be also clearly identified due to the fact that the frame work of the LCI analysis depends on this selection. The selection of FU in waste-LCA studies must be done in accordance with the goal of the study. Unitary-based, generation-based, input-based, and output-based FU are commonly used in waste-LCA studies. To perform a comprehensive LCA study, all relevant impact categories must be evaluated before claiming the environmental superiority of one alternative. Multi-functionality and decision over allocation methods are also important issues in this context. While system expansion would be the best option, it is not always possible to solve the multi-functionality problem by system expansion. Allocation methods, especially when dealing with recycling waste management strategies, must be wisely performed considering if either an open-loop or closed-loop system is under consideration. Eventually, it must be kept in mind that without considering technical, economic, and social information, the most sustainable waste management strategy would never be achieved.

References

Ahbe S, Braunschweig A, Müller-Wenk R (1990) Methodik für Oekobilianzen auf der Basis Ükologischer Optimierung, Schriftenreihe Umwelt No. 133, Bundesamt für Umwelt. Wald und Landschaft (BUWAL), Bern, Switzerland

Bisinella V, Götze R, Conradsen K, Damgaard A, Christensen TH, Astrup TF (2017) Importance of waste composition for life cycle assessment of waste management solutions. J Clean Prod 164:1180–1191

Cheremisinoff NP (2003) Handbook of solid waste management and waste minimization technologies. Butterworth-Heinemann

Christensen TH (2011) Introduction to waste management. Solid Waste Technol Manage 1&2: 1–16

Christiansen K, Hoffman L, Virtanen Y, Juntilla V, Rønning A, Ekvall T, Finnveden G (1995) Nordic guidelines on life-cycle assessment. Nordic Council of Ministers

Consoli F (1993) Guidelines for life-cycle assessment. In: A code of practice

Curran MA (2017) Overview of goal and scope definition in life cycle assessment. In: Goal and scope definition in life cycle assessment. Springer, pp 1–62

Ekvall T, Azapagic A, Finnveden G, Rydberg T, Weidema BP, Zamagni A (2016) Attributional and consequential LCA in the ILCD handbook. Intern J Life Cycle Assess 21(3):293–296

Finkbeiner M (2016) Special types of life cycle assessment. Springer

Goedkoop M, Heijungs R, Huijbregts M, De Schryver A, Struijs J, Van Zelm R (2009) ReCiPe 2008. In: A life cycle impact assessment method which comprises harmonised category indicators at the midpoint and the endpoint level, p 1

Goedkoop M, Spriensma R (2000) The Eco-indicator 99: a damage oriented method for life cycle assessment, methodology report. PreConsultans BV, The Netherlands

Guinée J, Gorree M, Heijungs R, Huppes G, Kleijn R, Udo de Haes H, Van der Voet E, Wrisberg M (2002) Life cycle assessment. In: An operational guide to ISO standards, vol 1, 2, 3. Centre of Environmental Science, Leiden University (CML), The Netherlands
Hauschild MZ, Huijbregts MA (2015) Introducing life cycle impact assessment. In: Life cycle impact assessment. Springer, pp 1–16
Heijungs R, Guinée JB (2007) Allocation and 'what-if'scenarios in life cycle assessment of waste management systems. Waste Manag 27(8):997–1005
Heijungs R, Guinée JB, Huppes G, Lankreijer RM, Udo de Haes HA, Wegener Sleeswijk A, Ansems A, Eggels P, Duin RV, De Goede H (1992) Environmental life cycle assessment of products: guide and backgrounds (part 1)
Heijungs R, Wiloso E, Wang L (2014) Life cycle assessment of bioenergy systems. Sustain Bioenergy Product
Hoornweg D, Bhada-Tata P (2012) What a waste: a global review of solid waste management
Horne R (2009) Life cycle assessment: origins, principles and context. CSIRO Publishing, Melbourne
Hunt RG (1974) Resource and environmental profile analysis of nine beverage container alternatives
ISO 14041 (1998) Environmental management-life cycle assessment-goal and scope definition and inventory analysis. International Organization for Standardization
ISO 14042 (2000) Environmental management—life cycle assessment—life cycle impact assessment. Geneva, Switzerland
ISO 14043 (2000) Environmental management—life cycle assessment—life cycle interpretation, p 18
Jolliet O, Saadé-Sbeih M, Shaked S, Jolliet A, Crettaz P (2015) Environmental life cycle assessment. CRC Press
Klöpffer W (2014) Introducing life cycle assessment and its presentation in 'LCA Compendium'. In: Background and future prospects in life cycle assessment. Springer, pp 1–37
Laurent A, Bakas I, Clavreul J, Bernstad A, Niero M, Gentil E, Hauschild MZ, Christensen TH (2014a) Review of LCA studies of solid waste management systems—part I: lessons learned and perspectives. Waste Manag 34(3):573–588
Laurent A, Clavreul J, Bernstad A, Bakas I, Niero M, Gentil E, Christensen TH, Hauschild MZ (2014b) Review of LCA studies of solid waste management systems—part II: methodological guidance for a better practice. Waste Manag 34(3):589–606
McDougall FR, White PR, Franke M, Hindle P (2008) Integrated solid waste management: a life cycle inventory. Wiley
Mercante IT, Bovea MD, Ibáñez-Forés V, Arena AP (2012) Life cycle assessment of construction and demolition waste management systems: a Spanish case study. Intern J Life Cycle Assess 17(2):232–241
Novick D (1959) The federal budget as an indicator of government intentions and the implications of intentions. Defense Technical Information Center
Pharino C (2017) Challenges for sustainable solid waste management: lessons from Thailand. Springer
Pizzol M, Laurent A, Sala S, Weidema B, Verones F, Koffler C (2017) Normalisation and weighting in life cycle assessment: quo vadis? Intern J Life Cycle Assess 22(6):853–866
Plevin RJ, Delucchi MA, Creutzig F (2014) Using attributional life cycle assessment to estimate climate-change mitigation benefits misleads policy makers. J Ind Ecol 18(1):73–83
Rajaeifar MA, Ghanavati H, Dashti BB, Heijungs R, Aghbashlo M, Tabatabaei M (2017) Electricity generation and GHG emission reduction potentials through different municipal solid waste management technologies: a comparative review. Renew Sustain Energy Rev 79: 414–439
Rajaeifar MA, Tabatabaei M, Ghanavati H (2015a) Data supporting the comparative life cycle assessment of different municipal solid waste management scenarios. Data Brief 3:189–194

Rajaeifar MA, Tabatabaei M, Ghanavati H, Khoshnevisan B, Rafiee S (2015b) Comparative life cycle assessment of different municipal solid waste management scenarios in Iran. Renew Sustain Energy Rev 51:886–898
Saner D, Walser T, Vadenbo CO (2012) End-of-life and waste management in life cycle assessment—Zurich, 6 December 2011. Intern J Life Cycle Assess 17(4):504–510
Van de Klundert A, Anschütz J, Scheinberg A (2001) Integrated sustainable waste management: the concept. Tools for decision-makers experiences from the urban waste expertise programme (1995–2001). WASTE
Weitz KA (2012) Life cycle assessment and end of life materials management. In: Life cycle assessment handbook: a guide for environmentally sustainable products, pp 249–266
Wentzel H, Hauschild M, Alting L (1997) Environmental assessment of products, vol 1. Chapman & Hall, London
Wolf MA, Pant R, Chomkhamsri K, Sala S, Pennington D (2012) The International reference life cycle data system (ILCD) handbook-JRC reference reports
Yıldız-Geyhan E, Altun-Çiftçioğlu GA, Kadırgan MAN (2017) Social life cycle assessment of different packaging waste collection system. Resour Conserv Recycl 124:1–12

Chapter 13
Techno-Economic Aspects of Biogas Plants

Marzieh Shafiei

13.1 Introduction

Production of biomass-derived transportation fuels has been promoted to reduce the negative environmental and social effects of fossil fuels. Advanced biofuels have several definitions most of which are focused on the production of biofuels from non-food based materials, e.g., lignocelluloses, municipal solid waste or industrial wastes, which have low CO_2 emissions, high GHG reduction, and using minimum (or zero) indirect land use change (ILUC) (Balan et al. 2013). The U.S. Energy Information Administration (EIA) has predicted a 48% increase in the global energy demand from 2012 to 2040 if the current energy policies do not change (EIA 2016). Limited fossil fuels resources and their environmental drawbacks, e.g., the greenhouse effect and climate change, have emerged using renewable fuels (Igliński et al. 2015). Biogas has a significant contribution among the future been highlighted as the main drivers of renewable energies (Mao et al. 2015; Pike Research 2012). The target of the European renewable energy directive is to replace 27–30% of the total energy consumption with renewable sources by 2030. It is estimated that 14–26% of this renewable energy target can be met by biogas from farming and forestry residues (Holm-Nielsen et al. 2009; Meyer et al. 2017). Traditionally, biogas is produced from excess sludge in wastewater treatment plants or in the landfills from municipal solid waste. Other potential feedstock includes agricultural wastes, e.g., animal manure; food industry wastes, e.g., wastes from slaughterhouses, sugar refining, alcohol generation; or industrial wastes, e.g., wastes from biotechnological industries and paper manufacturing (Werner et al. 1989). A research by the US Energy Information Administration (EIA) revealed that if all of these types of raw materials, except lignocellulosic materials and lipids, are utilized in the US, the

M. Shafiei (✉)
Department of Chemical Engineering, Faculty of Engineering,
University of Isfahan, 81746-73441 Isfahan, Iran
e-mail: m.shafiei@eng.ui.ac.ir

M. Tabatabaei and H. Ghanavati (eds.), *Biogas*, Biofuel and Biorefinery Technologies 6, https://doi.org/10.1007/978-3-319-77335-3_13

produced amount of biogas would be about 420 billion (11.9×10^9 m^3) of methane. However, this amount of methane could only replace 5% of the natural gas used for electricity generation and cover 56% of the natural gas used in the transportation sector (EIA 2013; NREL 2013). Moreover, this report also reveals that if lignocellulosic materials are used, the produced methane would cover 46% of the natural gas demand in the electricity generation sector and all of the natural gas used for transportation.

Biogas is currently produced and used in Europe. In 2007, Germany was the largest biogas producer in Europe mainly from energy crops, while the UK was the second producer of biogas mainly from landfill sources (AEBIOM, 2009). Nevertheless, the main raw material for biogas production in Germany has been energy crops which has environmental and economical drawbacks (Meyer et al. 2017; Poeschl et al. 2010). Furthermore, environmental policies discourage waste landfilling and it is projected to be decreased due to its environmental drawbacks (Scharff 2014; European Commission 2016; Li et al. 2015; Kormi et al. 2017). Therefore, in order to meet the high energy demands of the future, use of more sustainable methods and feedstocks are recommended. Lignocellulosic agricultural residues, e.g., rice and wheat straw, bagasse, and corn stalks are the potential recommended feedstocks for the future. The co-digestion of these feedstocks with animal manure is recommended to reduce the demand for additional fertilizers (Meyer et al. 2017; NREL 2013).

Lignocellulosic feedstocks have a highly crystalline and recalcitrant structure. This structure resists to microbial attacks and diminishes the biogas yield (Karimi et al. 2013; Shafiei et al. 2015). Among the four steps of anaerobic digestion, i.e., hydrolysis, acidogenesis, acetogenesis, and methanogenesis, hydrolysis is the rate-limiting step when biogas is produced from lignocelluloses (Taherzadeh and Karimi 2008; Kabir et al. 2015). The biogas production process from these feedstocks requires an extra pretreatment step which is cost and energy intensive. In spite of high potential for biogas production from lignocellulosic materials, due to technological and economic problems, the industrial production of biogas from sole lignocellulosic materials has not been practiced yet (Shafiei et al. 2013; Sárvári Horváth et al. 2016). Biogas is currently produced in small farm scale or large industrial scale. The volume of digesters may vary from a few hundreds to thousands of cubic meters (Mussoline et al. 2012; Li et al. 2015). However, larger plants are economically more favorable both for the biogas production and for upgrading (Turton et al. 2009; Jalalzadeh-Azar 2010; Sun et al. 2015; Yliopisto 2009; Patterson et al. 2011; Persson 2003). The necessity of considering an economically viable scale is more highlighted when lignocellulosic materials are considered for biofuel production. However, before any kinds of industrial construction and investment, the economic profitability of the plant should be studied. Since biofuel production from lignocelluloses is a developing technology and has not been proved yet, this profitability analysis should be performed after the development and design of these new processes. Techno-economic analysis is a combination of these studies (Lauer 2017; Humbird 2011; Swanson 2010; Tao 2014; McAloon 2000).

A techno-economic analysis is a method for comparison of different process designs, to find the main bottlenecks of the processes, to direct the research studies towards removing these bottlenecks, and to estimate the absolute manufacturing cost of the lignocellulosic biofuels compared with those of the first generation biofuels and fossil fuels (Humbird 2011; Aden and Foust 2009; Aden 2002; Dutta et al. 2010; Eggeman and Elander 2005; Hamelinck et al. 2005; Humbird and Aden 2009; Iranmahboob et al. 2002; Kazi 2010; Lynd 1996; Klein-Marcuschamer et al. 2010; Piccolo and Bezzo 2009; Sassner et al. 2008; Sendich et al. 2008; von Sivers and Zacchi 1996; Wooley et al. 1999a; Yu et al. 2008). Other typical results of techno-economic analysis are estimation of total capital investment (TCI); manufacturing costs and minimum selling price of the products, and comparison of the economic profitability of different processes (McAloon 2000; Tao et al. 2013).

13.2 Techno-Economic Analysis: Definition and Methods

Although biofuels still require financial supports from the governments, it is predicted that their prices will be able to compete with those of the fossil fuels in the future. This is said to be because of "learning effects, large scale operations, and efficient system integrations". (Balan et al. 2013). Techno-economic analysis is a technique for further boosting of this commercialization. When a techno-economic assessment for a biofuel plant is considered, many types of feedstocks, products, and process configurations are possible. Detailed design for all of these possible configurations is a time consuming process and not applicable. Thus, it is very important to address the most important factors and choose those processes which seem to be the most economically feasible ones. In fact, techno-economic analysis enables the evaluation and comparison of the newly developed research findings, performed in the lab and pilot scales. Furthermore, the manufacturing cost of the biofuel produced by these new processes can be estimated and compared with those of conventional available biofuels, as well as gasoline and natural gas (Kazi 2010; Anex et al. 2010; Brown 2014, 2015; Chovau et al. 2013; von Sivers and Zacchi 1995; Wooley et al. 1999b).

Techno-economic analysis includes conceptual development of process flow diagrams (PFDs) and economic assessment of these processes. The economic assessment for biofuel production plants is basically similar to the established methods in the chemical and petrochemical industries. There are several criteria for comparison of the economic profitability of plants which will be discussed later in this section (Tao et al. 2013; Chovau et al. 2013; Tan et al. 2016).

13.2.1 General Design Considerations

Although techno-economic analysis involves design and economic assessment of a process, there are others factors which could affect the overall viability of the process. These factors include environmental, health and safety considerations, structural design, utilities, patents and legal restrictions, materials handling, and storage. Furthermore, the plant location, capacity, operation and control, and layout are also factors which should be considered (Turton et al. 2009; Peters et al. 2003). In this regard, the most important factors are briefly discussed here.

13.2.1.1 Health, Safety and Environmental Considerations

When a plant is designed, the potential hazards and dangers for both the labor and plant equipment should be considered. In chemical processes, usually severe process conditions, e.g., pressure higher than 5 bar and temperature more than 700 °C, or hazardous chemicals dictate the considerations concerning costly equipment (Peters et al. 2003; Douglas 1988). Furthermore, prevention of leakages of these hazardous liquids or gasses imposes additional investment expenses. Safety hazards are highly toxic substances with immediate negative health impacts while "industrial health and hygiene hazard" are those chemicals that long term exposure to them at low concentrations causes injuries. When designing a plant, the usage of these types of materials should be avoided as much as possible (Peters et al. 2003). Fortunately, production of biogas and other types of biofuels, involves moderate process conditions and less toxic or explosive materials compared with many other chemical or petrochemical industries (Taherzadeh and Karimi 2007, 2008; Karimi et al. 2013; Smith 2005). However, biogas production plants also face their typical health and safety issues.

The most important safety issues are explosion, hydrogen sulfide poisoning, asphyxiation, and disease. Dilution of biogas with 10–30% air produces an explosive gas. Around the biogas digesters, gas pipes, combined heat and power (CHP) units, gas flares, and gas storage tanks, collectively called Ex-Zones, all types of safety measures related to explosion should be considered including the installation and usage of acceptable devices. Accumulation of gasses at the feedstock storage area is a common safety hazard. Hydrogen sulfide, which is heavier than air at concentrations higher than 700 ppm could cause immediate death. Furthermore, accumulated biogas, carbon dioxide, and ammonia could cause unconsciousness and suffocation. Thus, considering ventilation equipment for closed storage areas and work places is necessary (Westenbroek and Martin 2017).

Possible infections by bacteria, viruses, or parasites should be avoided by addition of pre-sanitation equipment for specific types of feedstock. These feedstock and their required sanitation facilities are specified by the European Regulations (EC 1774/2002). The pre-sanitation involves pasteurization, i.e., maintaining materials at 70 °C for 1 h, or pressure sterilization, i.e., minimum temperature of 133 °C for 20 min (Findeisen 2015; Maciejczyk 2014). Regulations for "Toxic Hazardous Substances" and "Occupational Noise Exposure" are the two most common regulations which must be followed in the original design of a biogas plant (Peters et al. 2003).

There are general environmental protection rules about waste disposal and hydrocarbon emissions. Examples of these rules are available in the US federal environmental regulations (Peters et al. 2003). Methane is considered 20-folds a more effective greenhouse gas compared with carbon dioxide. Thus, special equipment and consideration would be required for minimizing emissions of this gas from biogas plants (Cakir and Stenstrom 2005; Börjesson and Mattiasson 2008). There are also regulations for the disposal of the wastewater and digested solids which must be considered in the process design.

13.2.1.2 Plant Location

Selection of the plant location mainly depends on the availability if the raw materials and the possible market for the products. Other most important factors are the availability of energy and water supplies, transportation facilities, as well as taxation and legal restrictions. These factors directly affect the results of a techno-economic analysis and thus, a preliminary estimation for plant location is necessary in order to minimize the costs associated with energy, water, and local income taxes. However, plant location is the most effective factor for the calculation of the TCI. The cost of construction materials, wage rates, and taxes may vary significantly based on the country. For instance, the TCI for a plant in the Middle East may be as low as 80% of the TCI for a plant in the US. There are other factors which could be important before and upon the completion of the "detailed estimate design" stage of the plant. These factors such as climate, waste disposal, labor supply, flood and fire protection, site characteristics, and community factors are not considered in the techno-economic analysis in the "study estimate" level of design (Peters et al. 2003; Smith 2005; Shafiei et al. 2014).

13.2.1.3 Plant Capacity

Plant capacity is an important factor which could affect the process economy. Techno-economic studies in general try to target economic scales usually leading to the design of large capacity plants (Anex et al. 2010). Typical plant capacities which were studied may vary from 50,000 to 600,000 ton biomass/year (Balan et al. 2013; Sassner et al. 2008; Anex et al. 2010; Brown 2015; Shafiei et al. 2011; Galbe et al. 2011). The possibility of the application of such large capacities depends on the availability of raw materials and the associated transportation costs, availability of the required capital investment, and availability of the product market (Pike Research 2012; Peters et al. 2003). For instance, when biogas is to be produced from municipal solid waste, maximum plant capacity would be the amount of waste which is produced in a specific city. Transportation of raw materials from other cities to the plant site involves additional costs and the economic feasibility for such transportation should be considered first.

13.2.1.4 Plant Operation and Control

Instrumentation and maintenance are two important factors in plant operation and control. Typical instrumentations in biogas plants involve instruments for online measurement of flow, level, temperature, pressure, pH, and redox potential (Wiese and König 2017). For process control in techno-economic analysis, digital, analogue or programmable logic controller (PLC) systems can be chosen. In large chemical industries, centralized automatic control facilities are used. These facilities, including computers are usually located in rooms which are separated from the rest of the compound for more labor safety. In small biogas plants, a few control boards and panels are often situated in a safe place inside the plant. The instrumentation charge is typically 2–8% of the fixed capital investment of the plant. Maintenance expenses are regarded as operating costs and may vary from 2 to 11% of the annual value of the fixed capital investment. This value is at maximum 6% for simple chemical processes while for a process with normal operating conditions, it would increase to a maximum of 9% (Peters et al. 2003).

13.2.2 Process Development and Simulation

The first step for a techno-economic analysis is the conceptual design of the processes involved in biofuel production, i.e., determination of the required equipment, their sequences, and many other details about the processes involved (Fig. 13.1) (Tao et al. 2013). These data will be summarized in a PFD. The next step is the simulation of the corresponding models in a simulation software, e.g., Aspen Plus. To do this, numerous types of data is required, and the most important ones are chemistry and kinetics of the main and side reactions. Typically, the reaction rates or yields/productivities are available based on experimental lab- or pilot-scale studies. Other necessary data are physical, chemical, and thermochemical properties of all of the important substances in the processes, e.g., molecular weight, boiling

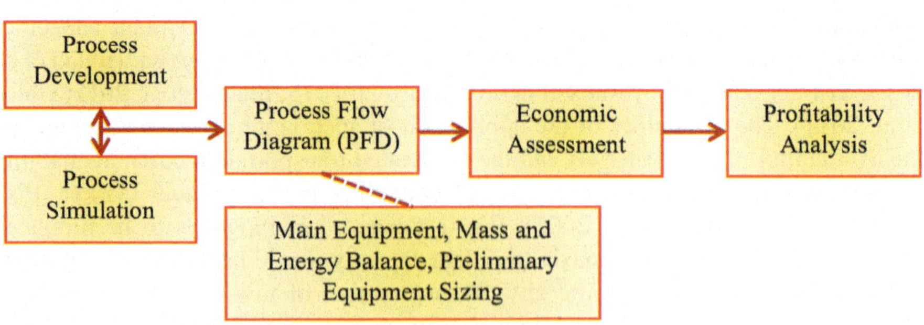

Fig. 13.1 General procedure for techno-economic analysis

point, triple point properties, vapor pressure, and viscosity. Additionally, a simulation is reliable only if the process models are as correct as possible. For instance, Aspen Plus® has many pre-built process models for reactors, heat exchanger, and distillation columns. These models have been proven by several expert engineers and companies. (AspenTech 2017). The results of process simulation are mass and energy balances for the main process equipment.

13.2.3 Total Capital Investment (TCI) and Operating Cost Estimation

The TCI is estimated based on the costs of purchased equipment. After simulation is completed, the next step is the sizing or selection of equipment based on design rules of chemical engineering. For instance, heat exchangers, towers, and reactors should be sized while pumps and compressors should be selected from the available types and sizes. Upon completion of this step, the expenses for purchased equipment are calculated using available databases, e.g., available literature or software. For instance, Aspen Process Economic Analyzer has a wide and reliable library for prices of most of the equipment used in chemical processes. The next steps involve calculation of the fixed and working capitals and consequently TCI for the plant. These investment costs together with operating costs are defined based on recommended factors which should be multiplied by the total value for purchased equipment (Turton et al. 2009; Peters et al. 2003) (Fig. 13.2).

TCI for a plant involves fixed and working capital investments. The fixed capital investment includes direct and indirect costs. The direct costs represent the expenditures required for purchasing of the main process equipment. Furthermore, they also include the expenses for installation of the equipment, purchase and installation of the instrumentations and controls for the equipment, piping, electricals, all types of buildings, service facilities, yard improvements, and land. The indirect costs include construction expenses, engineering and supervision, contingencies, and contractors' fees. The above charges are estimated as recommended percentages of the purchased equipment cost (Fig. 13.2). If these values are used for the estimation, the ranges of uncertainties would increase up to 25–30%, however, the cost estimation would be much faster and without requiring site-specific data (Turton et al. 2009; Tao et al. 2013; Peters et al. 2003).

Working capital involves the credit for 1 month of raw materials, supplies, finished and stored products, and semi-finished products in the process. Furthermore, it also includes 1 month cash for the payable operating expenses, taxes, and accounts (Turton et al. 2009; Peters et al. 2003).

Total product cost is estimated based on TCI and other expenses of the project. These costs involve manufacturing costs and the general expenses. Manufacturing costs include fixed charges, direct production costs, and plant overheads. The direct manufacturing costs include the costs for raw materials, utilities, operating labor

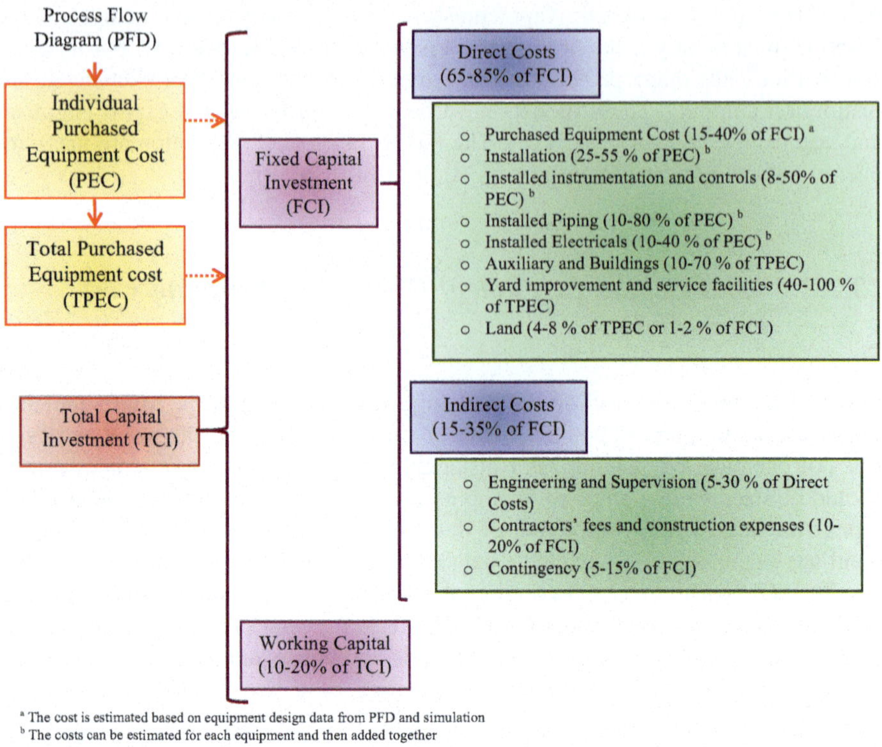

Fig. 13.2 Recommended values for calculation of Total Capital Investment (TCI) (Turton et al. 2009; Peters et al. 2003)

and direct supervisory, operating charges, maintenance, and repairs, as well as patents and royalties. The fixed charges include income taxes, rent (if the cost for land is not considered as a part of TCI), insurance, and depreciation. General expenses include the costs for research and development, financing and administrative costs, and the costs for distribution and selling of the product. The amount of raw materials and utilities are determined based on the simulation and experimental data. Depending on the process type, i.e., batch or continuous, the number of operating labors required for each major equipment is estimated and the labor cost is predicted by multiplying the total number of labor by the labor wage rate. The rest of these expenses are mainly estimated based on recommended percentages of total or fixed capital investment. These recommended percentages may vary slightly depending on the plant type. However, for the preliminary estimations, it is possible to use the recommended values which are available in the literature (Turton et al. 2009; Peters et al. 2003) (Fig. 13.3).

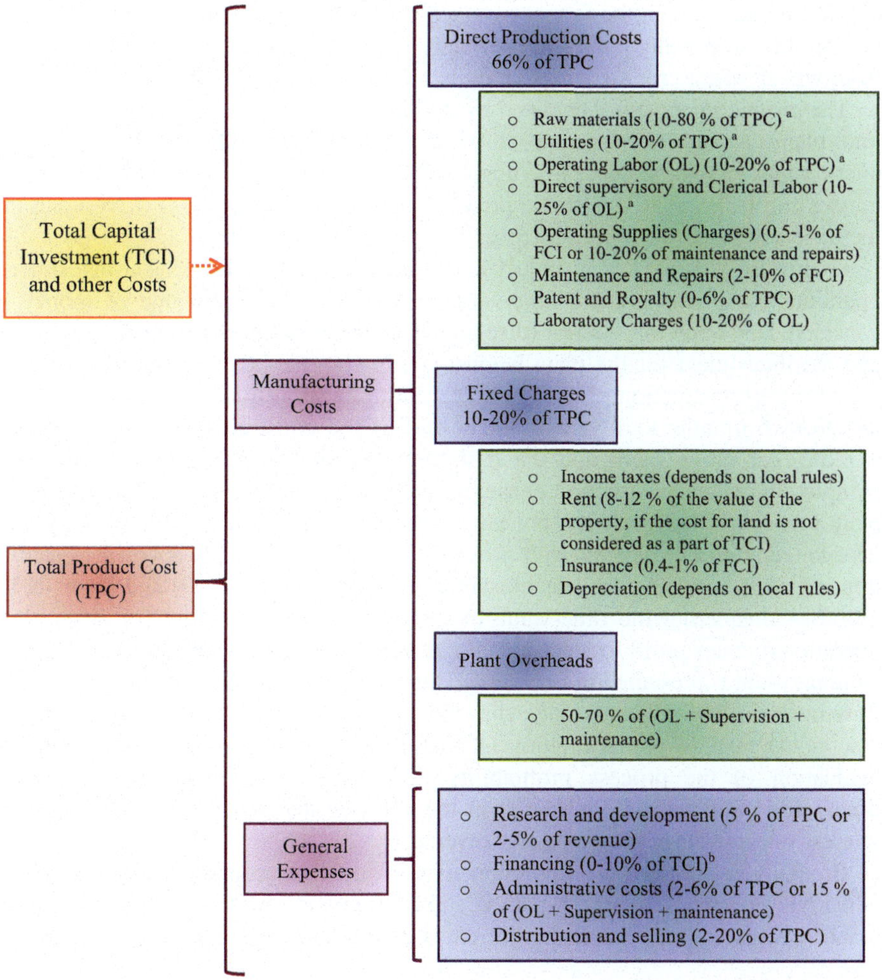

[a] These costs can be estimated directly using values obtained from simulation, e.g., mass and energy balance and unit costs, e.g., for unit price of materials or utilities.

[b] The financing cost is calculated as the sum which should be paid back if the capital is borrowed. Thus, it would include the amounts of borrowed money and the related interest. If the capital is 100% equity financed from the existing founds of the company, this parameter may be excluded or reduced to the interest rate of the invested money.

Fig. 13.3 Recommended values for the calculation of Total Product Cost (TPC) (Turton et al. 2009; Peters et al. 2003)

13.2.4 Profitability Analysis

After individual design and economic analysis for each process, the economic profitability of the processes involved should be compared. There are two approaches for designing of biofuel production plants. The lignocellulosic biofuel plant can be added to an existing plant or built as "grass root and clear field" or "green field" plants. For example, in the former case, a plant for biogas production from straw can be added to an existing plant from sewage sludge. In the latter case, the biogas plant from whole wheat crop or wheat straw is built from the beginning.

The production cost is calculated by dividing the total product cost (TPC) by the total plant capacity (Turton et al. 2009; Peters et al. 2003). For instance, the production cost of biomethane, i.e., upgraded and pressurized biogas, from lignocelluloses can be compared with the prices of compressed natural gas (CNG) and the methane produced by other processes.

Cash flow diagram for a project is depicted based on total project investment, operating expenses, and the time-value of money. The time-value of money is considered by applying interest (discount) rate. At zero time, the land is purchased and the investment for the plant begins. Gradually, the fixed capital investment is used for the construction of the plant, i.e., during the construction phase which may last from 6 months to 3 years. At the end of the construction phase, fixed and working capital is spent and the cash flow position is at minimum. Upon the completion of the construction phase, it may take several months to 2 years for the plant to work in full capacity. For new processes, the start up to full-capacity may take longer times compared with proven processes. The details about the calculations of cash flow diagram was presented by Turton et al. (2009) and Peters et al. (2003). Considering the time-value of money, any delays before full-scale plant operation dictates profit losses and is not desirable. A sample cash flow diagram for a biogas plant is presented in Fig. 13.4. The concepts of discounted cumulative cash flow position (net present value), discounted payback period (payout period), and discounted cash flow rate of return (DCFROR) are common parameters used for the evaluation of the process profitability. The term "discounted" is used when time-value of money is considered in the calculations through the application of interest rate for all of the costs and revenues.

Break-even point is an expression used in different conditions. In a situation when Net Present Value (NPV) at the end of a project equals zero, a break-even is reached. This indicates that the current assumptions for the project will lead to

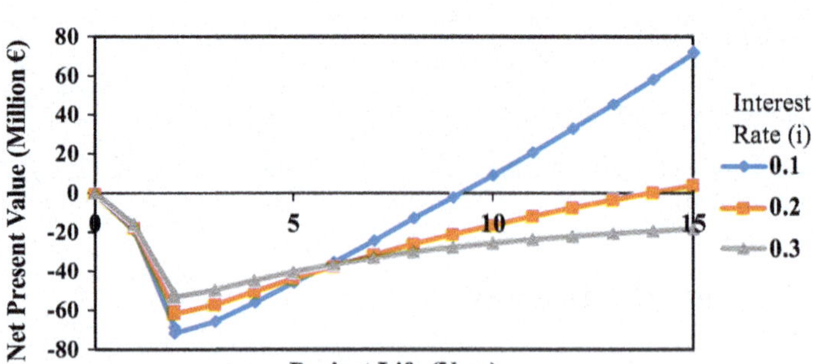

Fig. 13.4 Cash flow diagram for a biogas plant using different values of 0.1, 0.2, and 0.3 as interest rate. When i = 0.21, the Net Present Value (NPV) would be zero [Calculated from Shafiei et al. (2014)]

neither profit nor loss. Any of the following assumptions may be considered as the break-even parameter, e.g., the break-even interest rate (internal rate of return, IRR) or the break-even selling price of the product (Turton et al. 2009; Peters et al. 2003). In this regard, a very common expression for comparing the profitability of projects is the minimum selling prices (MSP) of the biofuel. Minimum ethanol selling price (MESP) is the price at which the NPV of the process would equal zero after a defined plant life time, e.g., 20 years (Humbird 2011; Swanson 2010; Tao 2014; McAloon 2000; Kazi 2010; Wooley et al. 1999a). When the profitability of two projects is to be compared, all possible factors which may affect the NPV should be eliminated. These investment parameters are presented in the following sections.

Sometimes the break-even point is defined as the minimum plant capacity at which the operating costs would be equal to earnings. Thus, at this capacity, which is a percentage of the plant full capacity, the net earnings are zero (Peters et al. 2003). Another definition for break-even point is when two additional plans should be compared for an existing plant. In this case, the required capital cost for these additions are calculated as well as the amounts of increased revenues. Then, the required time taken for these revenues to cover all the capital cost is calculated. At this time which is called the "breakeven point", all the expenses are recovered and from that time on the plant will produce extra profits. Therefore, the project which requires shorter time to payback the investment, is more profitable (Tao et al. 2013). However, this method is not suitable for the comparison of investments which have big differences in the amounts of capital cost or involve differences in the investment assumptions, e.g., rate of return on investment or taxation rate. A more general and accurate method for the comparison of the profitability of new or additional projects is the incremental analysis which has been widely discussed by Turton et al. previously (2009).

Profitability index (PI) or present value ratio (PVR) is calculated by dividing the present value of benefits (all positive cash flows) by the present value of costs (all negative cash flows). A profitable process has PI values greater than 1. If the NPV is equal to zero, then the PI would be 1 and a break-even situation has occurred (Turton et al. 2009; Shafiei et al. 2014).

13.2.5 Investment Parameters

In every techno-economic analysis, the investment parameters should be clearly stated. The economic viability of the processes may change significantly by changing these parameters such as taxation rate and interest rate (desired rate of return). The method for calculation of depreciation depends on the regulations in the plant location. For instance, the interest rate should be at least equal to the interest rate enforced by the banks. This value is as low as few percentages in the European countries while in the Middle East, it can be as high as 20%. The calculated IRR or DCFROR should be always more than the desired rate of return. If longer plant lives are considered, the NPV would be higher while shorter time for the depreciation of the plant would help the economy of the plant. The year at which the analysis was performed could also affect the capital costs significantly. Thus, a comparison of the results of techno-economic analyses is not easy and accurate (Chovau et al. 2013; Gnansounou et al. 2015).

13.3 Comparison of Techno-Economic Research Studies

A comparison of various techno economic analyses conducted for biogas production is presented in Table 13.1. Barta et al. (2010) compared the economic viability of the addition of different anaerobic digestion system to an ethanol production plant in Sweden. Aspen Plus was used for process simulation and Aspen Icarus Process Evaluator was applied for the economic evaluation. The ethanol plant utilized spruce wood and was based on steam explosion pretreatment catalyzed by SO_2. The anaerobic digestion scenarios were compared with a case of using an evaporation system for treatment of the stillage. The study compared the energy efficiency and ethanol production costs of the processes. The alternative scenarios included four scenarios for biogas production from the liquid fraction of the stillage (A scenarios) and one scenario for anaerobic digestion of the whole stillage (B). In the four former scenarios, i.e., A1–A4, in addition to ethanol and heat provided to the district heating system, the by-products could be upgraded biogas and pellets; pellets; upgraded biogas; or no other by-product. The results indicated that the energy efficiency of all the alternative scenarios were 87–92% while the base case had an energy efficiency of 81%. Scenarios A3 and B were associated with the lowest ethanol production costs of 4.01 and 4.00 (SEK[1]/L), respectively. These values were less than 5.14 SEK/L for the base case.

In a study by Akbulut (2012), the economy of a biogas plant as an addition to a farm for milk production was investigated. In addition to cow manure, which was available for free, sheep manure would be transported at a cost of 5.8 €/ton. In addition to ordinary operating costs, animal feed was purchased for cows. The main products of the plant were electricity, solid and liquid fertilizers, and milk. The final plant NPV was € 9.88 million using the most optimistic parameters. Although this study used a discounted cash flow analysis, the estimation method or reference for the equipment prices was not mentioned. Furthermore, mass and energy balance calculations and the cash flow analysis were performed manually. The taxation rate was too low and the contingencies and working capital were not included in the total capital investment.

Wheat straw and paper tube residuals were used as a feedstock for biogas production via solid state fermentation (Shafiei et al. 2013). Steam explosion at 180–190 °C was used for the pretreatment of the lignocelluloses. The process was developed based on the experimental data and simulated by Aspen Plus®.

[1]Swedish kronor

Table 13.1 Comparison of techno economic analysis for biogas production[a]

Reference	Barta et al. (2010)	Akbulut (2012)	Shafiei et al. (2013)	Shafiei et al. (2014)	Rajendran et al. (2014)	Kabir et al. (2015)	Mel et al. (2015)	Larsson et al. (2015)
Year	2009	2010	2013	2014	2012	2014	–	–
Feedstock	Ethanol plant stillage	Cow manure and Sheep dung	Paper tube residuals/ wheat straw	Pinewood	Municipal solid waste and biogas from WWTP	Forest residues	Fruits and vegetable wastes	Kraft pulp mill waste
Feedstock cost	0	0 and 5.8 €/ ton	80/50 €/ton	60 €/ton	0 $/kg + ? $/L	0.4 $/kg	RM 1,062,335/ y[d]	0
Pretreatment	SO_2-catalysed steam pretreatment	–	Steam explosion	Steam explosion/ NMMO	–	Methanol	–	–
Main product	Ethanol	Milk	Raw biogas	Biomethane	Biomethane	Biomethane	Biomethane	Liquefied methane
Main product price (production cost)[b]	4.01/4.00 SEK/L[c]	0.28 €/kg	0.57 €/m^3 (0.48 €/m^3/ 0.36 €/m^3)	1.15 €/m^3 (1.35/1.17 €/m^3)	1.81 $/L (0.76 $/ L)	1.81 $/L (gasoline equivalent)/ 3.0 $/kg	RM 5.0/kg (RM 4.4/ kg)[d]	119 €/MWh (80 €/MWh)
Byproduct	Biomethane and heat	Electricity/ solid and liquid fertilizer	Solid fuel and steam	Solid residue and sludge	Fertilizer	Lignin	–	–
Byproduct price	600 and 280 SEK/MWh[c]	€0.1/kWh, €120/t, €32/t	0.02 and 0.003 €/kg	0.04 €/kg	–	3.0 $/kg	–	–

(continued)

Table 13.1 (continued)

Reference	Barta et al. (2010)	Akbulut (2012)	Shafiei et al. (2013)	Shafiei et al. (2014)	Rajendran et al. (2014)	Kabir et al. (2015)	Mel et al. (2015)	Larsson et al. (2015)
Plant capacity	200,000 ton/year dry spruce chips	28,105 ton/year (DM 10.8)%	200,000 ton/year (dry basis)	100,000 tons/year (dry basis)	110,000 m^3 MSW/year + 3500 m^3/day raw biogas	20,000 ton/year	83.79 m^3/h	1.3, 9 and 13 ton/day of primary sludge, Secondary sludge and methanol
Contingency (%)	–	–	15	15	–	–	–	–
Working capital	2.6/2.5% of FCI	–	5% of TCI	5% of TCI	5% of TCI	–	11.7% of TCI	–
Interest rate (%)	7	14	10	10	10	7	6	6
IRR (DCFROR)	–	–	34%/34%	0%/0%	34%/34%	11.8%	9.61%	–
Tax rate	–	5% calculated	30%	30%	33%	33%	–	–
Project life (years)	15	20	15	15	20	15	–	15
Location	Sweden	Turkey	Sweden	Sweden	Sweden	Sweden	Malaysia	Sweden
Depreciation	Straight line	–	Straight line	Straight line	Straight line	Straight line	–	–
Depreciation period (years)	15	10	7	7	–	–	–	15

(continued)

Table 13.1 (continued)

Reference	Barta et al. (2010)	Akbulut (2012)	Shafiei et al. (2013)	Shafiei et al. (2014)	Rajendran et al. (2014)	Kabir et al. (2015)	Mel et al. (2015)	Larsson et al. (2015)
Payback period (year)	–	3.4	5	7.6/8.3	6	5.34	8.2	7.8
Total capital investment	1268/1305 MSEK[c]	10.26 M€	63.9 M€/ 61.8 M€	65.1 M€/ 69.7 M€	49.2 M$	60.5 M$	MRM 25.7[d]	15 M€
NPV	–	9.88 M€	34.7 M€/ 65.6 M€	82.2 M€/ 71.9 M€	106 M$	19.9 M$	MRM 6.8[d]	–

[a]In each case, the results of the best scenarios are presented, M€: million Euro; M$: million US dollar
[b]Product price is the value at which the product is sold and depends on the market; production cost is the calculated value that the financier should pay to produce the product
[c]MSEK: Million Swedish Kronor, US $ ≈ 7.3 SEK (2010)
[d]MRM: million RM, RM = MYR = Malaysian Ringgit (1 MYR ~ $ 0.31, 2014)

The economic evaluation was performed with Aspen Process Economic Analyzer. Manufacturing cost of raw biogas was 0.36 or 0.48 €/m^3 of methane for paper tubes and the straw while 46 and 56% of this cost was contributed by the raw material expenses. The respective total project investment of the processes were 63.9 and 61.8 million €, respectively.

Techno-economic analysis of steam explosion and Nmethylmorpholine-N-oxide (NMMO) pretreatment for biomethane production from pinewood was performed by Shafiei et al. (Shafiei et al. 2014). Aspen Plus® and Aspen Process economic analyzer were used for this analysis. Total capital investment of the processes were 65.1 and 69.7 million € for steam explosion and NMMO pretreatment, respectively. The respective manufacturing costs of the products were 0.77 and 0.97 €/m^3 of biomethane, respectively, excluding taxes as well as selling and distribution prices.

Rajendran et al. (2014) investigated the techno-economic aspects of biogas production from organic fraction of the municipal solid waste (OMSW) in Sweden. Aspen Plus and Aspen Process Economic Analyzer were used for process simulation and economic evaluation, respectively. The base case included the processing of 55,000 m^3 of the feedstock and required 34.6 million USD for its capital cost. The NPV after 20 years of plant life was 27.2 million USD. The scenarios included two types of biogas upgrading technologies, i.e., carbon dioxide absorption by monoethanolamine and water scrubbing method. Furthermore, the economy of the plants with double capacity, i.e., the application of two digesters, was also investigated. In order to improve the economy of the biogas upgrading system, buying additional raw biogas from nearby wastewater treatment plant was included in the scenarios and compared with the other ones. The minimum selling price for compressed biogas was 0.76 USD/L. This value was corresponding to the scenario with two digesters which imported biogas from the wastewater treatment plant. This scenario also included both scrubbing and absorption technologies for biogas upgrading.

Experimental and economic evaluation of biomethane production from forest residues were performed by Kabir et al. (2015). The processes included pretreatments with acetic acid, ethanol, or methanol at 190 °C for 60 min using 50% (V/V) of the organic solvent. The processes were simulated and evaluated using Superpro® Designer program. Among the processes investigated, the one which used methanol was the most promising process due to the lower cost of methanol.

SuperPro Designer software was used for the simulation and economic evaluation of a process for production of biomethane from fruits and vegetable wastes (Mel et al. 2015). The process involved a single concrete digester with a volume of about 2400 working at mesophilic conditions and retention time of 25 d. Water-scrubbing was used for the purification of biogas to 95%. The process was too simple and lacked compressors, biogas drying system, storage tanks for feed and product, and wastewater treatment unit. Water was not recycled back and thus, the desorption column was not considered for this process. The wastewater from the anaerobic digestion was not further processed to meet standard quality as wastewater either.

Larsson et al. (2015) performed a techno-economic analysis for the addition of a unit for biogas production for waste streams of a pulp and paper mill in Sweden. They considered two scenarios: case 1 evaluated the addition of a high rate (upflow anaerobic sludge bed, UASB) thermophilic anaerobic digester prior to the activated sludge treatment. This digester treated the condensate from Kraft evaporator methanol and alkaline filtrates from the bleaching process. The other scenario included a completely stirred tank reactor (CSTR) fed by the primary and secondary sludge of the aerobic wastewater treatment and Kraft evaporator condensate. The main product of these processes was liquefied methane. The most profitable case was the second one with the production cost of 80 €/MWh.

13.4 Conclusions

Considering the tremendous amounts of waste generated on a daily basis, biogas production from these waste streams is a promising strategy to move towards the substitution of fossil fuels. Agricultural and forest residues are the most available raw materials for biogas production in terms of quantity. However, the challenges of microbial treatment of these lignocellulosic feedstock have raised economic concerns for their commercial conversion to biogas. Many studies have investigated the economics of biogas production from such types of feedstock; however, different results were obtained and comparison of the results did not come into a good conclusion. There have been two main approaches for economic assessment of the processes involved; via calculation of cash flow diagram, and by calculation of production cost. In each method, there are many parameters which can affect the final results. Cost of raw materials, tax rates, contingencies, selling price of the products, plant location and capacity, and interest rate are among the most important parameters affecting the final results of economic assessments. While calculation of NPV may not show the whole picture of the economic feasibility of a process, it can be concluded that simultaneous application of both methods, i.e., cash flow analysis and calculation of production cost, could facilitate the comparison of the results from different studies. While some studies revealed that biogas, biomethane or electricity production was economically feasible, some studies concluded very low potentials for the commercialization of biogas production from lignocelluloses in the near future. This was concluded due to the payback periods of more than 5 years and low return rates compared with what expected by the industrial stakeholders. Furthermore, issues such as process instabilities and maintenance problems are hardly ever discussed in techno-economic analysis, while these concerns are very important as risk factors of a process. Other risk factors which are not discussed in techno-economic analysis are escalation factors for raw materials and the products. These factors have drastic effects on the revenues and raw material expenses in the cash flow analysis, and thus, a deep market analysis is required to have a correct picture of the future prices. Finally, in addition to economic viability of a process, issues such as energy efficiency and analysis of life cycle assessment are also essential to have a real green process which reduced GHG emissions.

References

Aden A et al (2002) Lignocellulosic biomass to ethanol process design and economics utilizing co-current dilute acid prehydrolysis and enzymatic hydrolysis for corn stover. National Renewable Energy Laboratory, US Department of Energy

Aden A, Foust T (2009) Technoeconomic analysis of the dilute sulfuric acid and enzymatic hydrolysis process for the conversion of corn stover to ethanol. Cellulose 16(4):535–545

Akbulut A (2012) Techno-economic analysis of electricity and heat generation from farm-scale biogas plant: Çiçekdağı case study. Energy 44(1):381–390

Anex RP et al (2010) Techno-economic comparison of biomass-to-transportation fuels via pyrolysis, gasification, and biochemical pathways. Fuel 89(Supplement 1):S29–S35

AspenTech (2017) Aspen Plus®, cited 2017; Available from http://home.aspentech.com/products/engineering/aspen-plus/

Balan V, Chiaramonti D, Kumar S (2013) Review of US and EU initiatives toward development, demonstration, and commercialization of lignocellulosic biofuels. Biofuels Bioprod Biorefin 7(6):732–759

Barta Z, Reczey K, Zacchi G (2010) Techno-economic evaluation of stillage treatment with anaerobic digestion in a softwood-to-ethanol process. Biotechnol Biofuels 3(1):21

Börjesson P, Mattiasson B (2008) Biogas as a resource-efficient vehicle fuel. Trends Biotechnol 26(1):7–13

Brown TR (2015) A techno-economic review of thermochemical cellulosic biofuel pathways. Biores Technol 178:166–176

Brown TR et al (2014) 2—Techno-economic assessment (TEA) of advanced biochemical and thermochemical biorefineries, In: Waldron K (ed) Advances in biorefineries. Woodhead Publishing, pp 34–66

Cakir FY, Stenstrom MK (2005) Greenhouse gas production: a comparison between aerobic and anaerobic wastewater treatment technology. Water Res 39(17):4197–4203

Chovau S, Degrauwe D, Van der Bruggen B (2013) Critical analysis of techno-economic estimates for the production cost of lignocellulosic bio-ethanol. Renew Sustain Energy Rev 26:307–321

Douglas JM (1988) Conceptual design of chemical processes. McGraw-Hill chemical Engineering series. McGraw-Hill, Singapour, p 601

Dutta A et al (2010) An economic comparison of different fermentation configurations to convert corn stover to ethanol using Z. mobilis and Saccharomyces. Biotechnol Prog 26(1):64–72

Eggeman T, Elander RT (2005) Process and economic analysis of pretreatment technologies. Bioresour Technol 96(18):2019–2025

EIA (2013) Monthly energy review: energy consumption by sector. Available from http://www.eesi.org/files/00351304.pdf

EIA (2016) International Energy Outlook 2016. Independent Statistics & Analysis 2016; Available from: https://www.eia.gov/outlooks/ieo/pdf/0484(2016).pdf

European Commission (2016) Waste, cited 2017; Available from http://ec.europa.eu/environment/waste/landfill_index.htm

Findeisen C (2015) The importance of safety standards, risk assessment and operators training for a successful biogas market development. In: UNIDO Biogas Workshop. German Biogas Association, Fachverband Biogas e.V., Vienna

Galbe M, Wallberg O, Zacchi G (2011) Techno-economic aspects of ethanol production from lignocellulosic agricultural crops and residues. In: Moo-Young M (ed) Comprehensive biotechnology, 2nd edn. Academic Press, Burlington, pp 615–628

Gnansounou E, Vaskan P, Pachón ER (2015) Comparative techno-economic assessment and LCA of selected integrated sugarcane-based biorefineries. Biores Technol 196:364–375

Hamelinck CN, van Hooijdonk G, Faaij APC (2005) Ethanol from lignocellulosic biomass: techno-economic performance in short-, middle- and long-term. Biomass Bioenerg 28(4): 384–410

Holm-Nielsen JB, Al Seadi T, Oleskowicz-Popiel P (2009) The future of anaerobic digestion and biogas utilization. Biores Technol 100(22):5478–5484

Humbird D, Aden A (2009) Biochemical production of ethanol from corn stover: 2008 state of technology model, NREL/TP-510-46214. National Renewable Energy Laboratory, US Department of Energy

Humbird D et al (2011) Process design and economics for biochemical conversion of lignocellulosic biomass to ethanol, dilute-acid pretreatment and enzymatic hydrolysis of corn stover, Technical Report NREL/TP-5100-47764. National Renewable Energy Laboratory, U.S. Department of Energy

Igliński B, Buczkowski R, Cichosz M (2015) Biogas production in Poland—current state, potential and perspectives. Renew Sustain Energy Rev 50:686–695

Iranmahboob J, Nadim F, Monemi S (2002) Optimizing acid-hydrolysis: a critical step for production of ethanol from mixed wood chips. Biomass Bioenerg 22(5):401–404

Jalalzadeh-Azar A (2010) Technoeconomic analysis of biomethane production from biogas and pipeline delivery (Presentation). National Renewable Energy Laboratory: United States. Department of Energy

Kabir MM, Forgács G, Horváth I Sárvári (2015a) Biogas from lignocellulosic materials. In: Karimi K (ed) Lignocellulose-based bioproducts. Springer International Publishing, Cham, pp 207–251

Kabir MM et al (2015b) Experimental and economical evaluation of bioconversion of forest residues to biogas using organosolv pretreatment. Biores Technol 178:201–208

Karimi K, Shafiei M, Kumar R (2013a) Progress in physical and chemical pretreatment of lignocellulosic biomass. In: Gupta VK, Tuohy MG (eds) Biofuel technologies: recent developments. Springer, Heidelberg, pp 53–96

Karimi K, Shafiei M, Kumar R (2013b) Progress in physical and chemical pretreatment of lignocellulosic biomass. In: Gupta VK, Tuohy MG (eds) Biofuel technologies. Springer, Heidelberg, pp 53–96

Kazi FK et al (2010) Techno-economic analysis of biochemical scenarios for production of cellulosic ethanol, NREL/TP-6A2-46588. National Renewable Energy Laboratory, U.S. Department of Energy

Klein-Marcuschamer D et al (2010) Technoeconomic analysis of biofuels: a wiki-based platform for lignocellulosic biorefineries. Biomass Bioenerg 34(12):1914–1921

Kormi T et al (2017) Estimation of landfill methane emissions using stochastic search methods. Atmos Pollut Res 8(4):597–605

Larsson M et al (2015) Techno-economic assessment of anaerobic digestion in a typical Kraft pulp mill to produce biomethane for the road transport sector. J Clean Prod 104:460–467

Lauer M (2017) Methodology guideline on techno economic assessment (TEA) in ThermalNet TEA Methodology Guideline. Intelligent Energy Europe

Li S et al (2015a) Assessing the role of renewable energy policies in landfill gas to energy projects. Energy Econ 49:687–697

Li J et al (2015b) Mass flow and energy balance plus economic analysis of a full-scale biogas plant in the rice–wine–pig system. Biores Technol 193:62–67

Lynd LR (1996) Overview and evaluation of fuel ethanol from cellulosic biomass: technology, economics, the environment, and policy. Annu Rev Energy Env 21(1):403–465

Maciejczyk M (2014) Safety of biogas plants. German Biogas Association, Fachverband Biogas e. V., Bangkok

Mao C et al (2015) Review on research achievements of biogas from anaerobic digestion. Renew Sustain Energy Rev 45:540–555

McAloon A et al (2000) Determining the cost of producing ethanol from corn starch and lignocellulosic feedstocks, NREL/TP-580-28893. National Renewable Energy Laboratory, U.S. Department of Energy
Mel M et al (2015) Simulation study for economic analysis of biogas production from agricultural biomass. Energy Procedia 65:204–214
Meyer AKP, Ehimen EA, Holm-Nielsen JB (2017) Future European biogas: animal manure, straw and grass potentials for a sustainable European biogas production. Biomass Bioenergy
Mussoline W et al (2012) Design considerations for a farm-scale biogas plant based on pilot-scale anaerobic digesters loaded with rice straw and piggery wastewater. Biomass Bioenerg 46: 469–478
NREL (2013) Biogas potential in the United States, NREL/FS-6A20-60178. National Renewable Energy Laboratory, The U.S. Department of Energy
Patterson T et al (2011) An evaluation of the policy and techno-economic factors affecting the potential for biogas upgrading for transport fuel use in the UK. Energy Policy 39(3):1806–1816
Persson M (2003) Evaluation of upgrading techniques for biogas
Peters MS, Timmerhaus K, West RE (2003) Plant design and economics for chemical engineers, 5th edn. McGraw-Hill Education
Piccolo C, Bezzo F (2009) A techno-economic comparison between two technologies for bioethanol production from lignocellulose. Biomass Bioenergy 33(3):478–491
Pike Research (2012) Worldwide power generation capacity from biogas will double by 2022. Available from: http://www.businesswire.com/news/home/20121107005284/en/Worldwide-Power-Generation-Capacity-Biogas-Double-2022-.U3NpFViSwwk
Poeschl M, Ward S, Owende P (2010) Prospects for expanded utilization of biogas in Germany. Renew Sustain Energy Rev 14(7):1782–1797
Rajendran K et al (2014) Uncertainty over techno-economic potentials of biogas from municipal solid waste (MSW): a case study on an industrial process. Appl Energy 125:84–92
Sárvári Horváth I et al (2016) Recent updates on biogas production—a review. Biofuel Res J 3(2): 394–402
Sassner P, Galbe M, Zacchi G (2008) Techno-economic evaluation of bioethanol production from three different lignocellulosic materials. Biomass Bioenergy 5
Scharff H (2014) Landfill reduction experience in The Netherlands. Waste Manag 34(11): 2218–2224
Sendich E et al (2008) Recent process improvements for the ammonia fiber expansion (AFEX) process and resulting reductions in minimum ethanol selling price. Biores Technol 99(17): 8429–8435
Shafiei M, Karimi K, Taherzadeh MJ (2011) Techno-economical study of ethanol and biogas from spruce wood by NMMO-pretreatment and rapid fermentation and digestion. Bioresour Technol 102(17):7879–7886
Shafiei M et al (2013) Techno-economical study of biogas production improved by steam explosion pretreatment. Bioresour Technol 148:53–60
Shafiei M et al (2014) Economic impact of NMMO pretreatment on ethanol and biogas production from pinewood. Biomed Res Int 2014:13
Shafiei M, Kumar R, Karimi K (2015) Pretreatment of lignocellulosic biomass. In: Karimi K (ed) Lignocellulose-based bioproducts. Springer International Publishing, Cham, pp 85–154
Smith R (2005) Chemical process design and integration, 3rd edn. Wiley, England
Sun Q et al (2015) Selection of appropriate biogas upgrading technology—a review of biogas cleaning, upgrading and utilisation. Renew Sustain Energy Rev 51:521–532
Swanson RM et al (2010) Techno-economic analysis of biofuels production based on gasification, Technical Report NREL/TP-6A20-46587. National Renewable Energy Laboratory, U.S. Department of Energy
Taherzadeh MJ, Karimi K (2007) Enzyme-based hydrolysis processes for ethanol from lignocellulosic materials. BioResources 2:707–738
Taherzadeh MJ, Karimi K (2008a) Pretreatment of lignocellulosic wastes to improve ethanol and biogas production: a review. Int J Mol Sci 9:1621–1651

Taherzadeh MJ, Karimi K (2008b) Pretreatment of lignocellulosic wastes to improve ethanol and biogas production: a review. Int J Mol Sci 9(9):1621–1651

Tan ECD et al (2016) Conceptual process design and economics for the production of high-octane gasoline blendstock via indirect liquefaction of biomass through methanol/dimethyl ether intermediates. Biofuels Bioprod Biorefin 10(1):17–35

Tao LA, Aden A, Elander RT (2013) Economics of pretreatment for biological processing. In: Wyman CE (ed) Aqueous pretreatment of plant biomass for biological and chemical conversion to fuels and chemicals. Wiley, UK

Tao L et al (2014) NREL 2012 achievement of ethanol cost targets: biochemical ethanol fermentation via dilute-acid pretreatment and enzymatic hydrolysis of corn stover, Technical Report NREL/TP-5100-61563. National Renewable Energy Laboratory, U.S. Department of Energy

Turton R, Bailie RC, Whiting WB (2009) Analysis, synthesis, and design of chemical processes, 3rd edn. Prentice Hall, Upper Saddle River

von Sivers M, Zacchi G (1995) A techno-economical comparison of three processes for the production of ethanol from pine. Bioresour Technol 1

von Sivers M, Zacchi G (1996) Ethanol from lignocellulosics: a review of the economy. Bioresour Technol 56:131–140

Werner U, Stoehr U, Hees N (1989) Biogas plants in animal husbandry. German Appropriate Technology Exchange (GATE) and German Agency for Technical Cooperation (GTZ) GmbH

Westenbroek PA, Martin J (2017) Anaerobic digesters and biogas safety, cited 2017; Available from http://articles.extension.org:80/pages/30311/anaerobic-digesters-and-biogas-safety

Wooley R et al (1999a) Lignocellulosic biomass to ethanol—process design and economics utilizing co-current dilute acid prehydrolysis and enzymatic hydrolysis—current and futuristic scenarios, NREL/TP-580-26157

Wooley R et al (1999b) Process design and costing of bioethanol technology: a tool for determining the status and direction of research and development. Biotechnol Prog 5

Wiese J, König R *Application report, laboratory analysis & process analysis, biogas plant monitoring*. Lellbach biogas plant

Yliopisto J (2009) Evaluation of potential technologies and operational scales reflecting market needs for low-cost gas upgrading systems. In: Seventh framework programme theme energy, Biowaste as feedstock for 2nd generation

Yu Y, Lou X, Wu H (2008) Some recent advances in hydrolysis of biomass in hot-compressed water and its comparisons with other hydrolysis methods. Energy Fuels 22(1):46–60

Chapter 14
Exergy-Based Performance Assessment of Biogas Plants: Application of Advanced Exergy and Exergoeconomic Analyses for Evaluating Biogas Upgrading Process

Hojat Ansarinasab and Mehdi Mehrpooya

14.1 Introduction

Nowadays, both energy and environmental problems call upon a reconsideration of the current energy infrastructure by improving the design and performance of energy conversion processes. Accordingly, research attempts have been propelled towards the use of advanced engineering approaches such as emergy, life cycle assessment (LCA), energy, and exergy for decision-making on the efficiency, productivity, and sustainability of energy conversion systems, particularly renewable energy systems. Among the methods developed so far, exergy concept is attractive because of its potency for locating and quantifying the irreversibility aspects of energy conversion systems. Moreover, all kinds of material and energy flows can be fairly evaluated/valued using this concept.

Simply speaking, exergy is the maximum amount of obtainable shaft work from an energy or material flow when it brings to equilibrium with its reference environment (Dadak et al. 2016). Due to unique conceptual features of the exergy-based analyses in designating both the energy quantity and quality more accurately compared with the conventional energy analysis, these approaches have recently become very popular for scrutinizing the efficiency, productivity, and sustainability aspects of biofuel production and utilization systems. In better words, exergy is the

H. Ansarinasab
Faculty of Energy Systems Engineering, Petroleum University of Technology (PUT), Abadan, Iran

M. Mehrpooya (✉)
Renewable Energies and Environmental Department, Faculty of New Sciences and Technologies, University of Tehran, Tehran, Iran
e-mail: mehrpoya@ut.ac.ir

M. Tabatabaei and H. Ghanavati (eds.), *Biogas*, Biofuel and Biorefinery Technologies 6, https://doi.org/10.1007/978-3-319-77335-3_14

confluence of energy, environment, and sustainable development (Dincer 2011) (Fig. 14.1).

Even though the reasons and sources of internal and external irreversibilities can be accurately identified by exergy analysis, it alone cannot be used as a perfect decision-making metric because of the fact that classical analysis does not consider economic and environmental constraints of energy systems. Fortunately, integrated exergy-based approaches like exergoeconomic and exergoenvironmental approaches can easily address this issue. In recent years, these approaches have found international reputation for micro/macro-level analysis, development, and optimization of energy systems from exergy/economic and exergy/environmental perspectives, respectively. Advanced exergy, exergoeconomic, and exergoenvironmental analyses respectively split the exergy destruction of a given component as well as its associated costs and environmental impact into four parts, i.e., avoidable endogenous, unavoidable endogenous, avoidable exogenous, and unavoidable exogenous. In better words, advanced exergy-based analyses can reveal technical, economical, and environmental interactions among components of energy systems.

In spite of the fact that biogas produced from waste organics is a renewable clean biofuel, its production and upgrading processes require huge amounts of energy, water, and materials. In light of that, there is an urgent need to find the most resource-efficient, cost-effective, and environmentally-friendly biogas production and upgrading processes in response to both energy crisis and environmental pollution. Biogas upgrading is one the most important step of biogas production plant

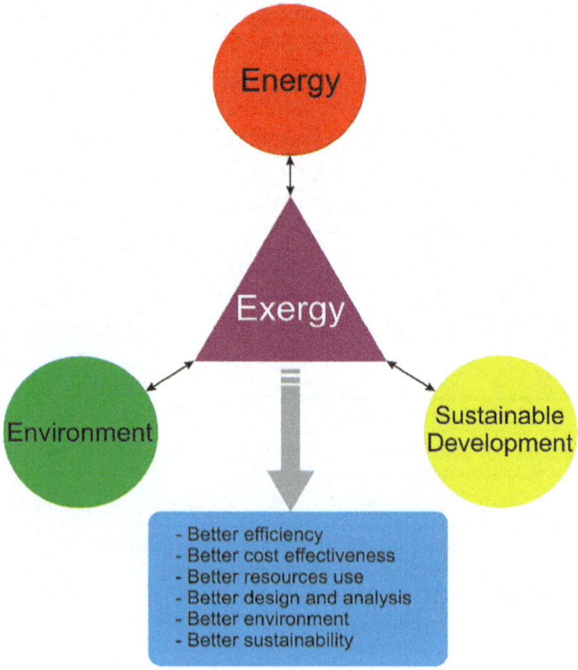

Fig. 14.1 The interdisciplinary triangle of exergy. Adopted from Dincer (2011)

in order to reduce its CO_2 and H_2S contents to a safe level as required by gas-powered vehicles and equipment. It should be noted that biogas upgrading process for removing CO_2 from evolved biogas can be carried out various technologies including high pressure water scrubbing, cryogenic separation, membrane separation, organic physical scrubbing, chemical scrubbing, and pressure swing adsorption. In this chapter, two biogas upgrading processes, i.e., high pressure water scrubbing (HPWS) and cryogenic separation (CS) were simulated and analyzed using advanced exergy and exergoeconomic analyses in order to pinpoint the breakthrough points for further thermodynamic and economic improvements. More specifically, the effects of influential parameters were investigated on the exergetic and exergoeconomic parameters of both processes.

14.2 Processes Description

14.2.1 HPWS Process

In this method, CO_2 and H_2S are separated from biogas and absorbed by the liquid absorbent. Figure 14.2 illustrates the process flow diagram of the HPWS simulated in this study. Table 14.1 presents thermodynamic properties and chemical compositions of the main streams of the HPWS process. In this process, raw biogas was compressed (to 8 bar) and then cooled down (to about 25 °C) by intermediate cooler (stream 5). The compressed biogas entered from the bottom of the absorption tower T-1. High pressure water with a flow rate of 71,820 kg/h was sprayed from top section of the tower. The purity of biogas (stream 6) at top outlet of the tower was higher than 90%. The remaining impurities like oxygen could not be removed simply because they were inert gases. The tower bottom outlet (stream 7) containing CO_2 and H_2S entered the water regeneration tower T-2. In this unit water entered from the top and contacted with air (stream 8) entering from the bottom. Acid gases were separated because of breaking their bonds with water and took away by air (stream 9). Finally, regenerated water transferred to the absorption tower (T-1) by pump P-1.

14.2.2 CS Process

In this method, biogas impurities like CO_2 and H_2S are liquefied and separated from the methane. Figure 14.3 shows the process flow diagram of the CS process simulated throughout this study. In this process, separation is done in three successive stages. Table 14.2 presents thermodynamic properties and chemical compositions of the main streams of the CS process. In each stage of separation, temperature and pressure of the streams were adjusted in order to maximize CO_2 separation from

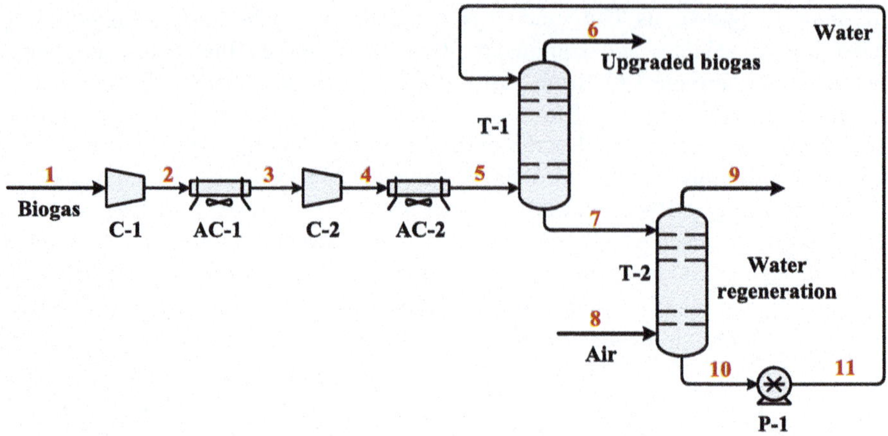

Fig. 14.2 Process flow diagram of the HPWS process

Table 14.1 Thermodynamic properties and chemical compositions of the main streams of the HPWS process

Stream name	Biogas (1)	(6)	(8)	(9)
Flow (kg/h)	533.32	226.07	2885.03	3217.80
Temperature (°C)	25.00	13.04	25.00	13.14
Pressure (bar)	2.00	6.00	1.20	1.20
Components (mol%)	–	–	–	–
CH_4	60.76	91.87	0.00	0.41
CO_2	36.49	4.28	0.00	6.21
H_2S	0.82	0.00	0.00	0.15
H_2O	0.00	0.27	0.00	1.27
O_2	0.97	1.69	21.00	19.30
N_2	0.97	1.90	79.00	72.65

methane. In this process, raw biogas compressed by three stage compression and cooled down by two intermediate coolers. The compressed biogas having temperature of 103.2 °C and pressure of 50 bar (stream 6) entered heat exchangers HE-1 and HE-2 and cooler E-1, respectively. The compressed biogas was cooled down to −45 °C (stream 9) and then transferred into two phase separator F-1 and a portion of its CO_2 was separated (stream 10). The partially purified biogas (stream 12) was flowed to heat exchangers HE-4 and HE-3 and was then cooled again by cooler E-2. The outlet stream was sent to valve V-1 and its pressure and temperature reached to −70 °C and 40 bar, respectively. This stream entered two phase separator F-2 and a portion of its CO_2 was liquefied and separated. Finally, stream 19 was cooled down in cooler E-2 and flowed into expansion valve V-1. The outlet

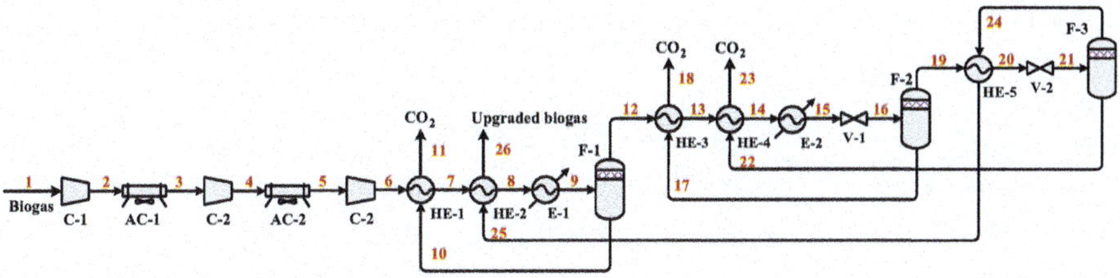

Fig. 14.3 Process flow diagram of the CS process

stream reached to −120 °C and 20 bar. In phase separator F-3, the remained CO_2 was separated from the methane.

14.3 Thermodynamic Modeling

Both HPWS and CS processes were developed using the Aspen HYSYS. The well-known Peng-Robinson equation of state was utilized for simulating these processes due to the nature of components available in the biogas. This equation of state can be written as below (Eq. 14.1) (Robinson et al. 1985):

$$p = \frac{RT}{v-b} - \frac{a}{v(v+b)+b(v-b)} \tag{14.1}$$

Here, the values of a and b can be expressed as follows (Eqs. 14.2 and 14.3):

$$a = \sum\sum z_i z_j (a_i a_j)^{0.5}(1-k_{ij}) \tag{14.2}$$

$$b = \sum z_i b_i \tag{14.3}$$

The power consumption rates of equipment involved in both biogas upgrading processes are presented in Table 14.3.

14.4 Exergy Analysis

14.4.1 Conventional Exergy Analysis

The total exergy flow rate associated with each stream could be computed as the summation of chemical, physical, potential and kinetic exergies. In this work, potential and kinetic exergies were ignored because of their insignificant contribution to total exergy. Physical and chemical exergies are written as below (Eqs. 14.4 and 14.5) (Mehrpooya et al. 2016):

Table 14.2 Thermodynamic properties and chemical compositions of the main streams of the CS process

Stream name	Biogas (1)	(11)	(18)	(23)	(26)
Flow (kg/h)	532.93	185.56	121.86	35.08	190.44
Temperature (°C)	25.00	60.00	−46.00	−49.00	22.00
Pressure (bar)	2.00	50.00	40.00	2.00	5.00
Components (mol%)	–	–	–	–	–
CH_4	61.10	23.91	29.51	1.48	92.67
CO_2	36.93	75.63	69.98	98.48	4.05
H_2S	0.01	0.02	0.02	0.02	0.00
H_2O	0.00	0.00	0.00	0.00	0.00
O_2	0.98	0.19	0.20	0.01	1.67
N_2	0.98	0.26	0.29	0.01	1.61

Table 14.3 Power consumption rates of equipment involved in both biogas upgrading processes

HPWS process		CS process	
Component name	Power (kW)[a]	Component name	Power (kW)[a]
C-1	16.23	C-1	26.37
C-2	11.72	C-2	27.48
AC-1	6.34	C-3	13.51
AC-2	7.61	AC-1	14.28
P-1	10.21	AC-2	17.14

[a]Mechanical efficiency = 0.75

$$\dot{E}^{PH} = H - H_o - T_o(S - S_o) \tag{14.4}$$

$$\dot{E}^{CH} = \sum x_i \dot{E}_i^o + G - \sum x_i G_i \tag{14.5}$$

where $\dot{E}_i^o$, x_i and G_i refer to the standard chemical exergy, molar fraction, and Gibbs free energy, respectively, for ith stream.

By defining the product exergy ($\dot{E}_{P,k}$) and fuel exergy ($\dot{E}_{F,k}$) associated with the kth process component, the exergy destruction ($\dot{E}_{D,k}$), exergy efficiency (ε_k) and exergy destruction ratio (y_k) of each element can be calculated as below (Eqs. 14.6, 14.7, and 14.8) (Ansarinasab and Mehrpooya 2017a):

$$\dot{E}_{D,k} = \dot{E}_{F,k} - \dot{E}_{P,k} \tag{14.6}$$

$$\varepsilon_k = \frac{\dot{E}_{P,k}}{\dot{E}_{F,k}} \quad \text{or} \quad \varepsilon_k = 1 - \frac{\dot{E}_{D,k}}{\dot{E}_{F,k}} \tag{14.7}$$

$$y_k = \frac{\dot{E}_{D,k}}{\dot{E}_{F,tot}} \quad (14.8)$$

Table 14.4 summarizes the fuel and product definitions for components of both HPWS and CS processes.

14.4.2 Advanced Exergy Analysis

The classical exergy method can only reveal the irreversibilities of each component, while the sources and inevitability of these irreversibilities are still important. Advanced exergy method can provide more information by splitting the exergy destruction of process components into endogenous/exogenous parts based on the origin of the irreversibilities and into avoidable/unavoidable parts based on the removing ability of the irreversibilities. Endogenous part of exergy destruction can be computed if it is assumed that other elements operate ideally. In better words, this value reveals the amount of exergy destruction of the *k*th element occurred due to poor efficiency of other elements. Engineering (graphical) method, as shown in Fig. 14.4 (Ansarinasab et al. 2017a; Kelly et al. 2009), can be used to compute endogenous exergy destruction based on the following equations (Eqs. 14.9 and 14.10) (Sadaghiani et al. 2017):

$$\dot{E}_{D,t} = \dot{E}_{F,t} - \dot{E}_{p,t} \quad (14.9)$$

$$\dot{E}_{D,t} = \sum_k \dot{E}_{D,k} = \dot{E}_{D,k} + \dot{E}_{D,other} \quad (14.10)$$

Furthermore, exogenous part of exergy destruction caused by the interaction among the considered element and other elements of the system can be computed by subtracting the endogenous exergy destruction from the total exergy destruction as below (Eq. 14.11) (Vatani et al. 2014):

$$\dot{E}^{EX}_{D,k} = \dot{E}_{D,k} - \dot{E}^{EN}_{D,k} \quad (14.11)$$

Due to economic and technical obstacles, all of the thermodynamic irreversibilities of a given element cannot be removed briefly called "inevitable" or "unavoidable" exergy destruction. In contrast, avoidable exergy destruction can be declined by technical improvements. The unavoidable conditions for both developed biogas upgrading processes, affecting their efficiency and performance significantly, are tabulated in Table 14.5.

The following relations can be utilized for calculating the unavoidable and avoidable exergy destructions (Eqs. 14.12 and 14.13) (Vatani et al. 2014):

Table 14.4 Fuel and product definitions for components of both HPWS and CS processes

HPWS process			CS process		
Component	Exergy of fuel	Exergy of product	Component	Exergy of fuel	Exergy of product
C-1	$\dot{E}_F = \dot{W}_{C\text{-}1}$	$\dot{E}_P = \dot{E}_2 - \dot{E}_1$	C-1	$\dot{E}_F = \dot{W}_{C\text{-}1}$	$\dot{E}_P = \dot{E}_2 - \dot{E}_1$
C-2	$\dot{E}_F = \dot{W}_{C\text{-}2}$	$\dot{E}_P = \dot{E}_4 - \dot{E}_3$	C-2	$\dot{E}_F = \dot{W}_{C\text{-}2}$	$\dot{E}_P = \dot{E}_4 - \dot{E}_3$
AC-1	$\dot{E}_F = (\dot{E}^{PH})_2 + \dot{W}_{AC\text{-}1}$	$\dot{E}_P = (\dot{E}^{PH})_3 - \dot{E}_{air,out}$	C-3	$\dot{E}_F = \dot{W}_{C\text{-}3}$	$\dot{E}_P = \dot{E}_6 - \dot{E}_5$
AC-2	$\dot{E}_F = (\dot{E}^{PH})_4 + \dot{W}_{AC\text{-}2}$	$\dot{E}_P = (\dot{E}^{PH})_5 - \dot{E}_{air,out}$	AC-1	$\dot{E}_F = (\dot{E}^{PH})_2 + \dot{W}_{AC\text{-}1}$	$\dot{E}_P = (\dot{E}^{PH})_3 - \dot{E}_{air,out}$
T-1	$\dot{E}_F = \dot{E}_5 - \dot{E}_6$	$\dot{E}_P = \dot{E}_7 - \dot{E}_{11}$	AC-2	$\dot{E}_F = (\dot{E}^{PH})_4 + \dot{W}_{AC\text{-}2}$	$\dot{E}_P = (\dot{E}^{PH})_5 - \dot{E}_{air,out}$
T-2	$\dot{E}_F = \dot{E}_7 - \dot{E}_{10}$	$\dot{E}_P = \dot{E}_9 - \dot{E}_8$	HE-1	$\dot{E}_F = \dot{E}_{10} - \dot{E}_{11}$	$\dot{E}_F = \dot{E}_6 - \dot{E}_7$
P-1	$\dot{E}_F = \dot{W}_{P\text{-}1}$	$\dot{E}_P = \dot{E}_{11} - \dot{E}_{10}$	HE-2	$\dot{E}_F = \dot{E}_{25} - \dot{E}_{26}$	$\dot{E}_F = \dot{E}_8 - \dot{E}_7$
			HE-3	$\dot{E}_F = \dot{E}_{17} - \dot{E}_{18}$	$\dot{E}_F = \dot{E}_{13} - \dot{E}_{12}$
			HE-4	$\dot{E}_F = \dot{E}_{23} - \dot{E}_{24}$	$\dot{E}_F = \dot{E}_{14} - \dot{E}_{13}$
			HE-5	$\dot{E}_F = \dot{E}_{25} - \dot{E}_{24}$	$\dot{E}_F = \dot{E}_{20} - \dot{E}_{19}$
			E-1	$\dot{E}_F = \dot{E}^{Q(E\text{-}1)}$	$\dot{E}_F = \dot{E}_9 - \dot{E}_8$
			E-2	$\dot{E}_F = \dot{E}^{Q(E\text{-}2)}$	$\dot{E}_F = \dot{E}_{15} - \dot{E}_{14}$

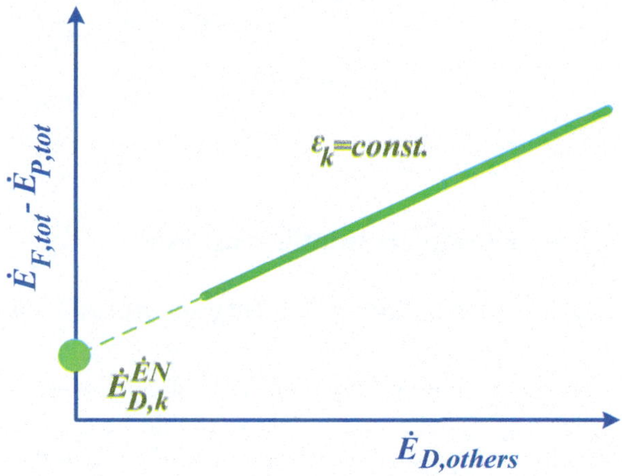

Fig. 14.4 Engineering (graphical) method used to compute endogenous exergy destruction (Ansarinasab et al. 2017a; Kelly et al. 2009). With permission from Elsevier, Copyright© 2018

Table 14.5 Assumptions for the unavoidable conditions

Component	Main aim for unavoidable conditions	Unavoidable conditions
Compressor	To maximize the isentropic efficiency	η_{is} = 90.0%
Pump	To maximize the isentropic efficiency	η_{is} = 90.0%
Tower	To maximize the exergetic efficiency	ε = 99.9%
Air cooler	To minimize the Min. Approach	ΔT_{min} = 5.0 °C, $\Delta P = \Delta P_{real}$
Heat exchanger	To minimize the Min. Approach	ΔT_{min} = 0.5 °C, $\Delta P = \Delta P_{real}$

$$\dot{E}^{UN}_{D,k} = \dot{E}_{P,k}\left(\frac{\dot{E}_{D,k}}{\dot{E}_{P,k}}\right)^{UN} \tag{14.12}$$

$$\dot{E}^{AV}_{D,k} = \dot{E}_{D,k} - \dot{E}^{UN}_{D,k} \tag{14.13}$$

To achieve more informative results regarding the role of each element on total exergy destruction, the above mentioned concepts can be integrated. Accordingly, hybrid destruction parts can be obtained by combining the unavoidable/avoidable and exogenous/endogenous destructions as below (Eqs. 14.14, 14.15, 14.16, and 14.17) (Vatani et al. 2014):

$$\dot{E}^{UN,EN}_{D,k} = \dot{E}^{EN}_{P,k}\left(\frac{\dot{E}_{D,k}}{\dot{E}_{P,k}}\right)^{UN} \tag{14.14}$$

$$\dot{E}^{UN,EX}_{D,k} = \dot{E}^{UN}_{D,k} - \dot{E}^{UN,EN}_{D,k} \tag{14.15}$$

$$\dot{E}_{D,k}^{AV,EN} = \dot{E}_{D,k}^{EN} - \dot{E}_{D,k}^{UN,EN} \quad (14.16)$$

$$\dot{E}_{D,k}^{AV,EX} = \dot{E}_{D,k}^{AV} - \dot{E}_{D,k}^{AV,EN} \quad (14.17)$$

14.5 Exergoeconomic Analysis

14.5.1 Conventional Exergoeconomic Analysis

In order to provide useful data for designing a cost-effective energy conversion system, exergoeconomic analysis should be applied. The first step in an exergoeconomic analysis is the development of an economic model.

14.5.1.1 Economic Model

An economic model was developed based on the Total Revenue Requirement (TRR) method proposed by the Electric Power Research Institute (Cao et al. 2016). TRR model considers the whole costs associated with a project such as expenses and carrying charges. Table 14.6 summarizes economic constants and assumptions used in this study (Ansarinasab et al. 2016).

The levelized annual total revenue requirement (TRR_L) can be determined by implementing the capital recovery factor (*CRF*) as below (Eq. 14.18) (Bejan and Tsatsaronis 1996):

$$TRR_L = CRF \sum_{1}^{BL} \frac{TRR_j}{(1 + i_{eff})^j} \quad (14.18)$$

The capital recovery factor (*CRF*) can be expressed as below (Eq. 14.19) (Bejan and Tsatsaronis 1996):

Table 14.6 Economic constants and assumptions (Ansarinasab et al. 2016)

Economic parameters	Value
Average annual rate of the cost of money (i_{eff})	10%
Average nominal escalation rate for the operating and maintenance cost (r_{OMC})	5%
Average nominal escalation rate for fuel (r_{FC})	5%
Plant economic life (book life)	25 years
Total annual operating hours of the system operation at full load	7300

$$CRF = \frac{i_{eff}(1 + i_{eff})^{BL}}{(1 + i_{eff})^{BL} - 1} \tag{14.19}$$

where i_{eff} and BL stand for the average annual rate of the cost of money and plant economic life, respectively. In both biogas upgrading processes, TRR_j as the revenue requirement in jth year of process performance, consisted of four annual costs as follow (Eq. 14.20) (Mehrpooya et al. 2015):

$$TRR_j = TCR_j + ROI_j + FC_j + OMC_j \tag{14.20}$$

where *ROI* and *TCR* stand for minimum return on investment and total capital recovery, respectively. *FC* and *OMC* refer to fuel costs and operation and maintenance costs, respectively. The levelized value of electrical power cost (electrical power consumed by compressors, pumps, and air coolers) as fuel costs can be computed by implementing the fuel cost (FC_0) at beginning of the first year of plant economic life as follows (Eqs. 14.21 and 14.22) (Mehrpooya et al. 2015):

$$FC_L = FC_0 \times CELF = FC_0 \frac{k_{FC}(1 - k_{FC}^{BL})}{(1 - k_{FC})} CRF \tag{14.21}$$

$$k_{FC} = \frac{1 + r_{FC}}{1 + i_{iff}} \quad r_{FC} = \text{constant} \tag{14.22}$$

where *CELF* and r_{FC} stand for the constant escalation levelization factor and average nominal escalation rate for fuel, respectively. The levelized annual operation and maintenance cost (OMC_L) can be achieved by the same procedure as following (Eqs. 14.23 and 14.24) (Mehrpooya et al. 2015):

$$OMC_L = OMC_0 \times CELF = OMC_0 \frac{k_{OMC}(1 - k_{OMC}^{BL})}{(1 - k_{OMC})} CRF \tag{14.23}$$

$$k_{OMC} = \frac{1 + r_{OMC}}{1 + i_{iff}} \quad r_{OMC} = \text{constant} \tag{14.24}$$

Similarly, r_{OMC} refers to nominal escalation rate of operation and maintenance cost. Finally, the levelized carrying charges (CC_L) can be achieved as follow (Eq. 14.25) (Mehrpooya et al. 2015):

$$CC_L = TRR_L - FC_L - OMC_L \tag{14.25}$$

The conventional exergoeconomic method was carried out at the component-level by using related principles for each equipment. Therefore, the cost rate associated with the kth equipment for *CI* (refers to capital investment) and *OMC* (refer to operation and maintenance cost) can be computed as follows (Eq. 14.26) (Mehrpooya et al. 2015):

$$\dot{Z}_k = \dot{Z}_k^{CI} + \dot{Z}_k^{OM} = \frac{CC_L + OMC_L}{\tau} \frac{PEC_k}{\sum_k PEC_k} \tag{14.26}$$

where τ and PEC_k refer to the operational hours and the purchased cost of the *k*th equipment, respectively. Table 14.7 depicts purchased-equipment cost functions.

14.5.1.2 Cost Balance Equations

The cost balance equation for each component of the process can be written as follows (Eq. 14.27) (Mehrpooya et al. 2017):

$$\sum_i (c_i \dot{E}_i)_k + \dot{Z}_k^{CL} + \dot{Z}_k^{OM} = \sum_o (c_o \dot{E}_o)_k \tag{14.27}$$

where c_i stands for average unit cost of exergy in \$/GJ unit.

Moreover, auxiliary equations can be obtained according to the fuel and product rules (Lazzaretto and Tsatsaronis 1997). The cost balance and auxiliary equations for each component of both biogas upgrading processes are presented in Table 14.8.

Table 14.7 Purchased cost equations of the process components

Component	Purchased equipment cost functions	Reference
Heat exchanger	$PEC_{HE} = 130\left(\frac{A}{0.093}\right)^{0.78}$	Ansarinasab et al. (2017b)
	$PEC_E = 8500 + 409(A)^{0.8}$	Ansarinasab et al. (2017b)
Compressor	$PEC_C = 7900(HP)^{0.62}$	Ansarinasab et al. (2017b)
Air cooler	$PEC_{AC} = 30(A)^{0.4}$	Ansarinasab et al. (2017b)
Pump	$PEC_P = 800\left(\frac{W_P}{10}\right)^{0.26}\left(\frac{1-\eta_P}{\eta_P}\right)$	Ansarinasab et al. (2017b)
Flash drum	$PEC_F = 1.218(42 + 163\ W)$	Ansarinasab et al. (2017b)
Tower	$PEC_T = 1.218[C_b + NC_t + C_p]$ $C_t = 457.7 \exp(0.1739D)$, N = number of trays $C_b = 1.218 \exp[6.629 + 0.1826(\ln W) + 0.02297(\ln W)2]$, $C_p = 300D^{0.7396}L^{0.7068}$	Cooper et al. (2010)

Table 14.8 Cost balances and auxiliary equations of the process components

HPWS process			CS process		
Component	Cost balance	Auxiliary equation	Component	Cost balance	Auxiliary equation
C-1	$\dot{C}_1 + \dot{C}_{\dot{W}(C\text{-}1)} + \dot{Z}_{C\text{-}1} = \dot{C}_2$	$c_1 = 0$	C-1	$\dot{C}_1 + \dot{C}_{\dot{W}(C\text{-}1)} + \dot{Z}_{C\text{-}1} = \dot{C}_2$	$c_1 = 0$
C-2	$\dot{C}_3 + \dot{C}_{\dot{W}(C\text{-}2)} + \dot{Z}_{C\text{-}2} = \dot{C}_4$	*None*	C-2	$\dot{C}_3 + \dot{C}_{\dot{W}(C\text{-}2)} + \dot{Z}_{C\text{-}2} = \dot{C}_4$	*None*
AC-1	$\dot{C}_2 + \dot{C}_{\dot{W}(AC\text{-}1)} + \dot{Z}_{AC\text{-}1} = \dot{C}_3$	*None*	C-3	$\dot{C}_5 + \dot{C}_{\dot{W}(C\text{-}3)} + \dot{Z}_{C\text{-}3} = \dot{C}_6$	*None*
AC-2	$\dot{C}_4 + \dot{C}_{\dot{W}(AC\text{-}2)} + \dot{Z}_{AC\text{-}2} = \dot{C}_5$	*None*	AC-1	$\dot{C}_2 + \dot{C}_{\dot{W}(AC\text{-}1)} + \dot{Z}_{AC\text{-}1} = \dot{C}_3$	*None*
T-1	$\dot{C}_5 + \dot{C}_{11} + \dot{Z}_{T\text{-}1} = \dot{C}_6 + \dot{C}_7$	$c_6 = c_7$	AC-2	$\dot{C}_4 + \dot{C}_{\dot{W}(AC\text{-}2)} + \dot{Z}_{AC\text{-}2} = \dot{C}_5$	*None*
T-2	$\dot{C}_7 + \dot{C}_8 + \dot{Z}_{T\text{-}2} = \dot{C}_9 + \dot{C}_{10}$	$c_9 = c_{10}, c_8 = 0$	HE-1	$\dot{C}_6 + \dot{C}_{10} + \dot{Z}_{HE\text{-}1} = \dot{C}_7 + \dot{C}_{11}$	$c_{10} = c_{11}$
P-1	$\dot{C}_{10} + \dot{C}_{\dot{W}(P\text{-}1)} + \dot{Z}_{P\text{-}1} = \dot{C}_{11}$	*None*	HE-2	$\dot{C}_7 + \dot{C}_{25} + \dot{Z}_{HE\text{-}2} = \dot{C}_8 + \dot{C}_{26}$	$c_{25} = c_{26}$
			HE-3	$\dot{C}_{12} + \dot{C}_{17} + \dot{Z}_{HE\text{-}3} = \dot{C}_{13} + \dot{C}_{18}$	$c_{17} = c_{18}$
			HE-4	$\dot{C}_{13} + \dot{C}_{22} + \dot{Z}_{HE\text{-}4} = \dot{C}_{14} + \dot{C}_{23}$	$c_{22} = c_{23}$
			HE-5	$\dot{C}_{19} + \dot{C}_{24} + \dot{Z}_{HE\text{-}5} = \dot{C}_{20} + \dot{C}_{25}$	$c_{24} = c_{25}$
			E-1	$\dot{C}_8 + \dot{Z}_{E\text{-}1} = \dot{C}_9 + \dot{C}_{Q(E\text{-}1)}$	*None*
			E-2	$\dot{C}_{14} + \dot{Z}_{E\text{-}2} = \dot{C}_{15} + \dot{C}_{Q(E\text{-}2)}$	*None*
			F-1	$\dot{C}_9 + \dot{Z}_{F\text{-}1} = \dot{C}_{10} + \dot{C}_{12}$	$c_{10} = c_{12}$
			F-2	$\dot{C}_{16} + \dot{Z}_{F\text{-}2} = \dot{C}_{17} + \dot{C}_{19}$	$c_{17} = c_{19}$
			F-3	$\dot{C}_{21} + \dot{Z}_{F\text{-}3} = \dot{C}_{22} + \dot{C}_{24}$	$c_{22} = c_{24}$
			V-1	$\dot{C}_{19} + \dot{C}_{16}$	*None*
			V-2	$\dot{C}_{20} + \dot{C}_{21}$	*None*

14.5.1.3 Exergoeconomic Variables

Based on the product and fuel rules presented in the previous section, cost rate related to the fuel ($\dot{C}_F$) and product ($\dot{C}_p$) can be determined for the process equipment. Accordingly, the average cost per unit exergy of the fuel (c_F) and product (c_P) can be defined as follows (Eqs. 14.28, and 14.29) (Mehrpooya et al. 2018):

$$c_{F,k} = \frac{\dot{C}_{F,k}}{\dot{E}_{F,k}} \tag{14.28}$$

$$c_{P,k} = \frac{\dot{C}_{P,k}}{\dot{E}_{P,k}} \tag{14.29}$$

The costs rate associated with the irreversibilities of the *k*th process component is exergy destruction cost. This hidden cost can be obtained using the following equation where the average cost per unit exergy of the fuel (c_F) is constant (Eq. 14.30) (Mehrpooya et al. 2018).

$$\dot{C}_{D,k} = c_{F,k}\dot{E}_{D,k} \tag{14.30}$$

Moreover, the exergoeconomic factor as an important relationship between the exergy related costs (exergy destruction cost) and non-exergy related costs (investment costs) is used to evaluate the process efficiency as follows (Eq. 14.31) (Mehrpooya et al. 2018):

$$f_k = \frac{\dot{Z}_k}{\dot{Z}_k + \dot{C}_{D,k}} \tag{14.31}$$

Relative cost difference (r_k) between unitary average cost of product and fuel exergies for the *k*th equipment can be calculated as below (Eq. 14.32) (Mehrpooya et al. 2018):

$$r_k = \frac{c_{P,k} - c_{F,k}}{c_{F,k}} = \frac{1 - \varepsilon_k}{\varepsilon_k} + \frac{\dot{Z}_k}{c_{F,k}\,\dot{E}_{P,k}} \tag{14.32}$$

14.5.2 Advanced Exergoeconomic Analysis

The cost rate associated with the irreversibilities of the *k*th component can be split into endogenous/exogenous and avoidable/unavoidable parts as follows (Eqs. 14.33, 14.34, 14.35, and 14.36) (Mehrpooya and Ansarinasab 2015):

$$\dot{C}_{D,k}^{EN} = c_{F.k}\,\dot{E}_{D,k}^{EN} \tag{14.33}$$

$$\dot{C}_{D,k}^{EX} = c_{F.k}\,\dot{E}_{D,k}^{EX} \tag{14.34}$$

$$\dot{C}_{D,k}^{UN} = c_{F.k}\,\dot{E}_{D,k}^{UN} \tag{14.35}$$

$$\dot{C}_{D,k}^{AV} = c_{F.k}\,\dot{E}_{D,k}^{AV} \tag{14.36}$$

Similar to the advanced exergy analysis, the mentioned concepts can be consolidated to attain more insights regarding the role of each element on total exergy destruction cost (Eqs. 14.37, 14.38, 14.39, and 14.40) (Mehrpooya and Ansarinasab 2015):

$$\dot{C}_{D,k}^{UN,EN} = c_{F.k}\,\dot{E}_{D,k}^{UN,EN} \tag{14.37}$$

$$\dot{C}_{D,k}^{UN,EX} = c_{F.k}\,\dot{E}_{D,k}^{UN,EX} \tag{14.38}$$

$$\dot{C}_{D,k}^{AV,EN} = c_{F.k}\,\dot{E}_{D,k}^{AV,EN} \tag{14.39}$$

$$\dot{C}_{D,k}^{AV,EX} = c_{F.k}\,\dot{E}_{D,k}^{AV,EX} \tag{14.40}$$

Furthermore, endogenous/exogenous and avoidable/unavoidable investment costs can be written as follows (Eqs. 14.41, 14.42, 14.43, and 14.44) (Mehrpooya and Ansarinasab 2015):

$$\dot{Z}_k^{EN} = \dot{E}_{P,k}^{EN}\left(\frac{\dot{Z}}{\dot{E}_P}\right)_k^{real} \tag{14.41}$$

$$\dot{Z}_k^{EX} = \dot{Z}_k - \dot{Z}_k^{EN} \tag{14.42}$$

$$\dot{Z}_k^{UN} = \dot{E}_{P,k}\left(\frac{\dot{Z}}{\dot{E}_P}\right)_k^{UN} \tag{14.43}$$

$$\dot{Z}_k^{AV} = \dot{Z}_k - \dot{Z}_k^{UN} \tag{14.44}$$

In addition, the integrated endogenous/exogenous and avoidable/unavoidable investment costs can be presented as follows (Eqs. 14.45, 14.46, 14.47, and 14.48) (Mehrpooya and Ansarinasab 2015):

$$\dot{Z}_k^{UN,EN} = \dot{E}_{P,k}^{EN}\left(\frac{\dot{Z}}{\dot{E}_P}\right)_k^{UN} \tag{14.45}$$

$$\dot{Z}_k^{UN,EX} = \dot{Z}_k^{UN} - \dot{Z}_k^{UN,EN} \quad (14.46)$$

$$\dot{Z}_k^{AV,EN} = \dot{Z}_k^{EN} - \dot{Z}_k^{UN,EN} \quad (14.47)$$

$$\dot{Z}_k^{AV,EX} = \dot{Z}_k^{EX} - \dot{Z}_k^{UN,EX} \quad (14.48)$$

In order to compare the performance of each element before and after modification, some parameters can be determined based on avoidable endogenous part of investment and exergy destruction costs as follows (Eqs. 14.49, 14.50, and 14.51) (Ansarinasab and Mehrpooya 2017b):

$$\varepsilon_{modified} = \frac{\dot{E}_{P,k}}{\dot{E}_{F,k} - \dot{E}_{D,k}^{UN} - \dot{E}_{D,k}^{AV,EX}} \quad (14.49)$$

$$f_k^{AV,EN} = \frac{\dot{Z}_k^{AV,EN}}{\dot{C}_{D,k}^{AV,EN} + \dot{Z}_k^{AV,EN}} \quad (14.50)$$

$$\dot{C}_{tot}^{AV,EN} = \dot{C}_{D,k}^{AV,EN} + \dot{Z}_k^{AV,EN} \quad (14.51)$$

where $\varepsilon_{modified}$ refers to the modified exergy efficiency and $f_k^{AV,EN}$ and $\dot{C}_{tot}^{AV,EN}$ refer to the available endogenous exergoeconomic factor and total operating cost, respectively.

14.6 Results and Discussion

14.6.1 Conventional Method

Tables 14.9 and 14.10 tabulate thermodynamic data for both HPWS and CS processes, respectively.

Tables 14.11 and 14.12 summarize the results of the conventional exergy and exergoeconomic analyses of the HPWS and CS processes, respectively. Clearly, the majority of irreversibility of the HPWS process belonged to column T-2 (33304.52 W), compressor C-1 (3304.58 W), and air cooler AC-2 (3200.61 W), respectively. Similarly, these components had the highest exergy destruction costs. The values of exergoeconomic factor of the HPWS varied from a minimum value of 67.21% (pump P-1) to a maximum value of 99.79% (column T-1). To decrease the total cost of the HPWS process, investment cost associated with column T-1 should be decreased even though this modification could increase the cost of exergy destruction. Moreover, the exergetic performance of pump P-1 must be improved even if this could increase the investment cost. Furthermore, compressor C-2 (4857.78 W) and compressor C-1 (4838.64 W) had the highest contributions to the total exergy destruction of the CS process. These components had the highest

Table 14.9 Thermodynamic data for the HPWS process

Stream no.	T (°C)	P (bar)	$\dot{m}$ (kg/h)	h (kJ/kg)	s (kJ/kg.C)	$\dot{E}^{PH}$ (kW)	$\dot{E}^{CH}$ (kW)	$\dot{E}^{TOT}$ (kW)	$\dot{C}$ ($/h)	c ($/GJ)
1	25.00	2.00	533.32	−7103.13	6.74	9.33	2887.18	2896.51	0.00	0.00
2	102.30	4.50	533.32	−6993.54	6.81	22.26	2887.18	2909.44	5.42	0.52
3	40.00	4.50	533.32	−7084.29	6.55	20.48	2887.18	2907.66	8.41	0.80
4	95.93	8.00	533.32	−7005.21	6.60	29.78	2887.18	2916.96	12.67	1.21
5	25.00	8.00	533.32	−7108.62	6.29	28.19	2887.18	2915.37	15.81	1.51
6	13.04	6.00	226.07	−4895.03	9.48	15.56	2720.45	2736.01	21.25	2.16
7	30.37	6.00	72153.29	−15833.57	3.06	20.24	13146.24	13166.48	102.30	2.16
8	25.00	1.20	2885.03	−0.28	5.21	11.65	3.60	15.24	0.00	0.00
9	13.14	1.20	3217.80	−953.97	5.19	12.87	160.90	173.77	1.39	2.22
10	30.33	1.20	71820.51	−15864.19	3.06	4.52	12970.14	12974.65	103.60	2.22
11	30.36	5.00	71820.51	−15863.68	3.06	13.50	12970.14	12983.63	104.90	2.24

Table 14.10 Thermodynamic data for the CS process

Stream no.	T (°C)	P (bar)	$\dot{m}$ (kg/h)	h (kJ/kg)	s (kJ/kg.C)	$\dot{E}^{PH}$ (kW)	$\dot{E}^{CH}$ (kW)	$\dot{E}^{TOT}$ (kW)	$\dot{C}$ ($/h)	c ($/GJ)
1	25.00	2.00	532.93	−7176.71	6.73	9.33	2870.87	2880.20	0.00	0.00
2	147.50	7.00	532.93	−6998.58	6.84	30.86	2870.87	2901.73	7.82	0.75
3	35.00	7.00	532.93	−7166.83	6.37	26.44	2870.87	2897.31	13.34	1.28
4	162.98	25.00	532.93	−6981.19	6.48	49.07	2870.87	2919.93	21.42	2.04
5	35.00	25.00	532.93	−7181.78	5.94	43.35	2870.87	2914.21	27.26	2.60
6	103.19	50.00	532.93	−7090.55	6.00	54.13	2870.87	2924.99	31.98	3.04
7	22.46	50.00	532.93	−7224.98	5.60	51.99	2870.87	2922.86	32.03	3.04
8	−15.92	50.00	532.93	−7295.78	5.34	52.85	2870.87	2923.72	32.08	3.05
9	−45.00	50.00	532.93	−7439.07	4.74	58.13	2870.87	2929.00	31.77	3.01
10	−45.00	50.00	185.56	−8863.21	2.53	15.24	295.92	311.17	3.74	3.34
11	60.00	50.00	185.56	−8477.10	3.98	12.81	295.92	308.74	3.71	3.34
12	−45.00	50.00	347.38	−6678.34	5.92	41.43	2576.40	2617.83	31.46	3.34
13	−48.63	50.00	347.38	−6706.21	5.80	42.28	2576.40	2618.68	31.49	3.34
14	−55.67	50.00	347.38	−6755.10	5.58	43.93	2576.40	2620.33	31.54	3.34
15	−63.45	50.00	347.38	−6807.35	5.34	45.93	2576.40	2622.33	31.48	3.33
16	−70.00	40.00	347.38	−6807.35	5.38	44.76	2576.40	2621.16	31.43	3.33
17	−70.00	40.00	121.86	−8785.68	2.46	11.28	246.14	257.42	3.42	3.69
18	−46.00	40.00	121.86	−8706.23	2.83	10.24	246.14	256.38	3.41	3.69
19	−70.00	40.00	225.52	−5738.26	6.95	31.91	2331.87	2363.78	31.43	3.69
20	−78.03	40.00	225.52	−5801.91	6.63	33.91	2331.87	2365.78	31.39	3.69
21	−120.00	2.00	225.52	−5801.91	7.65	14.93	2331.87	2346.80	31.39	3.72
22	−120.00	2.00	35.08	−9472.21	1.09	3.04	7.19	10.23	0.15	4.12
23	−49.00	2.00	35.08	−8988.10	3.60	0.47	7.19	7.66	0.11	4.12
24	−120.00	2.00	190.44	−5125.86	8.85	10.50	2326.08	2336.58	34.66	4.12
25	−77.49	5.00	190.44	−5050.47	8.87	14.23	2326.08	2340.31	34.71	4.12
26	22.00	5.00	190.44	−4852.34	9.69	11.82	2326.08	2337.90	34.67	4.12

Table 14.11 Results of the conventional exergy and exergoeconomic analysis of the HPWS process

Component	$\dot{E}_F$ (W)	$\dot{E}_P$ (W)	$\dot{E}_D$ (W)	c_F ($/Gj)	c_P ($/Gj)	$\dot{C}_D$ ($/h) × 10^3	$\dot{Z}$ ($/h) × 10^3	ε (%)	y_D (%)	r (%)	*f* (%)
C-1	16234.66	12930.08	3304.58	19.72	101.84	234.60	3587.94	79.64	6.34	416.43	93.86
C-2	11715.03	9305.74	2409.29	19.72	112.31	171.04	2930.74	79.43	4.62	469.52	94.49
AC-1	28597.42	26477.15	2120.27	19.72	45.19	150.52	2277.54	92.59	4.07	129.18	93.80
AC-2	37392.89	34192.28	3200.61	19.72	40.07	227.22	2277.54	91.44	6.14	103.19	90.93
T-1	179364.11	178285.85	1078.27	1.51	5.91	5.85	2822.09	99.40	2.07	292.57	99.79
T-2	191827.83	158523.31	33304.52	2.16	7.31	258.74	2680.12	82.64	63.91	238.63	91.20
P-1	10208.49	8981.00	1227.49	19.72	27.94	87.14	178.63	87.98	2.36	41.68	67.21

Table 14.12 Results of the conventional exergy and exergoeconomic analysis of the CS process

Component	$\dot{E}_F$ (W)	$\dot{E}_P$ (W)	$\dot{E}_D$ (W)	c_F ($/Gj)	c_P ($/Gj)	$\dot{C}_D$ ($/h) × 10^3	$\dot{Z}$ ($/h) × 10^3	ε (%)	y_D (%)	r (%)	*f* (%)
C-1	26370.52	21531.88	4838.64	19.72	86.67	343.50	4846.35	81.65	4.90	339.52	93.38
C-2	27481.82	22624.04	4857.78	19.72	85.00	344.86	4972.32	82.32	4.92	331.06	93.51
C-3	13505.71	10780.09	2725.62	19.72	107.18	193.50	3200.68	79.82	2.76	443.51	94.30
AC-1	45142.11	41443.33	3698.78	19.72	47.65	262.58	3903.87	91.81	3.74	141.61	93.70
AC-2	66207.37	62349.11	3858.26	19.72	38.33	273.91	3903.87	94.17	3.91	94.39	93.44
HE-1	2428.43	2138.41	290.02	3.34	24.25	3.49	157.50	88.06	0.29	626.48	97.84
HE-2	2410.37	862.59	1547.78	4.12	62.23	22.96	157.50	75.79	1.57	1410.49	87.28
HE-3	1038.20	852.75	185.45	3.69	55.80	2.47	157.50	82.14	0.19	1411.00	98.46
HE-4	2568.81	1645.46	923.35	4.12	33.02	13.70	157.50	74.06	0.93	701.47	92.00
HE-5	3734.98	1997.62	1737.36	4.12	29.60	25.77	157.50	73.48	1.76	618.55	85.94
E-1	6504.48	5276.64	1227.84	2.66	9.92	11.76	126.11	81.12	1.24	272.84	91.47
E-2	2125.68	1998.76	126.91	2.66	20.35	1.22	126.11	94.03	0.13	665.21	99.05

exergy destruction cost as well. Additionally, the higher values of exergoeconomic factor for the components of the CS process indicated that their investment costs must be minimized in order to decrease total cost of the process.

14.6.2 Advanced Exergy Analysis

Figure 14.5 depicts the results of advanced exergy analysis for the HPWS process. Obviously, the endogenous portion of exergy destruction was larger than its

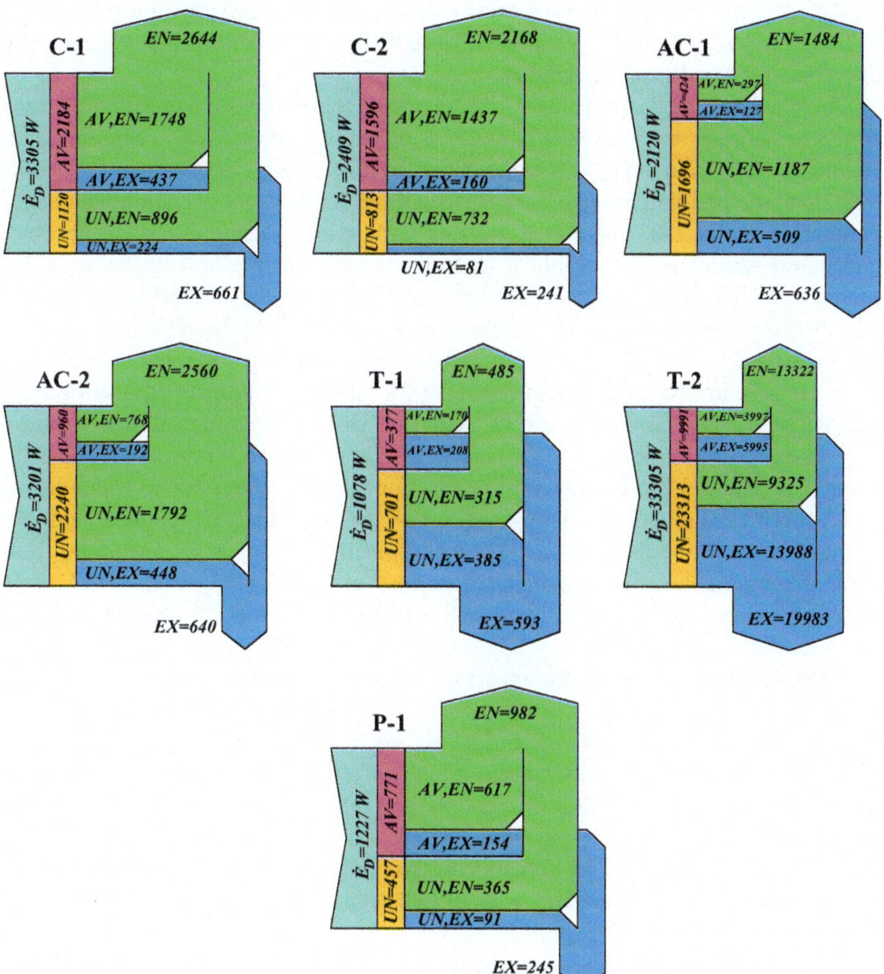

Fig. 14.5 Breakdown of exergy destruction rate of the components of HPWS process

corresponding exogenous part for all the process equipment except columns T-1 and T-2. Therefore, technical relationship between the process equipment was not considerable. According to the avoidable part of exergy destruction, potential improvement of compressors C-1 and C-2 and pump P-1 were higher than the other process components. However, the total irreversibility rate associated with pump P-1 was very low, showing its trivial importance on the overall performance. Based on the avoidable endogenous part of exergy destruction, compressors C-1 (1748 W) and C-2 (1437 W) should be first modified, respectively.

Figure 14.6 shows the results of advanced exergy analysis for the CS process. The endogenous part of exergy destruction showed a significant contribution to the total irreversibilities compared with its exogenous part for all the components. This indicated weak, interactions between the CS process equipment from technical viewpoint. The avoidable part of exergy destruction was large for the compressors C-1 (3187 W), C-2 (2241 W), and C-3 (1806 W), while this value was small for the air coolers and heat exchangers.

Figures 14.7, 14.8, 14.9 and 14.10 manifest the results of advanced exergoeconomic analysis for the HPWS and CS processes, respectively. Similar to the advanced exergy analysis, the endogenous cost of exergy destruction was higher than its corresponding exogenous cost for all the HPWS process equipment except columns T-1 and T-2. The cost associated with exergy destruction of compressors and pump was avoidable, while it was unavoidable for the heat exchangers, columns, and air coolers of the HPWS process. Based on the avoidable endogenous part of exergy destruction cost, compressors C-2 and C-1 of the HPWS process and compressors C-1 and C-3 of the CS process should be first modified, respectively.

According to Fig. 14.9, investment cost for all the equipment of HPWS process was endogenous except columns T-1 and T-2. Unlike the exergy destruction cost, investment cost for the air coolers and heat exchangers was avoidable, while this value was unavoidable for the compressor, pump, and column due to technological and economic limitations. This manifested the fact that air coolers and heat exchangers could be improved exergoeconomically. Furthermore, according to the avoidable endogenous part of investment cost, air coolers AC-2 and AC-1 of the HPWS process and heat exchangers HE-4 and HE-5 of the CS process should be first modified, respectively.

Table 14.13 summarizes a comparison between the results of conventional and advanced analyses for both processes on the basis of exergetic efficiency, total cost, and exergoeconomic factor. Clearly, exergetic efficiencies of all the process equipment could increase after performing the required modifications. Moreover, exergoeconomic factor of the compressors and pump could decrease after carrying out modifications for both processes, while exergoeconomic factor of the heat exchangers, columns, and air coolers could increase. Overall, air coolers should be first modified since their total costs were higher than the other equipment, followed by compressors.

Table 14.14 presents three different strategies for reducing the avoidable exergy destruction cost. These strategies should be performed either on the process or its units. Strategy A could be applied when the cost rate is high due to high

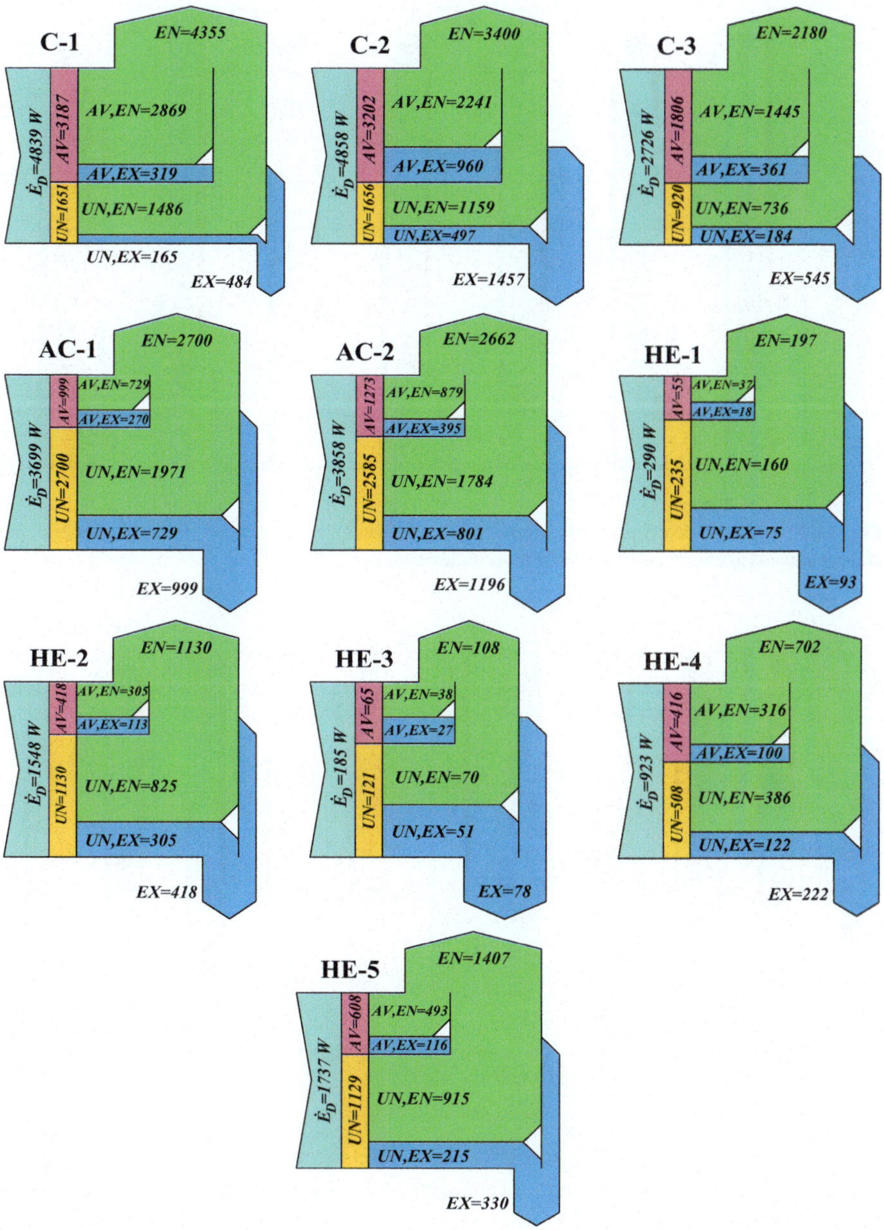

Fig. 14.6 Breakdown of exergy destruction rate of the components of CS process

irreversibilities in the elements involved. Clearly, strategy A should be applied to all the equipment. Strategy B could be applied for some equipment such as air cooler AC-1 and columns of the HPWS process and heat exchangers HE-1, HE-2, HE-3,

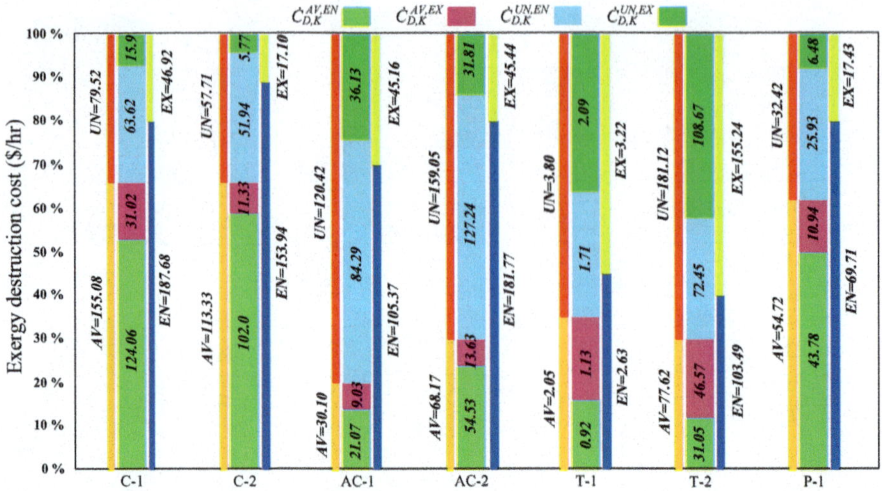

Fig. 14.7 Splitting exergy destruction costs of the components of HPWS process into endogenous/exogenous and avoidable/unavoidable parts

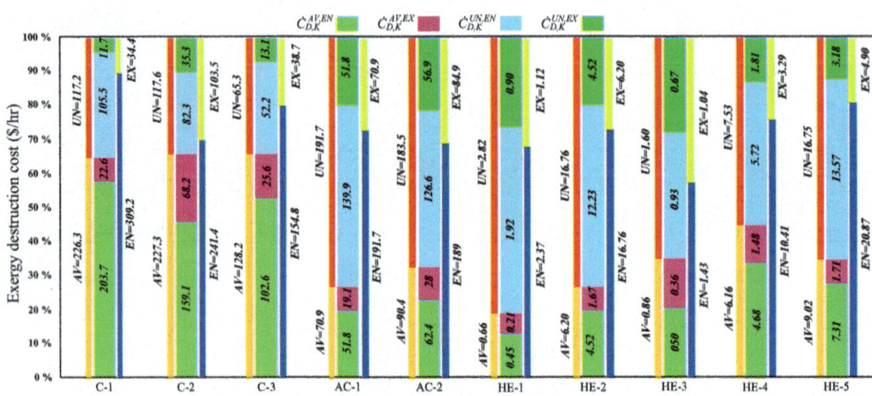

Fig. 14.8 Splitting exergy destruction costs of the components of CS process into endogenous/exogenous and avoidable/unavoidable parts

HE-4, and air coolers of the CS process where the avoidable exogenous part of exergy destruction cost was large compared with its avoidable endogenous portion. Finally, strategy C could be an inevitable option if the avoidable exogenous cost had a high contribution to the total avoidable exergy destruction cost. This strategy could be used for some equipment such as columns of the HPWS process and air cooler AC-1 and heat exchangers HE-1 and HE-3 of the CS process.

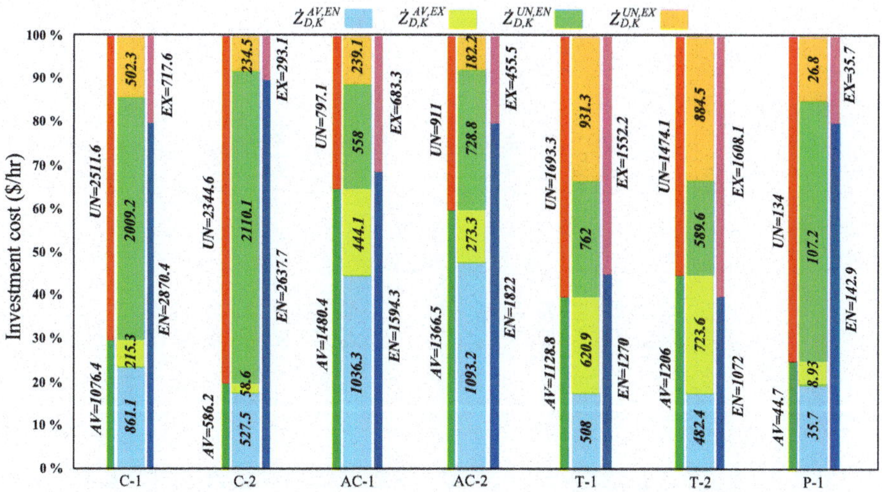

Fig. 14.9 Splitting investment costs of the components of HPWS process into endogenous/ exogenous and avoidable/unavoidable parts

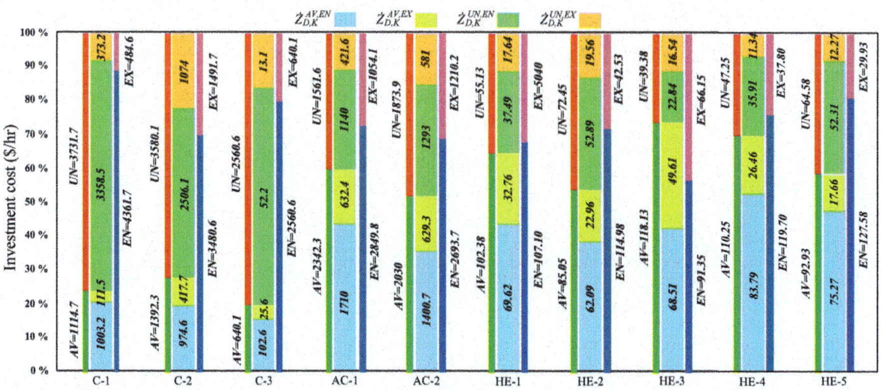

Fig. 14.10 Splitting investment costs of the components of CS process into endogenous/ exogenous and avoidable/unavoidable parts

14.6.3 Cost Sensitivity Analysis

Various operating variables such as pressure drop in valves and pressure ratio of compressors and pumps can profoundly affect the process efficiency. In order to reduce costs and improve the process performance, these operating variables should be elaborately adjusted. Figure 14.11 illustrates the variations of avoidable endogenous exergoeconomic factor and advanced exergy destruction cost of compressor C-2 and air cooler AC-1 of the HPWS process by changing the pressure

Table 14.13 Comparison between the results of conventional and advanced analyses for both HPWS and CS processes

Process	Component	Conventional			Advanced		
		ε	f	$\dot{C}_{tot}$	$\varepsilon_{modified}$	$f^{AV,EN}$	$\dot{C}_{tot}^{AV,EN}$
		(%)	(%)	(\$/hr) × 10^3	(%)	(%)	(\$/h) × 10^3
HPWS							
	C-1	79.64	93.86	3822.54	88.09	87.41	985.17
	C-2	79.43	94.49	3101.78	86.63	83.80	629.53
	AC-1	92.59	93.80	2428.06	98.89	98.01	1057.35
	AC-2	91.44	90.93	2504.75	97.80	95.25	1147.75
	T-1	99.40	99.79	2827.94	99.90	99.82	508.90
	T-2	82.64	91.20	2938.86	97.54	93.95	513.47
	P-1	87.98	67.21	265.77	93.57	44.94	79.50
CS							
	C-1	81.65	93.38	5189.85	88.24	83.13	1206.85
	C-2	82.32	93.51	5317.19	90.99	85.97	1133.67
	C-3	79.82	94.30	3394.18	88.18	83.31	614.68
	AC-1	91.81	93.70	4166.46	98.27	97.06	1761.65
	AC-2	94.17	93.44	4177.78	98.61	95.74	1463.08
	HE-1	88.06	97.84	160.99	98.28	99.36	70.07
	HE-2	75.79	87.28	180.46	83.87	93.21	66.61
	HE-3	82.14	98.46	159.97	95.77	99.27	69.01
	HE-4	74.06	92.00	171.20	83.90	94.71	88.47
	HE-5	73.48	85.94	183.27	80.22	91.15	82.58

Table 14.14 Three different strategies for discounting avoidable cost of exergy destruction

Process	Component	Cost of exergy destruction categories ($/hr)				The part should be focused	Possible strategies to reduce cost of exergy destruction		
		$\dot{C}_D$	$\dot{C}_{D,k}^{AV}$	$\dot{C}_{D,k}^{AV,EN}$	$\dot{C}_{D,k}^{AV,EX}$		Strategy A[a]	Strategy B[b]	Strategy C[c]
HPWS									
	C-1	234.60	155.08	124.06	31.02	EN.	*		
	C-2	171.04	113.33	102.00	11.33	EN.	*		
	AC-1	150.52	30.10	21.07	9.03	EN./EX.	*	*	
	AC-2	227.22	68.17	54.53	13.63	EN.	*		
	T-1	5.85	2.05	0.92	1.13	EN./EX.	*	*	*
	T-2	258.74	77.62	31.05	46.57	EN./EX.	*	*	*
	P-1	87.14	54.72	43.78	10.94	EN.	*		
CS									
	C-1	343.50	226.28	203.65	22.63	EN.	*		
	C-2	344.86	227.29	159.10	68.19	EN.	*		
	C-3	193.50	128.21	102.57	25.64	EN.	*		
	AC-1	262.58	70.90	51.76	19.14	EN./EX.	*	*	
	AC-2	273.91	90.39	62.37	28.02	EN./EX.	*	*	*
	HE-1	3.49	0.66	0.45	0.21	EN./EX.	*	*	*
	HE-2	22.96	6.20	4.52	1.67	EN./EX.	*	*	
	HE-3	2.47	0.86	0.50	0.36	EN./EX.	*	*	*
	HE-4	13.70	6.16	4.68	1.48	EN./EX.	*	*	
	HE-5	25.77	9.02	7.31	1.71	EN.	*		

[a]Strategy A: Improving the efficiency of the *k*th component or replacing the component with efficient devices
[b]Strategy B: Improving the efficiency of the remaining components
[c]Strategy C: Structural optimization of the overall system

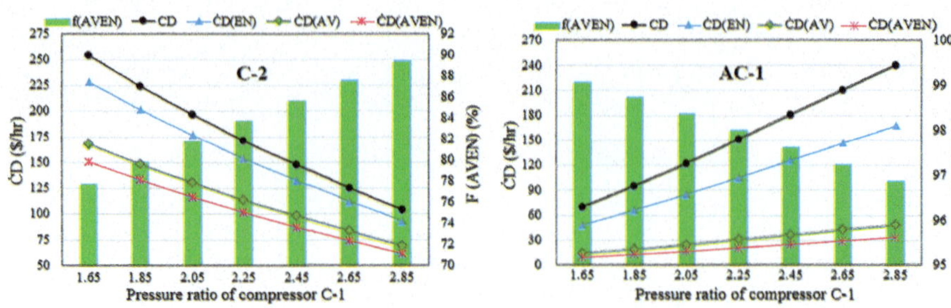

Fig. 14.11 Variations in advanced exergy destruction cost and exergoeconomic factor of the compressor C-2 and air cooler AC-1 of the HPWS process by changing the pressure ratio of the compressor C-1

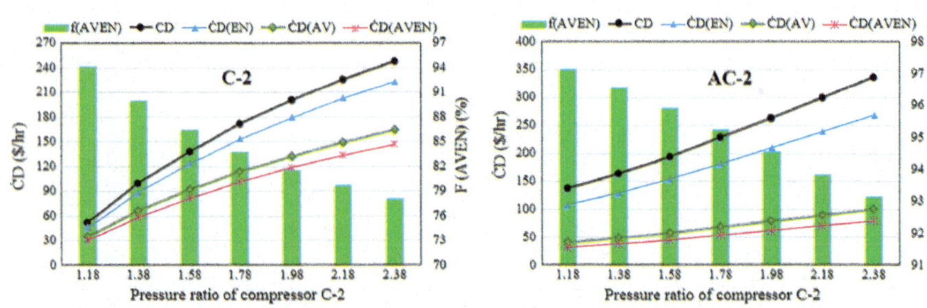

Fig. 14.12 Variations in advanced exergy destruction cost and exergoeconomic factor of the compressor C-2 and air cooler AC-2 of the HPWS process by changing the pressure ratio of the compressor C-2

ratio of compressor C-1 from 1.65 to 2.85. Obviously, avoidable endogenous exergoeconomic factor of the compressor C-2 increased by increasing the pressure ratio of the compressor C-1, while its advanced exergy destruction cost decreased. However, an opposite behavior was found for the air cooler AC-1. This could be attributed to the fact that power consumption of the compressor C-1 increased by increasing the pressure ratio of compressor C-1, while it decreased for the compressor C-2.

Figure 14.12 explains the variations of avoidable endogenous exergoeconomic factor and advanced exergy destruction cost of the compressor C-2 and air cooler AC-2 of the HPWS process by modifying the pressure ratio of compressor C-2. It is clear from this figure that avoidable endogenous exergoeconomic factor of the compressor C-2 and air cooler AC-2 decreased by increasing the pressure ratio of the compressor C-2, while their advanced exergy destruction costs decreased. This could be ascribed to more electrical power consumption of the compressor C-2 and more cold energy demand of the air cooler AC-2 by increasing the pressure ratio of the compressor C-2.

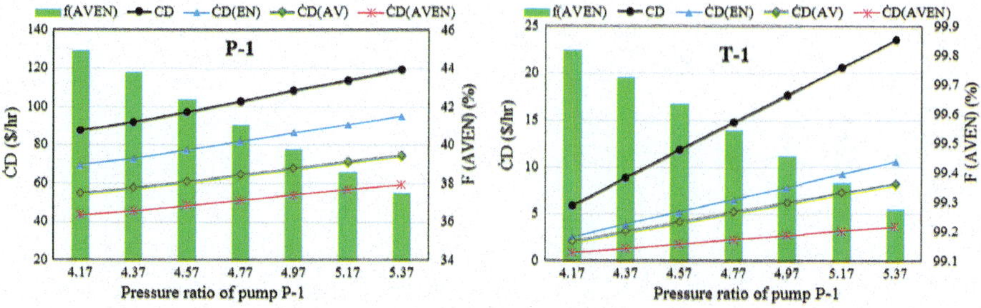

Fig. 14.13 Variations in advanced exergy destruction cost and exergoeconomic factor of the pump P-1 and column T-1 of the HPWS process by increasing the pressure ratio of pump P-1

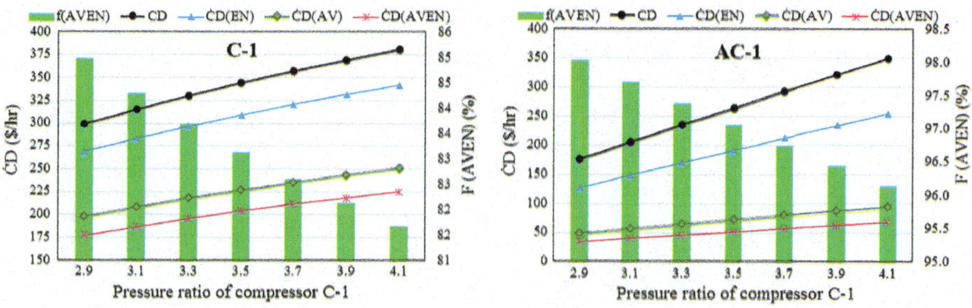

Fig. 14.14 Variations in advanced exergy destruction cost and exergoeconomic factor of the compressor C-1 and air cooler AC-1 of the CS process by increasing the pressure ratio of the compressor C-1

Figure 14.13 reveals the variations of avoidable endogenous exergoeconomic factor and advanced exergy destruction cost of the pump P-1 and column T-1 of the by HPWS process by changing the pressure ratio of pump P-1. Clearly, advanced exergy destruction cost of the pump P-1 increased as the pressure ratio of pump P-1 increased from 4.17 to 5.37, while avoidable endogenous exergoeconomic factor of the column T-1 decreased. This occurred due to the fact that the irreversibilities increased in column T-1 by increasing the pressure ratio of pump P-1.

Figure 14.14 displays the variations in advanced exergy destruction cost and exergoeconomic factor of the compressor C-1 and air cooler AC-1 of the CS process by increasing the pressure ratio of the compressor C-1. It is obvious from this figure that avoidable endogenous exergoeconomic factor of the compressor C-1 and air cooler AC-1 decreased as the pressure ratio of compressor C-1 increased, while advanced exergy destruction cost showed a slight increment. The reason for this behavior was that more electrical power was required by compressor C-1 and more cold energy was demanded by air cooler AC-1 as the pressure ratio of the compressor C-1 increased.

Figure 14.15 indicates the variations in advanced exergy destruction cost and exergoeconomic factor of the compressors C-2 and C-3 of the CS process by

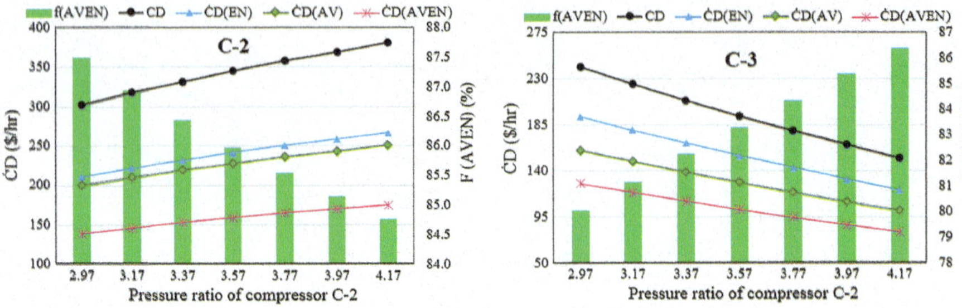

Fig. 14.15 Variations in advanced exergy destruction cost and exergoeconomic factor of the compressors C-2 and C-3 of the CS process by increasing the pressure ratio of compressor C-2

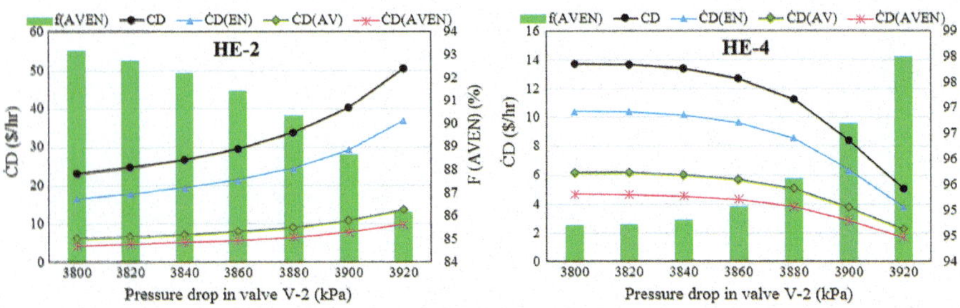

Fig. 14.16 Variations in advanced exergy destruction cost and exergoeconomic factor of the heat exchangers HE-2 and HE-4 of the CS process by increasing the pressure drop in the expansion valve V-4

increasing the pressure ratio of the compressor C-2. Clearly, increasing the pressure ratio of compressor C-2 from 2.97 to 4.17 led to a decrease in avoidable endogenous exergoeconomic factor of the compressor C-2, while this change increased its advanced exergy destruction cost. Unlike the compressor C-2, avoidable endogenous exergoeconomic factor of the compressor C-3 increased by increasing the pressure ratio of compressor C-2, while its advanced exergy destruction cost decreased. This occurred since the power consumption of the compressor C-2 increased by enhancing the pressure ratio of compressor C-2, whereas the power consumption of the compressor C-3 decreased.

Figure 14.16 demonstrates the variations in advanced exergy destruction cost and exergoeconomic factor of the heat exchangers HE-2 and HE-4 of the CS process by increasing the pressure drop in the expansion valve V-4. Clearly, avoidable endogenous exergoeconomic factor of the heat exchanger HE-2 decreased by increasing the pressure drop in the expansion valve V-2, while its advanced exergy destruction cost increased. Moreover, opposite findings were obtained for the heat exchanger HE-4.

Overall, exergy-based methods, particularly advanced approaches, can be very powerful tools with a wide variety of applications for various biogas production, upgrading, and utilization systems. Future works should include conventional and advanced exergoenvironmental analyses for component-level analysis of biogas plants from exergy/environmental viewpoint. The application of advanced exergoeconomic and exergoenvironmental approaches together with elaborated evolutionary-and knowledge-based optimization tools should be also considered in future works in order to make decisions on operational conditions of biogas plants.

14.7 Conclusions

Conventional and advanced exergy and exergoeconomic analyses were successfully applied for two biogas upgrading processes, i.e., HPWS and CS processes. Conventional method showed that the majority of irreversibility of the HPWS process belonged to column T-2 (33304.52 W), compressor C-1 (3304.58 W), and air cooler AC-2 (3200.61 W), respectively, Moreover, the majority of irreversibility of the CS process originated from from compressor C-2 (4857.78 W) and compressor C-1 (4838.64 W). These components had the highest exergy destruction costs as well. Therefore, to minimize total irreversibilities and to mitigate inefficient costs of both processes, these components must be first considered. The main results of advanced exergy and exergoeconomic analyses are summarized as below:

- Unlike the air coolers, columns, and heat exchangers, the cost rates of exergy destruction of the pump and compressors were avoidable.
- The investment cost rates of the pump and compressors were unavoidable, while these cost rates were avoidable for the air coolers, columns, and heat exchangers.
- The endogenous part of exergy destruction and investment costs dominated over exogenous counterpart, indicating weak economic and technical interactions between the processes components.
- Improving the efficiencies of the components or replacing them with more efficient ones can be a useful strategy for reducing the exergy destruction cost rate.

References

Ansarinasab H, Mehrpooya M (2017a) Advanced exergoeconomic analysis of a novel process for production of LNG by using a single effect absorption refrigeration cycle. Appl Therm Eng 114:719–732

Ansarinasab H, Mehrpooya M (2017b) Evaluation of novel process configurations for coproduction of LNG and NGL using advanced exergoeconomic analysis. Appl Therm Eng 115:885–898

Ansarinasab H, Afshar M, Mehrpooya M (2016) Exergoeconomic evaluation of LNG and NGL co-production process based on the MFC refrigeration systems. Iran J Oil Gas Sci Technol 5(3):45–61

Ansarinasab H, Mehrpooya M, Parivazh MM (2017a) Evaluation of the cryogenic helium recovery process from natural gas based on flash separation by advanced exergy cost method–Linde modified process. Cryogenics 87:1–11

Ansarinasab H, Mehrpooya M, Mohammadi A (2017b) Advanced exergy and exergoeconomic analyses of a hydrogen liquefaction plant equipped with mixed refrigerant system. J Clean Prod 144:248–259

Bejan A, Tsatsaronis G (1996) Thermal design and optimization. Wiley

Cao L, Liu J, Xu X (2016) Robustness analysis of the mixed refrigerant composition employed in the single mixed refrigerant (SMR) liquefied natural gas (LNG) process. Appl Therm Eng 93:1155–1163

Cooper JR, Penney WR, Fair JR, Walas SM (2010) Chemical process equipment–selection and design. Butterworth-Heinemann

Dadak A, Aghbashlo M, Tabatabaei M, Najafpour G, Younesi H (2016) Exergy analysis as a tool for decision making on substrate concentration and light intensity in photobiological hydrogen production. Energy Technol 4(3):429–440

Dincer I (2011) Exergy as a potential tool for sustainable drying systems. Sustain Cities Soc 1(2): 91–96

Kelly S, Tsatsaronis G, Morosuk T (2009) Advanced exergetic analysis: approaches for splitting the exergy destruction into endogenous and exogenous parts. Energy 34(3):384–391

Lazzaretto A, Tsatsaronis G (1997) On the quest for objective equations in exergy costing. In Conference on the quest for objective equations in exergy costing, AES, vol 37, pp 197–210

Mehrpooya M, Ansarinasab H (2015a) Advanced exergoeconomic evaluation of single mixed refrigerant natural gas liquefaction processes. J Nat Gas Sci Eng 26:782–791

Mehrpooya M, Ansarinasab H (2015b) Exergoeconomic evaluation of single mixed refrigerant natural gas liquefaction processes. Energy Convers Manag 99:400–413

Mehrpooya M, Sharifzadeh MMM, Rosen MA (2015) Optimum design and exergy analysis of a novel cryogenic air separation process with LNG (liquefied natural gas) cold energy utilization. Energy 90:2047–2069

Mehrpooya M, Dehghani H, Moosavian SA (2016) Optimal design of solid oxide fuel cell, ammonia-water single effect absorption cycle and Rankine steam cycle hybrid system. J Power Sources 306:107–123

Mehrpooya M, Ansarinasab H, Sharifzadeh MMM, Rosen MA (2017) Process development and exergy cost sensitivity analysis of a hybrid molten carbonate fuel cell power plant and carbon dioxide capturing process. J Power Sources 364:299–315

Mehrpooya M, Sharifzadeh MMM, Ansarinasab H (2018) Investigation of a novel integrated process configuration for natural gas liquefaction and nitrogen removal by advanced exergoeconomic analysis. Appl Therm Eng 128:1249–1262

Robinson DB, Peng D-Y, Chung SY (1985) The development of the Peng-Robinson equation and its application to phase equilibrium in a system containing methanol. Fluid Phase Equilib 24(1–2):25–41

Sadaghiani MS, Mehrpooya M, Ansarinasab H (2017) Process development and exergy cost sensitivity analysis of a novel hydrogen liquefaction process. Int J Hydrogen Energy 42:29797–29819

Vatani A, Mehrpooya M, Palizdar A (2014) Advanced exergetic analysis of five natural gas liquefaction processes. Energy Convers Manag 78:720–737

Chapter 15
Advanced Soft Computing Techniques in Biogas Production Technology

Fatemeh Almasi, Salman Soltanian, Soleiman Hosseinpour, Mortaza Aghbashlo and Meisam Tabatabaei

15.1 Introduction

The increasing of global energy demand and environmental concerns caused by using fossil fuels have pleaded in favor of using alternative renewable and sustainable energy resources. Among the various renewable energy resources developed to date, biofuels obtained from biomass via thermochemical and biochemical routes are excellent energy sources for completely/partially replacing the conventional fossil fuels. Anaerobic digestion is a promising biochemical pathway for converting organic matter into biogas using bacterial consortiums in the absence of oxygen. The obtained gas contains 50–70% methane, 30–40% carbon dioxide, and trace amounts of other gases such as hydrogen sulfide, ammonia, and hydrogen. In this process, solid and liquid biofertilizers having valuable agronomical features are also generated.

Fast and accurate modeling, optimization, control, and fault diagnose of anaerobic digestion systems are extremely critical aspects of biogas production

F. Almasi · S. Hosseinpour (✉) · M. Aghbashlo (✉)
Department of Mechanical Engineering of Agricultural Machinery, Faculty of Agricultural Engineering and Technology, College of Agriculture and Natural Resources, University of Tehran, Karaj, Iran
e-mail: shosseinpour@ut.ac.ir

M. Aghbashlo
e-mail: maghbashlo@ut.ac.ir

S. Soltanian
Biofuel Research Team (BRTeam), Karaj, Iran

M. Tabatabaei (✉)
Biofuel Research Team (BRTeam) / Agricultural Biotechnology Research Institute of Iran (ABRII), Agricultural Research, Education and Extension Organization (AREEO), Karaj, Iran
e-mail: meisam_tabatabaei@abrii.ac.ir; meisam_tab@yahoo.com

M. Tabatabaei and H. Ghanavati (eds.), *Biogas*, Biofuel and Biorefinery Technologies 6, https://doi.org/10.1007/978-3-319-77335-3_15

technology. However, anaerobic digestion process is very complex as it depends on many endogenous and exogenous factors. A considerable amount of studies have been conducted to assess the effects of influential factors like temperature, total solid, volatile solid, ultimate and proximate characterizations of the input feedstocks, carbon/nitrogen ratio, chemical oxygen demand, alkalinity, organic loading rate, hydraulic retention time, volatile fatty acid, pH, co-substrate inputs (co-digestion), pretreatment type, and reactor structure on the biogas production process (Carrère et al. 2010; Khalid et al. 2011; Horváth et al. 2016). Despite the extensive attempts have been made to scrutinize the details of the anaerobic digestion process, it is impossible to consider all the influential parameters simultaneously. In addition, it is very difficult or even impossible to identify the relationships between inputs and outputs of such complex biochemical processes using the conventional mathematical and phenomenological approaches. Unlike the conventional modeling techniques, heuristically soft computing methods can easily deal with the complexity, nonlinearity, and uncertainty of ill-defined biogas production systems using experimental data. It is worth mentioning that these techniques do not need any information about the mechanisms and principles of digestion process. Therefore, using advanced soft computing techniques is vital to control the anaerobic digestion process, to diagnose faults occurring in the process, to discount energy and operating costs, to optimize the operating conditions of the digesters, and to enhance biogas or methane production rate. Accordingly, this chapter is devoted to the applications of soft computing techniques for modeling and optimization of biogas production processes.

This chapter is arranged into four main sections. In the first section, after a brief description of artificial neural network (ANN) and its learning algorithms, an overview is provided of the most important investigations relating to the application of this approach for modeling and predicting the biogas production processes. In the second parts, fuzzy logic systems are introduced to predict and control anaerobic digestion processes. In the third section, evolutionary algorithms like genetic algorithm (GA), ant colony optimization (ACO), and particle swarm optimization (PSO) used for optimization purposes are introduced. In the last part, hybrid models are briefly presented and discussed and their applications in biogas production processes are summarized. It should be noted that it is impossible to introduce and comprehensively explain all soft computing techniques used in this domain in a single book chapter. For details, the readers are referred to the textbooks like (Nguyen and Sugeno 2012; Gazi and Passino 2011; Dorigo and Stützle 2010).

15.2 Artificial Neural Network (ANN)

Figure 15.1 shows the main components of a biological neural network and manifests its similarity with an ANN. The ANN model is an extended version of the first artificial neuron proposed by McCulloch and Pitts in the year 1943 (McCulloch and Pitts 1943). Cell body of neuron is called "Soma" which is connected to fibers

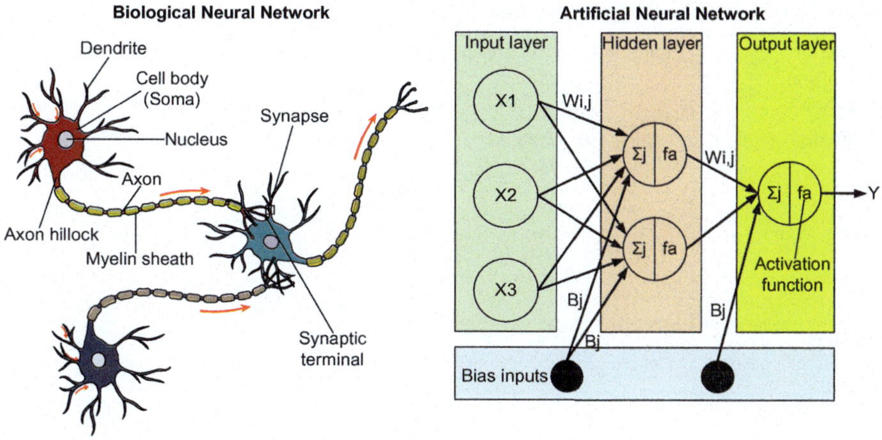

Fig. 15.1 The main components of a biological neural network and an ANN

branches called "Dendrites". Dendrite has a connection rule for bringing input signals to the neuron. The neuron is activated when the intensity of the received signals reaches to a specific excitation threshold. Consequently, a new signal is transferred to the "Synaptic terminals" via a connecting line named "Axon". Then, the signal is sent to other neurons by "Synapses" located on their dendrites or their cell body.

There are many different types of ANN models used for predicting scientific and engineering problems. The multi-layer perceptron artificial neural network (MLP ANN) model is one of the most used ANN architectures. This network is composed of at least three layers, i.e., input layer, hidden layer, and output layer. The interconnections between artificial neurons are called weight and biases. The inputs of *j*th neuron ($X_{i,j}, i = 1, 2, \ldots, D$ where D is the number of neurons connecting to the *j*th neuron) are multiplied by their weights ($W_{i,j}$) and summed together and then bias input of *j*th neuron (B_j) is added to it to form Σ_j (Eq. 15.1)

$$\Sigma_j = B_j + \sum_{i=1}^{D} W_{i,j} X_{i,j} \tag{15.1}$$

The obtained value (Σ_j) is then fed into a activation function, i.e., $f_a(\Sigma_j)$. Table 15.1 lists some important activation functions. Each neuron is linked to others by weighted connections adjusted by training the network (Dibaba et al. 2016; Sewsynker-Sukai et al. 2017).

Training is the process of changing the weights between layers to reduce the differences between computational outputs and desired outputs at an allowable level. After the training, the model is often validated by the set of data which is never seen before by network. Then, the output of a new set of data can be predicted by the trained network (Almasi et al. 2014). The best model is chosen according to

Table 15.1 Some important activation functions

Name of activation function	Equation of activation function	Plot
Unit step	$f_a(x) = \begin{cases} 0 & for\, x < 0 \\ 1 & for\, x \geq 0 \end{cases}$	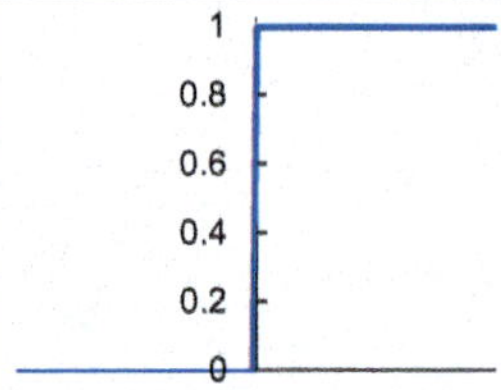
Identity	$f_a(x) = x$	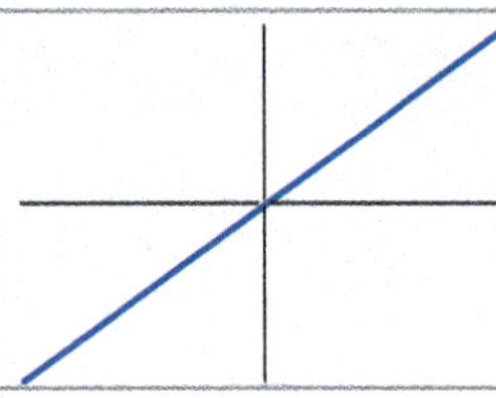
Piecewise linear	$f_a(x) = \begin{cases} 1 & for\, x \geq x_{max} \\ \frac{x - x_{min}}{x_{max} - x_{min}} & for\, x_{min} < x < x_{max} \\ 0 & for\, x \leq x_{min} \end{cases}$	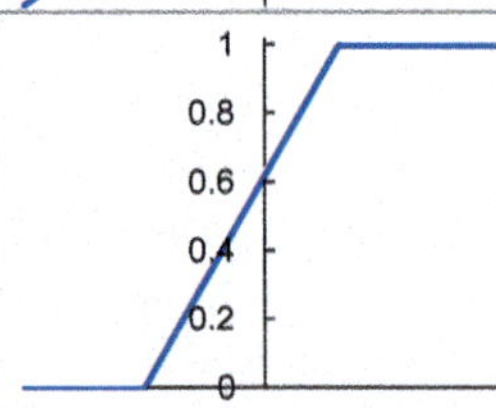
Sigmoid	$f_a(x) = \frac{1}{1 + e^{-\alpha x}}$	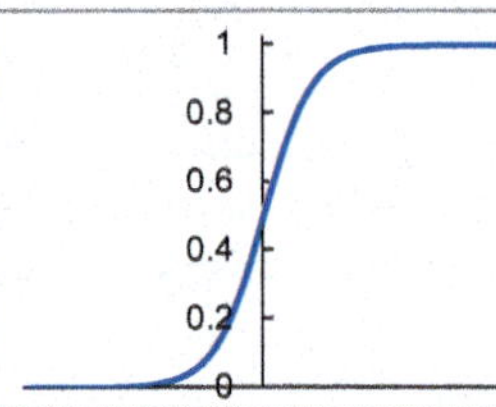
Tangent	$f_a(x) = \tan(ax)$	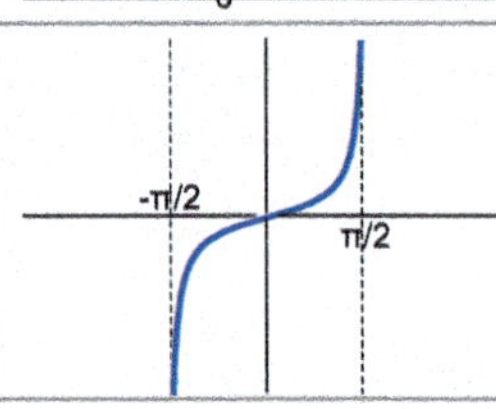
Hyperbolic tangent	$f_a(x) = \tanh(x) = \frac{e^{\alpha x} - e^{-\alpha x}}{e^{\alpha x} + e^{-\alpha x}}$	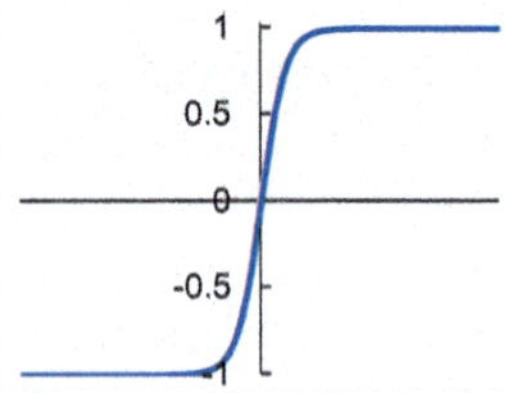

(continued)

Table 15.1 (continued)

Name of activation function	Equation of activation function	Plot
Gaussian	$f_a(x) = e^{-\alpha x^2}$	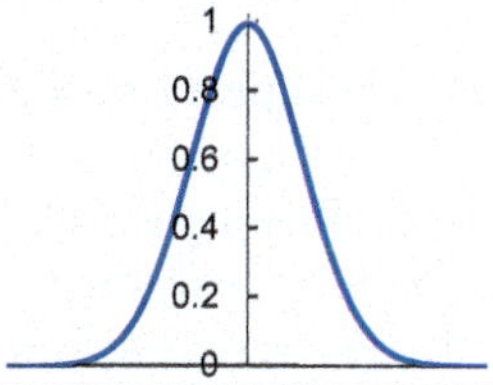

the statistical indicators and model simplicity. The main statistical parameters often used for assessing the performance of ANN models are Mean Absolute Error (MAE), Mean Square Error (MSE), Root Mean Square Error (RMSE) and Regression Coefficient (R^2) (Table 15.2).

15.2.1 Neural Networks Learning

The network can learn by changing the connection weights and biases so-called "training algorithm". Learning techniques in ANNs include supervised learning and unsupervised learning.

15.2.1.1 Supervised Learning Rule

In the supervised learning rule, the inputs and the desired outputs are available. The predicted outputs are compared with desired outputs and the error value is then computed. The error is minimized by adjusting the weights and biases repeating process over and over (Sewsynker-Sukai et al. 2016). This learning method include

Table 15.2 Main statistical parameters often used for assessing the performance of ANN models

Statistical parameter for prediction accuracy	Equation
Mean Absolute Error (MAE)	$\frac{1}{n}\sum_{i=1}^{n}\left\|y_i^{predicted} - y_i^{observed}\right\|$
Mean Square Error (MSE)	$\frac{1}{n}\sum_{i=1}^{n}\left\|y_i^{predicted} - y_i^{observed}\right\|^2$
Root Mean Square Error (RMSE)	$\sqrt{\frac{1}{n}\sum_{i=1}^{n}\left\|y_i^{predicted} - y_i^{observed}\right\|^2}$
Regression Coefficient (R^2)	$1 - \frac{\sum_{i=1}^{n}\left(y_i^{observed} - y_i^{predicted}\right)^2}{\sum_{i=1}^{n}\left(y_i^{observed} - y_i^{meanobserved}\right)^2}$

Widrow–Hoff rule, Gradient descent, Delta rule, Back propagation (BP) rule, Cohen–Grossberg learning rule, and Adaptive conjugate gradient model of Adeli and Hung (Siddique and Adeli 2013).

15.2.1.2 Unsupervised Learning Rule

Unlike the supervised rule, there is no desired output(s) in the unsupervised rule. It searches for characteristics and features of the inputs to find a pattern between inputs. Unsupervised learning rules are used for clustering and affinity methods. Hebbian's learning rule and Kohonen's rule belong to the unsupervised training rule (Siddique and Adeli 2013). It is noted that this learning rule is rarely used in biogas production problems domain.

15.2.2 Applications of ANN Technology in Biogas Production

Table 15.3 shows some important applications of ANN technology in order to model the anaerobic digestion process for biogas production. According to Table 15.3, prediction of the evolved biogas compositions was the main issue attracting researchers' attention to use ANN technology in this domain.

15.3 Fuzzy Logic

Traditional models and control laws for controlling systems are linear while not all real-world systems can be solved using linear models. Therefore, fuzzy logic systems have been invented, developed, and applied to numerous non-linear systems, fitting mathematical modeling and human thinking for controlling systems. These approaches provide a model of reasoning propositions of human having an approximate value. Fuzzy logic is a multivalued logic and relies on the theory of fuzzy sets achieved by generalization and expansion of crisp sets in a natural way.

A fuzzy set A is mapped onto X with a real number in range of [0, 1] by a membership function (MF) $\mu_A(x)$. It means that the quantity of $\mu_A(x)$ at X illustrates the membership degree of X in A. A fuzzy set A in X is represented as a set of arranged pairs that X is a universe of discourse and its elements are defined by x (Eq. 15.2).

$$A = \{(x, \mu_A(x)) | x \in X\} \tag{15.2}$$

Table 15.3 Some important applications of ANN technology in order to model the anaerobic digestion process for biogas production

ANN model	Aim(s)	Model input(s)	Model output(s)	Best model(s)	Result(s)	References
Batch gradient decent with momentum algorithm	To predict the trace compounds	In H_2S model: Sulfate loading rate [gSO_4Sm^{-3} d^{-1}], OLR [$kgCODm^{-3}$ d^{-1}], and H_2S in biogas [ppm] In NH_3 model: Total nitrogen loading rate [gNm^{-3} d^{-1}], OLR [$kgCODm^{-3}$ d^{-1}], NH_3 in biogas [ppm], biogas-productivity [m^3 biogas m^{-3} d^{-1}], pH, ammonia in reactor [$mgN\text{-}NH_3$ l^{-1}]	In H_2S model: Hydrogen sulfide in biogas [ppm] In NH_3 model: Ammonia in biogas [ppm]	In H_2S model: 5 hidden neurons and tangent sigmoid and pure linear transfer functions in the hidden and output layers, respectively In NH_3 model: 7 hidden neurons and tangent sigmoid and pure linear transfer functions in the hidden and output layers, respectively	This model could be successfully used to control the process and avoid the harmful trace compounds production	Strik et al. (2005)
MLP ANN with BP training algorithm	To model and control biogas production in a thermophilic upflow anaerobic sludge blanket (UASB) reactor	OLR, VFA of the effluent, influent-effluent ALK, influent–effluent pH, and temperature of the reactor	Biogas production	Sigmoid transfer function	The developed model satisfactorily estimated biogas production in a thermophilic UASB	Kanat and Saral (2009)
Feed-forward BP ANN model, radial basis function-based neural network (RBF), and generalized regression neural network (GRNN)	To predict COD and total soluble solid (TSS) concentrations of the effluent and biogas	Flow (m^3/day), volumetric load (kg/m^3), COD (mg/l), and TSS (mg/l) concentrations in the influent	COD and TSS concentrations of the effluent and biogas production (m^3/h)	Feed-forward BP	The feed-forward BP ANN model was found to be the best approach for prediction of the desired outputs	Elnekave et al. (2012)

(continued)

Table 15.3 (continued)

ANN model	Aim(s)	Model input(s)	Model output(s)	Best model(s)	Result(s)	References
Free forward BP ANN model	To study the effects of different factors on biogas production	pH, moisture content (MC), VS, VFAs, and CH_4 fraction	Biogas production	–	ANN model predicted the biogas production with an $R^2 > 0.915$	Nair et al. (2016)
Quasi-Newton method and conjugate gradient BP algorithms	To model and optimize biogas and methane yield	pH, COD, ammonium, ALK, total Kjeldahl nitrogen, total phosphorus, VFA and HRT	Biogas and methane yield	10 hidden neurons MLP ANN model and Tangent sigmoid and linear functions in the hidden and output layers, respectively	The developed ANN model could satisfactorily predict and optimize biogas and bio-methane production	Antwi et al. (2017)

The function of $\mu_A(x)$ that specifies the fuzziness of a fuzzy set A in X in the fuzzy logic are Triangular, Trapezoidal, Gaussian, Bell-shaped, and Sigmoidal MFs (Table 15.4).

The linguistic variables used as appellation of fuzzy subsets are variables whose quantities are words or sentences. The specifications of systems which cannot be defined by numerical values are described by linguistic variables. A fuzzy logic system has four main steps including fuzzification, fuzzy rule base, fuzzy inference mechanism, and defuzzification. The fuzzy logic steps are shown in Fig. 15.2.

Fuzzification is the process of converting a numeric values (or crisp value) into fuzzy inputs (linguistic variables) by the determined membership degrees. In fuzzy If-Then rules, the conditional statements are formulated which contain fuzzy logic.

Table 15.4 Important MFs

Type of MF	Function	Parameters	Relation between parameters
Triangular	$\mu(x) = \max\left\{\min\left\{\frac{x-a}{b-a}, \frac{c-x}{c-b}\right\}, 0\right\}$	a, b, c	$a < b < c$
Trapezoidal	$\mu(x) = \max\left\{\min\left\{\frac{x-a}{b-a}, 1, \frac{d-x}{d-c}\right\}, 0\right\}$	a, b, c, d	$a < b < c < d$
Gaussian	$\mu(x) = \exp\left[-\frac{1}{2}\left(\frac{x-m}{\sigma}\right)^2\right]$	m, σ	m and σ are the center and width of the Gaussian MF
Bell-shaped	$\mu(x) = \frac{1}{1+\left\lvert\frac{x-m}{\sigma}\right\rvert^{2a}}$	m, σ, a	m and σ are the center and width of the bell-shaped MF Parameter a, usually positive, controls the slope of the MF
Sigmoidal	$\mu(x) = \frac{1}{1+\exp[-a(x-c)]}$	a, c	Parameter a controls the slope of the MF at cross point $x = c$

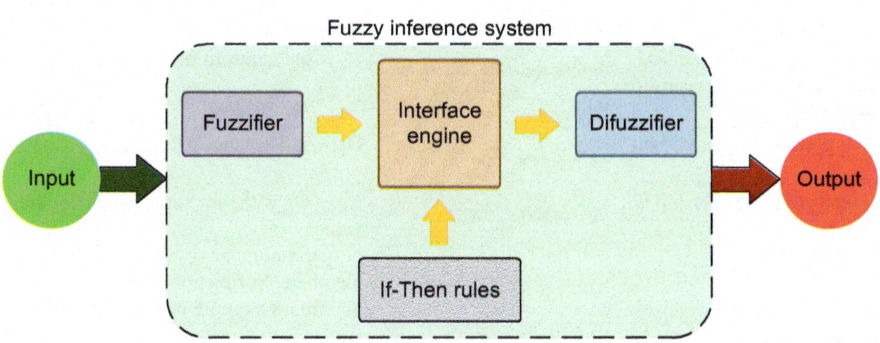

Fig. 15.2 Fuzzy logic steps

The relationships between variables are demonstrated by fuzzy If-Then rule as follows:
If ⟨fuzzy antecedent proposition⟩ Then ⟨fuzzy consequent proposition⟩

In contrast with the fuzzification, the defuzzification is a process that converts the fuzzy value of a fuzzy set into a crisp value. It is important to be done when the crisp value is prepared for the user. Some of the most important used defuzzification methods are presented in Table 15.5.

15.3.1 *Inference Mechanism*

In the fuzzy logic inference process, a nonlinear mapping is formulated from an input space to output space by using fuzzy logic. Inference mechanisms, also called fuzzy models, can be categorized as Mamdani, Sugeno, and Tsukamoto fuzzy inferences. The main difference among these models lies in the defuzzification procedure.

15.3.1.1 Mamdani Fuzzy Inference

A typical rule in a Mamdani fuzzy model is defined by fuzzy If-Then rules as bellow:

k: If x is A_i^k and y is B_j^k then z is C_l^k
k = 1, 2, …, R (R is the maximum number of rules, $R \subset N \times M$)
i = 1, 2, …, N, j = 1, 2, …, M and l = 1, 2, …, L. N, M and L are the numbers of membership functions for inputs and outputs, respectively.

Table 15.5 Some of the most important defuzzification methods

Defuzzification method	Equation	Parameter
Max-membership method (height method)	$x^* = \frac{\sum_{k=1}^{m} c_k h_k}{\sum_{i=1}^{n} h_k}$	c_k is the peak value of the fuzzy sets and h_k is the height of the clipped fuzzy sets
Centre of gravity method (center of area or centroid method)	$x^* = \frac{\sum_{i=1}^{m} \mu_c(X).X_i}{\sum_{i=1}^{m} \mu_c(X_i)}$	x^* is the centroid of the area
Weighted average method	$x^* = \frac{\sum \mu_c(X).X'}{\sum \mu_c(X_i)}$	Defuzzified value x^*
Mean-max membership	$x^* = \frac{\sum_{i=1}^{N} \mu_{max}(X_i)}{N}$	$\mu_{max}(x_i)$ is the maximum membership value N is the number of times the membership function reaches the maximum support value
Centre of sums	$x^* = \frac{\sum_{i=1}^{m} X_i . \sum_{k=1}^{n} \mu_k(X_i)}{\sum_{i=1}^{m} \sum_{k=1}^{n} \mu_k(X_i)}$	Defuzzified value x^*

In this model, crisp values are used as inputs and defuzzification is done to convert a fuzzy set to a crisp value. The overall output is gained by employing the center of gravity defuzzification method.

15.3.1.2 Sugeno Fuzzy Inference

The Sugeno fuzzy inference or TSK fuzzy model was introduced by Takagi, Sugeno, and Kang to create fuzzy rules from a given input/output data set. A typical rule in the Sugeno fuzzy model is similar to Mamdani model but the output in Sugeno model is a crisp function $z = f(x, y)$ that is usually polynomial of the inputs.

$$k : \text{If x is } A_i^k \text{ and y is } B_j^k \text{ then } Z_k = f(x, y)$$

Another difference between Sugeno and Mamdani models is the process of obtaining the overall output. In Sugeno model, it is obtained by weighted average of crisp outputs Z_k that prevents wasting time for defuzzification.

15.3.1.3 Tsukamoto Fuzzy Inference

Similar to Mamdani model, a typical rule in a Tsukamoto fuzzy model is defined by rules as bellow:

$$k : \text{If x is } A_i^k \text{ and y is } B_j^k \text{ then z is } C_l^k$$

Fuzzy consequent proposition of each if-then fuzzy rule is defined by a monotonic MF. The crisp output value of each rule is used to obtain the overall output by the weighted averages defuzzification method (Siddique and Adeli 2013).

15.3.2 Applications of Fuzzy Logic in Biogas Production

In recent years, there has been an increasing interest in using fuzzy logic for predicting the behavior of anaerobic digestion systems to maximize biogas production and optimize their performance. Table 15.6 shows some important applications of fuzzy logic systems in order to model the anaerobic digestion systems.

15.4 Evolutionary Optimization Algorithms

The evolution of nature has immensely impressed the scientists to use the ideology behind them for developing evolutionary optimization techniques. Several evolutionary optimization algorithms have been emerged by converting the behavior of

Table 15.6 Some important applications of fuzzy logic systems in order to model the behavior of anaerobic digestion systems

Fuzzy logic model	Aim(s)	Model input(s)	Model output(s)	Result(s)	References
Mamdani-type fuzzy inference system	To develop a fast predicting MIMO (multiple inputs and multiple outputs) fuzzy-logic-based model for the estimation of biogas and methane production rates in a pilot-scale mesophilic UASB reactor	Volumetric OLR, volumetric COD removal rate (RV), influent ALK, influent, and effluent pH	Biogas and methane production	The fuzzy logic control system forecasted biogas and methane production rates with an R^2 value over 0.98	Turkdogan-Aydınol and Yetilmezsoy (2010)
Mamdani-type fuzzy system	To predict the biogas production rate in a large-scale UASB reactor treating distillery wastewater	Biogas production of one day before, COD reduction of reactor one and two day (s) before (Kg/d), COD loading (Kg/d), equalized and effluent pH	Biogas production	The developed model was able to assess the performance of the large-scale UASB reactor treating distillery spent	Varne and Macwan (2012)
Mamdani method	To predict the effect of microwave pretreatment on the waste activated sludge (WAS) solubility and anaerobic digestibility or biogas production	Microwave temperature, microwave intensity, WAS concentration, and volume percentage of WAS pretreated	WAS solubility and anaerobic digestibility or biogas production	Fuzzy model predicted the response variables, i.e., WAS solubilisation and cumulative biogas production with R^2 values of 0.81 and 0.98, respectively	Saha et al. (2014)
Takagi-Sugeno structure	To control methane production and VFA concentration in order to avoid acidification process at closed-loop	Influent flow rate and VFA concentration	Gaseous outflow rate of methane	The fuzzy logic controller was able to achieve the process stability and the maximum methane production at higher VFA concentrations without acidification process	Robles et al. (2017)

different creatures in the nature to mathematical models. In this section, the relatively successful evolutionary algorithms including GA, ACO, and PSO are briefly introduced.

15.4.1 Genetic Algorithm (GA)

GA was initially introduced by John Holland in 1975. The GA has been inspired by focusing on the hypothesis of Darwinian evolution. The nature has stringent selection rules to survive individuals. Consequently, the individuals who have superior fitness are more likely to survive and spread their genetic material in next generation. As shown in Fig. 15.3, the variables of a possible solution are encoded into a chromosome which is a DNA molecule with part or all of the genetic material (genome). A gene is a locus of DNA representing a bit of the encoded variables, typically a digit value of 0 or 1. Therefore, a population of the possible solutions or chromosomes forms a generation which is manipulated based on the evolutionary rule of survival of the superior fitness through genetic operations such as reproduction (selection), crossover, and mutation. Over the time, generations will evolve into the best fitness, i.e., optimizing the problem.

Let's assume that the optimization problem depends on *n* variables. The algorithm of the GA for achieving the optimum solution is as follows:

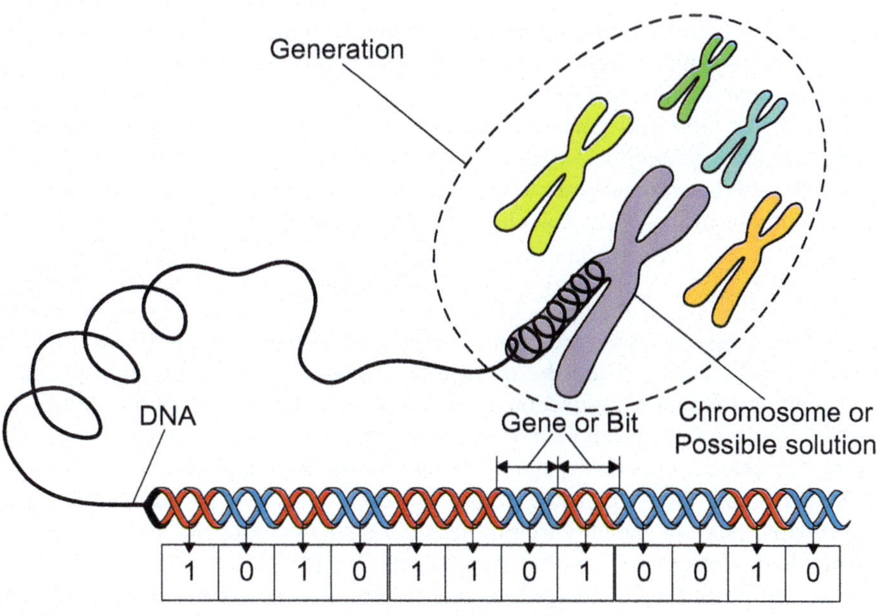

Fig. 15.3 A schematic representation of the genetic algorithm

Step 1: initially, a population of possible solutions is randomly selected in the search domain. The variables of each possible solution are separately encoded and then stuck together to form a binary string, representing a chromosome. The population of the chromosomes indicating a generation is put to a gene pool. The fitness values of the chromosomes are evaluated using a pre-specified fitness function.

Step 2: the reproduction (selection) genetic operator is performed to the gene pool of chromosomes under a given reproduction rate, p_r. The production rate is the probability of the chromosomes' presence in the next generation via the reproduction implementation. Thus, the number of chromosomes is transformed by reproduction operator, N_r, obtained by the following equation (Eq. 15.3).

$$N_r = p_r \times N_{chrom} \tag{15.3}$$

where N_{chrom} is the number of chromosomes of the gene pool.

Two types of reproduction operator have been typically introduced, i.e., roulette wheel selection and elitist selection. In the Roulette wheel selection, the chromosomes are chosen according to their probabilities obtained by the following equation (Eq. 15.4).

$$p_i = \frac{F_i}{\sum_{i=1}^{N_{chrom}} F_i} \tag{15.4}$$

where F_i is the fitness of the *i*th chromosome. The roulette wheel selection is unfortunately based on the stochastic selection, disrupting the higher fitness chromosomes. The elitist selection transforms the superior chromosomes having better fitness compared with the others.

Step 3: the crossover genetic operator is applied to the gene pool of the chromosomes under a specified crossover rate, p_c. The crossover rate is defined as the probability of the chromosomes to be selected to combine. Therefore, the number of mixed chromosomes, N_c, is calculated by the following equation (Eq. 15.5).

$$N_c = p_c \times N_{chrom} \tag{15.5}$$

Various types of crossover operator have been defined. Single-point crossover and two-point crossover are the most popular crossover operator. Either random selection or tournament selection is used to pick the parent chromosomes for mating. The random selection chooses pairs of parent chromosomes stochastically. In the tournament selection, pairs of chromosomes are randomly chosen from the gene pool and then the more fitness of each pair has permission to mate. Afterwards, pairs of parent chromosomes are also stochastically picked from the allowable chromosomes. As can be seen from Fig. 15.4, the single-point crossover selects one point (bit) of the pair of parent chromosomes (strings) randomly and then the right hand side bits of the parents are exchanged together. The two-point crossover operator considers two points of the pair of the parent chromosomes;

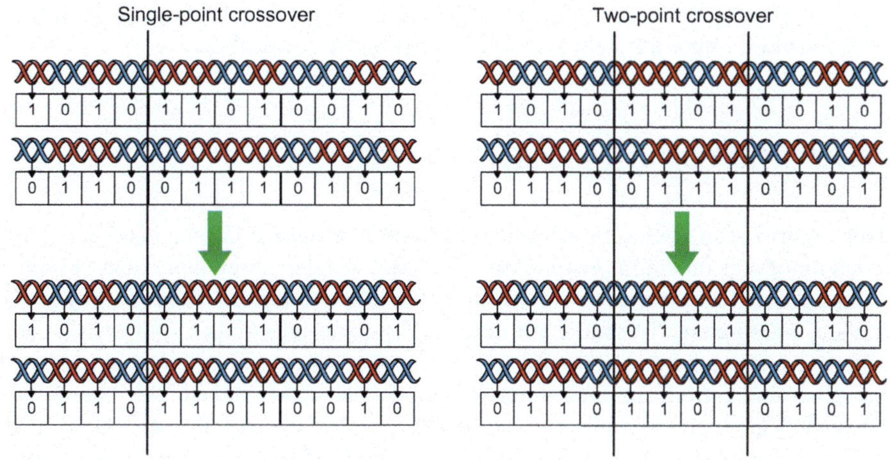

Fig. 15.4 The single-point and two-point crossover techniques

accordingly the bits among these points are interchanged between parent strings. In the mentioned crossover techniques, each pair of the parent chromosomes creates two new chromosomes (offspring) which are transformed to the next generation.

Step 4: the mutation genetic operator is implemented to the gene pool of the chromosomes under a given mutation rate, p_m. The mutation rate is described as the probability of chromosomes' transformation into the next generation due to mutation. Similar to the reproduction and crossover, the number of mutated chromosomes, N_m, is specified by the following equation (Eq. 15.6).

$$N_m = p_m \times N_{chrom} \tag{15.6}$$

It is noted that the number of chromosomes should be constant during the generations. Here, two types of mutation operator are introduced, i.e., single-point mutation and bit-wise mutation. Single point is the traditional mutation method in which a gene of a selected chromosome is stochastically chosen and then its digit value is changed to opposite value, i.e., zero is converted to one and vice versa. In the bit-wise mutation, a value is stochastically selected in the range (0, 1] for each gene of a picked out chromosome. If the value is lower than a pre-specified bit-wise mutation rate $p_{bw} \in (0, 1]$, then digit value of bit is reversely altered. This work is continued until the mutated chromosomes in the next generation reach to N_m. The aim of the mutation operation is to prohibit the chromosomes in the new generation to trap into a local optimum.

Step 5: After performing the above-mentioned steps, the fitness value of each chromosome of the new generation is evaluated. If the convergence criteria such as a given number of generation or a given iterations time is reached, the chromosome with the overall best fitness value is then presented as the optimum solution. Otherwise, steps 2–5 should be iterated again. The genetic parameters of the GA algorithms can profoundly influence the final solution. Accordingly, these parameters should be appropriately selected by trial-and-error method.

15.4.2 Ant Colony Optimization (ACO)

This optimization paradigm has been developed by mimicking the complex social behavior of ants in natural world. Ants randomly wander about their colony in order to find food. They return to colony just after discovering the food while leaving trails of pheromone. The trails are followed by other ants if they find such a path. Afterward, if other ants eventually achieve the food, they strengthen the intensity of the pheromone trails in their return path. As the time passed, the pheromone evaporates gradually, diminishing the attractiveness of the path for ants. As the path is longer, the more time it takes for ants to get the food and return to the nest. Thus, the pheromone trail of longer path is faster evaporated than the pheromone trail of shorter ones. Therefore, a shorter path has a higher chance to be found by ants. Over the time, the number of the ants travelling on the shorter path increases and hence, more pheromone reinforcements are applied to the path. Consequently, the shorter path (best solution) is unanimously chosen by the ants (Fig. 15.5).

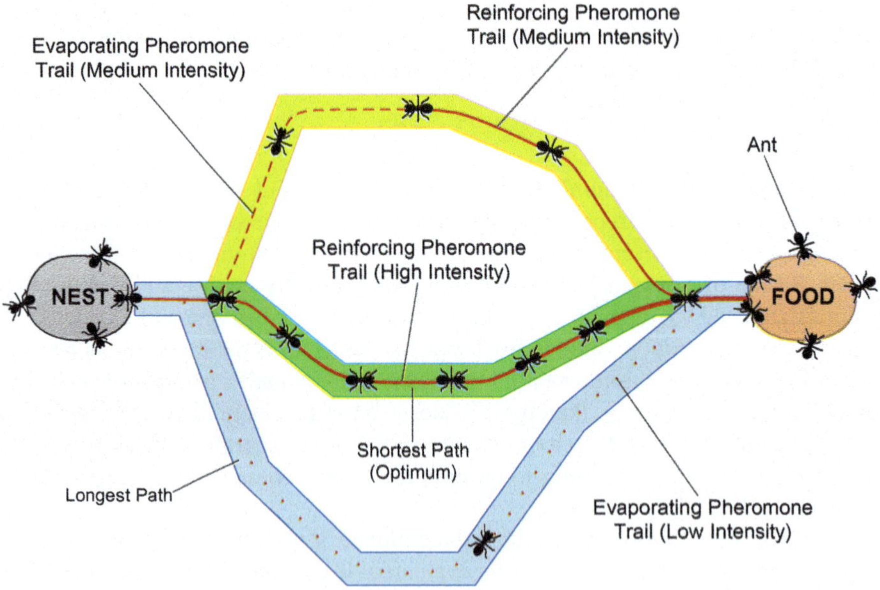

Fig. 15.5 A schematic illustration of ACO algorithm

15.4.2.1 Ant System (AS)

Ant System (AS) is the preliminary meta-heuristic algorithm of the ACO proposed to tackle the combinatorial optimization problem such as Traveling Salesman Problem (TSP). Assuming there are n towns and the problem is finding the shortest length of closed journey (tour) that touch all the towns just once. m artificial ants deployed for obtaining the best path. As an initial, the ants randomly distribute over the towns and then each ant chooses the next town with a probability obtained by the following equation (Eq. 15.7).

$$p_{i,j}^{k}(t) = \frac{[\tau_{i,j}(t)]^{\alpha}[\eta_{i,j}]^{\beta}}{\sum_{l \in N_i^k}[\tau_{i,l}(t)]^{\alpha}[\eta_{i,l}]^{\beta}} \; if \; j \in N_i^k \tag{15.7}$$

where $p_{i,j}^{k}(t)$ is the probability value of selecting jth town by the ant located in ith town at iteration t. $\tau_{i,j}(t)$ is pheromone value of the path connecting the ith town to the jth town. $\eta_{i,j}$ is the heuristic information defined as inverse of the distance between the ith town to the jth town. α and β are constant values specified by user. N_i^k is the allowed towns for kth ant located in ith town. It is noted that the amounts of pheromone are initially guessed as τ_0 for all paths at the first iteration. After finishing the tour by all ants, the pheromone values of their paths in the next iterations are updated by the following equation (Eq. 15.8) where the effect of the pheromone evaporation is considered in the first term.

$$\tau_{i,j}(t+1) = (1-\rho)\tau_{i,j}(t) + \sum_{k=1}^{m} \Delta\tau_{i,j}^{k}(t) \tag{15.8}$$

where ρ is the evaporation rate of the pheromone trails. $\Delta\tau_{i,j}^{k}(t)$ is the amount of the added pheromone by kth ant on the path of ith town to jth town which is defined by the following equation (Eq. 15.9).

$$\Delta\tau_{i,j}^{k}(t) = \begin{cases} \frac{1}{L^k(t)} & (i,j) \in Y^k(t) \\ 0 & (i,j) \notin Y^k(t) \end{cases} \tag{15.9}$$

where $Y^k(t)$ and $L^k(t)$ are the path and its length from which the kth ant passed at iteration t, respectively. After some iterations, the shortest path becomes unanimous choice which could be a global optimum. Nevertheless, many ants are able to travel on good path but sub-optimum paths called “stagnation” especially when lengths of the tour are close together. The appropriate values of pre-specified parameters in the AS algorithm vary in different problems, often set by trial and error.

15.4.2.2 Ant System with Elitist Strategy

The first popular strategy for improving the AS is elitist approach for ant system. This method emphasizes the more pheromone trail on the overall shortest path found during all iterations besides the considerations of the other ants' pheromone reinforcement on their path. Using the elitist strategy, the ants are efficiently focused on the optimum solution instead of dispersion on lower significant paths. For this purpose, the pheromone trails are updated using the following equation (Eq. 15.10)

$$\tau_{i,j}(t+1) = (1-\rho)\tau_{i,j}(t) + \sum_{k=1}^{m} \Delta\tau_{i,j}^{k}(t) + \gamma\Delta\tau_{i,j}^{best} \tag{15.10}$$

where $\Delta\tau_{i,j}^{best}$ is the amount of added pheromone by the elitist ants during all iteration which is obtained by the following equation (Eq. 15.11). γ is a pre-specified constant value.

$$\Delta\tau_{i,j}^{best} = \begin{cases} \frac{1}{L^{best}} & (i,j) \in Y^{best} \\ 0 & (i,j) \notin Y^{best} \end{cases} \tag{15.11}$$

where $Y^{best}(t)$ and $L^{best}(t)$ are the best path and its length in all iterations, respectively.

15.4.2.3 Ant System with Elitist Strategy and Ranking

The main controversy of the AS arises over the danger of over-emphasizing of pheromone trails on the sub-optimum path caused by many ants. In order to remove this obstacle, a modified algorithm of the AS has been introduced which is based on the overall elitist ant and the high rank ants at the end of a tour (Bullnheimer et al. 1997). The modification is simply applied to the AS. After completing a tour, the ants are sorted in terms of shorter tour length and then σ number of higher rank ants is selected as the elitist ants of tth iteration. Moreover, the overall elitist ant (the shortest tour length) during all iterations is obtained as well. According to this strategy, the pheromone trails are updated by the following equation (Eq. 15.12).

$$\tau_{i,j}(t+1) = (1-\rho)\tau_{i,j}(t) + \sum_{r=1}^{\sigma-1} (\sigma - r)\Delta\tau_{i,j}^{r}(t) + \sigma\Delta\tau_{i,j}^{best} \tag{15.12}$$

where r is the rank of ants, $\Delta\tau_{i,j}^{best}$ is determined by Eq. (15.11), and $\Delta\tau_{i,j}^{r}(t)$ is found as follows (Eq. 15.13).

$$\Delta\tau_{i,j}^{r} = \begin{cases} \frac{1}{L^{r}(t)} & (i,j) \in Y^{r} \\ 0 & (i,j) \notin Y^{r} \end{cases} \quad (15.13)$$

where $Y^r(t)$ and $L^r(t)$ are the path and its length of the ant with rank r at iteration t, respectively.

15.4.2.4 Max-Min Ant System (MMAS)

The problem of premature stagnation is the most important disadvantage of the greedy search strategies of ACO. Therefore, Max-Min Ant System (MMAS) algorithm has been proposed to keep balance between an exploitation of the best solutions found during the search, and avoidance of the early search stagnation (Stützle and Hoos 1997; Stützle and Hoos 2000) MMAS has three differences compared with the AS, i.e., (1) after finishing a tour, only one single ant is allowed to add pheromone which may be either the global-best ant or the iteration-best ant, (2) to escape from stagnation of the search, the pheromone trails are bounded between the lower and upper thresholds i.e. $\tau_{i,j}(t) \in [\tau_{min}, \tau_{max}]$, and (3) to reach a higher exploration of solutions at the beginning of the algorithm, the initial values of the pheromone trails are set to τ_{max}. Consequently, the pheromone trails are updated using the following equation (Eq. 15.14).

$$\tau_{i,j}(t+1) = (1-\rho)\tau_{i,j}(t) + \Delta\tau_{i,j}^{best} \quad (15.14)$$

where $\Delta\tau_{i,j}^{best}$ is the amount of added pheromone defined depend on choosing global-best ant or iteration-best ant. Accordingly, Eq. (15.11) can be used if global-best ant is selected. However, this equation can be simply modified if iteration-best ant is chosen.

The global-best ant choice is caused the search may be rapidly focused around the path which is not necessarily the optimum and therefore, limits the exploration of possibly better ones. On the other hand, the iteration-best ant reduces the danger of stagnation because the iteration-best solutions may vary during iterations, resulting in the extensive search over different solutions. It is proved that the combination of the global-best ant and iteration-best ant strategies is more efficient in convergence of the solutions into the optimum. It is demonstrated that the maximum and minimum pheromone limitations in the MMAS algorithm are obtained by the following equations (Eqs. 15.15 and 15.16).

$$\tau_{max}(t) = \frac{1}{\rho}\frac{1}{L^{best}} \quad (15.15)$$

$$\tau_{min}(t) = \frac{1-\sqrt[n]{p_{best}}}{(avg-1)\sqrt[n]{p_{best}}}\tau_{max}(t) \quad (15.16)$$

where p_{best} is the probability by which an ant chooses the best path (optimum). *avg* is the average number of choices of each ant. According to these equations, the upper and lower pheromone thresholds vary at each iteration until final solution.

15.4.2.5 Application of ACO in Biogas Production Technology

As an example of the application of ACO in biogas production, an optimal management of anaerobic co-digestion process using ACO is given in details in the following paragraphs. In an anaerobic co-digestion two or more substrates are simultaneously digested. The advantages of anaerobic co-digestion include: (1) it dilutes the potential toxic chemicals like ammonia and improves the stability of digestion (Sosnowski et al. 2003; Dai et al. 2013), (2) it prevents the inhibition of digestion process caused by volatile fatty acid (VFA) accumulation and a decrease in pH (Lin et al. 2014), (3) it promotes the quality of waste management in order to achieve the environmental sustainability, (4) it enhances mass conversion and reduces the weight and volume of digested residual (Macias-Corral et al. 2008), and (5) it increases the biogas and consequently methane production (Jafari et al. 2014).

A key issue of anaerobic co-digestion is management of real-time discharge of the organic waste from different sources (e.g., sewage sludge, agricultural residues, forest residues, food residues and etc.) for producing biogas. The ACO algorithms can be applied to the waste management in order to optimize the discharge of different organic waste sources for maximizing the biogas production (Fig. 15.6). The volume of different waste sources is discharged to produce maximum volume of biogas considering the significant characterizations of each feedstock such as: Total Solid (TS), Volatile Solid (VS), Chemical Oxygen Demand (COD), Total Nitrogen (TN), Alkalinity (Alk), Toxicity level (Tox), Volatile Fatty Acid (VFA), pH, etc.

An optimal management of anaerobic co-digestion of sewage sludge and food residue has been performed using the MMAS algorithm with the aim of maximizing biogas production (Verdaguer et al. 2016). The co-substrates contributions to input feed of anaerobic digester were optimized by considering their COD, TN, Alk and Tox as well as the availability of each waste source over time. The fitness function has been defined by the following equation (Eq. 15.17).

$$I = \sum_{w=1}^{n} \sum_{s=1}^{l_w} y_w^s V_w^s C_w^4 \sum_{d=1}^{3} C_w^d \tag{15.17}$$

where n and l_w are the number of the waste types and sources, respectively. w and s are the waste number and source number, respectively. V_w^s is the volume of substrate discharged from each source as a part of input feed. y_w^s is the decision variable which is 1 if the source is selected and otherwise is 0. C_w^d are coefficients related to COD for $d = 1$, COD/TN for $d = 2$, alkalinity for $d = 3$, and toxicity level for $d = 4$. These values have been expressed by the following equation

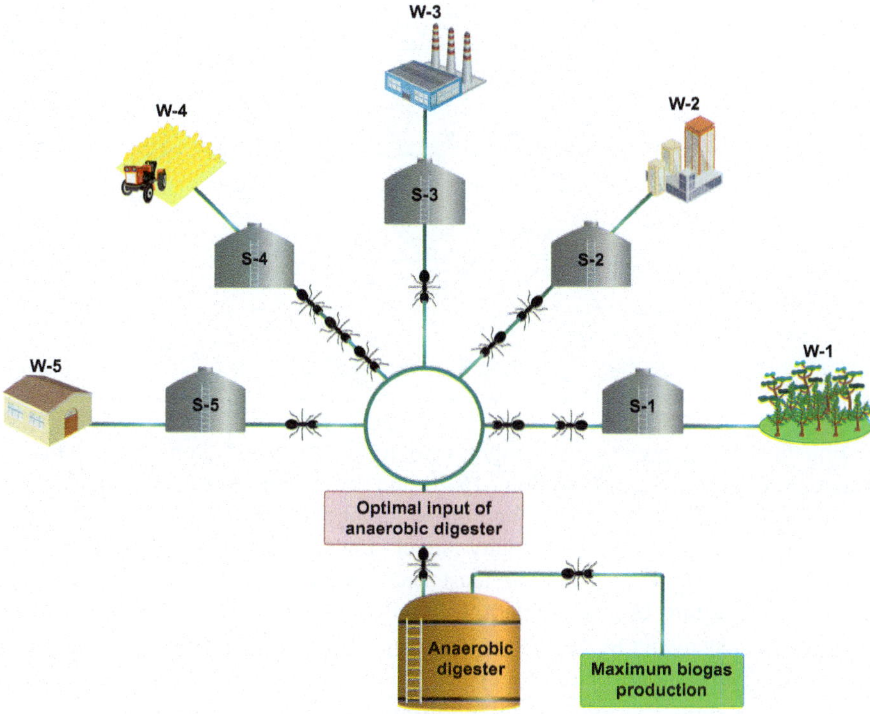

Fig. 15.6 A schematic illustration of waste management for an anaerobic co-digestion process

(Eq. 15.18) according to the previous experimental studies about the effect of feedstock characterizations on the biogas production.

$$C_w^d = \begin{cases} C_w^1 = 0.000035Z_w^1 - 0.5 \\ C_w^2 = e^{-\frac{(Z_w^2-40)^2}{2\times 15^2}} \\ C_w^3 = e^{-\frac{(Z_w^3-4500)^2}{Z_w^3\times 2000^2}} \\ C_w^4 = e^{-\frac{(Z_w^4)^2}{2\times 0.13^2}} \end{cases} \tag{15.18}$$

where Z_w^d are the COD, COD/TN, alkalinity, and toxicity contents of the substrate for d = 1, 2, 3 and 4, respectively.

This problem has been solved under constraints G_1–G_5 defined by the following equations (Eqs. 15.19–15.23).

$$G_1 = \sum_{w=1}^{n}\sum_{s=1}^{l_w} y_w^s V_w^s \leq V_{AnD} \tag{15.19}$$

$$G_2 = \frac{\sum_{w=1}^{n}\sum_{s=1}^{l_w} y_w^s V_w^s Z_w^2}{\sum_{w=1}^{n}\sum_{s=1}^{l_w} y_w^s V_w^s} \in \left[Z_{min}^2, Z_{max}^2\right] \quad (15.20)$$

$$G_3 = \frac{\sum_{w=1}^{n}\sum_{s=1}^{l_w} y_w^s V_w^s Z_w^3}{\sum_{w=1}^{n}\sum_{s=1}^{l_w} y_w^s V_w^s} \in \left[Z_{min}^3, Z_{max}^3\right] \quad (15.21)$$

$$G_4 = \frac{\sum_{w=1}^{n}\sum_{s=1}^{l_w} y_w^s V_w^s Z_w^4}{\sum_{w=1}^{n}\sum_{s=1}^{l_w} y_w^s V_w^s} \le Z_{max}^4 \quad (15.22)$$

$$G_5 = \sum_{s=1}^{l_w} y_w^s = 1, \quad w = 1, 2, 3, \ldots, n \quad (15.23)$$

Equation 15.19 prevents the input feed to be exceeded from the volume of anaerobic digester, V_{AnD}. Equations 15.20 and 15.21 bind the mean values of the COD/TN and alkalinity in the appropriate range, respectively. Equation 15.22 keeps the mean value of toxicity level lower than the allowable value. Equation 15.23 represents that at least one source should be involved in the anaerobic digestion. In this problem, the ants have chosen the solutions with the probabilities calculated by Eq. 15.24. This equation is obtained using $V_w^s \sum_{d=1}^{4} Z_w^d$ as heuristic information in the Eq. 15.7. The other processes of the optimization algorithm are similar to the MMAS algorithm as mentioned before.

$$p_{w,s}^k(t) = \frac{\left[\tau_{w,s}(t)\right]^{\alpha}\left[V_w^s \sum_{d=1}^{4} Z_w^d\right]^{\beta}}{\sum_{l=1}^{l=l_w}\left[\tau_{w,l}(t)\right]^{\alpha}\left[V_w^l \sum_{d=1}^{4} Z_w^d\right]^{\beta}} \quad (15.24)$$

15.4.3 Particle Swarm Optimization (PSO)

PSO has been inspired by collective behavior of social intelligent of some organisms such as schools of fishes (or flocks of birds). During developments of particle swarm concept, it was observed that movement behavior of the agents is more similar to a swarm than a flock or a school. The term of "particle" is due to the fact that velocities and accelerations are suitably applied to particles. Therefore, the name of "particle swarm" was chosen to introduce optimization concept.

PSO is a stochastic population-based problem, comprising primitive mathematical related to positions and velocities of particles in the hyperspace. For a better understanding of the algorithm, it is initially worth describing how fishes (or birds) participate in their school (or flock) in order to discover bait. As shown in Fig. 15.7, each probable solution for an optimization problem is treated as a fish, called a particle. At the initial time, the fishes randomly commence seeking the search

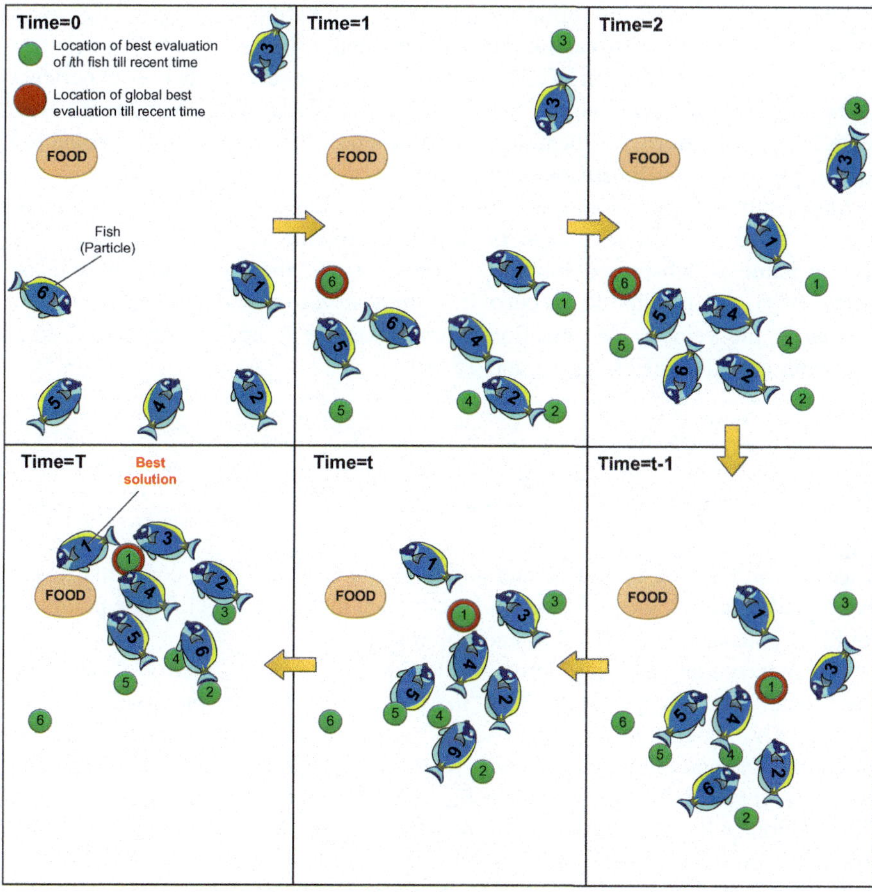

Fig. 15.7 A schematic representation of collaboration among fishes to attain food illustrated for understanding PSO algorithm

domain. Over time, each fish remembers its own nearest position to food and shares generously to the others. Therefore, the fishes tend to the target (food) with a change in their speed rely on their best experienced position and the best position of the most successful fish. By following this process, the school movement trajectory is improved till the fishes reach in the vicinity of final destination which is food or the best solution of the investigating optimization problem.

The PSO algorithm described originally by Kennedy and Eberhart is known as the Standard (global) PSO (Eberhart and Kennedy 1995). The main steps for implementing the PSO are subsequently discussed.

Step 1: the positions of the particles are randomly chosen and their velocities are arbitrary initialized on n-dimensions in the problem domain. The position and velocity of ith fish (particle) in hyperspace at time t are represented by $X_i^t = \left(x_{i,1}^t, x_{i,2}^t, x_{i,3}^t, \ldots, x_{i,j}^t, \ldots, x_{i,n}^t\right)$ and $V_i^t = \left(v_{i,1}^t, v_{i,2}^t, v_{i,3}^t, \ldots, v_{i,j}^t, \ldots, v_{i,n}^t\right)$ respectively.

Step 2: the value of optimization fitness function corresponding to the n-dimensional position of each particle is obtained.

Step 3: fitness evaluation of each particle is compared with its best experienced fitness ($P_{Best,i}$). If current value is more suitable than $P_{Best,i}$, then the current value is set as $P_{Best,i}$. Therefore, the location of particle's best fitness ($x_j[G_{Best}]$) is equal to its current position in n-dimensional domain.

Step 4: fitness evaluation of each particle is compared with overall previous best fitness of the population (G_{Best}). If current evaluation is better than G_{Best}, then current value is substituted for G_{Best}. Consequently, the location of overall best fitness ($x_j[G_{Best}]$) is equal to its current position in n-dimensional space.

Step 5: the velocity and position of each particle is updated according to the following equations (Eqs. 15.25 and 15.26).

$$v_{i,j}^t = v_{i,j}^{t-1} + C_c R_c^t \left\{ x_j \left[P_{Best,i} \right] - x_{i,j}^{t-1} \right\} + C_s R_s^t \left\{ x_j [G_{Best}] - x_{i,j}^{t-1} \right\} \tag{15.25}$$

$$x_{i,j}^t = x_{i,j}^{t-1} + v_{i,j}^t \tag{15.26}$$

where C_c and C_s are cognitive and social positive acceleration constants, respectively, and R_c^t and R_s^t are two random functions in the range [0, 1]. These parameters could be constant or modified at each time step.

The first, second, and third terms of the Eq. 15.25 are the "inertia", "cognition", and "social" parts, respectively. The "inertia" term introduces previous velocity of particle. The "cognition" and the "social" terms represent the effects of the individual particle and the particles group, respectively, contributing to the change of a particle's velocity. Without considering these two terms, the particles will retain their motion in the same speed and direction until they exceed the boundary of the space (e.g., motion of dead fish in its school).

It is noted that particle's velocity of each dimension are restricted to a maximum value called $v_{MAX,j}$. If the particle's velocity of each dimension exceed from the user-specified $v_{MAX,j}$, then the velocity is set to $v_{MAX,j}$.

Step 6: the steps of 2–5 are iterated until a criterion is satisfied which could usually be a sufficiently good fitness or a maximum number of iterations.

The essential merits of the PSO compared to the other optimization methods are its facile implementation and few adjustable parameters.

15.5 Hybrid Models and Applications in Biogas Production

Hybrid models such as adaptive neural-fuzzy interface system (ANFIS), Fuzzy-neural, Geno-Fuzzy, and Neuro-genetic have been developed to benefit from the advantages of all the involved approaches. Among the hybrid models developed to date, ANFIS is the most attractive paradigm proposed by Jang in the year 1993.

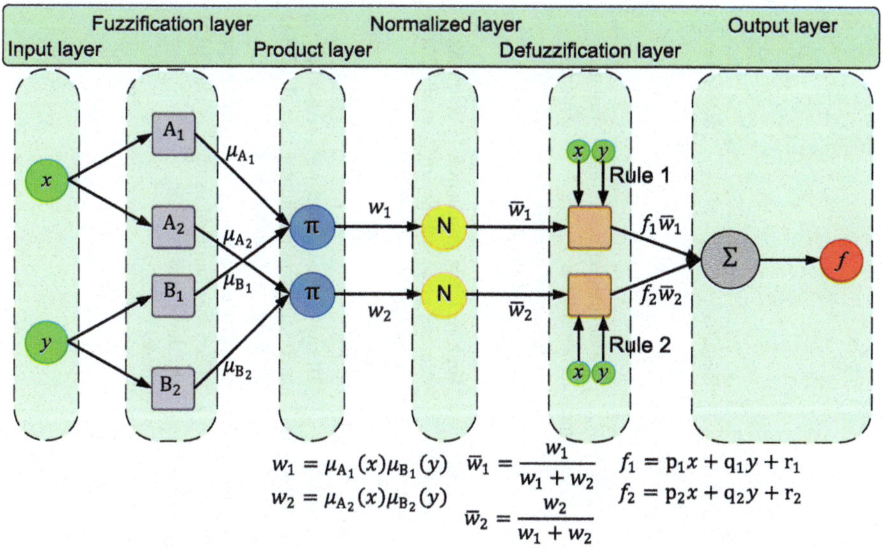

Fig. 15.8 Schematic structure of ANFIS

This intelligent technique has merged the merits of an artificial neural network, i.e., learning capability with Takagi, Sugeno and Kang (TSK) fuzzy inference system, i.e., human decision-making ability (Fig. 15.8 illustrates a concise structure of ANFIS). Some applications of ANFIS approach in biogas production technology are summarized in Table 15.7.

ANN technology consolidated with evolutionary optimization algorithms like GA, ACO, and PSO are considered among the most successful hybrid soft comptuing approches. These hybrid models can applied to model complex engineering problems like biogas production processes that cannot be estimated and optimized by conventional approaches. In such approaches, evolutionary algorithms utilize the outlet of the developed ANN models as fitness functions to select the optimal input variables, leading to the optimum biogas production or methane yield. Table 15.8 lists some important applications of hybrid models for optimizing the anaerobic digestion processes.

Table 15.7 Some important applications of ANFIS approach to model the anaerobic digestion process

Type of biogas production system	Aim(s)	ANFIS model input(s)	ANFIS model output(s)	Result(s)	References
High-rate anaerobic wastewater treatment systems	To predict and simulate the response and performance of high-rate anaerobic digestion systems after two-fold OLR with two-fold hydraulic loading rate (HLR) shock	OLR, HLR, alkalinity loading rate (ALR), volumetric methane production rate (VMP), total organic carbon (TOC), and VFA at time (n)	VMP, TOC, VFA at time (n + 1)	The developed model has an acceptable performance to predict the response of the systems	Tay and Zhang (2000)
Anaerobic hybrid reactor	To control an anaerobic hybrid reactor	pH, ALK and total volatile acids (TVA)	Influent feed flow rate	High methane production rate and great stability in the process was achieved using the developed model	Waewsak et al. (2010)
Landfill	To simulate and predict the landfill gas production rate	Time, sludge addition rate, and leachate recirculation rate	Biogas production rate	The developed model was successfully in predicting the biogas production rate	Abdallah et al. (2011)
Two-phase AD model	To predict biogas production	HRT, pH, and OLR	Cumulative gas production	The process was effectively modeled using the developed ANFIS model	Arumugam et al. (2015)

Table 15.8 Some prominent applications of ANN technology integrated with evolutionary algorithms to predict and optimize the anaerobic digestion process

Waste type	Aim(s)	Prediction model	Optimization algorithm	Model input(s) and its optimal value	Model output(s) and its optimal value	References
Organic waste	Modeling and optimizing the biogas production process in an industrial digester	Two-hidden layers BP trained MLP ANN model with sigmoid function	GA	Temperature (36 °C), TS (6.6%), VS (52.8%), and pH (6.4)	Methane fraction (77%)	Qdais et al. (2010)
Co-substrate of cow dung, paper waste, rice bran, banana stem, and saw dust	Modeling and optimizing the co-digestion process	One-hidden layer BP trained MLP ANN model with 2 hidden neurons and sigmoid transfer function	GA	Concentrations of cow dung (25% w/w), banana stem (25% w/w), rice bran (5% w/w), paper waste (25% w/w), and sawdust (20% w/w)	Biogas performance (10.144 ml)	Kana et al. (2012)
Co-digestion of potato waste with Pistia stratiotes	Modeling and optimizing the co-digestion process	One-hidden layer BP trained MLP ANN model with 12 hidden neurons, and tangent sigmoid and pure linear transfer functions in the hidden and output layers, respectively	GA	Substrate (potato waste) concentration (7 g TS/L), proportion of co-substrate (Pistia stratiotes) (69.08% TS, w/w), and inoculum concentration (78% VS/VS)	Methane yield (449.4 L/kg VS_{fed})	Jacob and Banerjee (2016)
Sludge	Modeling and optimizing the methane percentage of biogas produced in an industrial wastewater digester	One-hidden layer BP trained MLP ANN model with 20 hidden neurons, and tangent sigmoid transfer function	PSO	Temperature (35.4 °C), TS (31751 mg/L), VFA (8.78 mg/L), ALK (2754 mg/L) and pH (6.86)	Methane percentage of the produced biogas (66.5%)	Akbaş et al. (2015)
Sludge	Modeling and optimizing the	One-hidden layer BP trained MLP ANN model with 15	PSO			

(continued)

Table 15.8 (continued)

Waste type	Aim(s)	Prediction model	Optimization algorithm	Model input(s) and its optimal value	Model output(s) and its optimal value	References
	biogas production rate in an industrial wastewater digester	hidden neurons and tangent sigmoid transfer function		Temperature (35 °C), SLR (403.79 m^3/day, SRT (18.3 day), and pH (6.85)	Biogas production rate (3459 m^3/day)	Akbaş et al. (2015)
Sludge	Modeling and optimizing the biogas quality in an industrial wastewater digester	One-hidden layer BP trained MLP ANN model with 30 hidden neurons and tangent sigmoid transfer function	PSO	Temperature (35.8 °C), TS (22090 mg/L), VS (18361 mg/L), VFA (17.04 mg/L), ALK (3934 mg/L), SLR (371.5 m^3/day), OLR (2.02 kg/m^3 day), SRT (18.05 day), and pH (6.87)	Methane percentage of the biogas produced (66.8%) and biogas production rate (3322 m^3/day)	Akbaş et al. (2015)

15.6 Conclusions

Various soft computing approaches have been used for modeling and optimizing biogas production systems because of their capability to deal with complexity, nonlinearity, and uncertainty associated with the digestion process. Interest in applying such advanced tools to biogas production technology is growing due to the need for fast and accurate control of ill-defined digestion systems. Overall, advanced soft computing approaches will serve as powerful tools for modeling, optimizing, and controlling biogas production systems. Modeling and optimizing the biogas production processes using various soft computing techniques have been reported in the literature, while these methods have rarely been used for real-time monitoring and control of digestion systems. Therefore, future work should be directed towards the application of advanced soft computing techniques for real-time monitoring and control of biogas production systems and exploring strategies to enhance the quantity and quality of the biogas evolved.

References

Abdallah M, Warith M, Narbaitz R, Petriu E, Kennedy K (2011) Combining fuzzy logic and neural networks in modeling landfill gas production. World Acad Sci Eng Technol 78:559–565

Akbaş H, Bilgen B, Turhan AM (2015) An integrated prediction and optimization model of biogas production system at a wastewater treatment facility. Biores Technol 196:566–576

Almasi F, Jafari A, Akram A, Nosrati M, Afazeli H (2014) New method of Artificial Neural Networks (ANN) in modeling broiler production energy index in Alborz Province. Int J Adv Biol Biomed Res 2(5):1707–1718

Antwi P, Li J, Boadi PO, Meng J, Shi E, Deng K, Bondinuba FK (2017) Estimation of biogas and methane yields in an UASB treating potato starch processing wastewater with backpropagation artificial neural network. Biores Technol 228:106–115

Arumugam T, Parthiban L, Rangasamy P (2015) Two-phase anaerobic digestion model of a tannery solid waste: experimental investigation and modeling with ANFIS. Arab J Sci Eng 40 (2):279–288

Bullnheimer B, Hartl RF, Strauss C (1997) A new rank based version of the Ant System. A computational study

Carrère H, Dumas C, Battimelli A, Batstone D, Delgenès J, Steyer J, Ferrer I (2010) Pretreatment methods to improve sludge anaerobic degradability: a review. J Hazard Mater 183(1):1–15

Dai X, Duan N, Dong B, Dai L (2013) High-solids anaerobic co-digestion of sewage sludge and food waste in comparison with mono digestions: stability and performance. Waste Manag 33 (2):308–316

Dibaba OR, Lahiri SK, T'Jonck S, Dutta A (2016) Experimental and artificial neural network modeling of a Upflow Anaerobic Contactor (UAC) for biogas production from Vinasse. Int J Chem Reactor Eng 14(6):1241–1254

Dorigo M, Stützle T (2010) Ant colony optimization: overview and recent advances. In: Handbook of metaheuristics. Springer, Berlin, pp 227–263

Eberhart R, Kennedy J (1995) A new optimizer using particle swarm theory. In: Proceedings of the sixth international symposium on Micro Machine and Human Science. MHS'95, IEEE, pp 39–43

Elnekave M, Celik SO, Tatlier M, Tufekci N (2012) Artificial neural network predictions of Up-Flow Anaerobic Sludge Blanket (UASB) reactor performance in the treatment of citrus juice wastewater. Pol J Environ Stud 21(1)

Gazi V, Passino KM (2011) Swarm stability and optimization. Springer Science & Business Media, Berlin

Horváth IS, Tabatabaei M, Karimi K, Kumar R (2016) Recent updates on biogas production—a review. Biofuel Res J 3(2):394–402

Jacob S, Banerjee R (2016) Modeling and optimization of anaerobic codigestion of potato waste and aquatic weed by response surface methodology and artificial neural network coupled genetic algorithm. Biores Technol 214:386–395

Jafari A, Rafiee S, Nosrati M, Almasi F (2014) Investigation yield and energy balances for biogas production from cow and poultry manure. Int J Renew Energy Res (IJRER) 4(2):312–320

Kana EG, Oloke J, Lateef A, Adesiyan M (2012) Modeling and optimization of biogas production on saw dust and other co-substrates using artificial neural network and genetic algorithm. Renew Energy 46:276–281

Kanat G, Saral A (2009) Estimation of biogas production rate in a thermophilic UASB reactor using artificial neural networks. Environ Model Assess 14(5):607–614

Khalid A, Arshad M, Anjum M, Mahmood T, Dawson L (2011) The anaerobic digestion of solid organic waste. Waste Manag 31(8):1737–1744

Lin Y, Ge X, Li Y (2014) Solid-state anaerobic co-digestion of spent mushroom substrate with yard trimmings and wheat straw for biogas production. Biores Technol 169:468–474

Macias-Corral M, Samani Z, Hanson A, Smith G, Funk P, Yu H, Longworth J (2008) Anaerobic digestion of municipal solid waste and agricultural waste and the effect of co-digestion with dairy cow manure. Biores Technol 99(17):8288–8293

McCulloch WS, Pitts W (1943) A logical calculus of the ideas immanent in nervous activity. Bull Math Biophys 5(4):115–133

Nair VV, Dhar H, Kumar S, Thalla AK, Mukherjee S, Wong JW (2016) Artificial neural network based modeling to evaluate methane yield from biogas in a laboratory-scale anaerobic bioreactor. Biores Technol 217:90–99

Nguyen HT, Sugeno M (2012) Fuzzy systems: modeling and control, vol 2. Springer Science & Business Media, Berlin

Qdais HA, Hani KB, Shatnawi N (2010) Modeling and optimization of biogas production from a waste digester using artificial neural network and genetic algorithm. Resour Conserv Recycl 54 (6):359–363

Robles A, Latrille E, Ruano M, Steyer J-P (2017) A fuzzy-logic-based controller for methane production in anaerobic fixed-film reactors. Environ Technol 38(1):42–52

Saha M, Eskicioglu C, Sadiq R (2014) A fuzzy rule-based approach for modelling effects of bench-scale microwave pre-treatment on solubilisation and anaerobic digestion of secondary sludge. Int J Environ Eng 6(2):183–204

Sewsynker-Sukai Y, Faloye F, Kana EBG (2016) Artificial neural networks: an efficient tool for modelling and optimization of biofuel production (a mini review). Biotechnol Biotechnol Equip 2818:1–15. https://doi.org/10.1080/13102818.2016.1269616

Sewsynker-Sukai Y, Faloye F, Kana EBG (2017) Artificial neural networks: an efficient tool for modelling and optimization of biofuel production (a mini review). Biotechnol Biotechnol Equip 31(2):221–235

Siddique N, Adeli H (2013) Computational intelligence: synergies of fuzzy logic, neural networks and evolutionary computing. Wiley, Hoboken

Sosnowski P, Wieczorek A, Ledakowicz S (2003) Anaerobic co-digestion of sewage sludge and organic fraction of municipal solid wastes. Adv Environ Res 7(3):609–616

Strik DP, Domnanovich AM, Zani L, Braun R, Holubar P (2005) Prediction of trace compounds in biogas from anaerobic digestion using the MATLAB neural network toolbox. Environ Model Softw 20(6):803–810

Stützle T, Hoos H (1997) MAX-MIN ant system and local search for the traveling salesman problem. In: IEEE international conference on evolutionary computation, IEEE, pp 309–314

Stützle T, Hoos HH (2000) MAX–MIN ant system. Future Gener Comput Syst 16(8):889–914

Tay J-H, Zhang X (2000) A fast predicting neural fuzzy model for high-rate anaerobic wastewater treatment systems. Water Res 34(11):2849–2860

Turkdogan-Aydınol FI, Yetilmezsoy K (2010) A fuzzy-logic-based model to predict biogas and methane production rates in a pilot-scale mesophilic UASB reactor treating molasses wastewater. J Hazard Mater 182(1):460–471

Varne AL, Macwan J (2012) Fuzzy rule based approach for modeling biogas production rate in a real scale UASB reactor treating distillery wastewater. J Environ Res Dev 6(3A)

Verdaguer M, Molinos-Senante M, Poch M (2016) Optimal management of substrates in anaerobic co-digestion: an ant colony algorithm approach. Waste Manag 50:49–54

Waewsak C, Nopharatana A, Chaiprasert P (2010) Neural-fuzzy control system application for monitoring process response and control of anaerobic hybrid reactor in wastewater treatment and biogas production. J Environ Sci 22(12):1883–1890

Chapter 16
New "Omics" Technologies and Biogas Production

Gholamreza Salehi Jouzani and Reza Sharafi

16.1 Introduction

The increasing concern about energy security, negative environmental effects of fossil fuels and improvements in living standards have forced national and international authorities to promote green and renewable energy alternatives. Application of renewable energies is among the most encouraging and environmentally-compatible ways to mitigate pollution, improve energy security, and diminish consumption of conventional fuels. Different waste streams, such as forestry and agricultural residues, sewage and industrial organic wastewater, municipal solid wastes, as well as livestock and poultry dung have been shown to hold economic potentials for production of different types of bioenergies, such as bioethanol, biobuthanol and biogas (Benato and Macor 2017; Mao et al. 2015).

Biogas production through anaerobic digestion, is a promising waste treatment strategy and is becoming increasingly important because of being clean, renewable and environmentally friendly. Furthermore, biogas energy production relies on a balanced carbon dioxide cycle (Bremges et al. 2015; Deng et al. 2016). Principally, biogas production process takes place under anaerobic conditions, in which complex consortia of microorganisms are involved in hydrolysis and fermentation of organic matter, resulting in biogas (methane) as the final product. The anaerobic digestion process of biomass for biogas production is generally performed in four steps, including hydrolysis, acidogenesis, acetogenesis and methanogenesis. Each step is catalyzed by the activity of a specific group of microorganisms, commonly named as lignocellulose degrading, acidogenic, acetogenic and methanogenic microorganisms (Maus et al. 2014). In spite of the huge number of research works

G. Salehi Jouzani (✉) · R. Sharafi
Department of Microbial Biotechnology, Agricultural Biotechnology Research Institute of Iran (ABRII), Agricultural Research, Education and Extension Organization (AREEO), Fahmideh Blvd, P.O. Box: 31535-1897, Karaj, Iran
e-mail: gsalehi@abrii.ac.ir

M. Tabatabaei and H. Ghanavati (eds.), *Biogas*, Biofuel and Biorefinery Technologies 6, https://doi.org/10.1007/978-3-319-77335-3_16

focused on the identification of the microorganisms and the genes involved in biogas production from different types of biomass, the majority of the participating microbes and genes are still unknown, due to the lack of optimized culture media and conditions needed to grow the majority of the microbial communities involved. It is well known that biogas production efficiency of biogas plants is significantly dependent on the microbial community composition and activity, type of input waste, and process parameters, and therefore, the technologies providing a more in-depth understanding about the correlation among these parameters are of great importance (Heyer et al. 2016).

New "omics" technologies have become available for different families of cellular molecules, including genes, RNAs, proteins, and metabolites (Raupach et al. 2016). Recent developments in "omics" technologies, such as genomics, metagenomics, transcriptomics, proteomics, metabolomics, and next generation sequencing (NGS) technologies have provided excellent opportunities for developing new high-throughput structural and functional genomic surveys which can be efficiently used in the identification of uncultivable microorganisms, unknown genes, and pathways involved in biogas production from different biomass resources (Fig. 16.1). For instance, sequencing of the total metagenomic DNA in biogas process could efficiently help to obtain unbiased insights into the composition and function of a microbial community under investigation (Bremges et al. 2015). Overall, these innovations have great potentials to improve the efficiency of biogas production from different agricultural and municipal wastes. In line with that, in this chapter, we have tried to briefly elaborate on the state of the art of the application of different "omics" technologies in biogas production (Fig. 16.1).

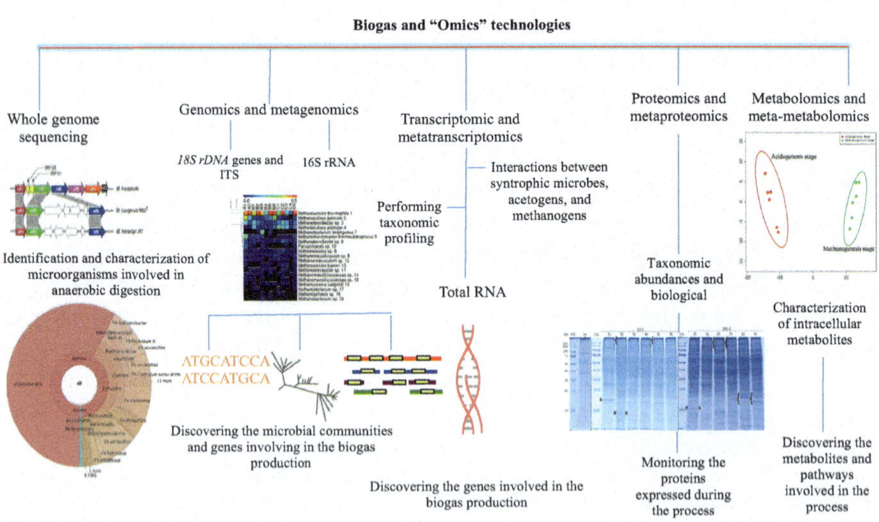

Fig. 16.1 Potential applications of new "omics" technologies in biogas production studies

16.2 Whole Genome Sequencing of Microorganisms Involving in Anaerobic Digestion

Identification and characterization of dominant bacteria and archaea involved in the four steps would be essential to target bioprocess optimization and metabolic engineering with an aim to enhance biogas production efficiency in biogas plants. Recent advances in NGS technologies for whole genome sequencing of diverse organisms have opened a wonderful way to explore adaptive genome features conferring competitiveness to different microorganisms contributing to one or more steps of anaerobic digestion within biogas reactor environments. During the last five years, the whole genome of many microbes, such as *Methanoculleus bourgensis*, *Clostridium bornimense, Clostridium ultunense, Ruminiclostridium cellulosi, Herbinix hemicellulosilytica* and *Peptoniphilus* sp. involved in different steps of anaerobic digestion have been sequenced and characterized (Table 16.1) (Maus et al. 2014, 2016; Hahnke et al. 2014; Koek et al. 2014a, b, 2015a, b; Tomazetto et al. 2014, 2016; Manzoor et al. 2013; Sun and Schnürer 2016).

Methanogens are members of the phylum Euryarchaeota. *Methanoculleus* species are known as one of the most abundant microorganisms involved in methanogenesis in biogas plants (Nettmann et al. 2010; Wirth et al. 2012; Maus et al. 2014). In particular, the species *M. bourgensis* is known as one of the key species in different biogas reactor systems. Maus et al. (2014) completely

Table 16.1 Application of whole genome sequencing in characterization of microorganisms involved in anaerobic digestion

Name	Type of life	Type of activity	Organic matter	References
Methanoculleus bourgensis	Mesophilic	Methanogenesis	Sewage sludge	Maus et al. (2014)
Porphyromonadaceae	Mesophilic	Hydrolysis and acidogenesis	Maize silage and pig and cattle manure	Hahnke et al. (2015)
Clostridium bornimense M2/40	Mesophilic	Biomass degradation	Maize silage and Wheat straw	Hahnke et al. (2015)
Ruminiclostridium cellulosi DG5	Thermophilic	cellulolytic	industrial-scale biogas plant	Koek et al. (2014a, b)
Peptoniphilus sp	Mesophilic	Protein breakdown and acidogenesis	Maize silage	Tomazetto et al. (2014)
Clostridium bornimense strain M2/40^T	Mesophilic	Acidogenic	Maize silage and wheat straw	Tomazetto et al. (2016)
Clostridium ultunense	Mesophilic	Syntrophic acetate-oxidizing	Sludge	Manzoor et al. (2013)
Clostridium sp	Mesophilic	cellulolytic	Slaughterhouse waste	Sun and Schnürer (2016)

sequenced the genome of the type strain *M. bourgensis* MS2T. The strain contained a chromosome with a size of 2.8 kbp, and showed significant genome similarities to *M. marisnigri JR1*. Specific genes contributing to methanogenesis and osmolyte production were detected in the genome of *M. bourgensis MS2T,* and the majority of the genetic information generally found in methanogenesis in biogas plants was detected in the genome of this strain. The same group of researchers in another work sequenced and annotated the genome of another strain BA1 belonging to this species (Maus et al. 2016).

Hahnke et al. (2015), by using Illumina MiSeq system, sequenced the whole genome of a mesophilic, anaerobic *Porphyromonadaceae* bacterium, previously isolated from a biogas-producing lab-scale continuously stirred tank reactor (CSTR) optimized for anaerobic digestion of co-fermented maize silage and pig/cattle manure. The genome sequencing results indicated that this bacterium may be involved in hydrolysis and acidogenesis during anaerobic digestion and biomethanation, as its genome contained diverse genes encoding proteins contributing to the degradation of complex carbohydrates and proteinaceous compounds, and catalyzing the production of volatile fatty acids (VFAs). This research group in another work, by using the same NGS technology sequenced the genome of a mesophilic and anaerobic bacterium *Clostridium bornimense* M2/40, previously isolated from a two-phase biogas reactor continuously fed with maize silage and 5% wheat straw. The results revealed two replicons, including a chromosome (containing 2613 putative genes) and a newly discovered secondary replicon (containing 680 putative genes). Koek et al. (2014a, b) sequenced the genome of the *Ruminiclostridium cellulosi* DG5 (formerly known as *Clostridium cellulosi*), a thermophilic, anaerobic and cellulytic bacterium, previously isolated from an industrial-scale biogas plant in Germany. This microbe is probably responsible for lignocellulose degradation during the anaerobic digestion process and its genome contained 2017 coding sequences, including 136 genes for carbohydrate-hydrolases. The characterized enzymes mainly belonged to different glycoside hydrolase classes which are mainly involved in hydrolysis and/or rearrangement of glycosidic bonds. In another work, same researchers also isolated and characterized a novel species *Herbinix hemicellulosilytica* from a thermophilic biogas reactor. The strain was efficiently able to degrade cellulose at thermophilic conditions (Koeck et al. 2015a). The results of the genome sequencing of the strain showed the presence of a total of 2681 protein coding sequences, including 155 genes encoding glycoside hydrolases (GH) and carbohydrate-binding modules (CBM) (Koeck et al. 2015b). The genes involved in the cellulytic system of the strains included three cellulases, one endoglucanase and two cellobiohydrolases, which presumably degrade cellulose (Koeck et al. 2015b).

Another recently sequenced genome belonged to a mesophilic and obligate anaerobic bacterium *Peptoniphilus sp.* strain ING2-D1G. This strain has high potentials for the hydrolysis of proteins and in acidogenesis during biomass degradation. The chromosome of the strain is 1.6 Mb in size, containing 1476 coding sequences, 53 tRNAs, and 4 ribosomal RNA (rrn) operons. The genome of the strains contained putative genes with potentials for production of acetate, lactate, and butyrate which are involved in the acidogenic metabolism. However, the strain did not possess any genes encoding cellulases (Tomazetto et al. 2014). Later in the year 2016, the same group of researchers annotated the genome sequence of another bacterium, *Clostridium bornimense* strain M2/40^T, isolated from a biogas reactor fed with maize silage and wheat straw. The genome of the strain consisted of two replicons, a chromosome with size of 2.9 Mb containing 2613 putative genes, and a chromid with the size of 700 Kb harboring 680 coding sequences. The genome sequence data indicated that the strain should be classified as acidogenic bacterium, as that it encodes all enzymes required for hydrogen, acetate, formate, lactate, butyrate, and ethanol production (Tomazetto et al. 2016).

Clostridium ultunense strain Esp was another bacterium, whose genome recently has been sequenced. This bacterium belongs to the syntrophic acetate-oxidizing bacteria which play a key role during biogas production from protein-rich materials. The total genome size was about 6.2 Mb containing a total of 6446 putative genes, 6296 of which were protein-encoding sequences, and the others were rRNA genes (Manzoor et al. 2013). Sun and Schnürer (2016) reported on the genome sequence of *Clostridium sp.* Bc-iso-3, a cellulolytic strain isolated from a Swedish industrial-scale biogas digester. The genome size was about 4.3 Mb which contained 3711 coding sequences. Totally, 181 carbohydrate metabolism enzymes (CAZyme) were detected in the genome of the strain, of which one had auxiliary activity, 23 carbohydrate-binding modules, nine carbohydrate esterases, 29 glycoside hydrolases, 16 glycosyltransferases, and three polysaccharide lyases. All these results reveal the high potential of NGS to explore genes and pathways involved in the biogas process.

16.3 Microbial Composition and Functions in Anaerobic Digestion: Genomics and Metagenomics Approaches

High throughput genomics and metagenomics sequencing has been widely used to more deeply investigate the microbial communities and genes involved in the biogas production process from different biomass. For improving the efficiency of biogas digesters, the presence of an active complex microbial communities, effectively hydrolyzing different polymers to methane, is essential. However, understandings on these communities is currently limited as a large proportion of these organisms are uncultivable (Stevenson et al. 2004; Tian et al. 2016). Moreover, only little knowledge exists on the variations in taxonomic, functional and metabolic patterns of microbiomes found in biogas digesters (Luo et al. 2016). Therefore, identification and characterization of microbial communities active during anaerobic digestion for biogas production is an essential task to enhance efficient substrate use and process stability (Stark et al. 2014).

Microbial community composition in the anaerobic digestion system can be identified by PCR amplification and analysis of conserved housekeeping genes as marker genes. The *16SrRNA* (for bacteria) as well as *18SrDNA* genes and ITS sequences (for fungi) have been widely used (Vanwonterghem et al. 2014; Su et al. 2012). Different traditional molecular fingerprinting markers, such as random amplified polymorphic DNA (RAPD; Dubey et al. 2014; Koeck et al. 2014a, b), ribosomal intergenic spacer analysis (RISA; Ciesielski et al. 2013), single-strand-conformation polymorphism (SSCP; Delbès et al. 2000; Leclerc et al. 2001), ribotyping-16SrDNA sequencing (Kröber et al. 2009; Koeck et al, 2014a, b), and variable number of tandem repeats (VNTR; Koeck et al. 2014a, b) have been widely used to study biogas digesters. However, these traditional techniques are culture dependent, time-consuming, and relatively low-throughput. The NSG and "omics" strategies have significantly decreased the cost and improved the yield and quality of the sequence data generated. These advantages make it feasible to rapidly sequence tens to hundreds of amplicon samples on a single run without need to purification and cultivation of each microorganism (Vanwonterghem et al. 2014; Delmont et al. 2012).

Metagenomics is the study of genetic material of all or a group of microbial communities recovered directly from environmental samples. In this science, studies are performed without isolation and purification of microbes. Hence, the genetic materials (DNA of all microbiome) are extracted directly from the environmental samples. Recent explosion in the application of NSG to explore metagenomic or *16S rRNA* and *18SrRNA/ITS* taxonomic diversity of microbial environments has provided a huge deal of data which has opened a new view on microbial communities involved in different processes, such as biogas production. These studies provide a genuine understanding and fingerprinting of structure, functions, and interactions of the microbial communities present in the biogas production process in a culture-independent manner. (Gilbert et al. 2011; Dudhagara et al. 2015; Campanaro et al. 2016). Different metagenomics technologies, such as denaturing/temperature gradient gel electrophoresis (DGGE/TGGE) (Connaughton et al. 2006; Liu et al. 2009), terminal-restriction fragment length polymorphism (T-RFLP) (Carballa et al. 2011; Ziganshin et al, 2013), clone library sequencing (Dong et al. 2015), fluorescence in situ hybridization (FISH) (Nettmann et al. 2010), and 454 pyrosequencing (Li 2013) have been used during the last decade to identify the composition, dynamics, and bioconversion functions of microbial communities in biogas digesters. These studies have mostly been carried out to detect microbial communities in large (Liu et al. 1999), lab-scale (Li 2013) biogas reactors or small-sized household digesters (Dong et al. 2015; Tian et al. 2016).

Based on the source biomass (manure, lignocellulosic materials, etc.), the microbial communities involved in the anaerobic digestion are different. Tsapekos et al. (2016) showed that during the process of anaerobic co-digestion of pig manure and ensiled meadow grass, species similar to *Coprothermobacter proteolyticus* and to *Clostridium thermocellum* with high proteolytic and cellulolytic activities were found firmly attached to the solid fraction of the digested feedstock, whereas liquid samples contained different microbial community composition, mainly dominated by Proteobacteria and archaea. Furthermore, an unclassified *Alkaliphilus* sp. was found in high relative abundance in all samples.

Tian et al. (2016) used a combination of *16S rRNA* gene PCR-based techniques to study community composition, richness, and evenness in different household biogas digesters in Yunnan Province, China. They showed that the composition of bacterial and archaeal community in different regions were significantly different, and was drastically affected by seasonal changes. The dominant bacterial phyla included *Chloroflexi, Bacteroidetes, Firmicutes,* and *Proteobacteria,* and the most dominant archaeal phylum was Euryarchaeota. The dominant genera were *Bacteroides, Clostridium, Smithella, Syntrophus* and *Bellilinea. Chloroflexi* and *Crenarchaeota* (species *Methanosaeta concilii*) were identified as the most dominant bacteria and archaea, respectively. Suksong et al. (2016) optimized different key parameters of biogas production in a solid-state anaerobic digestion (SS-AD) fed by oil palm biomass. The most dominant bacteria were *Ruminococcus* sp. and *Clostridium* sp., while *Methanoculleus* sp. was detected as the most dominant archaea. A combination of Illumina MiSeq platform sequencing and DGGE were used to study the effect of straw microbial community on thermophilic biogas process. The results obtained revealed that *Methanosarcina* was the most efficient in producing biogas (Li et al. 2017). Sun et al. (2016) by using 16S rRNA gene amplicons could show that *Bacteroidetes* and *Firmicutes* were the most dominant microorganisms involved in the degradation of lignocellulos to enhance biogas production. Zhao et al. (2016) investigated the microbial community of a biogas reactor fed by *Pennisetum sinese* Roxb as mono-substrate by *16S rRNA* gene analysis. The dominant bacteria were *Treponema porcinum, Eubacterium limosum, Clostridium, Bacteroidetes, Firmicutes,* and *Anaerofilum,* while the dominant archaea were *Methanobacterium curvum, Methanosarcina barkeri, Methanobacterium bryantii,* and *Methanofollis ethanolicus*. Luo et al. (2016) used metagenomic analysis to elucidate the structure, function, and metabolic patterns of the core microbial consortium existing in 14 full-scale biogas reactors fed by sludge or manure, operated under various conditions. The metagenomics analysis showed higher relative abundance of hydrogenotrophic methanogens, and that the sludge-based samples were clearly distinct from the manure-based samples from both taxonomic and functional patterns perspectives. Treu et al. (2016a, b) also used high throughput metagenomics to characterize the structure and function of the microbial community in biogas reactors. They found that a complex consortium of unknown species was involved in the anaerobic digestion process, but it could only be characterized at high taxonomic levels (Treu et al. 2016a, b).

Another application of metagenomics in biogas production is the monitoring of inoculated microorganisms during the biogas production process. For instance, Tsapekos et al. (2017) evaluated the efficiency of two hydrolase producing bacteria, *C. thermocellum* and *Melioribacter roseus* in degrading lignocellulosic matter in biogas production process in batch reactors and CSTRs. The results confirmed that the presence of *C. thermocellum* could enhance biogas production up to 34%, whereas another bacterium did not show significant effect on methane production. They used high-throughput *16S rRNA* amplicon sequencing to evaluate the effect of bacterial inoculation on bacterial and archaeal populations. The results confirmed that both strains were not markedly resided into biogas microbiome, and did not significantly alter the microbial communities.

Another application of metagenomic studies in anaerobic digestion is the characterization of functions of those microorganisms involved in the breakdown of polysaccharides and methanogenesis during biogas production (Vanwonterghem et al. 2014). For instance, Jaenicke et al. (2010) used 454 pyrosequencing to elucidate the structure and function of the microbial consortium contributing to the anaerobic digestion of a mixture of maize silage, green rye, and chicken manure in a production-scale biogas plant. They found that *Firmicutes* and *Bacteroidetes* were dominant in the process and were likely responsible for the breakdown of polysaccharides and fermentation. Moreover, metagenomics studies have been used to study metabolic pathways in the anaerobic digestion process. In line with that, Wirth et al. (2012) could evaluate the regulatory role of hydrogen metabolism in a biogas plant fed periodically with a mixture of maize silage and pig manure, and tried to enhance biogas production by optimizing the balance between H_2-producers and H_2-consumers. In addition to the wide application of metagenomic sequencing in phylogenetic and functional diversity in anaerobic digestion systems, it can also be used to evaluate the effects of reactor set-up, pretreatment methods, operational conditions, and feedstock composition on the community composition and function (Vanwonterghem et al. 2014).

16.4 Microbial Functions in Anaerobic Digestion: Transcriptomics and Metatranscriptomics Approaches

Commonly, the transcriptome includes a set of all RNA molecules in one cell or a population. It is sometimes used to refer to all RNAs, or just mRNA, depending on the particular experiment. Therefore, transcriptomics is the study of the whole transcriptome of an organism. Another term is metatranscriptomics which refers to the sequencing of reverse transcribed mRNA extracted from microbial communities of different environmental samples. The advantage of this technology is the reduction of the level of complexity seen in metagenomics by only focusing on those microorganisms and genes that are metabolically active (Sue et al. 2012; Vanwonterghem et al. 2014). Previously, microarray technologies have been widely used to explore gene expression profiles for different organisms. However, this technology is time consuming and expensive, and cannot detect novel genes, as the probes are designed and used based on known genes, whereas, metatranscriptomics is very efficient and does not need any information about the genes of interest.

In investigating biogas production from different feedstock, transcriptomics analysis could play an important role in finding the right microorganisms with appropriate genes and metabolites. Transcriptomics technologies can be used not only for exploring active microorganisms, new genes, and pathways involved in lignocellulose degradation and methane production, but also can help to find markers for monitoring industrial biogas production to prevent failures or to model the whole process (Stark et al. 2014). Zakrzewski et al. (2012) for the first time used metatranscriptome sequencing of 16S ribosomal sequence tags to perform taxonomic profiling of the active part of the microbial community in an anaerobic digestion system. Their results showed Euryarchaeota and Firmicutes as dominant and the most active phyla during the process, whereas a small part was assigned to the *Bacteroidetes*, *Actinobacteria* and *Synergistetes* phyla. Maus et al. (2016) used metatranscriptome sequencing analysis to determine the active microbial flora in an exemplary thermophilic biogas plant. The meta-transcriptomic *16SrRNA* analysis showed that the genera *Defluviitoga* (9.2%), *Clostridium* cluster III (4.8%), and *Tepidanaerobacter* (1.1%) as well as *Methanoculleus* (5.7%) were the most transcriptionally active microorganisms during the process, whereas *Hallocella* (1.8%), *Tepidimicrobium* (0.5%), and *Methanothermobacter* (<0.1%) were less active.

This approach can be used to determine the role and function of high or low abundant individual microbes in maintaining the efficiency and stability of the process, and also to evaluate the functionally versatile lineages with the capability to drive diverse processes from hydrolysis to acetate oxidation (e.g. *Clostridia*), and even to assess the role of different strains and species of a specific genus under different environmental conditions or presence of syntrophic or competitive community members. By using this "omics" technology, it has been shown that *Clostridia* were the dominant class of hydrolytic organisms in the biogas fermenters, playing a key role during the initial biomass degradation (Gulert et al. 2016; Sundberg et al. 2013; Vanwonterghem et al. 2014; Ziganshin et al. 2013).

The metatranscriptome analysis also has the potential to study microbe-microbe interactions between syntrophic microbes, acetogens, and methanogens. For instance, electron transfer mechanisms occurring as result of interspecies microbe-microbe interactions (e.g., between *Geobacter* and *Methanosaeta* populations) during the anaerobic digestion have been explored (Morita et al. 2011; Liu et al. 2012). Moreover, metatranscriptomics holds the potential to be applied to quickly assess regulatory reactions in biogas plants to monitor shifts in metabolic pathways and profiles, and changes in the balance of functional guilds. This would help to determine the optimized conditions in which pathways of interest would be active, allowing a defined and special microbial community towards efficient biodegradation and fermentation of biomass. For instance, Gulert et al. (2016) used metagenomics and metatranscriptomics to evaluate hydrolysis rates in a commercial biogas plant fed with maize silage, cow manure, chicken manure, and feces samples from herbivores. The maximum active cellulolytic GHs genes were observed in the biogas reactor fed by elephant feces. RNA-Seq results indicated that highly transcribed cellulases of Firmicutes were four times more than that of *Bacteroidetes* in the biogas fermenter, while in the elephant feces samples, the distribution of these enzymes was very similar. Based on these results, they suggested that increase of *Bacteroidetes* and *Fibrobacteres* populations may enhance hydrolytic performance in the anaerobic digesters.

It is clear that the application of metatranscriptomics in combination with other "omics" technologies and chemical, rheological, and physical parameters could assist with achieving the best results. However, transcriptomics and metatranscriptomics studies are faced with some technical challenges, such as impurity of the final extracted RNA from environmental samples, fast degradation of RNA due to its short half-life, difficulties in enriching mRNA and bias related to cDNA synthesis and amplification (Vanwonterghem et al. 2014). Therefore, optimizing total RNA extraction and cDNA synthesis protocols is of importance in transcriptome studies in biogas process. In line with that, attempt have made to overcome these challenges. For instance, Stark et al. (2014) tried to set up different RNA-extraction protocols which could be efficiently used for metatranscriptomics studies in biogas plants.

Anaerobic fungi have been known as efficient degraders of lignocellulosic biomass in the digestive tracts of their host animals and are regarded as a promising reservoir for bioaugmentation in biogas production processes (Dollhofer et al. 2017; Gruninger et al. 2014). Dollhofer et al. (2017) by using genomics (PCR) and transcriptomics (cDNA synthesis) approaches could detect anaerobic fungi GH5 endoglucanase gene in two digesters fed by lignocellulosic biomass. In addition, they isolated and characterized a new *Piromyces* species from a PCR-positive digester. However, the cellulytic activity was at low level in the digesters, and it was proposed that anaerobic fungi from digesters sludge should be added to those digesters to enhance cellulytic activities. Bremges et al. (2015) used deep sequencing of metagenome and meta-transcriptome to explore the genes involved in methanogenic pathways in an agricultural production-scale biogas plant, and reconstructed the major genes contributing to the methane metabolism in the process.

16.5 Microbial Functions in Anaerobic Digestion: Proteomics and Metaproteomics Approaches

Commonly, proteomics is considered as the large-scale characterization of all protein components of an organism, whereas metaproteomics is a set of all proteins expressed by complex microbial communities under a given set of conditions at a specific point in time in one ecosystem. In order to perform metaproteomics studies, at the first step, proteins are extracted from an environmental sample (e.g., soil, water, and fermentation process), followed by fractionation and separation using liquid chromatography, and finally detection and sequencing are performed using tandem mass spectrometry (MS/MS) (Vanwonterghem et al. 2014; Langley et al. 2013). By using these approaches, it is possible to organize proteome analysis on an organism or environmental samples, and also to detect the presence and quantity of enzymes and overall metabolic activity of microbial communities in different environmental samples, such as anaerobic digesters.

One of the application of metaproteomics is proteotyping of microbial flora involved in anaerobic digestion process. Heyer et al. (2016) used a metaproteomics approach (liquid chromatography coupled to tandem MS) to determine microbial community of 35 different industrial biogas plants. The obtained proteome datasets showed that *Bacillales, Enterobacteriales, Bacteriodales, Clostridiales, Rhizobiales* and *Thermoanaerobacteriales* as well as *Methanobacteriales, Methanosarcinales* and *Methanococcales* were playing key roles in biogas production process. These results were very similar to those data achieved from metagenomics studies. Recently, Kohrs et al. (2017) used tandem MS-based proteotyping to identify taxonomic abundances and biological processes in some biogas plants. Proteotyping and T-RFLP fingerprinting indicated significant differences in the composition of individual microbial communities, indicating multiple steady-states.

Proteomics and metaproteomics studies allow to monitor all proteins expressed during the process of lignocellulose degradation and fermentation regardless of the microorganisms producing them. Hence, these approaches have evolved as powerful methods to detect catalytic enzymes, metabolic pathways and even novel functional proteins expressed by microbial communities in biogas plants (Stolze et al. 2015; Montag and Schink 2016; Lin et al. 2016; Vanwonterghem et al. 2014).

Abram et al. (2011) used metatranscriptomics approach to determine key metabolic pathways during anaerobic digestion process in an anaerobic industrial-like wastewater treatment bioreactor. Their results showed that the core microbiome in the process was very diverse and complex, and the archaeal population was mainly composed of *Methanocorpusculum*-like (76%) microorganisms. They identified 33 different proteins which were excised from 2-D gels. The detected proteins were related to methanogenesis process, such as CO_2 and acetate, glycolysis, and the pentose phosphate pathways. In addition, this approach using protein assignments indicated the presence of some specific microorganisms in the bioreactor. Hanreich et al. (2013) performed integrated metagenomic and meta-proteomic studies to evaluate the hydrolase activity of different groups of microorganisms involved in the process of an anaerobic digesting system. Their results indicated that *Bacteroidetes* expressed a high number of hydrolases and sugar transporters, whereas few glycoside hydrolases were expressed by *Firmicutes*. In addition, the key enzymes of the methanogenesis were highly expressed during the process.

Metaproteomics can also be used to enhance the efficiency and stability of the anaerobic digestion process, by characterizing regulatory elements contributing to enzyme production, through identification of biomarkers as predictive indicators of process failure in anaerobic digesters following perturbation, and to elucidate critical steps in the conversion of biomass to methane by generating functional data (Vanwonterghem et al. 2014; Heyer et al. 2016).

In spite of the above-mentioned attractive features, metaproteome analyses are faced with different shortcomings as well, such as sample impurities and complexity and limited availability of genome sequences. Therefore, developing new suitable extraction and fractionation methods for metatranscriptomics studies is of importance (Heyer et al. 2016; Kohrs et al. 2015). Kohrs et al. (2014) showed that sample pre-fractionation with liquid isoelectric focusing improved deep metaproteome analysis of different types of biogas plants. This group of researchers also developed a fast and easy centrifugal fractionation method for metaproteome extraction from biogas sludge samples. This method uses a centrifugal fractionation, and gel-based separation followed by LC-MS/MS identification (Kohrs et al. 2015).

Ortseifen et al. (2016) used a combined strategy based on metagenomics and metaproteomics approaches to elucidate the composition of microbial communities and proteins involved in the anaerobic digestion in an agricultural biogas plant. The highly abundant proteins of the biogas microbiome were related to the pathways involved in methanogenesis, transport, and carbon metabolism. In addition, by integration of metagenome and metaproteome data, it was possible to construct a lineage link between special microorganisms and the detected genes encoding special proteins contributing to the process. For instance, meta-proteome data showed that 3 dominant proteins contributing to the methanogenesis process belonged to *Methanoculleus* sp.

Overall, it can be concluded that integration of metaproteomics data with advanced imaging techniques, metagenomics, metatranscriptomics, and metabolomics could be used for efficient exploration of metabolic pathways involved in biogas production from organic materials.

16.6 Microbial Functions in Anaerobic Digestion: Metabolomics and Meta-Metabolomics Approaches

Characterization of intracellular metabolites involved in anaerobic fermentation processes could help to delineate metabolic pathways of microbial communities for the optimization of the process to enhance the efficiency of biogas production (Yang et al. 2014). Metabolomics and meta-metabolomics are the science of total metabolites of an organism or all organisms present in a specific environment under a specific condition. To do these analyses, commonly different chromatography techniques, such as GC-MS are used. This method provides the opportunity to reconstruct metabolic pathways and to discover genes and enzymes. On the another hand, metabolomics helps to distinguish post-translational modifications (Vanwonterghem et al. 2014). Furthermore, metabolomics analysis can be used to elucidate microbial and metabolite distribution in the anaerobic digestion under different conditions (Beale et al. 2016).

Yang et al. (2014) optimized a meta-metabolite extraction protocol based on acetonitrile/methanol/water (2:2:1, by vol.) to evaluate microbial community and their metabolites involved in the anaerobic fermentation of corn stalk in a biogas digester using GC–MS. The comparison of metabolite profiles during the process indicated a significant increase in the levels of sugars and sugar alcohols during the methanogenesis and fatty acids during acidogenesis. Beale et al. (2016) used an integrated metagenomics and metabolomics analysis in the laboratory scale digesters. Their results indicated that the ratio of oxidizing bacteria (methane, sulphide, and sulphate) to sulphate reducing bacteria had a significant effect on the efficiency of biogas production. Ina addition, increases in short chain fatty acids improved the biogas production efficiency.

Sasaki et al. (2014) compared the metabolite profiles of the microbiomes during methanogenic process between a stable methanogenic reactor and a deteriorated reactor. Their results showed that the carbon flux was higher during the stable methanogenesis process, as the concentrations of the intracellular metabolites involved in the Embden–Meyerhof and pentose phosphate pathways were higher in this period. In addition, the concentrations of the intermediate metabolites in the reductive branch of tricarboxylic acid cycle, malate, fumarate, and succinate were higher in the deteriorated reactor, whereas the glutamate levels were higher in the stable methanogenic reactor. In another study, Kučera et al. (2017) studied the metabolites changes during the process of anaerobic digestion of wine waste in a microscale discontinuous fermenter using GC-MS in combination with principal component analysis and orthogonal projection. Their findings showed that particular polyphenolic structures were dominant during the process of anaerobic digestion. A group of dihydro-flavonoids appeared at early stages of the process (acidogenic phase), however, they were degraded in the next steps. As some dihydroflavonoids (e.g., taxifolin) are very toxic, so, the application of un-stabilized digestate as a fertilizer would represent a potential environmental risk. These studies indicated that the application of metabolomics could help scientists to better understand biogas production processes, especially when used in combination with other "omics" technologies.

16.7 Conclusions

Biogas production from organic matters is becoming an increasingly promising bioenergy source owing to being clean, renewable, and environmentally compatible. The most important factor during the process of anaerobic digestion for biogas production is the presence and activity of a complex consortia of microorganisms contributing to an efficient decomposition of organic matter and the subsequent stages, i.e., acidogenesis and methanogenesis. The majority of the participating microbes, their genes, and metabolic pathways during this process are not still well known, as most of the biogas microbial communities are non-cultivable. Recent developments in "omics" technologies, such as genomics, metagenomics, transcriptomics, proteomics, metabolomics, and NGS have provided excellent opportunities for identification and characterization of microbial flora, their genes, encoded transcripts, proteins, and metabolites, as well as the metabolic pathways contributing to the anaerobic digestion process. By such new findings, it will be possible to optimize the process conditions for efficient and economic production of biogas from different organic materials, especially from lignocellulosic biomass and wastewaters. Recent studies have indicated that co-application of two or more different "omics" technologies could result in deeper insights into the different steps of anaerobic digestion process. Finally, it could be concluded that these recent revolutionary innovations and developments in "omics" and NGS technologies will be instrumental in enhancing the efficiency of biogas production from a diverse range of organic matters with complex structures.

Acknowledgements The authors would like to thanks Dr. Meisam Tabatabaei for his assistance with carefully reviewing the manuscript and the improvements made as a result.

References

Abram F, Enright A-M, O'Reilly J, Botting CH, Collins G, O'Flaherty V (2010) A metaproteomic approach gives functional insights into anaerobic digestion. Appl Microbiol 110:1550–1560

Anjum R, Grohmann E, Krakat, N (2017) Anaerobic digestion of nitrogen rich poultry manure: impact of thermophilic biogas process on metal release and microbial resistances. Chemosphere

Beale DJ, Karpe AV, McLeod JD, Gondalia SV, Muster TH, Othman MZ, Palombo EA, Joshi D (2016) An 'omics' approach towards the characterization of laboratory scale anaerobic digesters treating municipal sewage sludge. Water Res 88:346–357

Benato A, Macor A (2017) Biogas engine waste heat recovery using organic Rankine cycle. Energies 10(3):327

Bremges A, Maus I, Belmann P, Eikmeyer F, Winkler A, Albersmeier A (2015) Deeply sequenced metagenome and metatranscriptome of a biogas-producing microbial community from an agricultural production-scale biogas plant. GigaScience 4(1):33

Campanaro S, Treu L, Kougias PG, De Francisci D, Valle G, Angelidaki I (2016) Metagenomic analysis and functional characterization of the biogas microbiome using high throughput shotgun sequencing and a novel binning strategy. Biotechnol Biofuels 9(1):26

Carballa M, Smits M, Etchebehere C, Boon N, Verstraete W (2011) Correlations between molecular and operational parameters in continuous lab-scale anaerobic reactors. Appl Microbiol Biotechnol 89:303–314

Ciesielski S, Bułkowska K, Dabrowska D, Kaczmarczyk D, Kowal P, Możejko J (2013) Ribosomal intergenic spacer analysis as a tool for monitoring methanogenic archaea changes in an anaerobic digester. Curr Microbiol 67(2):240

Connaughton S, Collins G, O'Flaherty V (2006) Development of microbial community structure and activity in a high-rate anaerobic bioreactor at 18 C. Water Res 40:1009–1017

Delbès C, Moletta R, Godon JJ (2000) Monitoring of activity dynamics of an anaerobic digester bacterial community using 16S rRNA polymerase chain reaction–single-strand conformation polymorphism analysis. Environ Microbiol 2(5):506–515

Delmont TO, Simonet P, vogel TMJ (2012) Describing microbial communities and performing global comparisons in the "omic" era. ISME 6(1625–1628):18

Deng L, Liu Y, Zheng D, Wang L, Pu X, Song L, Wang Z, Lei Y, Chen Z, Long Y (2016) Application and development of biogas technology for the treatment of waste in China. Renew Sustain Energy Rev

Dollhofer V, Callaghan TM, Griffith GW, Lebuhn M, Bauer J (2017) Presence and transcriptional activity of anaerobic fungi in agricultural biogas plants. Biores Technol 235:131–139

Dong M, Wu Y, Li Q, Tian G, Yang B, Li Y, ZhangL Wang Y, Xiao W, Yin F, Zhao X, Zhang W, Cui X (2015) Investigation of methanogenic community structures in rural biogas digesters from different climatic regions in Yunnan, south west China. Curr Microbiol 70(5):679–684

Dubey SK, Meena RK, Sao S, Patel J, Shukla P (2014) Isolation and characterization of cellulose degrading bacteria from biogas slurry and their RAPD profiling. Curr Res Microbiol Biotechnol 2(4):416–421

Dudhagara P, Ghelani A, Bhavsar S, Bhatt S (2015) Metagenomic data of fungal internal transcribed Spacer and 18S rRNA gene sequences from Lonar lake sediment, India. Data Brief 4:266–268

Gilbert JA, Meyer F, Bailey MJ (2011) The future of microbial metagenomics (or is ignorance bliss?). ISME J 5(5):777

Gruninger RJ, Puniya AK, Callaghan TM, Edwards JE, Youssef N, Dagar SS, Fliegerová K, Griffith GW, Forster R, Tsang A, McAllister T, Elshahed MS (2014) Anaerobic fungi (phylum Neocallimastigomycota): advances in understanding their taxonomy, life cycle, ecology, role and biotechnological potential. FEMS Microbiol Ecolo 90:1–17

Güllert S, Fischer MA, Turaev D, Noebauer B, Ilmberger N, Wemheuer B, Alawi M, Rattei T, Daniel R, Schmitz RA, Grundhoff A (2016) Deep metagenome and metatranscriptome analyses of microbial communities affiliated with an industrial biogas fermenter, a cow rumen, and elephant feces reveal major differences in carbohydrate hydrolysis strategies. Biotechnol Biofuels 9(1):121

Hahnke S, Maus I, Wibberg D, Tomazetto G, Pühler A, Klocke M, Schlüter A (2015) Complete genome sequence of the novel *Porphyromonadaceae* bacterium strain ING2-E5B isolated from a mesophilic lab-scale biogas reactor. J Biotechnol 193:34–36

Hanreich A, Schimpf U, Zakrzewski M, Schluter A, Benndorf D, Heyer R (2013) Metagenome and metaproteome analyses of microbial communities in mesophilic biogas-producing anaerobic batch fermentations indicate concerted plant carbohydrate degradation. System App Microbiol 36(5):330–338

Heyer R, Benndorf D, Kohrs F, De Vrieze J, Boon N, Hoffmann M (2016) Proteotyping of biogas plant microbiomes separates biogas plants according to process temperature and reactor type. Biotechnol Biofuels 9(1):155

Heyer R, Kohrs F, Benndorf D, Rapp E, Kausmann R, Heiermann M (2013) Metaproteome analysis of the microbial communities in agricultural biogas plants. New Biotechnol 30 (6):614–622. https://doi.org/10.1016/j.nbt.2013.01.002

Jaenicke S, Ander C, Bekel T, Bisdorf R, Droge M, Gartemann K-H, Junemann S, Kaiser O, Krause L, Tille F et al (2010) Comparative and joint analysis of two metagenomic datasets from a biogas fermenter obtained by 454-pyrosequencing. PLoS ONE 6:1–15

Koeck DE, Ludwig W, Wanner G, Zverlov VV, Liebl W, Schwarz WH (2015a) *Herbinix hemicellulosilytica* gen. nov., sp. nov., a thermophilic cellulose-degrading bacterium isolated from a thermophilic biogas reactor. Int J Syst Evol Microbiol 65(8):2365–2371

Koeck DE, Maus I, Wibberg D, Winkler A, Zverlov VV, Liebl W, Pühler A, Schwarz WH, Schlüter A (2015b) Draft genome sequence of *Herbinix hemicellulosilytica* T3/55 T, a new thermophilic cellulose degrading bacterium isolated from a thermophilic biogas reactor. J Biotechnol 214:59–60

Koeck DE, Wibberg D, Maus I, Winkler A, Albersmeier A, Zverlov VV, Liebl W, Pühler A, Schwarz WH, Schlüter A (2014a) Complete genome sequence of the cellulolytic thermophile *Ruminoclostridium cellulosi* wild-type strain DG5 isolated from a thermophilic biogas plant. J Biotechnol 188:136–137

Koeck DE, Zverlov VV, Liebl W, Schwarz WH (2014b) Comparative genotyping of *Clostridium thermocellum* strains isolated from biogas plants: genetic markers and characterization of cellulolytic potential. Syst Appl Microbiol 37(5):311–319

Kohrs F, Heyer R, Bissinger T, Kottler R, Schallert K, Püttker S, Behne A, Rapp E, Benndorf D, Reichl U (2017) Proteotyping of laboratory-scale biogas plants reveals multiple steady-states in community composition. Anaerobe

Kohrs F, Heyer R, Magnussen A, Benndorf D, Muth T, Behne A, Rapp E, Kausmann R, Heiermann M, Klocke M, Reichl U (2014) Sample prefractionation with liquid isoelectric focusing enables in depth microbial metaproteome analysis of mesophilic and thermophilic biogas plants. Anaerobe 29:59–67

Kohrs F, Wolter S, Benndorf D, Heyer R, Hoffmann M, Rapp E et al. (2015) Fractionation of biogas plant sludge material improves metaproteomic characterization to investigate metabolic activity of microbial communities. Proteomics 15(20):3585–3589. https://doi.org/10.1002/pmic.201400557

Kröber M, Bekel T, Diaz NN, Goesmann A, Jaenicke S, Krause L, Miller D, Runte KJ, Viehöver P, Pühler A, Schlüter A (2009) Phylogenetic characterization of a biogas plant microbial community integrating clone library 16S-rDNA sequences and metagenome sequence data obtained by 454-pyrosequencing. J Biotechnol 142(1):38–49

Kučera L, Kurka O, Barták P, Bednář P (2017) Liquid chromatography/high resolution tandem mass spectrometry–tool for the study of polyphenol profile changes during micro-scale biogas digestion of grape marcs. Chemosphere 166:463–472

Langley SR, Dwyer J, Drozdov I, Yin X, Mayr M (2013) Proteomics: from single molecules to biological pathways. Cardiovasc Res 97:612–622

Leclerc M, Delbes C, Moletta R, Godon JJ (2001) Single strand conformation polymorphism monitoring of 16S rDNA Archaea during start-up of an anaerobic digester. FEMS Microbiol Ecol 34(3):213–220

Li JM (2013) Biogas production in China: current status and future development. www.epa.gov/agstar/documents/conf13/Biogas Production in China—current status. Accessed June 2017

Li X, Liu YH, Zhang X, Ge CM, Piao RZ, Wang WD, Cui ZJ, Zhao HY (2017) Evaluation of biogas production performance and dynamics of the microbial community in different straws. J Microbiol Biotechnol 27:524–534

Lin YW, Tuan NN, Huang SL (2016) Metaproteomic analysis of the microbial community present in a thermophilic swine manure digester to allow functional characterization: A case study. Int Biodeterior Biodegradation 115:64–73

Liu F, Rotaru AE, Shrestha PM, Malvankar NS, Nevin KP, Lovley DR (2012) Promoting direct interspecies electron transfer with activated carbon. Energy Environ Sci 5(10):8982–8989

Liu Y, Balkwill DL, Aldrich HC, Drake GR, Boone DR (1999) Characterization of the anaerobic propionate-degrading syntrophs *Smithella propionica* gen. nov., sp. nov. and *Syntrophobacter wolinii*. Int J Sys Bacteriol 49:545–556

Liu FH, Wang SB, Zhang JS, Zhang J, Yan X, Zhou HK, Zhao GP, Zhou ZH (2009) The structure of the bacterial and archaeal community in a biogas digester as revealed by denaturing gradient gel electrophoresis and 16S rDNA sequencing analysis. J Appl Microbiol 106(3):952–966

Luo G, Fotidis IA, Angelidaki I (2016) Comparative analysis of taxonomic, functional, and metabolic patterns of microbiomes from 14 full-scale biogas reactors by metagenomic sequencing and radioisotopic analysis. Biotechnol Biofuels 9(1):51

Manzoor S, Müller B, Niazi A, Bongcam-Rudloff E, Schnürer A (2013) Draft genome sequence of *Clostridium ultunense* strain Esp, a syntrophic acetate-oxidizing bacterium. Genome announcements 1(2):e00107–e00113

Mao C, Feng Y, Wang X, Ren G (2015) Review on research achievements of biogas from anaerobic digestion. Renew Sustain Energy Rev 45:540–555

Maus I, Stantscheff R, Wibberg D, Stolze Y, Winkler A, Pühler A, König H, Schlüter A (2014) Complete genome sequence of the methanogenic neotype strain *Methanobacterium formicicum* MFT. J Biotechnol 192:40–41

Maus I, Wibberg D, Winkler A, Pühler A, Schnürer A (2016) Complete genome sequence of the methanogen *Methanoculleus bourgensis* BA1 isolated from a biogas reactor, 4(3):3–4

Montag D, Schink B (2016) Biogas process parameters—energetics and kinetics of secondary fermentations in methanogenic biomass degradation. Appl Microbiol Biotechnol 100(2):1019–1026

Morita M, Malvankar NS, Franks AE, Summers ZM, Giloteaux L, Rotaru AE, Rotaru C, Lovley DR (2011) Potential for direct interspecies electron transfer in methanogenic wastewater digester aggregates. MBio 2(4):e00159–11

Nettmann E, Bergmann I, Pramschüfer S, Mundt K, Plogsties V, Herrmann C, Klocke M (2010) Polyphasic analyses of methanogenic archaeal communities in agricultural biogas plants. Appl Environ Microbiol 76(8):2540–2548

Ortseifen V, Stolze Y, Maus I, Sczyrba A, Bremges A, Albaum SP, Jaenicke S, Fracowiak J, Pühler A, Schlüter A (2016) An integrated metagenome and-proteome analysis of the microbial community residing in a biogas production plant. J Biotechnol 231:268–279

Raupach MJ, Amann R, Wheeler QD, Roos C (2016) The application of "-omics" technologies for the classification and identification of animals. Organisms Diversity Evol 16(1):1–12

Sasaki D, Sasaki K, Tsuge Y, Morita M, Kondo A (2014) Comparison of metabolomic profiles of microbial communities between stable and deteriorated methanogenic processes. Biores Technol 172:83–90

Stark L, Giersch T, Wünschiers R (2014) Efficiency of RNA extraction from selected bacteria in the context of biogas production and metatranscriptomics. Anaerobe 29:85–90

Stevenson BS, Eichorst SA, Wertz JT, Schmidt TM, Breznak JA (2004) New strategies for cultivation and detection of previously uncultured microbes. Appl Environ Microbiol 70:4748–4755

Stolze Y, Zakrzewski M, Maus I, Eikmeyer F, Jaenicke S, Rottmann N et al. (2015) Comparative metagenomics of biogas-producing microbial communities from production-scale biogas plants operating under wet or dry fermentation conditions. Biotechnol Biofuels 8(1):14

Su C, Lei L, Duan Y, Zhang K-Q, Yang J (2012) Culture-independent methods for studying environmental microorganisms: methods, application, and perspective. Appl Microbiol Biotechnol 93:993–1003

Suksong W, Kongjan P, Prasertsan P, Imai T, Sompong O (2016) Optimization and microbial community analysis for production of biogas from solid waste residues of palm oil mill industry by solid-state anaerobic digestion. Biores Technol 214:166–174

Sun L, Schnürer A (2016) Draft genome sequence of the cellulolytic strain Clostridium sp. Bc-iso-3 isolated from an industrial-scale anaerobic digester. Genome Announcements, 4(5):e01188–16

Sun L, Liu T, Müller B, Schnürer A (2016) The microbial community structure in industrial biogas plants influences the degradation rate of straw and cellulose in batch tests. Biotechnol Biofuels 9(1):128

Sundberg C, Al-Soud WA, Larsson M, Alm E, Yekta SS, Scvensson BH, Sorenson SJ, Karlsson A (2013) 454 pyrosequencing analyses of bacterial and archaeal richness in 21 full-scale biogas digesters. FEMS Microbiol Ecol 85:612–626

Tian G, Li Q, Dong M, Wu Y, Yang B, Zhang L et al. (2016) Spatiotemporal dynamics of bacterial and archaeal communities in household biogas digesters from tropical and subtropical regions of Yunnan Province, China. Environ Sci Poll Res 23(11):11137–11148

Tomazetto G, Hahnke S, Koeck DE, Wibberg D, Maus I, Pühler A, Klocke M, Schlüter A (2016) Complete genome analysis of Clostridium bornimense strain M2/40 T: A new acidogenic *Clostridium* species isolated from a mesophilic two-phase laboratory-scale biogas reactor. J Biotechnol 232:38–49

Tomazetto G, Hahnke S, Maus I, Wibberg D, Pühler A, Schlüter A, Klocke M (2014) Complete genome sequence of *Peptoniphilus sp.* strain ing2-D1G isolated from a mesophilic lab-scale completely stirred tank reactor utilizing maize silage in co-digestion with pig and cattle manure for biomethanation. J Biotechnol 192:59–61

Treu L, Campanaro S, Kougias PG, Zhu X, Angelidaki I (2016a) Untangling the effect of fatty acid addition at species level revealed different transcriptional responses of the biogas microbial community members. Environ Sci Technol 50(11):6079–6090

Treu L, Kougias PG, Campanaro S, Bassani I, Angelidaki I (2016b) Deeper insight into the structure of the anaerobic digestion microbial community; the biogas microbiome database is expanded with 157 new genomes. Biores Technol 216:260–266

Tsapekos P, Kougias PG, Frison A, Raga R, Angelidaki I (2016) Improving methane production from digested manure biofibers by mechanical and thermal alkaline pretreatment. Biores Technol 216:545–552

Tsapekos P, Kougias PG, Vasileiou SA, Treu L, Campanaro S, Lyberatos G, Angelidaki I (2017) Bioaugmentation with hydrolytic microbes to improve the anaerobic biodegradability of lignocellulosic agricultural residues. Biores Technol 234:350–359

Vanwonterghem I, Jensen PD, Ho DP, Batstone DJ, Tyson GW (2014) Linking microbial community structure, interactions and function in anaerobic digesters using new molecular techniques. Curr Opin Biotechnol 27:55–64

Wirth R, Kovács E, Maróti G, Bagi Z, Rákhely G, Kovács KL (2012) Characterization of a biogas-producing microbial community by short-read next generation DNA sequencing. Biotechnol Biofuels 5(1):41

Yang D, Fan X, Shi X, Lian S, Qiao J, Guo R (2014) Metabolomics reveals stage-specific metabolic pathways of microbial communities in two-stage anaerobic fermentation of corn-stalk. Biotech Lett 36(7):1461–1468

Zakrzewski M, Goesmann A, Jaenicke S, Junemann S, Eikmeyer F, Szczepanowski R, Al-Soud WA, Sorensen S, Puhler A, Schluter A (2012) Profiling of the metabolically active community from a production-scale biogas plant by means of high-throughput metatranscriptome sequencing. J Biotechnol 158:248–258

Zhao C, Ai C, Li Q, Yang C, Zhou G, Liu B (2016) Diversity of archaea and bacteria in a biogas reactor fed with *Pennisetum sinese* Roxb by 16S rRNA sequence analysis. Tropic J Pharmaceutical Res 15(12):2659–2667

Ziganshin AM, Liebetrau J, Pröter J, Kleinsteuber S (2013) Microbial community structure and dynamics during anaerobic digestion of various agricultural waste materials. Appl Microbiol Biotechnol 97:5161–5174

Chapter 17
Small Scale Biogas Production

Ram Chandra Poudel

17.1 Global Challenges for Sustainable Development

Fossil fuels are in many cases inaccessible and/or unaffordable in some rural areas of developing countries. People are thus forced to rely on locally available traditional resources like firewood, dried cattle dung, coal, agricultural wastes, etc. for cooking, lighting and heating purposes. Therefore, deforestation, environmental degradation, ecological imbalance, sanitation and other public health issues and socioeconomic difficulties are the prime concern with traditional energy dependency. According to IEA and World Bank (2015) report, more than 2.7 billion people rely on conventional energy resources of which 1.2 billion populations are living without electricity. The report notes that the indoor air pollution caused by burning such fuels takes premature lives of four million annually and 946 million people do not have access to toilet forcing to defecate in open areas. Moreover, the number of people lacking improved sanitation, drinking water sources and surface water stand at 2.4 billion, 663 million and 159 million, respectively. Surprisingly, around 90% of these affected people live in rural areas of Sub-Saharan Africa and South Asia (IEA and World Bank 2015). Implementing small-scale biogas (SSB) plant at household levels has positive and synergistic effect in mitigating such integrated global problems. As such, it is undoubtedly true that small household biogas system is the integral part of sustainable development.

R. C. Poudel (✉)
Department of Microbiology, National College
(Tribhuvan University), Kathmandu, Nepal
e-mail: poudelram07@gmail.com

M. Tabatabaei and H. Ghanavati (eds.), *Biogas*, Biofuel and Biorefinery Technologies 6, https://doi.org/10.1007/978-3-319-77335-3_17

17.2 Small-Scale Biogas Technology and Its Benefits

SSB technology is predominant in rural areas of the developing world. Therefore, it is also named as rural household biogas technology. Countries with highest number of small-size biogas plant installed are Thailand, India, China, Malaysia and Philippines (Daniel et al. 2009). By the end of 2010, China had more than 40 million small family-sized (6, 8 and 10 m^3) biogas plants, with a target of adding 400,000–500,000 plants annually (Chen et al. 2010; Zhang and Wang 2014). Similarly, countries like Nepal, Vietnam, Pakistan, Kenya, Uganda, etc. has adopted similar simple technologies which has contributed to significant improvements in the number of their installations in recent years.

SSB system facilitates circular economy by providing a clean and self-sufficiency renewable energy and nutrients rich fertilizers in rural areas (Fig. 17.1). It offers alternative solutions to proper management of such organic wastes. Therefore, this technology is well-suited in rural areas to improve livelihoods and to maintain economical, ecological and agricultural sustainability.

It should be noted that the extent of the benefits expected from SSB depends upon different factors such as efficiency of biogas production, availability (quantity) and quality of resource material, size and design of biogas plants, etc. Nevertheless, the general advantages are as follows:

- Reduces greenhouse gas (GHG) (methane, carbon dioxide and nitrous oxide) emissions in three ways: (i) improving manure and other organic waste management system, (ii) substituting fossil fuels and non-renewable energy for

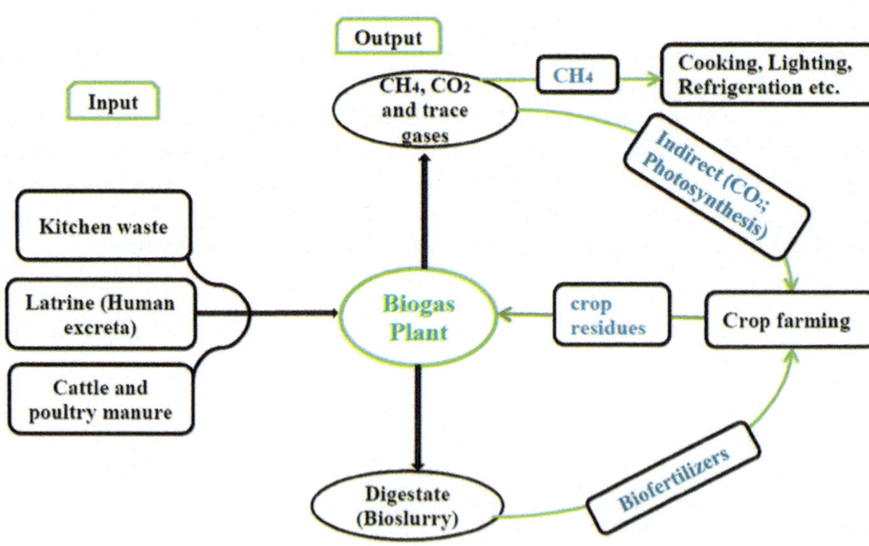

Fig. 17.1 Overview of the benefits of small-scale biogas technology

cooking and lighting and (iii) replacing synthetic fertilizer with biofertilizer (Daniel et al. 2009).

- Reduces demands for wood, charcoal, agricultural residue and dried cow dung for cooking and therefore, prevents deforestation and land use change as well as landslides while maintaining ecological diversity.
- Provides nutrients rich bio-fertilizers thereby closing the nutrient cycles and consequently improves soil fertility, soil microbial biodiversity and reduces soil erosion.
- Mitigate indoor air pollution as well as the associated infections such as respiratory infections and lung diseases, low birth weight, asthma, cataract, etc.
- Attaching latrines to biogas plants improves household sanitation and eventually the surrounding society while improving the surface water quality.
- Increases the rural household economy by generating secured energy, substituting synthetic fertilizers and enhancing crop productivity.
- Diminishes the workloads mostly for women and school going girls, in cutting and collecting firewood and in preparing cattle dung cakes and this extra time can be utilized for rest, education or other income generating activities.
- Rural public health system is improved through the access to reliable energy sources for cold storage of blood and other samples, vaccines, etc.
- Contributes to maintaining food quality through refrigeration run by biogas.
- With sustainable energy, business will grow, jobs and markets are created.

17.3 Design of Biogas Plant

As a matter of fact, the main technical prerequisite of an efficient biogas plant encompasses an air tight reactor with ample space for microbial activities to take place. However, implementing a single size or type of digester for all households is impractical. During the last few decades, SSB plants have been developed in different sizes and shapes/types depending upon constraints like simplicity, availability of feedstock, geographical scenario, climatic condition, operational and maintenance skills, efficiency of digestion process and energy yields, cost value and stability.

17.3.1 Sizing of Biogas Plant

Along with the above mentioned parameters, the dimensioning of the reactor volume largely depends upon the hydraulic retention time (HRT) and duplication time of the microbial populations, in order to avoid the loss of slow-growing methanogenic population from the digester. In general, digester ranges from small-scale household digesters, i.e., 1 m^3 to large commercial scale bioreactor, i.e., 10,000 m^3.

Table 17.1 The differences between small-scale and medium/large-scale biogas plants

	Small-scale biogas plants	Medium/Large-scale biogas plants
1.	More common in rural parts of developing and under developed countries. For instance: India, China, Nepal, Thailand, Nigeria, Vietnam etc.	More prevalent in developed countries like Germany, Denmark, Netherland, USA, UK etc.
2.	Feedstock include locally available organic materials like kitchen waste, cattle manure, human excreta from latrines	Agricultural wastes, manures from animal farms like pig, cow and poultry farms, dairy wastes etc. are the input materials
3.	Heating and mixing mechanisms are often absent	Heating and mixing mechanisms are necessary
4.	Biogas is used for cooking, lighting and heating purposes at household levels	Biogas is used to generate electricity (on-grid or off-grid), as transport fuels and for district heating benefitting large populations
5.	Uses simple, low-tech design and is cost-effective and therefore, low efficient process	Use advanced and high-tech design with high investments which maximize the efficiency of the digestion process
6.	Low-risks and less-expensive in operation and maintenance	High risks and expensive in operation and maintenance
7.	Upgrading of biogas is not essential	Upgrading of biogas is crucial

Therefore, depending upon the size of a digester, plants can be broadly divided as follows.

(1) Small-scale household plant (1–20 m^3) and
(2) medium (community based)/large scale commercial plant (>20 to 10,000 m^3).

Although, the general principles applies in all types of digesters; small-scale differs from medium and large-scale in many ways as shown in Table 17.1.

17.3.2 Types of Small-Scale Biogas Plant

Biodigesters are of two operation modes: continuous-flow and batch digester depending upon the feeding nature. In continuous-flow type, new substrate is frequently fed into a digester subsequently pushing the digestate (bioslurry) out from the system. In contrast to this, batch digesters are fed once with an intention to remove the digestate upon the completion of the digestion and the process will then be repeated. Unlike, continuous-flow type, batch type digesters do not supply gas on a regular basis. However, they benefit a simple and cost-effective technology. By fitting inlet and outlet tubes, batch digesters can be converted into semi-continuous type.

The basic standard of micro-digesters falls under the following design types. However, other digester designs also exist which are developed by individuals, private institutions, or government agencies depending upon the geographical and socio-economic conditions of the user. In general, the main structure of micro-digesters falls in three categories with some variations: floating drum/tank design (batch or semi-continuous), fixed/in ground dome style (continuous flow) and tunnel/tube design (continuous flow).

17.3.2.1 Floating Tank/Drum Design

This design was originated in India in 1950 (Marchaim 1992) and is widely popular in Indian households. The tank size normally used in practice is 1000 L (1 m^3). Examples of floating drum are Pragati model, ARTI model, etc.

This design consists of two tanks which are slightly different in diameter (Fig. 17.2). In general, both are made up of plastic or metal but the larger tank can also be built with brick and concrete under or above the ground. The larger reactor is kept facing upwards into which the smaller one is fitted upside down forming a floating clamper. As a principle, the open side of the smaller tank (floating tank) in always submerged in the substrate mixtures under digestion. The produced gas is stored under the floating tank which slides up and down depending on the volume of the gas produced. Pressure can be applied by adding extra weight on the top of the smaller tank. This pressure allows the gas to flow through the pipeline for use.

The main disadvantage of this type of digester is that there is a high chance of corrosion of the floating tank and therefore, the system has a short life span. Nevertheless, this period can be prolonged by painting the inner and outer wall of the drum but the painting could be toxic to methane producing bacteria.

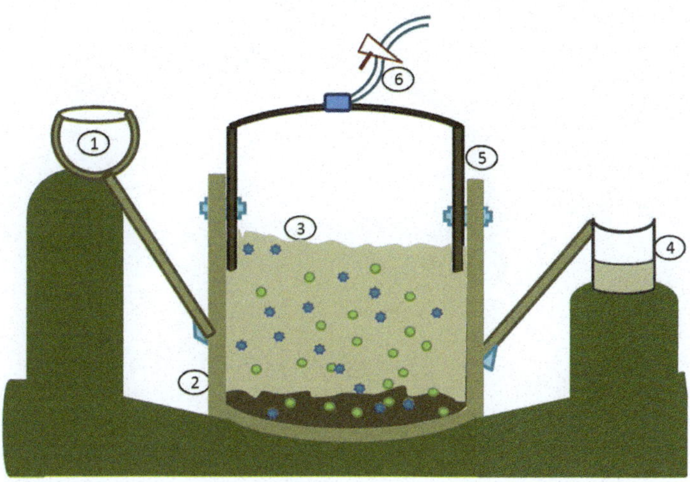

Fig. 17.2 Design of floating tank digester: (1) inlet; (2) reactor (larger tank); (3) digesting substrate mixtures; (4) outlet (digestate holding tank); (5) floating/inner tank, and (6) gas pipe

17.3.2.2 Fixed-Dome Digester Design

This design, a modified form of septic tank, was originally developed in China in 1936 and is the most popular design in the developing world in 6, 8, and 10 m^3 sizes (Zhang and Wang 2014; Chen et al. 2010; Marchaim 1992). The design consists of an underground digesting pit with a dome-shaped cover on the top built of brick, stone or concrete (Fig. 17.3). A second pit called the expansion or slurry reservoir is built higher on the side of the digester pit and is open to atmospheric pressure. The gas produced from the mixture under digestion is accumulated in the top space of the dome-shaped chamber. The pressure exerted by the gas then displaces some of the slurry into the expansion chamber. Upon the use of biogas, the slurry flows back from the reservoir to the chamber. Given the geo-thermal activity of the Earth, this type of underground dome-style digester can maintain a constant temperature even in colder climates.

Examples of these types of digesters are China fixed dome, Indian-fixed dome (Deenbandu), Nepalese fixed dome (GGC-2047) etc.

17.3.2.3 Tunnel/Tube Digester Design

This digester is based on a very simple design developed in Taiwan which consists of a thick polyethylene tube serving as digester with an inlet at one

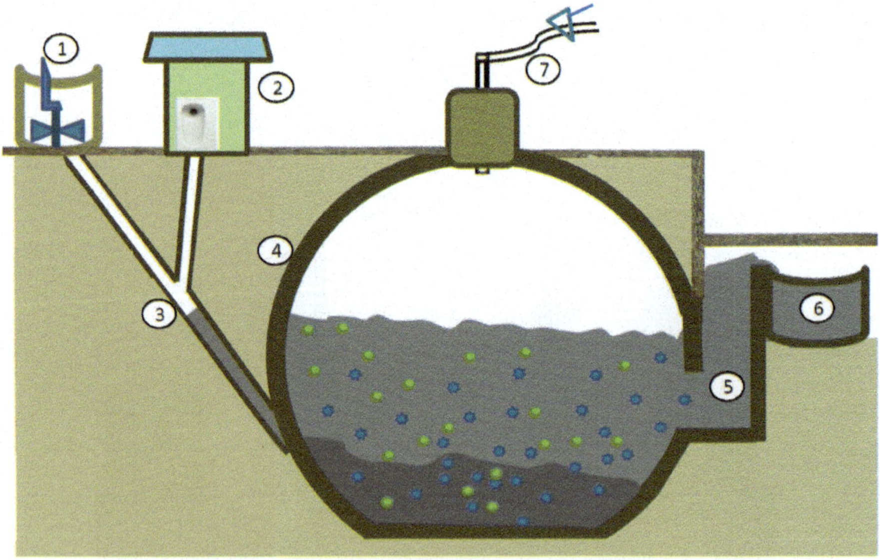

Fig. 17.3 Fixed-dome digester (Chinese model): (1) inlet; (2) Latrine; (3) inlet pipe; (4) fixed-dome (bioreactor); (5) expansion chamber (reservoir); (6) outlet (slurry holding tank), and (7) gas pipe

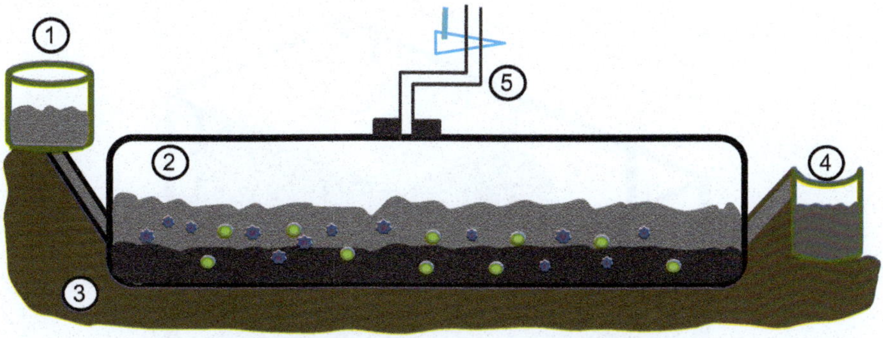

Fig. 17.4 Tunnel biodigester design: (1) inlet; (2) tubular digester; (3) underground soil; (4) outlet; and (5) gas pipe

end and an outlet at the other end while a gas outlet is fitted on the top (Fig. 17.4).

This design applies the displacement principle meaning that there is a simultaneous movement of the slurry in and out of the tunnel shaped chamber. The gas holder space is approximately ¼ of the total digester volume. The system is built underground and therefore, uses geothermal energy to maintain the temperature. In order to reduce the low flow rate of the produced gas, an additional tubular polytheylene could be set up where the gas is to used, e.g., kitchen.

17.3.2.4 Modern Portable Design

With the aim of using the vast potentials of biogas as an energy carrier at small scale, there are a few companies worldwide offering various modernized versions of the above-mentioned biogas production systems. Using these kinds of user-friendly, easy-to-operate, portable biogas production systems, users can produce a proportion of their household energy requirements (e.g., for cooking, warming, etc.). It should be mentioned that the conventional small scale biogas production systems suffer from a range of shortcomings which could be effectively overcome using such modern designs; a typical example is presented in Fig. 17.5. In line with that, commercially available biogas production systems have been introduced into the market such as the design by Homebiogas Inc. This technology well suits rural areas where access to sustainable sources of heat is a challenge while waste-oriented feedstock such as food wastes and/or animal manure are available in ample amounts on a daily basis. Being a portable apparatus, it is possible to operate the system in cold weather by using either a simple heating device or by placing it in a greenhouse or an indoor space. They also offer the advantage of low maintenance costs compared with the conventional small scale systems.

(a)

(b)

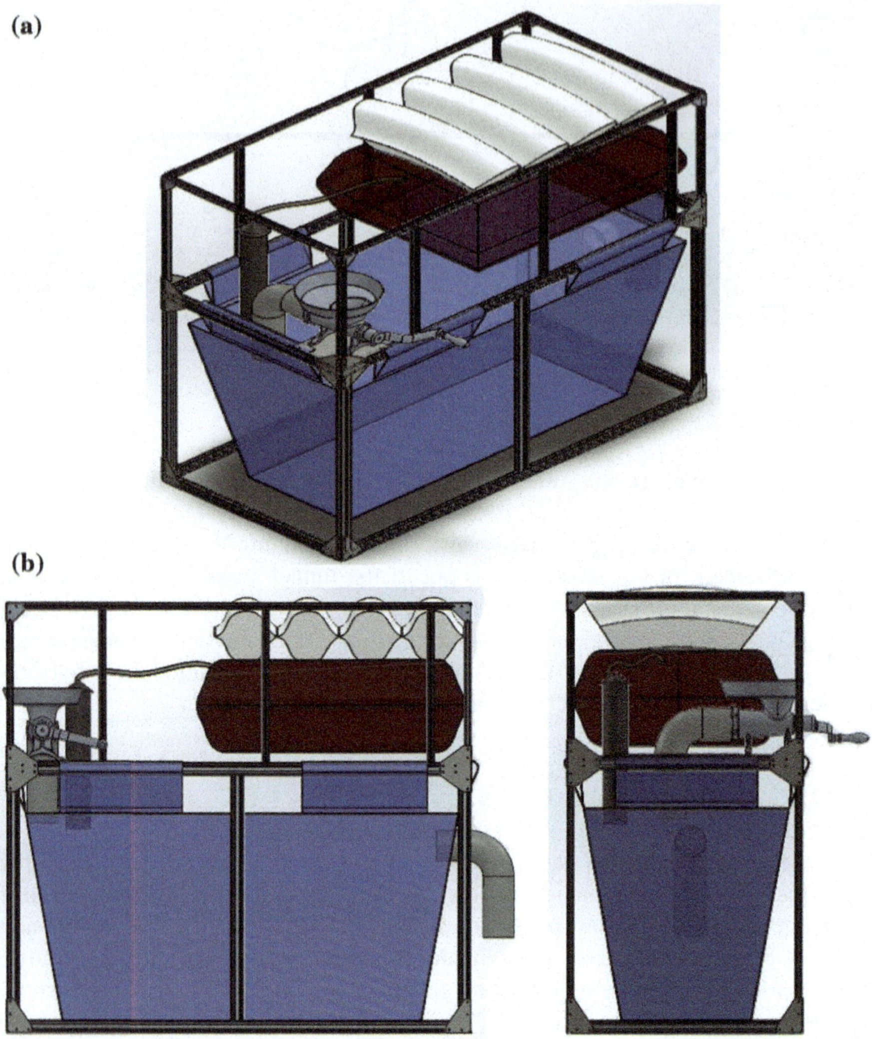

Fig. 17.5 A typical example of user-friendly, easy-to-operate, portable biogas production systems; **a** 3D view and **b** cross-section views

17.4 Activation, Operation and Maintenance

Before the start-up process, it is of utmost importance to inspect the whole plant system to certify that its construction and technical part is well efficient. Like large-scale biogas plant, the activation of SSB plants is possible by feeding the digester with animal manure or human excreta which are potential sources of methanogenic bacteria. It may take the digester a few weeks to adjust to the new environment and once the process is started, the gas will be produced continuously depending upon the type and amount of the substrates fed on daily basis.

The operation and maintenance of small-scale plants is simple and less risky compared with large-scale ones. Operation of biogas plants is linked to safety issues as well and potential risks and hazardous situations should be diminished by following safety measures to ensure safe and stable operation (Paterson et al. 2015). Animal manure and human excreta are partially digested substrates so their digestion alone would lead to low methane yields. Therefore, to enhance the efficiency of the digestion process, animal manure should be co-digested with crop residues and other plant materials having low cellulose, hemicelluloses and lignin contents. This also allows to balance the C:N ratio and reduce the failure risks of the digestion process by high ammonia production. However, some feeding materials need pre-treatment for continuous gas production. In this case, the materials are pre-treated by chopping into fine particles before being fed into the digester. In addition, they can be treated with Sodium Hydroxide (NaOH). For instance, pre-treatment of corn stover at ambient temperature with specific dose and loading rate of NaOH could enhance the biodegradability and increase substrate availability for digestion and finally boost the biogas production yield (Pang et al. 2008). As the pretreatment of substrates with high lignin, hemicelluloses and lignin contents requires would be energy-intensive and costly; therefore, it is beneficial to avoid using such materials in SSB plants.

A properly built biogas plant requires less maintenance and can generate gas for at least 15–20 years without major problems and additional cost (Daniel et al. 2009). A trained consumer is able to easily maintain a household digester. It is necessary to frequently check gas pipes to avoid leaking problems and corrosion of pipes and to perform other small repairs. Depending upon the design and feedstock, the undigested materials deposited at the bottom of the digester should be removed at certain intervals varying from 1 to 5 years. In the floating drum design, maintenance is required to avoid rusting of the drum reactor.

17.5 Small Biogas Systems and Sustainability

17.5.1 Biogas for Sustainability

Sustainability has now become a global concern. SSB system is a paradigm shift for sustainability in rural areas as it endorses the stability of environmental, economic and social elements. This is due to the fact that this energy system applies a simple technology using locally available resource materials to supply clean and efficient renewable energy and at the same time, it cracks the major environmental problems like indoor air pollution, soil degradation, deforestation, desertification, global warming, etc. while it also contribute to solving public health problems such as respiratory infections, water-borne and air-borne diseases, and to resolving social issues like some gender related problems.

17.5.2 Sustainability of Biogas Production Scheme

SSB technology is installed mainly for its household use; therefore, the rate of its implementation depends on family decisions and economic conditions. These families or the potential users are usually of low literacy rate while have low investment capacity and low access to communication and transport. As such, the obstacles for sustainable biogas promotion and production in rural areas are economical, technical, and social barriers. Rural people generally can not afford the high initial cost for biogas installation. Similarly, it may be difficult for them to obtain proper technical assistance during operation and maintenance of biogas plants. On the other hand, it may sometimes difficult to convince the uneducated people living in rural areas regarding the overall benefits of biogas technology. Moreover, the currently available SSB technology is not suitable for consistent biogas production and the biogas yields are largely affected by the climatic conditions. The reality is that such SSB technology produces extra biogas in summers and inadequate amounts in winters. As a result, this conventional technology fails to enhance digestion process and biogas yields efficiently in rural regions with cold climates. Therefore, improvements may include insulation of the underground design digesters with compost heaps or green house as well as heating the digester with external supply of energy, the latter one is costly though and may be unsuitable in rural areas. SSB systems do not have direct fossil fuels inputs yet some extra energy of such may be required at some points during the execution process.

Overall, to achieve sustainability in biogas production social, environmental, and economic elements should be observed while effective local/national policies and organizational capacity building are also of crucial importance. It is significant to evaluate the cost-benefits of biogas production including sanitary and health benefits as well as the reduction of GHG emissions. This will assist with more efficiently promoting biogas as well as with moving towards policies in favour of SSB development like subsidies, incentives, technical support, and awareness activities.

17.6 Small-Scale Biogas in Urban Areas

Generation and improper treatment of organic municipal wastes in the urban areas of developing countries is of great concern from environmental, socio-economic and aesthetic point of views. To solve this global problem faced in urban areas, biogas production at household level could be an efficient solution to not only target resources recovery but also to reduce demands for fossil fuels. Moreover, it can also contribute to reducing the waste volumes and associated transportation and disposal cost. Production of biogas in urban areas can also create a new job market by establishing new enterprises, manufacturing technical equipments, construction, operation and maintenance of biogas plants, etc. Furthermore, the digestate could provide a good source of biofertilizer for urban farming. Unlike large-scale biogas plants, small-scale household biogas plants have gained no attention in developed countries. Therefore, there is no investment at all in terms of research and development on small-sale biogas production either. As a result, this technology has failed to gain popularity in urban areas.

The successful implementation of family size biogas plants involves local and national policies including alternative organic waste management strategy, subsidies in biogas plants installation and research and development with reference to biogas yields and digester design. This propels SSB technology to cross the threshold of competitive markets with multiple benefits.

17.7 Conclusions

The integrated global environmental, economical, health and social problems associated with using conventional fuels and the global needs for sustainable development have persuaded the governmental, institutional and private agencies and policies makers to step forward towards sustainable energy carriers. In line with that, biogas technology has been shown to hold enormous potentials in fulfilling sustainable energy demands in rural and urban areas.

Promoting and executing SSB technology in both rural and urban areas, using freely available local organic resources, will lead to reduced more sustainable national energy demands and will also reduce the use of traditional biomass and imported fossil fuels. By doing so, the targets of renewable energy generation, mitigation of GHGs emissions and sustainable waste management could be fulfilled.

It is important to invest in research and development to improve the various designs of the SSB systems based on specific geographical scenarios, climatic conditions, energy needs, waste management and biofertilizers requirements, etc. In this context, it is also important to obtain competitive prices with those of fossil-oriented energy systems.

References

Chen Y, Yang G, Sweeney S, Feng Y (2010) Household biogas use in rural China: a study of opportunities and constraints. Renew Sustain Energy Rev 14:545–549

Daniel B, Ksenia B, Martin B et al (2009) PoA CDM manual—"Mini biogas plants for households" UNEP. CD4CDM Working Paper No. 8, 2nd version: pp 2–16

International Energy Agency (IEA) and the World Bank (2015) "Sustainable energy for all 2015—Progress toward sustainable energy" (June), World Bank, Washington, D.C. https://doi.org/10.1596/978-1-4648-0690-2. License: Creative Commons Attribution CC BY 3.0 IGO

Marchaim U (1992) Biogas process for sustainable development. FAO Corporate Document Repository, ISBN: 92-5-103126-6

Pang YZ et al (2008) Improving biodegradability and biogas production of corn stover through sodium hydroxide solid state pretreatment. Energy Fuels 22(4):2761–2766

Paterson M et al (2015) Implementation guide for small-scale biogas plants. BioEnergy Farm II Publication, KTBL, Germany, pp 36–38

Zhang L, Wang C (2014) Energy and GHG analysis of rural household biogas systems in China. Energies 7:767–784

Chapter 18
Current State and Future Prospects of Global Biogas Industry

Karan Sehgal

18.1 Introduction: Snapshot of the Global Biogas Industry

Biogas provides a renewable and environmentally friendly process that supports sustainable agriculture. Best suited for household and small-scale farms, especially those located in warm climates, the use of biogas energy has increasingly become recognized as a suitable technology for industrial purposes as well, both for thermal generation and cogeneration (Kothari et al. 2010). From an environmental perspective, biogas integrated in farming systems can directly lower greenhouse gas (GHG) emissions by recovering the methane produced by manure (which is 22 times more effective than carbon dioxide at trapping heat and thus a more potent GHG).

In developing countries, the use of biogas has been treated as a clean energy source and as a replacement to firewood and charcoal—the predominant source of cooking fuel for over 2.7 billion people, or 38% of the world's population, mostly residing in Sub-Saharan Africa and Southeast Asia (WEO 2016). Additionally, the by-products of the 'digester' have been proven to be a potential organic waste of high nutrient quality (Arthur and Baidoo 2011). This model has been widely adopted in rural areas to address two of the major problems for rural development: agricultural production and the availability of clean fuel.

In developed nations, biogas has been primarily promoted as a technological solution that utilizes waste products from agricultural or agro-processing activities to generate electrical power for processing equipment, internal electricity requirements and other enterprise needs, including vehicular application. For the latter, biogas must be compressed and bottled, i.e., through the removal of carbon dioxide and hydrogen sulphide. This type of model is also garnering attention in developing nations as it can be expanded into a "public utility" to provide a clean source of fuel

K. Sehgal (✉)
International Fund for Agricultural Development (IFAD), Rome, Italy
e-mail: k.sehgal@ifad.org

M. Tabatabaei and H. Ghanavati (eds.), *Biogas*, Biofuel and Biorefinery Technologies 6, https://doi.org/10.1007/978-3-319-77335-3_18

Table 18.1 Different approaches to biogas promotion

	Donor support approach (starting in 1970s and 1980s)	Market creation approach (starting 1990s)	'Sustainable energy' approach (starting in 2000s)
Actor(s)	Homogenous: bilateral development relationships between country and donor	Heterogeneous: multiple government agencies, donors, and market participants	Polycentric: multiple public sector, private market, and community development stakeholders
Provision	Household cooking and small lamps for illumination	Household cooking (beginning to use biogas in rural industry)	Household cooking and fertilizer, electricity and rural industry, integrated with broader economic development benefits
Ownership	Public: often given away	Private: sold to consumers or intermediaries	Tailored: use of cost-sharing, public-private partnerships, in-kind community contribution, etc.
Capacity building	Rare: often limited to technical assistance and maintenance	Emergent: some focus on after sales service and business model development, strengthening regulatory capacities	Integrative: efforts centered on maintenance and business model development coupled with strengthening public and private institutions

Source Sovacool 2016

to residents living in rural and/or peri-urban areas or at a community level (schools, hospitals, and apartment blocks) where access to the electric grid is too costly or intermittent. However, although the set-up of a micro-network with a centralized distribution centre has great future prospects globally, the current regulatory framework for bottling biogas and storing it in high-pressure cylinders is still not properly developed, especially in many developing countries.[1] Table 18.1 tabulates different approaches to biogas promotion.

This chapter presents a series of case studies to illustrate the factors necessary for scaling up biogas programmes from the point at which the sector was largely donor-driven to a market-driven, commercial scale. The first section will provide an overview of the biogas sector in India and Nepal and seeks to demonstrate the importance of combining policy measures and innovative technologies. In these countries, biogas is now widely integrated with animal husbandry and is recognized

[1]In Kenya for example, ISO standards developed through the Association of Biogas Contractors (ABC-K) and the Kenya National Domestic Biogas Program (KENDBIP) do not have regulations for bottling biogas nor for feeding scrubbed (purified) biogas/biomethane into the national gas grid.

as a means of manure treatment in the agricultural sector, thus advancing other environmental goals, namely waste removal, environmental management (reduced deforestation and soil restoration) and energy production.

The second section seeks to answer the question of whether an institutional support structure, relevant policies and a regulatory framework are the basic fundamentals required to even consider any advances in the biogas sector. This section highlights how the lack of fiscal incentives for supporting a market-driven approach still hampers the growth of the biogas sector. In addition, it recognizes how uptake has slowed significantly in developing countries, a development attributable to the complex logistics (requiring large volumes of sand, gravel, cement, and stone), need for skilled expertise for proper operation and maintenance, and large volumes of animal manure and water for daily feeding of the biogas digester (which is cumbersome given the open-grazing cattle raising practices in certain communities).

The last section provides a summary of what can be considered the optimal blend of government and market-driven policies. Given that biogas is still largely promoted through a donor-driven agenda, the chapter concludes by examining whether the commercialization of a market-based biogas sector entails national government strategies to reduce subsidies and embrace the necessary blend of public-private sector support required to advance the future prospects of the biogas sector.

18.2 Different Biogas Applications in Developed and Developing Countries

Although commercial scale biogas projects have been predominant in developed countries, large-scale models are now also expanding in Africa. A number of biogas applications are widely available in the market—from biogas pumps, fans, refrigerators, and air conditioners, to power tools and other applications that can be modified to run on petrol or diesel engines, such as chaff cutters for chopping fodder, water pumps and milking machines (Oosterkamp 2013).

Over the past four decades, large scale programmes have financed traditional fixed dome biogas systems. Most, adopting the basic principles of the first biogas digesters implemented in India, China and Nepal—Indian Deenbandhu and Chinese conventional fixed dome or floating drum models. The failures have been evidenced over the years due to the inherent labour work (construction and digging required), complex logistics (especially in relation to the transportation of materials) and the lack of land tenure rights. As a result, numerous domestic biogas models have been advanced technically—from modifications to the traditional underground fixed-dome model, to fibreglass floating top and plastic tubular (plug-flow) systems. The differences and efficiencies are dependent on specific climatic contexts, the availability of spare parts, technical expertise and cost.

18.2.1 Case Study: India's Biogas Sector

Back in the 1880s, in a sewage plant in Mumbai, it was realized that there was the potential to produce a combustible gas—methane. From then onwards, India has been a pioneer in developing the biogas sector which is exemplified by the creation of state-run renewable energy agencies under the India Renewable Energy Development Agency (IREDA).[2] Furthermore, the National Biogas and Manure Management Programme (NBMMP), a central scheme of the government under the Ministry of New and Renewable Energy (MNRE), is the institutional agency that provides household biogas digesters for rural and semi-urban/households. About 4.75 million biogas digesters have been installed in the country up to 31st March, 2014, and an annual target of approximately 1 million additional biogas plants has been set (Virendra 2014).

Returning to the present day, India has made impressive steps towards developing the biogas sector and exporting the *Deenbandhu* technology to other countries. In particular, India's development of the biogas sector has been established alongside livestock manure management and with an emphasis on the application of bioslurry for vegetable production.

18.2.1.1 Biogas Compression and Bottling Model

India is a pioneer in bottling and compression for mini-grids and vehicular application. Removing carbon dioxide and hydrogen sulphide yields biomethane. Compressing this purified 98% methane gas into cylinders makes it easily usable for transport applications (three wheelers, cars, pick up vans, etc.) and also for stationary applications at various sites. Compressed Natural Gas (CNG) has become easily available and, therefore, biomethane (enriched biogas), which retains identical chemical characteristics to CNG, can be used for all applications for which CNG is used (Virendra 2005). However, storing methane in a given limited volume is still a technical and financial challenge for both gas transportation and storage. Moreover, cylinders for storing compressed biogas (which can be at high pressures, e.g., 200 psi) are not readily available in many developing countries, not to mention the lack of policy regulations for the use of this technology.

The Indian Institute of Technology (IIT), Delhi, is one of the institutes created as centres of excellence for training, research and development in science, engineering and technology in India. The "Biogas Development and Training Centre" (BDTC) under IIT, which was opened in 2008, has become more deeply engaged in the research and development of biogas compression and bottling since 2012. Ongoing research activities at BDTC, IIT Delhi include:

---

[2]For further information see: http://www.ireda.gov.in/ and Renewable Energy and Green Growth in India: http://www.teriin.org/projects/green/pdf/National-RE.pdf.

- Biogas production performance from non-edible oil seed cakes and other biodegradable raw materials;
- Enrichment of methane content in biogas by removal of CO_2 and H_2S through water scrubbing technology and membrane separation;
- Hydrogen sulphide removal from biogas through biological methods;
- Biogas bottling into CNG cylinders for automotive applications and biogas testing on engine performance and emissions.

In December 2012, the Indian Convention on Biogas took place in Delhi and provided a platform for many state government representatives and stakeholders in the biogas sector to highlight their achievements (Table 18.2). IIT-Delhi hosted the

Table 18.2 Highlights from the IIT convention on biogas

Entity	Achievements	Challenges	Opportunities
Government of Kerala	50,000 biogas plants deployed	For a 400–600 m^3 biogas plant, subsidies start decreasing. Assessing the production capacity and technology required for packaging biogas	For bigger plants, subsidies should not be needed, and biogas plants with daily capacities of 5,000–10,000 m^3 require performance based subsidies
Ashoka Bio Green	55,000 ton generated from sugarcane to obtain about 9000 kg of CNG	Promoting the conversion of the plant into an industrial plant for the use of CBG and vehicular application	Subsidies from 10 to 40% of total costs are financed by MNRE. Other partners are involved, such as the Bank of Baroda
Jain Irrigation Systems Ltd	In the process of promoting a "one-stop agri shop" for producers of bananas	Generating electricity from waste through a multiple benefits assessment of biogas production	Horticulture farmers in Maharashtra are known to be the banana belt, and account for 20% of the annual production of bananas
State of Punjab (Ganganagar)	Selling biogas at USD1.27/kg	Commercial LPG is expensive (USD1.05/kg) and thus its replacement with CBG is a profitable business, but the regulatory framework is absent	Confronted with technical problems relating to the compressor and methods for carbon dioxide removal i.e. absorption in water, absorption using chemicals, pressure swing adsorption and membrane separation
INSEDA	Installed 4265 biogas plants between 2006 and 2008 and saved 24,000 ton of CO_2 every year	The volatility of the carbon credits market	Project beneficiaries have applied for Carbon credits through INSEDA

Fig. 18.1 Demonstration model installed at IIT-Delhi; **a** biogas scrubbing system and **b** biogas operated three wheeler

event and showcased their demonstration model including biogas scrubbing system, compression and bottling (Fig. 18.1).

The compressed biogas (CBG) model at IIT-Delhi indicated that purified biogas can be compressed up to 200 bar in CNG cylinders (certified by the government) and that a standard CNG cylinder of 60 L can store 9 kg biogas at 200 bar. The average mileage of a car is around 23 km per kg (Virendra 2014). In 2012, there were 150 biogas upgrading plants and the CNG market rate was USD 0.61 cents/kg. The CNG subsidized rate was USD 0.41 cents/kg and the CNG purchase rate was USD 0.40 cents/kg (Virendra 2006).

18.2.1.2 Indian Council of Agricultural Research (ICAR)'s Model

The Indian Council of Agricultural Research (ICAR) is an autonomous organisation under the Department of Agricultural Research and Education (DARE), Ministry of Agriculture and Farmers Welfare, Government of India. The Council is the apex body for co-ordinating, guiding and managing research and education in agriculture, including horticulture, fisheries and animal sciences, in the entire country. With 101 ICAR institutes and 71 agricultural universities spread across India, this is one of the largest national agricultural schemes in the world. A number of ICAR institutes are advancing R&D in the biogas sector. The National Rice Research Institute (NRRI), Cuttack Centre, in the state of Odisha, developed a 5 m^3 paddy straw based biogas plant for running a 3 kW CNG engine for power generation (Fig. 18.2).

The Maharana Pratap University of Agriculture and Technology (MPUA&T), based in Udaipur, developed an aloe vera waste (40–50 L of juice are produced daily) biogas system generating 100–120 kg of waste/d (Panjabrao 2016). The waste generated consists of leaf peel, tentacles, froth of gel, etc. The non-edible aloe

Fig. 18.2 Biogas units at National Rice Research Institute (NRRI), Cuttack Centre

Fig. 18.3 Hybrid solar/biogas dryers for button mushrooms

vera waste cannot be used as cattle feed. The institute has also been conducting research on hybrid solar/biogas dryers for button mushrooms using 37.5 kg of dung, 12.5 kg of spent mushroom waste and 75 L of water (Fig. 18.3) (Jena 2016).

In summary, India's biogas sector is a proof of how various institutional, economic, environmental, technical and sociocultural factors come into play in the dissemination of biogas programmes as an alternative to firewood and charcoal. At the same time, the case studies in India highlight the policy gaps and difficulties in implementing these, for example in the CNG market, as vested interests amongst the political classes influence the decision making of the sector at national level.

18.2.2 Case Study: Nepal's Prospects on Use of Biogas

The first biogas programme in Nepal was initiated in 1974 by the Government of Nepal with the support of the Agricultural Development Bank of Nepal (ADBN). This was followed in 1977 by the establishment of the Gobar Gas Company, a

Table 18.3 Installations of biogas plants in Nepal

Number of biogas plants	Location/Area
4000	Mountainous areas
120,000	Hilly areas
150,000	Terai location
300	Institutional level (community scale)

state-owned enterprise responsible for advancing the development and dissemination of biogas technology (Mendis and van Nes 1999). Yet, it was not until the early 1990s that uptake radically increased:

- In July 1992, the Biogas Sector Partnership[3] (BSP) began operations with funding from the Directorate General for International Cooperation of the Netherlands (DGIS) through the Netherlands Development Organization (SNV);
- From 1992 to 2007, the BSP followed four implementation phases, resulting in the installation of 172,505 biogas plants (World Bank 2012);
- The German Development Bank (KfW) also started funding the BSP and it continued until July 2012. Moreover, this biogas model was replicated in 22 countries across Asia and Africa;
- Other donors also played a role in developing the national biogas sector, such as DANIDA, the Norwegian Government, World Bank and the UK's DFID (World Bank 2012).

Today, the Alternative Energy Promotion Centre (AEPC) maintains overall responsibilities for rural energy projects. In the fiscal year 2011/12, AEPC/BSP reported that out of a total technical potential of 1.3 million biogas digesters, around 300,000 biogas plants were already installed (AEPC 2015). These have been installed in 75 districts and more than 2800 village development committees (VDCs) in Nepal (Table 18.3).

Although the outreach of biogas is high in remote, rural areas (Bajgain 2005), the biogas sector has been largely stimulated by donor-driven support. Unlike in India, where government funding and institutional support were central, in Nepal, the biogas sector lacks the capacity to continue growing exponentially. In India, subsidized programmes ensure greater ownership and long-term sustainability by reducing the transaction costs for the private sector and at the same time allowing them to expand their business in areas they otherwise would not enter. In Nepal, the fading financial support from donor driven projects (including the volatility of carbon credits market) has led to the realization that the private sector needs to play a more prominent role and, for this, requires financial incentives to operate in rural areas.

[3]BSP is currently managed by the Ministry of Population and Environment (previously managed by the Ministry of Science and Technology) and provides subsidy support to promote cooking and lighting using biogas.

Fig. 18.4 Portable biogas installation in the hilly areas of Nepal

The country is also pushing to pilot test innovative biogas technologies. Through the support of International Fund for Agricultural Development (IFAD),[4] the Adaptation for Smallholder Agriculture in Hilly Areas (ASHA) project piloted portable flexi biogas digesters adapted to cold weather in the hilly areas (Fig. 18.4). The digesters were placed in three cluster areas at altitudes ranging from 1946 m in Dailekh district to 642 m in Surkhet district. All systems are functioning well, with an average of 3 h daily use after being fed 20–30 kg of dung/d. Beneficiary farmers consumed 5 kg of wood/d for cooking in their individual households and required 2 h/d to collect the firewood in nearby forest areas.

The use of biogas in cold climate areas is an area in which the International Center for Numerical Methods in Engineering (CIMNE) has excelled in Bolivia, Peru and Ecuador (Marti-Herrero et al. 2014). In all three countries, CIMNE has worked jointly with private biogas companies to construct tubular models currently operating at 4000 m (Perrigault 2012). The Endev-Bolivia project, supported by The Deutsche Gesellschaft für Internationale Zusammenarbeit (GIZ), has installed a total of 750 digesters in Bolivia. Hivos installed 30 digesters, 14 demonstration units (in universities and selected families as demonstrations) and 16 digesters

[4]The International Fund for Agricultural Development (IFAD) is a specialized United Nations agency dedicated to eradicating poverty and hunger in rural areas of developing countries. IFAD's Adaptation for Smallholder Agriculture Program (ASAP) is a climate finance window, created to support smallholder farmers cope with increasing climate change related effects. More information here: https://www.ifad.org/topic/asap/overview.

already sold on a commercial basis to families in the Bolivian altiplano linked to small-scale dairy farms (Marti-Herrero 2014). Over the past five years, a new wave of portable, prefabricated biogas digesters has arisen, such as the Flexi biogas technology manufactured by Biogas International in Kenya (Rota and Sehgal 2015), The Gesi Shamba model developed by SimGas in Tanzania, BioBolsa in Mexico or the Fibreglass model developed by Puxin in China (Table 18.4).

Table 18.4 Shortlist of portable, prefabricated biogas digesters

Renewable energy technology (RET)	Estimated unit cost (USD)	Innovative aspect	Validations from field testing
FLEXI BIOGAS  Kenya	USD 600 for domestic model (3 m^3)	– A flexible above-ground PVC envelope housed in a greenhouse tunnel – Pre-fabricated kit can be installed in a half-day by trained technicians – The system is portable (weighing 50 kg in total), simple to assemble and can be (re)moved	– Installed over 3000 units within projects in Rwanda co-financed by IFAD – Indian Institute of Technology (IIT) in Delhi conducted feasibility studies on the technology – Designer of the technology awarded the Renewable Energy Innovator of the year award (2012) in Washington D.C.
GESI SHAMBA  Tanzania	USD 790 for 6 m^3	– Off-the shelf domestic biogas digesters that are mass produced with recycled plastic – The digester consists of multiple removable parts, creating a modular system that is adaptable and scalable	– Finalist at the U.S. International Design Excellence Awards (IDEA®) and Ashden Awards 2014 – Pilot testing in Tanzania and within an IFAD funded GEF Project (LUSIP) in Swaziland in 2014

(continued)

Table 18.4 (continued)

Renewable energy technology (RET)	Estimated unit cost (USD)	Innovative aspect	Validations from field testing
FIBREGLASS  China	USD 570 for 6 m^3 model	– Simple to build and operate relative to fixed dome biogas digesters – The mould and the gas-holder can be built easily within 2 days – Efficient 'water trap' method for managing gas pressure and optimal operation	– Puxin Biogas won the "Blue Sky Award" given by United Nations Industrial Development Organization (UNIDO) and Shenzhen International Technology – In China, Puxin biogas system has been extended to over ten provinces and exported to over 20 countries, including use in UNDP and SNV projects
BIOBOLSA  Mexico	USD 600 for domestic model (3 m^3)	– Pre-fabricated kit, modular, scalable and adaptable for a variety of biodegradable material and for human waste management – Pre-fabricated kit can be installed in 1 day by trained technicians – Efficient zero-waste packaging and shipping for regional distribution	– Technology has been institutionalized with the International Renewable Resources Institute (IRRI) in Mexico – Installed over 3000 systems in Mexico, Nicaragua, Costa Rica, Haiti, Colombia, El Salvador and Panama – Endorsed by SNV and currently implementing programmes funded by Kellogg Foundation, USAID and the Mexican government

18.2.3 *Case Study—Rwanda and Kenya's Use of Biogas*

In both Kenya and Rwanda, biogas was introduced through the Kenya National Domestic Biogas Program (KENDIP) and the National Domestic Biogas Programme (NDBP), respectively.

In Rwanda, the biogas program was implemented by the energy sector of the Ministry of Infrastructure, with technical assistance from the Netherlands Development Organisation (SNV) and funding from the GIZ. Currently, the average price per digester is RWF 700,000 (USD 848), towards which the NDBP provides an investment subsidy of RWF 300,000 (USD 365) and the Banque Populaire du Rwanda a loan of RWF 300,000. The remaining amount is covered in a cash contribution by the user. Following the SNV experience in some Asian countries, the country is working to take advantage of the reduction in carbon emissions. NDBP has also implemented integrated biogas systems in Kirehe and Ngoma by constructing 76 fiberglass biodigesters, made in China, in order to reduce the construction time and minimize quality assurance issues.

18.2.3.1 Biogas in Prisons

The Kigali Institute of Science, Technology and Management (KIST) has developed and installed large-scale biogas plants in prisons in Rwanda to treat toilet wastes and generate biogas for cooking. After the treatment, the bio-effluent is used as fertiliser for the production of crops and fuelwood (Wheldon 2005). The scale of these biogas systems is enormous: a prison with a population of 5000 people produces between 25 and 50 m^3 of toilet wastewater each day. Using a 500 m^3 system (five linked digesters), this produces a daily supply of about 250 m^3 of biogas for cooking (Wheldon 2005). The program was funded by the UNDP in an effort to curb the difficult conditions for human waste management as well as the huge expenditures on firewood for cooking. The KIST program won an Ashden award in 2005.

18.2.3.2 Biogas in Slaughterhouses

Biogas can cater to rural energy needs by supplying a decentralized source of power with uninterrupted supply. The model provides waste management in a manner that simultaneously produces energy and reduces the environmental load. In Kenya, on the outskirts of Nairobi, Keekonyoike slaughterhouse installed a fixed dome biogas plant that has a 250 m^3 capacity and a biogas storage tank with a storage capacity of 200 m^3, piped to a 16 kW generator set (Njuguna 2013). At present, the slaughterhouse's biogas plant can handle 3100 kg of digestible waste out of the 14,460 kg available daily. Increasing the digestible amount (thereby increasing gas production) could allow for a conversion of the current plant to an industrial plant and potentially produce a net energy level of 2590 kWh, which could be used for operating machinery, powering the cold meat room, internal electricity requirements, or by piping the excess gas to nearby homes (Njuguna 2013).

18.2.3.3 Biogas for Household and Farm Level Use

Under the 'One Cow for Every Poor Family' or Girinka Programme, an IFAD supported Kirehe Community-based Watershed Management Project (2009–2016), implemented by the Ministry of Agriculture and Animal Resources, approximately 1500 household biogas systems have been installed. In Kagogo cowshed in Kirehe district, Rwanda, a dairy cooperative comprising 32 farmers with funding support from Heifer International implemented the first biogas unit (48 m^3) at a total cost of USD 32,000, operating a dual fuel engine that is running a water pump (2400 L/d) for livestock watering needs, a 3 phase motor for running a milking machine, and a chaff cutter for chopping fodder (2 h daily) (Fig. 18.5). The biogas generated is also being used directly as gas for cooking. This demonstrates demand for two forms of energy that cannot be fulfilled at present: one for direct heating (pasteurizing) and another for refrigeration.

In summary, the above three examples demonstrate how the potential of biogas is curtailed due to the lack of technical know-how, lack of policy support and limited access to finance. Most of these programmes, although largely donor-funded, aspire to reach a market based approach—that is, biodigester construction companies reaching an autonomous and profitable level so that they are able to sell digesters to farmer households. These programs are also multi-stakeholder, where financial institutions provide credit to the households for purchasing the unit and government or NGO extension services provide training on operation, maintenance and the use and application of the bioslurry.

Fig. 18.5 Installation of a 48 m^3 digester at a communal cow shed in Rwanda

18.2.4 Case Study: Snapshot of Biogas Sector in the EU

A total of 80% of the cars in the EU running on methane are manufactured in Italy (Chiaramonti 2014). By converting abundant domestic biomass resources—such as leaves, husks or stalks from corn—into transportation fuel, this biomass conversion technology continues to mature toward commercialization. Opportunities across the industry are growing exponentially for providers of raw materials, technology developers, refiners, and fuel distributors.

Germany and the Netherlands have the most auspicious programmes for renewable energy and in particular have emphasized on the important role that research and development (R&D) play in shaping and igniting the business/commercial side of the biogas sector. In Italy, Research Center for Alternative and Renewable Energies (CREAR) has dedicated activities on renewable energy, with particular attention to biomass and biofuels. The annual budget for R&D is a total of USD 6 million per year. The project BIOSYNG, supported by the Italian Ministry of Agriculture, Food and Forestry (MIPAAF) and spearheaded by students from the RE-CORD (a non-profit research institute associated with the University of Florence), has set-up a gasification plant (not for commercialization) with 1 MW capacity (or approximately 500 m^3/h) producing gas from gasification of ligno-cellulose biomass. On the outskirts of Florence, RE-CORD laboratories house a handful of demonstration plants such as:

- Microturbines convertible on biofuels (20–30 kW);
- Biomass gasifiers;
- Charcoal briquettes;
- Engines running on vegetable oil;
- Biomass pyrolysis;
- Thermochemical methane.

From a policy angle, many developed nations have developed feed-in tariffs (FiT), which are a good way (but not the only way) to provide incentives in the biomethane sector. In the UK, FiT is GBP 0.71/kWh. In France, FiTs are dependent on the producing capacity of the plant but range from EUR 45–95/MWh for landfill plants to EUR 69–125/MWh in anaerobic digestion plants (CREAR 2016). In Sweden, Germany and Austria, biomethane producers are entitled to receive a technology bonus, tax reliefs and other incentives. However, under current market conditions, biomethane cannot compete with natural gas. The biomethane sector can grow only if the costs of storage and transportation come down and thus become cost-competitive with natural gas. Methane is a crucial and key method of curbing GHG emissions—the documentation of the GHG reduction in CO_2 equivalent could be another driver for achieving stronger policy support at a national and regional level.

The 2050 European Low Carbon Economy Roadmap suggests that a 40% reduction in emissions by 2030 compared with 1990 would be cost-effective (Chiaramonti 2014). A reduction of less than 40% would increase the long-term costs of decarbonising the economy. The future prospects rely on the ability of the European countries to capitalize on the experiential knowledge and innovative applications of biogas that have already been tested and demonstrated. Looking to the future, by 2030, the European biogas industry will produce as much "green gas" as "green electricity" by using the natural gas distribution network for generating electricity, heating and cooling and as a fuel for vehicular application.

18.3 Factors Shaping the Biogas Sector

There are a number of factors that shape and develop the biogas sector. There is no one-fits-all solution. However, there are certain fundamental elements that need to exist in order to better align biogas programmes with national government strategies. Firstly, there must be political support from line ministries, i.e., Ministry of Environment, Energy and Livestock that are pushing forward a thematic agenda involving climate mitigation, waste-to-energy projects, health and sanitation, access to energy and others. At the same time, given that biogas is a cross cutting theme touching upon numerous thematic areas, there must be a realization of the inter-ministerial dialogue and cooperation required. Over the past forty years, there have been numerous programmes introducing trials of biogas on a wide range of scales and with various types of feedstock, but few have been maintained for more than a few years. The success of biogas has been limited by a combination of factors: poor institutional framework and infrastructure; inadequate planning policies; lack of coordination and linkages in biogas programmes; pricing distortions that have placed renewable energy at a disadvantage; high initial capital costs; weak dissemination strategies; lack of skilled human resources; and weak maintenance, after-sales services and infrastructure.

The prevalence of biogas among rural agricultural households and a lack of distribution systems to enhance commercialization have resulted in a failure to reinforce the concept of biogas for agricultural applications. The general perception of biogas still associate it with rural households that own a few cattle that can produce cow dung to be used as input substrate. The case studies in the previous section highlighted why this sector has not achieved a scaling-up process, especially in its potential to promote integrated farming systems. Issues at the policy level are one culprit but the other is the distribution network and financial sustainability for post-sales services. Likewise, many governments subsidize conventional fuel prices for the poorest, thus creating a barrier to entry for biogas technologies. At the technical level, the quality in installation and operation has been ambiguous, with many donor-funded projects lacking the capacity to monitor and backstop projects that have promoted biogas. Although biogas systems are cost-effective on a life-cycle basis, they are often not affordable or capable of being scaled up without the support of government and/or financial institutions. In addition, the capacity for the rural poor to pay is low and, once the operation and maintenance costs begin, there is limited sustainability in the financial support or other mechanisms put in place.

18.3.1 Ecological Barriers: Changing Climate and Volatile Fossil Fuel Prices

In this chapter, we started by underscoring predominantly the benefits of biogas for cooking, but also emphasized the potential of the effluent from the digester, the so called bioslurry. This is a perfect solution for organic farms that suffer from soil fertility shortages. The business case for investing in a biodigester is very positive and biodigesters fit seamlessly into most mixed farming systems (Warnars and Oppenoorth 2014). In general, there has been a shift in prioritizing the use of biogas energy primarily for the slurry rather than the energy-related benefits (cooking and lighting). Numerous studies have been undertaken demonstrating the 'magic' of bioslurry (De Groot and Bogdanski 2013).

China exemplifies how to bring forward an agenda with multi-faceted benefits for rural and peri-urban areas, which are also the main consumers of charcoal (a leading cause of deforestation). West Guangxi Poverty Alleviation Project has provided biogas tanks to about 30,000 poor households, saving 56,000 ton of firewood annually, which is equivalent to the recovery of 7470 ha of forest (IFAD 2011).

The experience of the cassava starch industry in Thailand also highlights the environmental benefits from the potential for GHG emissions reduction given that (i) fossil fuel oil is replaced with renewable biogas; (ii) the methane that previously escaped into the atmosphere during fermentation of wastewater in open lagoons is now captured as biogas; and (iii) part of the grid electricity produced mainly from coal and gas is replaced with on-site electricity produced from biogas (Hansupalak 2015).

18.3.2 Policy Barriers: Subsidy Schemes and the Role of the Private Sector

At the macro-level, promoting South-South cooperation (knowledge and technology transfer) for training and distribution as well as for material sourcing is key to making biogas technologies more affordable to the lower quartile rural populations (i.e., more resource-poor and marginalized smallholder farmers), mostly resident in Sub-Saharan Africa and South East Asia.

In developed countries, a lack of policy regulations on the use of biogas energy (especially in the transport sector) combined with resistant lobbies such as the fossil fuel industry impeding the biogas industry's growth and, therefore, the central role biogas could play in the energy and transport sector. Lessons learnt from other countries with more advanced biogas programmes (such as Sweden and India) show that there is also a large potential for biomethane in vehicle transportation (buses, tractors, cars, auto rickshaws). To this end, political decision-makers must consider the complex cultural constraints dictating institutional sustainability, subsidies that distort market prices and subsidies on fossil fuels, including chemical fertilizers.

18.3.3 Technical: Capacity Building, Quality Control Systems and Training

In Rwanda, SNV and Rwanda Energy Group (REG) have provided training on installation, operation and maintenance of all types of biogas units. This is part of a wider SNV programme that seeks to train over 800 youth in each sector of Rwandan industry. The goal is to train rural youth on group formation and capacity building in financial management. In relation to biogas, certificates have been handed out by the selected companies under the guidance of REG and SNV. These certified biogas installers now receive RWF 15,000 (USD 18) per installation from active biogas companies and have been trained in becoming self-sustaining. In turn, these certified biogas installers have trained beneficiary farmers on the operation and management (O&M) and troubleshooting of biogas digesters, such as routine feeding of the biogas digester and appropriate mixing of cow dung (in appropriate water to manure ratios).

In Vietnam, the biogas programme is currently implemented by the Biogas Project Division (BPD) under the Department of Livestock and Production (DLP), Ministry of Agricultural and Rural Development (MARD). The Vietnam program has been part of SNV's broader Asia Biogas Programme, and is currently funded by the EnDev Energising Development (2013–2017), the Blue Moon Fund (2013–2014), and the sales of Voluntary Gold Standard Credits (VERs)—the programme has been registered as a carbon project since 2012. It is estimated that 6 million ton of CO_2e are released annually by medium-scale pig farmers as a result of an estimated 73 million ton of pig waste disposed of improperly into ponds, channels and sewages (Teune 2007). In Vietnam, 1 million digesters have been constructed, including 500,000 medium scaled, 150,000 industrial level units that have covered lagoons for treating wastewater (UASB model), and 20,000 medium-sized pig farms, growing at an annual rate of 10% (Zwebe 2013). At present, under the SNV's programme, 140,000 biogas plants have been constructed and over 800 technicians and 1400 masons have been trained (Zwebe 2013).

In Cambodia, more than 300 trained and certified masons are available to construct biogas systems. 46 experienced Biodigester Construction Companies (BCCs) in 11 provinces can fulfil an order in less than two weeks. The National Biogas Programme (NBP) under the Ministry of Agriculture Fisheries and Forestry

(MAFF) is working closely with MFIs such as Amret and Prasac. A Czech non-governmental, non-profit organization, People in Need (PIN), is also supporting NBP and the private sector in building the capacities of BCCs to manage and expand their businesses, which includes support for promotion and marketing activities, but have also been directly involved in the provision of after sales services through the establishment of village-based local technicians who sell basic spare parts and provide technical advice to biogas users. NBP plans to install 3000 units under a recently approved Global Environment Fund (GEF) supported project: Building Adaptive Capacity for the Scaling up of Renewable Energy Technologies in Rural Cambodia (S-RET). The ADB project: Climate-Friendly Agribusiness Value Chains Sector Project, beginning in 2017, is also refurbishing approximately 11,000 biogas units and constructing an additional 5000 units during the period 2017–2020 under the supervision and management of NBP.

The experience of the cassava starch industry in Thailand shows that since 2002 most cassava starch factories have switched from fuel oil to biogas generated from their wastewater. The energy recovered enables factories to cover 100% of their thermal energy needs, including starch drying (previously done with fuel oil), and in some cases to also produce on-site electricity and reduce the use of grid electricity. The most efficient factories generate excess biogas and have invested in generators to produce electricity, covering 20–40% of their electricity needs. Factories report a return on investment of 2–5 years thanks to the savings from not needing fuel oil any longer (Hansupalak 2015). The reduction in GHG emissions is significant as well. Most cassava starch factories have adopted biogas technology in the past 10–12 years, using their wastewater as feedstock for biogas production. As a result, the factories can save on production costs and recover their investment within 5–10 years.

Biogas technology works well at large scales (cassava starch factories typically process 800 ton of cassava roots every 24 h). At small or medium scales, variations in the quality and concentration of wastewater during the day make it more difficult to stabilize the fermentation reaction. This requires monitoring the fermentation as it needs constant adjustment (e.g., increasing or decreasing the flow rate of organic material into the tank; adding starters or nutrients when necessary) and the equipment needs maintenance.

18.3.4 Financial: Role of Microfinance Institutions (MFIs)

In developing countries, the energy lending portfolio, especially in rural areas, remains negligible. The possibility for rural poor farmers to purchase biogas technologies that can reduce monthly energy expenditures, improve health conditions, increase agricultural productivity and increase income generation opportunities is still rare.

The development challenge is to seek modalities for engaging with private sector companies that promote environmental goods and services and to identify finance mechanisms (grants, equity loans etc.) to support capacity building (both human and financial). There are many other factors that need to be taken into consideration, including cost, capacity to deliver the technology in large quantities, a training support mechanism and timely post-sale services.

In Cambodia, a typical system costs US$500, of which the end user pays USD 350. NBP provides a USD 150 grant. For households that are not able to pay the USD 350 upfront but are creditworthy, three leading microfinance institutions (MFIs) provide payment schemes with an average payment of US$20 per month for up to 2 years. To date, 20,000 households have purchased this system. The latest consumer survey shows that 96% of these systems are still operating satisfactorily. A private company, Kamworks, is leading the solar energy sector and is a pioneer in the mobile technology PAYGO—a software for paying, monitoring, tracking and troubleshooting solar energy installations. Established in 2006 by Dutch solar engineers, to date 150,000 rural households have been reached, with 96% coverage of the PAYGO market in Cambodia. This model has been applied only for solar energy products but addresses a multitude of issues that are also relevant for private companies and clients in the biogas sector: (i) pre-financing the customer; (ii) managing and monitoring sales; (iii) collecting payments; (iv) controlling operations and installations/repairs; and (v) inventory management and financing.

Another programme being promoted through Hivos is the Africa Biogas Partnership Program (ABPP) in 6 countries in Africa (Ethiopia, Zimbabwe, Tanzania, Uganda, Burkina Faso, and Kenya), Indonesia, Bolivia and Guatemala. The program is well-aligned with national government strategies and, to date, 60,000 biogas units have been installed In Africa. 12,000 units have already been installed in Tanzania and another 10,000 biogas plants are planned in the period 2017–2019. In Indonesia, through the ABPP, a total of 15,000 units are planned for the next two years.[5]

18.3.5 Social: Trends of Urban Migration

There are a number of elements that curtail the spread of biogas programmes (Table 18.5). Another daunting trend is that of increasing rural to urban migration, with implications for the sustainability of biogas. For example, in rural areas, promoting biogas energy is complicated by years of habit using traditional methods that are perceived as cheap and inexpensive. Elderly people at home have a lower capability to deal with technological problems and properly manage livestock manure (collecting and feeding the digester). On top of this, over recent years biogas energy has more and more been recognized as a 'past' technology and synonymous with an inferior social status that (a) handles animal manure and (b) cannot afford

[5]For more information on the Africa Biogas Partnership Program see here: http://www.africabiogas.org/#.

Table 18.5 Summary of obstacles and solutions to the dissemination of biogas

Obstacles	Specific challenges	Possible solutions
Ecological barriers	– Large water requirements – Mitigating potential pathogen dispersal	– Support livestock numbers, veterinary services, fodder and water availability – Sustainable groundwater extraction – Quantification of reductions in deforestation and GHG emissions – Bioslurry application with influence on soil fertility
Policy barriers	– Lack of an enabling policy framework for accelerating take up – Lack of fiscal incentives for the private sector – Poor institutional arrangements (customs, regulations and import taxes) – Cost reductions through pursuit of economies of scale and linking with national rural energy subsidies – Subsidies for low-quality traditional biomass resources and kerosene	– Imposition of standards and provision of training for operating biogas digesters – Certification of biogas installers/ masons – Identification of impacts on other livelihood systems such as charcoal and firewood suppliers and defining alternatives
Technical barriers	– Small-scale farmers with small herds not able to get sufficient feedstock – Complex logistics in transportation of material components (especially in remote, rural areas) – Technical skills required for operation and maintenance – Scum formation in small sized plants – Corrosion of gas holder – Improper preparation in treatment of solids leading to blockage	– Adaptation of biogas digester design to the needs of the users – Capturing user-knowledge for design improvements – Optimising feeding methods and types of organic substrate – Appropriate delivery mechanisms (technical training manuals and facilities and post-sales services) – Pre-soaking, adjustment and scum formation of carbon to nitrogen ratio – Removal of inert particles such as sand and rocks
Financial barriers	– Initial costs too high – High transportation and maintenance costs of plant requiring skilled expertise – Marketing of organic fertilizer (bioslurry) limited – Construction from cheap construction materials (glass fibre, clay, jute-fibre reinforced plastic)	– Provision of a line of credit to lower the initial investment costs (possibly in combination with microfinance services) – Engaging women in self-help groups for multiplying finance and sharing knowledge – Quantification of economic benefits from substitution of chemical fertilizers with bioslurry, and use of free time (derived from

(continued)

Table 18.5 (continued)

Obstacles	Specific challenges	Possible solutions
		time saved in collecting firewood) on other income generating activities
Sociocultural barriers	– Open-grazing practices – De-stigmatizing livestock manure management and introduction of hygienic methods of feeding biodigester (integration of human waste over longer term) – Limited knowledge on the potential of biogas energy	– Stalling livestock for dung collection – Engaging youth in the deployment process for creating the next generation of users – Provision of information, awareness-raising and promotion regarding the benefits of RETs – Farmer-to-farmer exposure for stimulating take up

LPG. Destigmatizing these perceptions is a challenge in itself that can only be tackled by demonstrating the economic benefits of biogas for the user, the community and the environment (as, for example, in India, which remains one of the few countries where the use of human waste is accepted as substrate for biogas digesters[6]).

18.4 Conclusions

The use of biogas has been promoted by virtue of its chemical characteristics (composed of carbon dioxide, methane and traces of hydrogen sulphide) and produce multi-faceted benefits, such as the reduction of GHG emissions, improved livestock manure management, and health and sanitation benefits (Rota and Sehgal 2012).

With its many ecological and societal benefits, biogas development for energy and fertilizer has gained much attention in recent years. However, most biogas applications are restricted to locations where biogas is produced. At present, biogas is associated with rural households that own a few cattle that can produce cow dung to be used as biomass. Compressing and bottling biogas (biomethane) would aid in commercialization and popularizing its ease of use as this would clearly segregate production from distribution.

Many countries already seek to promote policies that create viable markets for biogas and, at the same time, use subsidies as a quality control mechanism to guarantee that markets function properly. Over the past decade, the uptake of biogas has slowed due to increased urbanization rates. Biogas is still largely promoted through subsidy programmes, and the commercialization of a market-based biogas

[6]Another impressive program in India is the Sulabh International movement. More than 200 biogas plants of 35–60 m^3 capacity have been constructed by Sulabh in different states of the country so far which are connected to human latrines (Pathak 2013).

sector requires the long-term vision to align this with national government strategies (Sovacool et al. 2015). However, to arrive at that juncture, significant development is required, such as quality control systems and training mechanisms. In addition, public-private partnerships with clear roles must be articulated to scale up the biogas sector. Private companies cannot do it alone and neither can government agencies.

Biogas and biomethane deserve particular attention and support among renewable energy sources as these low-carbon technologies promote closed loop waste-energy systems. The industry can significantly contribute to further development of rural areas. The use of biogas in stationary engines for different agricultural operations (milling, grinding, powering water pumps and chaff cutters, etc.) also shows that it has the capacity to be a profitable business that can generate ample opportunities for employment in rural areas.

Finally, a number of take-home messages can be presented as follows:

1. *Engaging private sector* entities is essential for achieving success in scaling up biogas technologies. In the long-run, it is important to facilitate platforms that can lead to rural youth entrepreneurship and agricultural development.
2. *Raising awareness on the potential of biogas digesters when integrated within farming systems.* The nature of the technology and simplicity to understand its operation allows for farmers to become technical service providers and to troubleshoot minor complications (clearing water from gas pipes due to condensation, feeding digester daily or not leaving gas valve open). Therefore, technical leaflets, exhibitions and demonstrations, community competitions, troubleshooting manuals and documentaries are all tools that need to be disseminated widely.
3. *There is a need for more sensitization and demonstrations on the use of biogas.* Farmer-to-farmer exchange visits and the generation of KM products are critical so that farmers can witness experiences of other farmers who are successfully operating biogas and disseminate knowledge relating to the opportunities to diversify livelihoods and increase revenue streams.
4. *Carve a role for the private sector in relation to the marketing, finance and after-sales services.* For now, Vietnam is the only viable market for commercially promoting biogas (as opposed to viability through Government programmes).
5. *Donor funded biogas has to stop.* Being largely promoted through subsidy programmes, the commercialization of the biogas sector is undoubtedly the long-term vision, but to arrive at that stage, a lot of development will need to take place (such as quality control systems and training mechanisms). A commercially viable biogas sector entails subsidies being used as a quality control mechanism to guarantee that markets function properly rather than distorting market prices and creating disincentives for potential clients to invest in the technology.

References

AEPC (2015) Market mapping study for biogas companies. Alternative Energy Promotion Centre, Final Report. Kathmandu, Nepal

Arthur R, Baidoo MF (2011) Harnessing methane generated from livestock manure in Ghana, Nigeria, Mali and Burkina Faso. Biomass Bioenerg 35:4648–4656

Bajgain S (2005) The Nepal biogas support program: a successful model of public private partnership for rural household energy supply. Full Report, SNV

Chiaramonti D (2014) L'insostenibile incertezza smart green: 98–102. Available online: http://www.qualenergia.it/sites/default/files/articolo-doc/98-103_QE_n2-2014_chiaramonti.pdf

CREAR (2016) BIO-MGT project full technical report. Available online: http://md.bluefactor.it/index.php?module=CMpro&func=listpages&subid=6&expsubid=6

De Groot L, Bogdanski A (2013) Bioslurry = brown gold? A review of scientific literature on the co-product of biogas production. Full report. Food and Agriculture Organization (FAO), Rome, Italy

Hansupalak N (2015) Biogas reduces the carbon footprint of cassava starch: a comparative assessment with fuel oil. J Clean Prod 1:8

IFAD (2011) Biogas in China. Full video. Available online: https://www.youtube.com/watch?v=bCRps9Jnwbk

Jena PK (2016) Energy management in agriculture, thermo-chemical and bio-chemical conversion technology. Annul Report. Central Institute of Agricultural Engineering (CIAE), Bhopal, India

Kothari R, Tyagi V, Pathak A (2010) Waste-to-energy: a way from renewable energy sources to sustainable development. Renew Sustain Energy Rev 14:3164–3170

Marti-Herrero J (2014) Improvement through low cost biofilm carrier in anaerobic tubular digestion in cold climate regions. Biores Technol 167:87–93

Marti-Herrero J et al (2014) Low cost tubular digesters as appropriate technology for widespread application: results and lessons learned from Bolivia. Renew Energy 71:156–165

Mendis MS, van Nes WJ (1999) The Nepal biogas support programme, elements for success in rural household energy supply. Policy and Best Practice Document, The Hague, The Netherlands

Njuguna A (2013) Business plan review for Keekonyoike Company Ltd. Melanel Consulting Company, Nairobi, Kenya

Oosterkamp WJ (2013) Engines on biogas for generators with a maximum power of 50 kWe. Technical Brief

Panjabrao DK (2016) Energy in agriculture and agro-based industries. Maharana Pratap University of Agriculture and Technology. Full Progress Report, Udaipur, India

Pathak B (2013) Sulabh international social service organization. Status report. Available online: http://indiagovernance.gov.in/files/sulab.pdf

Perrigault T (2012) Towards thermal design optimization of tubular digesters in cold climates: a heat transfer model. Biores Technol 124:259–268

Rota A, Sehgal K (2012) Livestock and renewable energy, livestock thematic papers IFAD

Rota A, Sehgal K (2015) How to do: mainstreaming portable biogas systems into IFAD-supported projects. Technical Brief. IFAD: Rome, Italy

Sovacool B (2016) Paradigms and poverty in global energy policy: research needs for achieving universal energy access. Environ Res Lett 11(6)

Sovacool B et al (2015) Scaling and commercializing mobile biogas systems in Kenya: a qualitative pilot study. Renew Energy 76:115–125

Teune B (2007) The biogas programme in Vietnam: amazing results in poverty reduction and economic development. Boiling Point 53:11–13

Virendra VK (2005) Biogas scrubbing, compression and storage: perspective and prospectus in Indian context. Renewable Energy Journal

Virendra VK (2006) Biogas purification and bottling into CNG cylinders: producing bio-CNG from biomass for rural automotive applications. Full report. Joint international conference on "sustainable energy and environment" Bangkok, Thailand

Virendra VK (2014) Upgraded bottled biogas: a green and low-cost fuel for automobiles. Akshay Urja. April 2014. Available online: http://mnre.gov.in/file-manager/akshay-urja/march-april-2014/EN/20-23.pdf

Warnars L, Oppenoorth H (2014) Bioslurry—a supreme fertiliser: a study on bioslurry results and uses. Full Report. Hivos, The Hague, The Netherlands

Wheldon A (2005) Biogas plants providing sanitation and cooking fuel in Rwanda. Technical Report, Kigali Institute of Science, Technology and Management (KIST), Rwanda, Ashden Awards

World Bank (2012) Nepal biogas support program IV (BSP). Implementation completion and results report, October

World Energy Outlook (2016) Biomass database. Full report. International Energy Agency

Zwebe (2013) Commercial development opportunities for small and medium scale biogas in Vietnam. PowerPoint Presentation at the International Workshop on Small- and Medium scale Biogas. Hanoi, Vietnam

The manufacturer's authorised representative in the EU is Springer Nature Customer Service Centre GmbH, Europaplatz 3, 69115 Heidelberg, Germany. If you have any concerns regarding our products, please contact ProductSafety@springernature.com

Printed and bound by CPI Group (UK) Ltd, Croydon, CR0 4YY
30/09/2026
02225438-0001